Intelligent Transportation Systems

Intelligent Transportation Systems

Edited by
Madison West

STATES
ACADEMIC PRESS
www.statesacademicpress.com

Published by States Academic Press,
109 South 5th Street,
Brooklyn, NY 11249, USA

ISBN: 978-1-63989-294-5

Cataloging-in-Publication Data

Intelligent transportation systems / edited by Madison West.
p. cm.
Includes bibliographical references and index.
ISBN 978-1-63989-294-5
1. Intelligent transportation systems. 2. Electronics in transportation.
3. Transportation--Technological innovations. I. West, Madison.
TE228.3 .I58 2022
388.312--dc23

For information on all States Academic Press publications
visit our website at www.statesacademicpress.com

STATES
ACADEMIC PRESS

Contents

Preface

This book has been an outcome of determined endeavour from a group of educationists in the field. The primary objective was to involve a broad spectrum of professionals from diverse cultural background involved in the field for developing new researches. The book not only targets students but also scholars pursuing higher research for further enhancement of the theoretical and practical applications of the subject.

Intelligent transportation system (ITS) refers to the integration of advanced data and communication technologies that are used in transportation and traffic management systems. It aims at improving the safety, efficiency and sustainability of transportation networks. ITS aims to enhance the driver's experience and reduce traffic congestion. Various ITS technologies are used for management of transport systems including basic systems such as car navigation, traffic signal control systems, container management systems, variable message signs, automatic number plate recognition, etc. Emergency vehicle notification systems, automatic road enforcement, collision avoidance systems, etc. are some of the ITS applications used in the present scenario. This book traces the progress of this discipline and highlights some of its key concepts and applications. It will help the readers in keeping pace with the rapid changes in this field. The topics covered in this extensive book deal with the core aspects of intelligent transportation systems.

It was an honour to edit such a profound book and also a challenging task to compile and examine all the relevant data for accuracy and originality. I wish to acknowledge the efforts of the contributors for submitting such brilliant and diverse chapters in the field and for endlessly working for the completion of the book. Last, but not the least; I thank my family for being a constant source of support in all my research endeavours.

Editor

1

Real-Time Smart Parking Systems Integration in Distributed ITS for Smart Cities

Muhammad Alam,[1] Davide Moroni,[2] Gabriele Pieri,[2] Marco Tampucci,[2] Miguel Gomes,[3] José Fonseca,[1] Joaquim Ferreira,[1] and Giuseppe Riccardo Leone[2]

[1] *Instituto de Telecomunicações, Universidade de Aveiro, Portugal*
[2] *Institute of Information Science and Technologies, National Research Council of Italy, Pisa, Italy*
[3] *Micro I/O, Sistemas Electrónicos, Lda, Aveiro, Portugal*

Correspondence should be addressed to Gabriele Pieri; gabriele.pieri@isti.cnr.it

Academic Editor: Eneko Osaba

Intelligent Transportation Systems (ITS) have evolved as a key research topic in recent years, revolutionizing the overall traffic and travel experience by providing a set of advanced services and applications. These data-driven services contribute to mitigate major problems arising from the ever growing need of transport in our daily lives. Despite the progress, there is still need for an enhanced and distributed solution that can exploit the data from the available systems and provide an appropriate and real-time reaction on transportation systems. Therefore, in this paper, we present a new architecture where the intelligence is distributed and the decisions are decentralized. The proposed architecture is scalable since the incremental addition of new peripheral subsystems is supported by the introduction of gateways which requires no reengineering of the communication infrastructure. The proposed architecture is deployed to tackle the problem of traffic management inefficiency in urban areas, where traffic load is substantially increased, by vehicles moving around unnecessarily, to find a free parking space. This can be significantly reduced through the availability and diffusion of local information regarding vacant parking slots to drivers in a given area. Two types of parking systems, magnetic and vision sensor based, have been introduced, deployed, and tested in different scenarios. The effectiveness of the proposed architecture, together with the proposed algorithms, is assessed in field trials.

1. Introduction

The advancements in urbanization and the dependence on transportation lead to the evident increase in the number of vehicles on the roads around the world, which have a negative impact on the quality of life. Congestion on the roads, CO2 emission, energy waste, road accidents, and increase stress level are some of the problems that are seriously affecting public health and wasting useful resources. It is clear that innovative solutions are needed, and that is where Intelligent Transportation Systems (ITS) should play the role in contributing to tangible results quickly and efficiently. Therefore, new disruptive technologies and techniques that ensure ITS to be more efficient, safer, and energetically sustainable need to be introduced.

The development of urban centers and the increasing number of vehicles on the roads around the world lead to the evident increase in demand for parking places in urban centers, which seriously affects the quality of the traffic. The search for the available parking place increases the traffic volume; it is estimated that more than 70% of the European population lives in urban centers under almost constant impacts of air pollution and noise caused by road traffic [1]. But, next to the pollution, the problem called *cruising for parking* causes more than 30% of the usual total traffic congestion resulting from the daily travels and traffic jams due to road constructions, traffic accidents, and related traffic problems. The average search for the available parking place is up to 20 minutes. According to other conducted studies this percentage could be much higher and so in a borough of

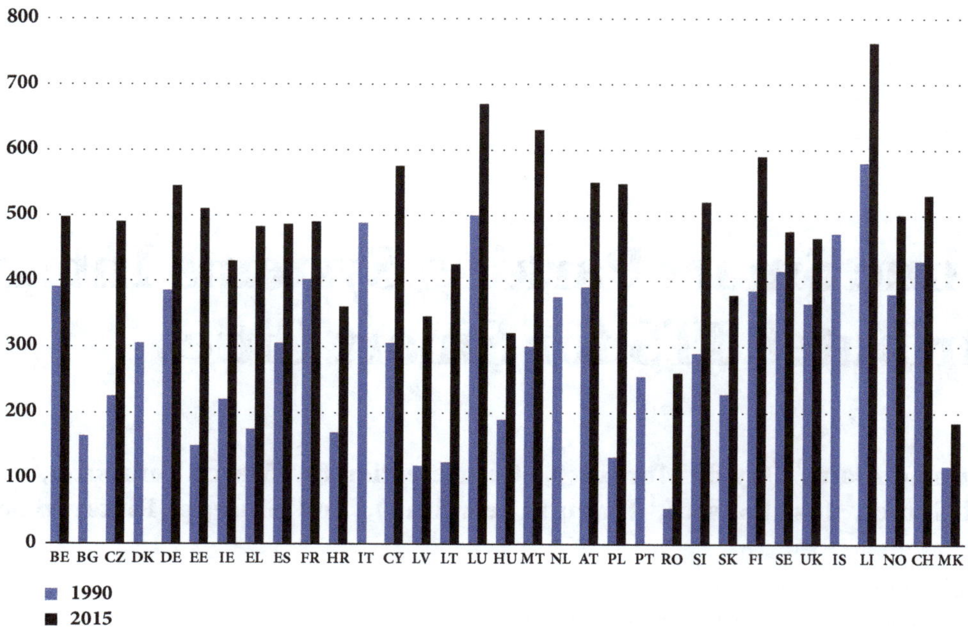

FIGURE 1: The increase in the number of passenger vehicles per 1000 inhabitants in EU [1].

Munich (Schwabing) even exceeds 44% and in a borough of Brooklyn, NY (Park Slope) exceeds 45%. Moreover, according to an earlier study [1], 24% of all traffic accidents are caused by careless parking (entry and exit from the parking lot), reckless departing from vehicles, sudden stops, etc.

In addition, from an economic point of view, in 2010, the specified congestion caused in America 4.8 billion hours of delay, 7.2 billion litres of additional fuel consumption, and the cost of 101 billion dollars. Only in Los Angeles the search for available parking places took 178,000 litres of fuel (which is enough to circle the world 38 times) and produced 730 tons of carbon dioxide [2].

The importance of applying the adequate strategy for the management of parking places can also be found in the estimate that during the day a typical vehicle stays in idle 95% of the time.

The driver's behavior, when choosing the type and the location of the parking place as well as making the final decision, depends on many factors and represents a considerably more complex process, which should be paid much more attention than before. The usage of advanced technologies is unavoidable in achieving selected combination of strategies.

Increased vehicle production in the last 30 years and higher purchasing power of citizens consequently induced extremely high number of vehicles which are daily or periodically using highways as shown in Figure 1. In addition, according to the report presented in [4], every car is parked for 23 hours (average) daily and thus makes it extremely difficult to find an empty parking. Today ITS are very complex systems, made of several subsystems working in isolation, which are often closed, i.e., they do not provide interfaces for direct access to third parties, and vertical, i.e., they only provide an end-to-end system from sensors/actuators to the human-machine interface (HMI) of the system. Hence, the fusion of data originated from subsystems operating in the same area is very difficult in practice and is bound to become worse in the future as the ICT infrastructure in transportation systems becomes bigger and more complex. Therefore, in this paper, we present an architecture which makes such interaction between heterogeneous data sources as seamless as possible, and which leverages such opportunity to shift the intelligence for decision making from humans in control centers to distributed machines. The proposed architecture tackles the problem of traffic management for the specific case of open parking space management, through local information diffusion on vacant parking slots to drivers. To this end, two types of parking systems' technologies are considered: using magnetic and vision sensor based. They were considered, deployed, and tested in different urban scenarios, and the effectiveness of the proposed architecture, along with the proposed algorithms, was assessed in field trials.

This paper is further organized as follows: the related work is presented in Section 2, Section 3 presents the proposed distributed architecture, and Section 4 presents vision based algorithm. The implementation of both magnetic and vision sensors are presented in Section 5 while Section 6 describes the equipments installation. The conclusions are presented in Section 7.

2. Related Work

As urban centers are becoming increasingly more populated and congested, the management of public parking places, according to local mobility policies, is a key strategic issue with high economical impact. Academia and industry are well aware of the potential gains that can be obtained taking advantage of information and communication technologies applied to this field. An efficient smart parking system will favor drivers, by reducing the time to find a parking place and by optimizing parking spaces' utilization rate, thus increasing the revenues for the urban parking operators.

The demand for the available parking place is a dynamic value, i.e., a value that changes over time, so logically there is a need to provide the information on the location and condition of parking offer in real time to the drivers of passenger cars. The systems for providing the information on parking offer to users already began to be used in some world cities more than 20 years ago and can be found under names like "A Guided Parking System" or "Dynamic Parking Guidance System" [5]. By the usage of Parking Guidance and Information (PGI) systems it is possible to improve to some extent the process of parking and increase its success and efficiency. The ecological effects of potential PGI systems application are evident in the reduction of noise (the decreasing number of vehicles in certain segments of the transport network) and in the reduction of fuel consumption and exhaust emissions (shorter travel times and the increase in the traffic flow).

Typically, a PGI consists of 4 components, namely, the parking lot monitoring, information dissemination, communication technology, and the control system [6]. The PGI systems use either smart sensors or cameras for monitoring lots occupancy. The sensors based monitoring is further classified as "On-Roadway" or "Off-Roadway". The on-road sensors are glued to the road surface; examples of such type of sensors are magnetic sensors and acoustic sensors. The off-road sensors are placed above the surface of the road; the examples are smart cameras for monitoring.

The body of knowledge in this area is vast, with large-scale field deployments since 1996. Lin *et al.* [7] provide an exhaustive, although not systematic, survey on smart parking solutions, from 2000 to 2016, and identify some challenges and open issues. The smart parking system can be designed in several ways and there are different technology options available. Regarding the technologies available for parking lots monitoring various sensor solutions such as magnetic sensor, acoustic sensors, RFIDs, ultrasounds, inductive loops, smart cameras, etc. can be used. Recently more advanced solutions for street parking detection and monitoring systems are proposed, for example, crowdsourcing [8], which is based on an ultrasonic range finder placed in each vehicle to detect free places near the parked vehicle. The collected information is then disseminated to the other vehicles' drivers to help them find these free slots.

In [9] the authors have used photoelectric sensors to detect the vehicles in a controlled zone. The sensors were installed at the entry/exit of the parking and connected with the server for information processing and dissemination. Such arrangement can provide real-time information about the number of free parking lots but does not provide exact location of the free space. In addition, the problem of finding exact free parking lot grows exponentially with the number of parking lots. An increased production of sensors and smart cameras has enhanced the scope of analysis and management of urban traffic activities. New disruptive techniques leveraging computer vision that provides an added value to cameras by automatically extracting relevant information have been proposed in several works. Therefore, these new techniques based on computer vision and video analytics become increasingly important for ITS. For instance, a smart camera based real-time car parking monitoring systems exploiting the convolutional neural network algorithms was presented in [10]. The system was tested and validated for robustness and effectiveness in various challenging scenarios such as low light, shadows, and occlusions. The presented system exhibits high performance; however it is not integrated with a fully distributed ITS architecture. In [11] the authors have used applied computational mechanisms and image processing techniques to monitor and authorize vehicles entering the parking area. However, the presented systems only detect and recognize the vehicles and verify if they are the registered users or not. Recently, a more interesting work on image processing for smart transport infrastructure is presented in [12] using drones. The presented work uses drone cameras to capture the videos and then analyze it for the available parking lots and congestion on the roads. However, the presented work shows less precision and also not fully integrated into a distributed ITS system. Other related works that use smart camera for parking lots monitoring can be found in [13–16].

The work presented in this paper addresses some of the challenges and openes issues identified in [7], notably at the sensing layer, guaranteeing personal privacy, and by using smart, low power, magnetic, and vision sensors. Moreover, with the aim of designing an architecture which can have high scalability and reliability, in the following section our fully distributed ITS is presented and described, along with the two different sensing technologies for covering different aspects of parking systems (e.g., outdoor and underground parking lots). Another issue, faced by our proposal, regards the conditions for proposing a low-cost solution, at least in comparison with other proposals mentioned above, and with regard to several aspects: from the sensors customization to the infrastructures and from the maintenance to the scalability.

3. Distributed ITS Architecture

In current information systems, supporting ITS applications, data flows from the edges of the system, i.e., sensors, road-side units, vehicles, and personal devices, towards a centralized data aggregation and control center. Automated or manual control and supervision decisions are then either taken and fed back to the edge actuation components or forwarded externally, e.g., to databases or web services. This architecture has a number of drawbacks, such as the single point of failure of the central subsystem (control center) and the increased end-to-end delay. Moreover, this legacy architecture has limited scalability and availability, as the addition of new road sections implies new configurations, parameterizations, and system restart. Therefore, based on these limitations, a different architecture was proposed, in the scope of the

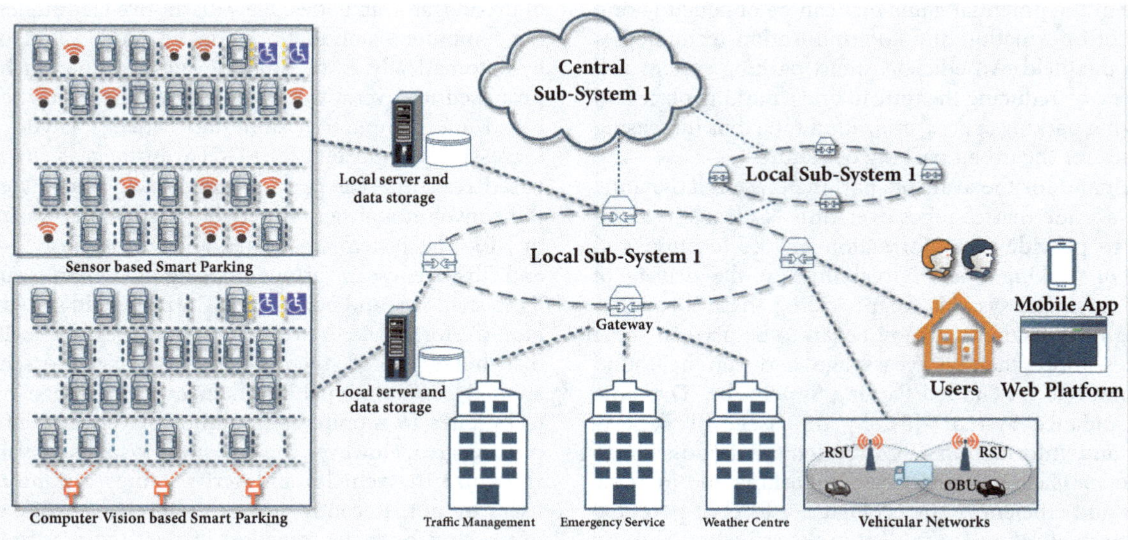

FIGURE 2: Distributed ITS architecture.

ICSI project [17], where the intelligence and storage are distributed across limited geographical areas, as depicted in Figure 2. In this architecture, data is processed in gateways located at each local subsystem without the intervention of the control center; hence the information is locally distributed. These gateways aggregate and store data, process events, issues warnings, etc., within a bounded geographical area. In turn, smart parking systems, vehicular networks via road-side units (RSUs), and other personal subsystems are connected with gateways. Additionally, each gateway is logically connected to the control center, for long-term data aggregation, optimizations, and big data processing. The proposed architecture provides higher dependability and reliability, as faults are confined within the respective gateway, while other gateways and the control center may continue to operate without failures. Furthermore, the new architecture is highly scalable, since the addition of new components of the system is automatically accommodated by a proportional increase of the number of gateways. Thus, thinking to the typical development in an urban environment, new clusters (e.g., an existing parking lot becoming monitored by smart cameras) could be added with only the cost of the infrastructure and the nodes, but at almost no cost for the whole architecture. In addition, reactions in bounded time are favored, because the complexity of interactions is limited to a local scope and the communication path is much shorter. A data distribution layer acts as a bridge between the cooperative sensing infrastructure (sensors, road-side units, vehicles, and personal devices), the local gateways, and the central subsystem. The data distribution layer is connected to the border routers of the vehicular network and the smart parking subsystems using the CoAP protocol. The border router communicates with On-Board Units (OBUs) using the ETSI ITS G5 and with the parking sensors and cameras using 6LoWPAN.

4. Design of Cooperative Computer Vision Algorithms for Real-Time Parking Lot Monitoring

For the analysis of parking lot occupancy status, the approach followed is based on an *ad hoc* computer vision pipeline capable of highlighting the changes in the Regions of Interest (RoI), with respect to an adaptive background reference image. The pipeline, described in the following, is particularly suitable to real-time deployment on resource-constrained embedded devices.

Change detection: in order to improve the computational efficiency, the frame differencing for detecting changes is performed only on predetermined RoI in the acquired frame. Each of the RoI corresponds to a specific parking slot, and for each of the regions the absolute accumulated pixel-wise difference is reported; such a difference is dynamically weighted in order to correct and improve the robustness of the algorithm with respect to environmental light changes. In order to perform this improvement, normalized versions of the images are computed and used, with respect to global illumination parameters (average and variance of both the current and reference image). The sum of the differences is scaled with respect to the size of the RoI, and finally it is stored in a buffer. At this point verification occurs in order to detect the possible change. In particular, a comparison of the stored actual value with the historical values allows filtering out possible spurious values (i.e., exceeding the threshold) due to, e.g., the presence of shadows. In the same way the stored value is compared with another threshold in order to detect possible changes with respect to the reference image. At this point the algorithm yields a first outcome which is a value regarding the occupancy status of the specific parking slot. **Intranode occupancy detection:** once the algorithm has computed values regarding the occupancy for each parking slot (corresponding to the RoIs), an intranode occupancy detection process occurs. In order to avoid transitory events (e.g., people crossing by and shadows casted by external objects), the occupancy status becomes effective and is transmitted to the smart camera network, only after being observed consecutively for a specific number of acquired frames. For each parking slot, the algorithm yields a confidence value in the range [0..255], meaning that values next to 0 represent an almost no-change detected with respect to the reference value, and thus the slot is likely to be free; higher values, on the other hand, indicate that main changes have occurred in the observed scene, and thus the slot is likely to be occupied.

Internode cooperative decision making: the confidence values produced by the single nodes as a 256 levels number should be transformed to binary values corresponding either to free or busy parking slots, thus taking a final decision regarding the parking availability. To this end, local confidence values will be propagated through the smart camera network thanks to its middleware. In particular when a parking space is monitored by more than one sensor node, the final decision regarding its occupancy is obtained in a cooperative way, at an internode level. This final decision is obtained aggregating all the confidence values produced by the different nodes (which are statically dislocated and have static tables of the monitored parking slots). If a slot k is monitored by $n = n(k)$ sensor nodes and $v_1^k(t), \ldots, v_n^k(t)$ are the confidence values measurements from each single sensor node at time t, then the aggregated measure is computed as

$$v^k(t) = \sum_{i=1}^{n} \omega_{i,k} v_i^k(t) \quad (1)$$

where $\omega_{i,k}$ are the nonnegative weights and

$$\sum_{i=1}^{n} \omega_{i,k} = 1 \quad (2)$$

Thus the final decision $st^k(t)$ regarding the slot k is obtained performing a comparison with a threshold ε:

$$st^k(t) = \begin{cases} 1 & \text{if } v^k(t) > \varepsilon \\ 0 & \text{if } v^k(t) \leqslant \varepsilon \end{cases} \quad (3)$$

For the implementation of a more robust algorithm that avoids meaningless oscillations, the above decision is further improved using two levels of thresholds $\varepsilon_1 < \varepsilon_2$ and considering the status of the slot at the previous measure obtained at time $t-1$:

$$st^k(t) = \begin{cases} 1 & \text{if } v^k(t) \geq \varepsilon_2 \text{ or } \left(v^k(t) > \varepsilon_1 \text{ and } st^k(t-1) = 1\right) \\ 0 & \text{if } v^k(t) \leq \varepsilon_1 \text{ or } \left(v^k(t) < \varepsilon_2 \text{ and } st^k(t-1) = 0\right) \end{cases} \quad (4)$$

Weights $\omega_{i,k}$ are determined heuristically for each physical configuration of the smart camera network, while the thresholds $\varepsilon_1, \varepsilon_2$ are set to a common value for all the nodes, the sensors, and the parking slots.

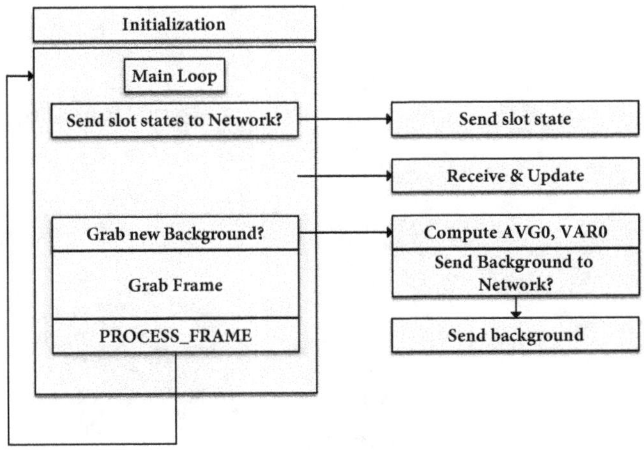

FIGURE 3: Parking lot occupancy flowchart in analysis software (this figure has been modified after [3] (Magrini et al., 2015)).

The general processing flow for the parking lot algorithm is presented in Figure 3. The main loop of the algorithm starts by possibly sending the status of the slot being monitored to the network. Afterwards, if needed, the new active background is acquired and stored, simultaneously with the computation of statistical data over the frame. Once the actual frame is acquired its real-time processing is performed. The processing follows the above described steps, from the normalization of the acquired frame to the final outcome evaluating the occupation status of the monitored slots. In the following cycle, the slot status are sent (if the communication is needed and foreseen to be performed at every single cycle) to the higher level of the network, where the computed values can be aggregated as previously described.

5. Implementation

5.1. Sensor Based Smart Parking Implementation

5.1.1. Architecture. Micro I/O has developed a smart parking system as depicted in Figure 4 that is suitable for both indoor and outdoor environments. The developed system is based on wireless magnetic sensors as depicted in Figure 5 that are deployed on the ground in each parking space for the respective occupancy detection. The system is compatible with multiple user platforms, for management and driver guidance to available spaces and interoperable with several complementary systems, such as LED panels, tags, indicators, alarms, and barriers. In terms of architecture, the system consists of three major layers:

(i) The endpoints layer, encompassing the parking sensors and actuators.

(ii) The communication layer, encompassing gateways and repeaters, to bridge the communications between the endpoints and the rest of the system.

(iii) The software layer, consisting on all cloud-based components responsible for data collection, parsing,

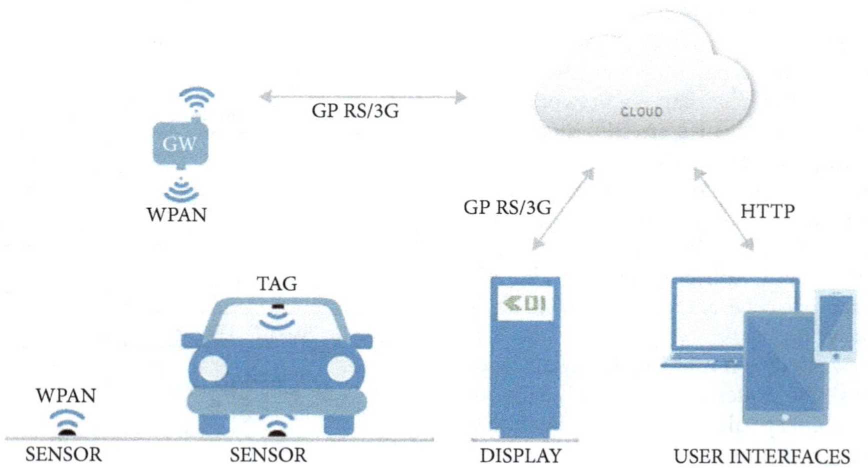

FIGURE 4: Sensor based smart parking overview.

FIGURE 5: Deployed sensors.

storage, and provision, from the web services and databases, to the multiple user platforms.

The system has the following features and specifications:

(i) Encapsulation: the sensor does not require cabling, as it features wireless communications and internal batteries for power. The device is installed inside a plastic capsule, which is then filled with epoxy, and after the epoxy has hardened, the device is completely isolated. The encapsulation enables complete impermeability and the ability to withstand extreme pressure (as the weight of the vehicle).

(ii) Deployment: after encapsulated, the sensor is of easy and fast deployment. It only required the application of asphaltic glue beneath the sensor and little pressure after its deployment on the ground, followed by few waiting minutes for the glue to dry and attain a robust fixation to the ground.

(iii) Programming and firmware updates: the sensor firmware can be remotely transferred to the device. The system features FOTA (Firmware Over The Air).

(iv) Initialization: in order to initialize the sensor, a provided magnet has to be approximated to the sensor. The device includes a reed sensor that will detect the presence of the magnet and activate the device.

(v) Detection: the sensor features a 3-axis digital magnetometer (Freescale MAG3110) for the vehicular detection, through the perception of the variations on the Earth's magnetic field, caused by the passage/presence of the vehicle above it. The sensor has a determined sampling rate and if the magnitude of the readings overcomes the algorithm's defined thresholds, the status change is communicated to the system.

(vi) Temperature compensation: the magnetometer sensibility varies with the temperature, being influenced both by the temperature of its surroundings and the exposure to sunlight. For proper operation, the sensor requires calibration and temperature compensation, in case of variations. For this purpose, the MAG3110 features a temperature sensor and samples are also taken along the magnetic readings, for compensation/recalibration, if necessary.

(vii) Communication: according to the application requirements, in terms of communications range and device autonomy, the chosen communications technology was LoRa, over 868MHz. This technology enables a communications range up to 200m and low power consumptions. There can be two types of communications:

(a) Status change notification: following a sampling event, the data is analyzed and, in case a status change is detected, the sensor sends the respective communication/notification to the system.

(b) I'm alive notification: if no status change is verified for a defined period of time, the sensor sends an "I'm alive" notification to the system, indicating it is operating properly.

(viii) Power management and autonomy: the device autonomy represents its lifetime. The sensors can feature up to 4 AA low self-discharge Li-SOCl2 batteries (SAFT LS14500, 3.6V, 2.6Ah); however, due to the permanent epoxy encapsulation, the batteries are not replaceable nor rechargeable; therefore power management is of great relevance. In order to expand the device's lifespan, the processes involving sampling, data processing, and communications were optimized:

(a) Sampling: in order to reduce the time the microprocessor is awake, but not compromise the quality of service levels, by ensuring prompt status updates, the defined sampling periodicity is of 4 seconds. After the device is activated, at every 4 seconds the microprocessor will wake up from sleep, take a sample and, if no change is detected, go back to sleep;

(b) Processing: in order to reduce power consumption, the selected microprocessor was the (Low Power) Microchip PIC18LF26K22-I/SS. Its power consumption in sleep mode (at which it is most of the time) is, approximately, 1μA.

(c) Communications: communications are the most power consuming processes. The chosen communications technology, LoRa on the 868MHz band, enables low power consuming communications, but the periodicity and duration (size) of the communications was also optimized, for greater power savings.

According to the defined periodicity of the "I'm alive" messages, for example, with intervals not smaller than 30 minutes, the sensors can achieve an autonomy beyond 8 years.

5.2. Vision Based Smart Parking. In this section, a smart camera prototype for vision based smart parking is presented. The prototype is based on previous work [3, 18]; choices that lead to the selection of specific hardware are motivated and discussed and technical facts are reported. Main emphasis is given to the vision board, which is a custom Printed Circuit Board (PCB) manufactured specifically for the sensor. Cameras, wireless module, batteries, photovoltaic panels, and housing are also briefly addressed.

5.2.1. Vision Board. For the implementation and development of the vision board, an embedded Linux architecture has been selected that provide enough computational power and can be easily programmed. A selection of ready-made Linux based prototyping boards had been evaluated with respect to computing power, flexibility, price, and support. The following candidates were considered: Raspberry Pi Model B (ARM11, 700 MHz), Phidget SBC (ARM9, 400 MHz), and BeagleBone–TI Sitara AM3359 (Cortex A8, 720 MHz). All the candidates have as common disadvantages high power consumption. It has been therefore decided to realize a custom vision component by designing, printing, and producing a new PCB. The new PCB has been designed to have the maximum flexibility of use while maximizing the performance/consumption ratio. A good compromise has been achieved by using a Freescale CPU based on the ARM architecture, with support for MMU-like operating

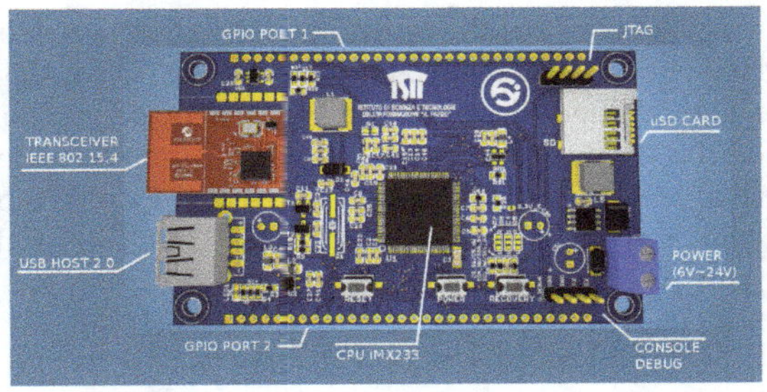

FIGURE 6: Top view design of the vision board.

FIGURE 7: Actual vision board.

systems GNU/Linux. This architecture has the advantage to integrate within it a PMU (Power Management Unit), in addition to numerous peripheral interfaces, thus minimizing the complexity of the board. Moreover, the CPU package of type TQFP128 has helped us to minimize the layout complexity, since it was not necessary to use multilayer PCB technologies for routing. Thus, the board can be printed also in a small number of instances. The choice has contributed to the further benefit of reducing development costs; in fact, the CPU only needs an external SDRAM, a 24MHz quartz oscillator, and an inductance for the PMU. The chosen architecture has been proved to have an average consumption measured at the highest speed (454MHz) less than 500mW. The system includes an on-board step-down voltage regulator type LM2576 featuring high efficiency to ensure a range of voltages between 6 and 25V, making it ideal for battery-powered systems, in particular for power supply by lithium batteries (7.2 V packs) and lead acid batteries (6V, 12V, and 24V packs). A microSD slot is present, which is essential for booting the system, booting the kernel and filesystem associated (EXT4); the board can be upgraded simply by changing the contents of the microSD. On board, there are two expansion connectors (strip 1x30), a specific connector for the serial console of the operating system and a serial connector for debugging. On the expansion connectors, SPI, I2C, and UART peripherals are provided as well as two low resolution ADC. Figure 6 shows the top a view of the board, rendered with SketchUp directly from source Eagle CAD. The main components including the CPU iMX233, the USB host port, and the microSD card slot have been noted. Also shown are the connectors for the GPIO and debugging via serial connection. Finally, in Figure 7 the realized operational vision board is shown.

5.2.2. The Camera. For the integration of a camera sensor on the vision board, some specific requirements were defined in the design stage for providing easiness of connection to the board itself and management through it, and capability to have at least a minimal performance in difficult visibility condition, e.g., in low light conditions. Thus the minimal constraints were as follows:

(i) An USB Video Class (UVC) device.

(ii) Possibility of removing IR filter or capability of Near-IR acquisition. Moreover, the selection of a low-cost device was the main requirement considered for the whole smart camera prototype.

The USB Device Class Definition for Video Devices, or USB Video Class (USB Implementers Forum, 2012), defines video streaming functionality on the Universal Serial Bus. Similarly, to most mass storage devices (USB flash disks, external SATA disk enclosures) that can be managed by a single driver because they conform to the USB Mass Storage specification; UVC compliant peripherals only need a generic driver. Among a very large list of UVC compliant devices (see e.g. http://www.ideasonboard.org/uvc/), an easy-to-buy and cheap camera (HP HD 2300 Webcam) was selected. Such camera is equipped with an IR filter, designed to reduce the noise from IR light sources, which is easily removable for our purposes of acquiring images even in low light conditions. The camera supports several colour spaces (such as YUYV and RGB) and image resolutions up to 1280x1024. The maximum achievable frame rate is 30 fps at 640x480 resolution. Through the Video For Linux 2 (V4L2) API this camera allows for several controls and commands, including brightness, contrast, saturation, hue, sharpness, and gamma correction.

5.2.3. The Wireless Communication Module. The smart camera was equipped with suitable interfaces for networking to connect to the distributed ITS architecture reported in Section 3. An approach in line with modern trends in the Internet of Things (IoT) has been followed [19] in order to be able to build a smart camera network in which each sensor provides its own interpretation of the scene. To this end, a transceiver compliant with IEEE 802.15.4 has been integrated directly on the PCB. More in detail the Microchip MRF24J40MB has been selected working in the 2.4GHz range and featuring data rate up to 250kbps. Typical power consumptions of the transceiver are 25mA in RX mode and 130mA in TX mode. The aforementioned transceiver has an antenna integrated in the PCB; in case an external antenna is needed, the similar transceiver MRF24J40MC can be considered instead. A suitable glue has been used to integrate the transceiver with the IPv6 stack, also containing the 6LoWPAN header compression and adaptation layer for IEEE 802.15.4 links [20]. The use of IEEE 802.15.4 is an excellent solution for building a network of smart cameras in a parking lot scenario, where the nodes are not excessively far and the mesh network, also by exploiting multihop mechanisms, can guarantee a good coverage. The whole prototype with energy harvesting system is depicted in Figure 8.

FIGURE 8: VSN sensor node prototype with energy harvesting system.

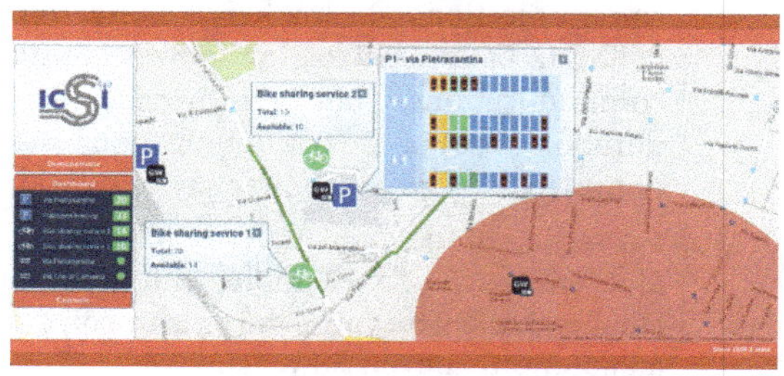

FIGURE 9: Web application.

5.2.4. Energy Harvesting and Housing. The other important component of the smart camera is the power supply and energy harvesting system that controls charging and permits to choose optimal energy savings policies. The power supply system includes the lead (Pb) acid battery pack and a module for harvesting energy through photovoltaic panel. A wide range of photovoltaic panels can be integrated, thus making possible to respond to the energy requirements in any application scenario. For the experimentation a 12V panel with a nominal power of 15W was used and managed by a controller outputting up to 4A DC current. The controller has protection against deep battery discharging and can deal with battery having capacity between 10 and 40Ah. A 12Ah battery has been normally used for tests.

5.3. Human-Machine Interface (HMI). A web platform and mobile applications has been developed, depicted in Figure 9, to validate the implementation and operational evaluation of the presented architecture taking into consideration integration with existing traffic management system. The main purpose of these applications is to integrate collected data from the developed wireless sensors and transmitting the aggregated results to a higher hierarchical level. A middleware has been designed and implemented over Low power Wireless Personal Area Networks (6LoWPAN) and

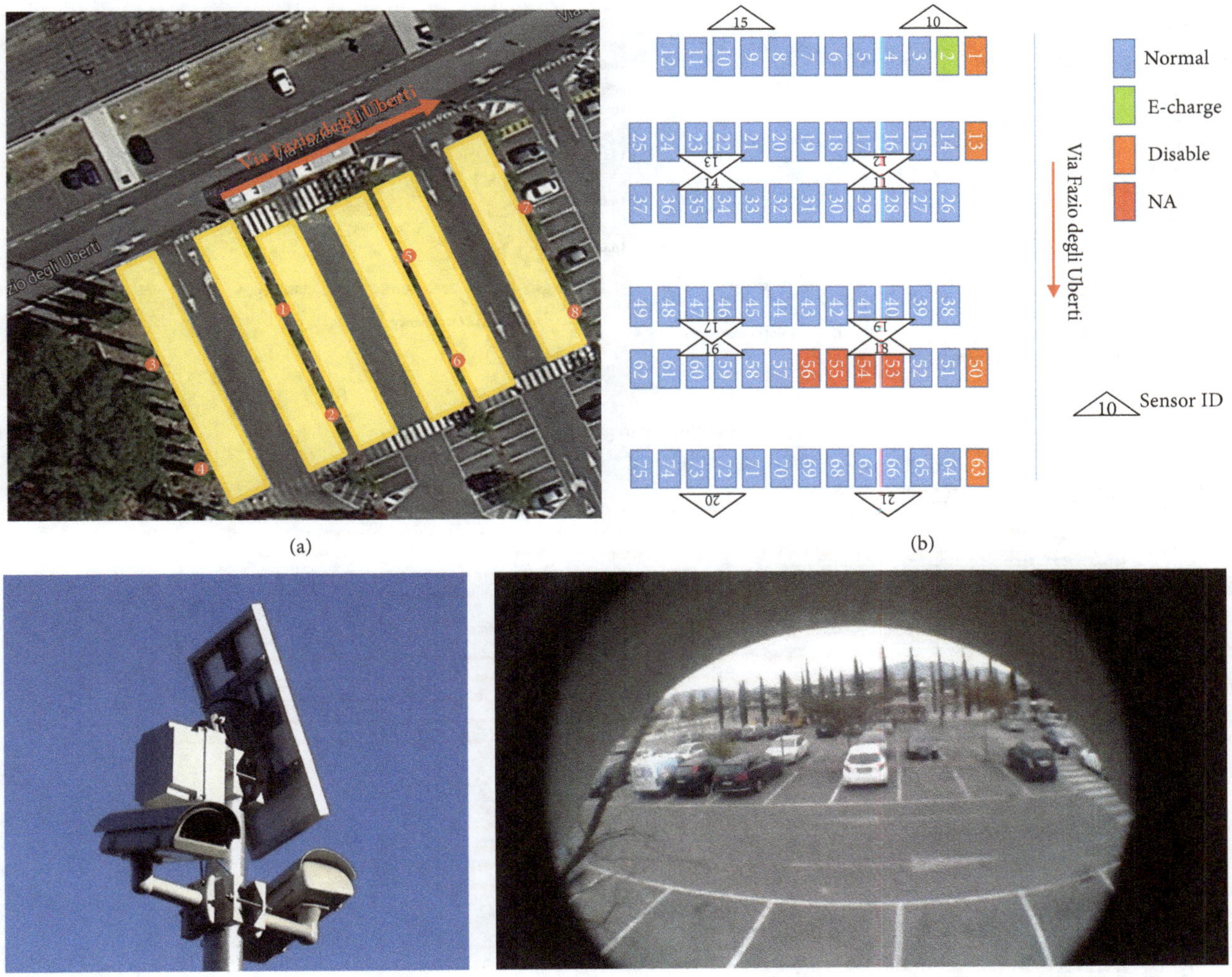

FIGURE 10: Results: (a) pole displacements for sensor installation, (b) schema of parking lot monitoring with sensor and spaces IDs, (c) example of installation of two sensors on a single pole, and (d) parking view from sensor ID = 16.

Constrained Application Protocol (CoAP), both specialized for tiny devices, to enable web services to manipulate simple resources (e.g., vision sensors) [21]. The basic tasks that were performed are as follows: sensors monitor the status of the parking area, the GWs manage the sensed data and process it and publish the event and information messages, and the information messages are displayed via both mobile and web application.

6. Equipment Installation

Tests were conducted in the north-west border of the city of Pisa, Italy. The parking area is very near to Pisa main touristic attractions, including the leaning tower, to the hospital premises and to several buildings and departments of the local university. The experimental area included two subregions consisting in the main access road (on which flow analysis is performed) and a parking area subregion, on which cooperative parking monitoring is deployed. A total of 12 sensors were deployed on 8 poles as shown in Figure 10(a). These sensors monitored a total of 75 parking lots consisting of 1 *Electric Cars slots*, 4 *Disabled slots* and 70 *Regular slots*. The schema of these parking slots with their ids is depicted in Figure 10(b) and the deployed sensors are shown in Figure 10(c). It must be noticed that there is a number of slots which are covered by more than one sensor. On these ones, the middleware performs aggregation of the results produced by each embedded vision node according to the model described in (4).

7. Results

7.1. Vision Based Results. The results of the monitoring of a typical working day (November 2, 2015) scenario is shown in Figure 11(a). The monitoring, due to wintertime (i.e., short daylight time) and to the need to save the battery at night-time, took place starting from 8:00 hours until 18:00 hours for each day. The occupancy status was recorded every 15 minutes

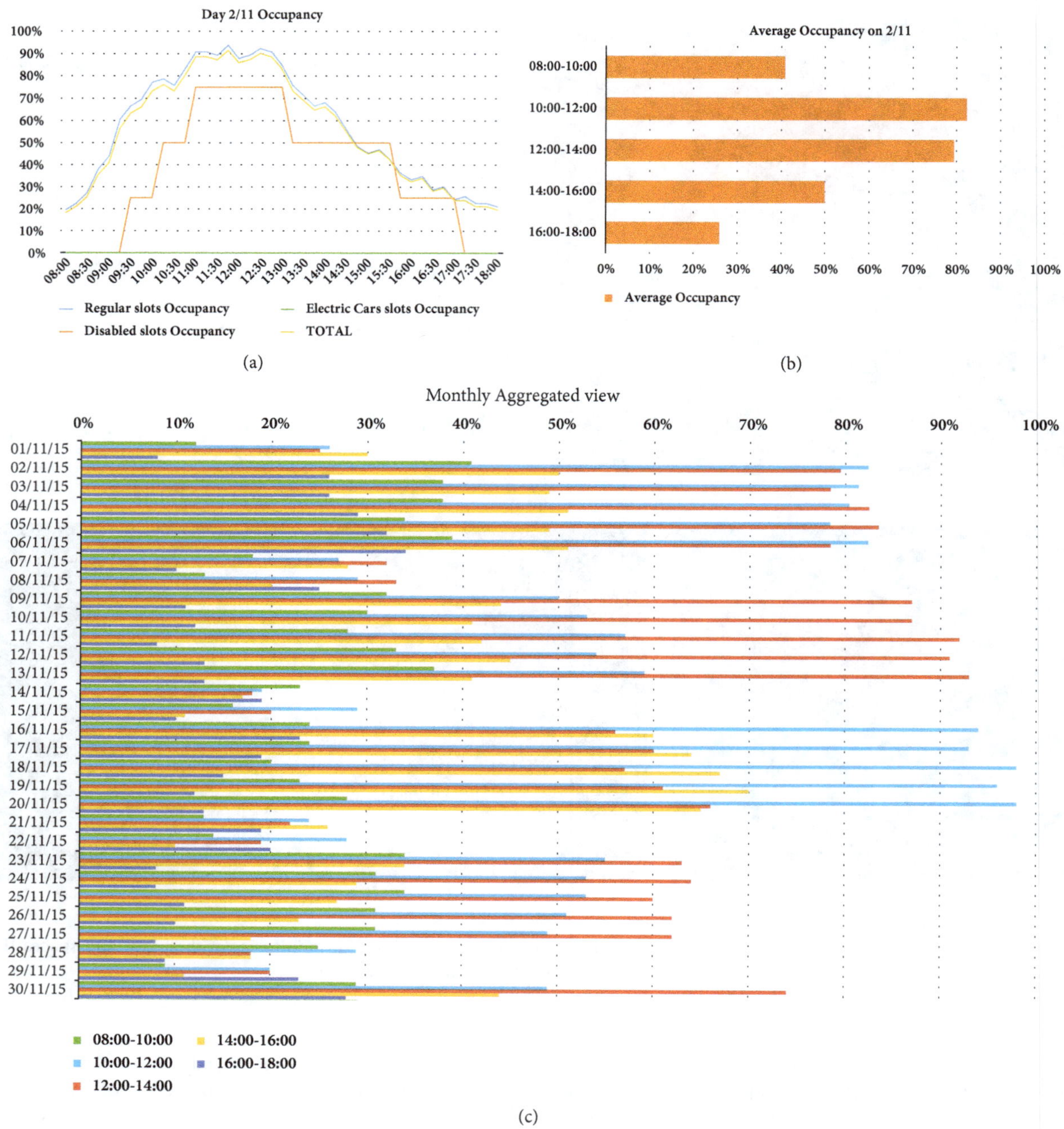

FIGURE 11: Results: (a) Occupancy status of the parking on November 2, 2015, (b) parking lot occupancy on November 2, 2015 aggregated by time slots, and (c) aggregated occupancy data by time slots for the month of November. Aggregated occupancy data by time slots for the month of December.

as depicted in Figure 11(a). The data are shown aggregated by time windows of 2 hours in Figure 11(b). The occupancy peak for that day was 92% of the capacity recorded at 11:45. As it is shown by these figures, the Electric Cars slot happened to be never occupied on that day, and this has happened several times during the monitoring; also for this reason, Pisa mobility agency had previously decided to decrease the Electric Cars available slots from two to only one.

A more extensive assessment of the results is given by a monthly view of the aggregated data and depicted in Figure 11(c). The figure shows the whole month of November 2015. The obtained results confirm the usage of Via Pietrasantina parking as a typical long-stay swapping parking used also for mobility exchange reasons (e.g., fast bus stop link to center and train station). In fact, the occupancy ratio reaches its highest values early and quickly in the morning and only decreases slowly in the afternoon to get to its minimum towards the offices closing time. Another relevant information regards weekends low usage of the parking (i.e., never exceeding 35% of the capacity).

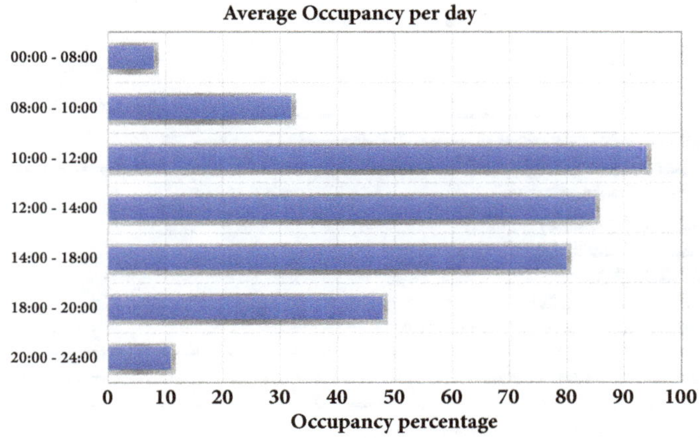

FIGURE 12: Average parking lots occupancy per day in outdoor parking.

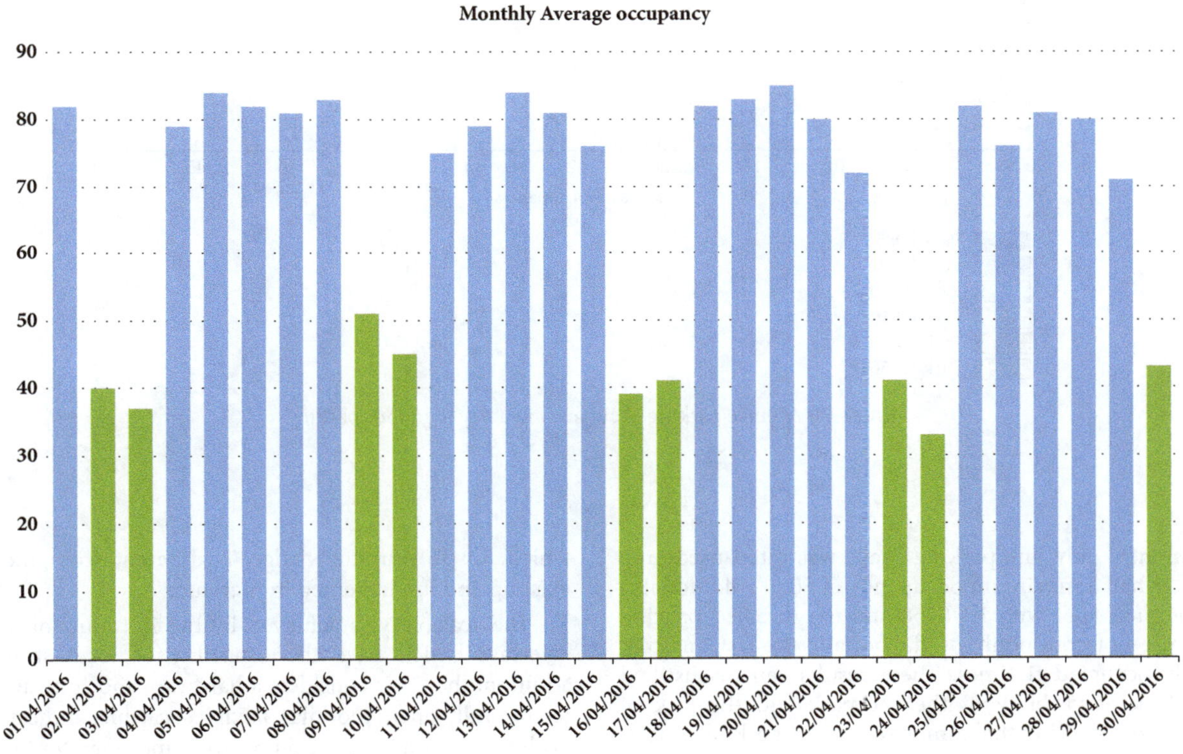

FIGURE 13: Parking occupancy for the whole month in outdoor parking.

7.2. Sensors Based Results. Sensors were deployed on an outdoor parking lot in order to avail the car parking experience on a typical working day. Figure 12 presents the average occupancy of the park on a daily basis. An overview on the graph shows a 12-hour window pattern, an higher occupancy rate between 8 and 20 followed by an abrupt decrease and almost vacancy in the remaining hours. Moreover, it can be seen that the parking lot is not used in full capacity having its maximum attendance between 10:00 and 12:00 (95%). After this peak, the number of occupied parking slots gradually descends until reaching the end of the 12 hour window. These results and subsequent pattern can be explained by the park vicinity, mostly composed by working spaces namely: offices and the university campus. This assertion was also verified by the low car park attendance during weekends showed in the average occupancy in the month of April depicted in Figure 13. The graph also shows a slight decrease on the number of cars parked on Fridays.

A different deployment was also conducted on a private underground parking lot and integrated with the system. Six sensors were used and placed on different parking spaces, one of which was reserved for persons with disabilities (Park ID: 5048). Figure 14 shows the occupancy rate of each space per hour. The parking lot usage was very reduced throughout the

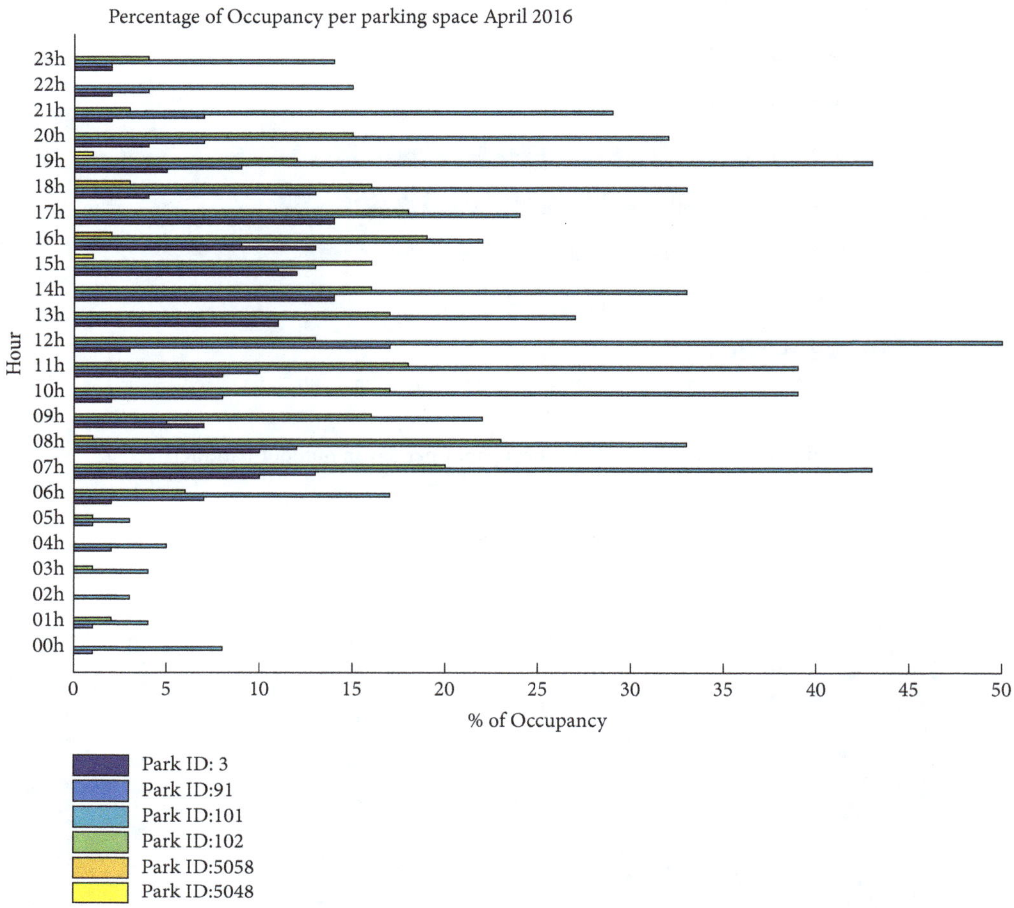

FIGURE 14: Average parking occupancy in a day of indoor parking.

entire month, only one parking space was often used and, even so, it had a maximum occupancy of 50%. Additionally, both the park spot with ID 5058 and the disabled parking space were not commonly used, specially the last one that was only occupied 11 times. These results can be justified by the lack of visibility of the car park and the lack of information regarding the number of vacant parking spaces. Thus, an application capable of providing real-time car park availability and free parking spaces for nearby drivers could improve both the parking lot usage as the entire parking experience.

8. Conclusions

In this paper, we have presented a fully scalable and distributed architecture that is easily integrated with multiple sensing sources.

The main novelty of the proposed approach lies in the context of scalability, and in particular the reduced costs for obtaining it and the ease of introducing a new element to the existing infrastructure. The joint design of these two aspects in our approach, in particular due to the distributed architecture designed, allows having a global complex system which, up to our knowledge, is different from the many existing and implemented in literature.

The scalability is achieved by the introduction of local gateways which have the capability of local storage. In addition, the GWs considers a hardware and software architecture fully open, modular, flexible, scalable, and capable of evolution that allows the authorities and operators to easily implement traffic management strategies and monitoring traffic based on ITS cooperative technologies.

The proposed architecture was then validated via real world deployment tests by integrating the two types of parking management systems. The magnetic sensors presented are fully encapsulated, dust/water resistant, enabling both indoor, and outdoor deployments and do not requiring any maintenance. The chosen communications technology is LoRa on the 868MHz band and enables low power consuming communications, but the periodicity and duration (size) of the communications was also optimized, for greater power savings. These sensors are suitable for mostly any type of car park and enable fast and simple deployments. Since they do not require infrastructure or cabling, the system becomes cheaper, involving less costs in terms of deployment and maintenance. The developed sensors were installed and tested in two different location: indoor parking and outdoor parking. The developed system was integrated in a novel distributed architecture in order to be able to provide scalability and an ease to the drivers in finding available parking lots, thus saving fuel and time. Scalability and reliability are granted by means of the designed distributed architecture.

Following the sensor based parking system, a smart camera prototype for vision based smart parking is also presented. Main emphasis is given to the vision board, which is a custom PCB manufactured specifically for the sensor. Cameras, wireless module, batteries, photovoltaic panels, and housing have also been described. In addition, a parking lot monitoring algorithm specifically designed for smart cameras and featuring an internode processing strategy for the final decision making is presented. 12 smart camera based sensors were deployed in on 8 poles in the city of Pisa, Italy. The parking lots were monitored for a whole month and various occupancy data were recorded: they proved to be useful both to disseminate real-time information to tourists and commuters as well to design and deploy adaptive policies for access and payment of parking lot facilities.

In the end, as already mentioned above our proposed approach seems to be quite effective in being applied to typical urban environments, in which the deployment of such infrastructures occurs gradually over time and with very different geographical dislocations.

Conflicts of Interest

The authors declare that they have no conflicts of interest.

Acknowledgments

The research leading to these results has received funding from the European Union's Seventh Framework Programme (FP7) under Grant Agreement no. 3176711.

References

[1] EU, Passenger cars in the EU - Statistics Explained - European Commission, http://ec.europa.eu/eurostat/, Accessed on: 15 May 2017.

[2] US. Data and Statistics — Bureau of Transportation Statistics, URL: https://www.rita.dot.gov/bts/data_and_statistics/index.html, Accessed on: 15 May 2017.

[3] M. Magrini, D. Moroni, G. Pieri, and O. Salvetti, "Smart cameras for ITS in urban environment," *Intelligent Transport Systems: Technologies and Applications*, pp. 167–188, 2015.

[4] SIEMENS, Future of Infrastructure: The Smart Way to Park, last visited: June 2016.

[5] M. Kodransky and G. Hermann, *Europe's Parking U-Turn: From Accommodation to Regulation*, Institute for Transportation and Development Policy, 2011, https://www.itdp.org/wp-content/uploads/2014/07/Europes_Parking_U-Turn_ITDP.pdf, Accessed on: 15 September, 2017.

[6] V. W. S. Tang, Y. Zheng, and J. Cao, "An intelligent car park management system based on wireless sensor networks," in *Proceedings of the SPCA 2006: 1st International Symposium on Pervasive Computing and Applications*, pp. 65–70, August 2006.

[7] T. Lin, H. Rivano, and F. Le Mouel, "A survey of smart parking solutions," *IEEE Transactions on Intelligent Transportation Systems*, vol. 18, no. 12, pp. 3229–3253, 2017.

[8] L. Ruizhi, R. Cristian, B. Peter, O. Shumao, and C. Liping, "Crowdsourcing on-street parking space detection," https://arxiv.org/abs/1603.00441, 2016.

[9] J. Vera-Gómez, A. Quesada-Arencibia, C. García, R. Suárez Moreno, and F. Guerra Hernández, "An Intelligent Parking Management System for Urban Areas," *Sensors*, vol. 16, no. 6, p. 931, 2016.

[10] G. Amato, F. Carrara, F. Falchi, C. Gennaro, and C. Vairo, "Car parking occupancy detection using smart camera networks and Deep Learning," in *Proceedings of the 2016 IEEE Symposium on Computers and Communication, ISCC 2016*, pp. 1212–1217, July 2016.

[11] E. C. Neto, E. S. Reboucas, J. L. De Moraes, S. L. Gomes, and P. P. R. Filho, "Development control parking access using techniques digital image processing and applied computational intelligence," *IEEE Latin America Transactions*, vol. 13, no. 1, pp. 272–276, 2015.

[12] G. Maria, E. Baccaglini, D. Brevi, M. Gavelli, and R. Scopigno, "A drone-based image processing system for car detection in a smart transport infrastructure," in *Proceedings of the 18th Mediterranean Electrotechnical Conference, MELECON 2016*, pp. 1–5, April 2016.

[13] Q. Wu, C. Huang, S.-Y. Wang, W.-C. Chiu, and T. Chen, "Robust parking space detection considering inter-space correlation," in *Proceedings of the IEEE International Conference on Multimedia and Expo, ICME 2007*, pp. 659–662, July 2007.

[14] S. Funck, N. Möhler and W. Oertel, "Determining car-park occupancy from single images," in *Proceedings of the 2004 IEEE Intelligent Vehicles Symposium*, pp. 325–328, June 2004.

[15] C.-C. Huang, Y.-S. Dai, and S.-J. Wang, "A surface-based vacant space detection for an intelligent parking lot," in *Proceedings of the 2012 12th International Conference on ITS Telecommunications, ITST 2012*, pp. 284–288, IEEE, November 2012.

[16] C. Gálvez del Postigo, J. Torres, and J. M. Menéndez, "Vacant parking area estimation through background subtraction and transience map analysis," *IET Intelligent Transport Systems*, vol. 9, no. 9, pp. 835–841, 2015.

[17] Intelligent Cooperative Sensing for Improved traffic efficiency, http://www.ict-icsi.eu/.

[18] M. Magrini, D. Moroni, G. Palazzese, G. Pieri, G. Leone, and O. Salvetti, "Computer Vision on Embedded Sensors for Traffic Flow Monitoring," in *Proceedings of the 18th IEEE International Conference on Intelligent Transportation Systems, ITSC 2015*, pp. 161–166, September 2015.

[19] Z. Shelby, "Embedded web services," *IEEE Wireless Communications Magazine*, vol. 17, no. 6, pp. 52–57, 2010.

[20] Z. Shelby and C. Bormann, *6LoWPAN: The wireless embedded Internet*, vol. 43, John Wiley & Sons, 2011.

[21] A. Azzara, M. Petracca, and P. Pagano, "The icsi m2m middleware for iot-based intelligent transportation systems," in *Proceedings of the 2015 IEEE 18th International Conference on Intelligent Transportation Systems*, pp. 155–160, 2015.

2

Highway Travel Time Prediction using Sparse Tensor Completion Tactics and K-Nearest Neighbor Pattern Matching Method

Jiandong Zhao,[1,2] Yuan Gao,[3] Jinjin Tang,[1] Lingxi Zhu,[1] and Jiaqi Ma[4]

[1]*School of Traffic and Transportation, Beijing Jiaotong University, Beijing, China*
[2]*MOE Key Laboratory for Urban Transportation Complex Systems Theory and Technology, Beijing Jiaotong University, Beijing, China*
[3]*School of Mechanical and Electronic Control Engineering, Beijing Jiaotong University, Beijing, China*
[4]*Department of Civil and Architectural Engineering and Construction Management, University of Cincinnati, Cincinnati, OH, USA*

Correspondence should be addressed to Jiandong Zhao; zhaojd@bjtu.edu.cn

Academic Editor: Steven I. Chien

Remote transportation microwave sensor (RTMS) technology is being promoted for China's highways. The distance is about 2 to 5 km between RTMSs, which leads to missing data and data sparseness problems. These two problems seriously restrict the accuracy of travel time prediction. Aiming at the data-missing problem, based on traffic multimode characteristics, a tensor completion method is proposed to recover the lost RTMS speed and volume data. Aiming at the data sparseness problem, virtual sensor nodes are set up between real RTMS nodes, and the two-dimensional linear interpolation and piecewise method are applied to estimate the average travel time between two nodes. Next, compared with the traditional K-nearest neighbor method, an optimal KNN method is proposed for travel time prediction. optimization is made in three aspects. Firstly, the three original state vectors, that is, speed, volume, and time of the day, are subdivided into seven periods. Secondly, the traffic congestion level is added as a new state vector. Thirdly, the cross-validation method is used to calibrate the K value to improve the adaptability of the KNN algorithm. Based on the data collected from Jinggangao highway, all the algorithms are validated. The results show that the proposed method can improve data quality and prediction precision of travel time.

1. Introduction

Travel time is a traffic parameter that can reflect traffic conditions effectively, and it is the most popular traffic information for travelers. How to predict travel time accurately and timely is a classic research question in intelligent transportation system (ITS) research.

In many countries, the travel time estimation and prediction studies are based on the traffic data collected by loop vehicle detectors. In China, although the highway mileage ranks first in the world currently, because of the limitation of early construction budget, no loop vehicle detectors or other traffic detection equipment was set on highways. In recent years, with the development of the remote transportation microwave sensor (RTMS) technology, some highways in China began to put an RTMS every 2 to 5 km to monitor the traffic flow conditions. RTMS is a fixed traffic information collection device installed by a side mount (the working process of RTMS is described in Section 2). This makes it possible to carry out highway travel time prediction in China.

Abnormal operation, noise interference, or transmission line failure can lead to incomplete traffic data collected by RTMSs, and the large distance between two sensors on a highway can lead to a sparse data problem. Data missing due to equipment failure and other hardware-related problems can be regarded as data missing at random. After introducing the concept of virtual sensor nodes, the sparse data problem caused by the large distance between microwave sensors can be regarded as data missing at the virtual node. These phenomena inevitably affect the quality of the traffic data collected by the remote transportation microwave sensor, thus affecting the accuracy of travel time prediction. In order to obtain accurate and reliable travel time prediction results, the first step is to solve the above two kinds of traffic data missing.

Aiming at repairing the traffic data missing at random, Coifman [1] proposed a creative interpolation algorithm based on kinetic-wave models to predict travel time. Li et al. [2] pointed out that Coifman's method is deviated, and proposed a new algorithm based on the temporal-spatial queueing model to describe the fast travel time variations using only the speed and headway time series that is measured at upstream and downstream detectors. Zhang et al. [3] pointed out that the mobility and scale limit of probe vehicles may lead to incomplete or inaccurate data and proposed a method to fill the missing data through the weighted average method and exponential smoothing method combined with the characteristics of road network. Zhang and Zhang [4] established a prediction model based on a vector autoregressive model and a general regression neural network with the spatial relationship of adjacent location traffic volume, and concluded that the latter is superior to the former when the missing data ratio is higher. Zhao et al. [5] adopted a two-dimensional linear time-space interpolation method to interpolate microwave vehicle data and predicted travel time based on the K-nearest neighbor pattern matching method. Qu et al. [6] used the matrix completion method to estimate the missing data. Tang et al. [7] estimated missing detector data based on fuzzy C-means and a genetic algorithm. Using weekly data similarity, the data vector structure is transformed into a matrix pattern for completion. Duan et al. [8] designed a denoising stacked autoencoder deep learning model and trained the model hierarchically using the full set of data from all vehicle detector stations to achieve traffic data imputation. Li et al. [9] compared the data completion effect both of the probability and of the kernel probability principal component analyses through the fusion of multidetection point data, showing that imputation errors can be reduced using temporal-spatial dependence. Li et al. [10] pointed out that the PCA trend brings about several advantages to traffic time series analysis and used PCA for missing data completion. Then, they used the multivariable linear regression method for traffic volume prediction. Tan et al. [11] introduced the tensor model and tensor decomposition to the issue of recovering missing traffic data. This model can solve the extreme situation where the data is completely lost. They used the dynamic tensor model to express the traffic data and predict short-term traffic volumes [12]. Considering the traffic volume time-space information, Ran et al. [13] reconstructed traffic volume data in the four-dimensional tensor mode and interpolated the missing data with a low-order tensor completion algorithm. They also converted traffic state estimation to a missing data estimation problem [14]. Zhou et al. [15] recovered the missing OD data by the tensor completion method, taking into account the low-dimensional potential structure and time-space characteristics of volume.

Through the above literature, it can be seen that the existing recovery theories and methods for missing traffic data at random can be attributed to the form of vector, matrix, and tensor, according to the combination of traffic data. The missing data recovery method based on the structure of "vector or matrix" can be mainly divided into statistical type, interpolation type, and predictive type. This method does not excavate the multimode traffic data characteristics such as time, space, week, and day, and the completion precision is limited. The tensor model is a new method to solve the missing data problem that can completely use the multimode characteristic of data and complete missing data in high-dimensional space [11].

Aiming at solving the sparse data problem, Quiroga and Bullock [16, 17] proposed a travel time estimation method based on the velocity integral model and the position interpolation model using expressway vehicle GPS. Wang et al. [18] proposed a model based on the naïve Bayesian method to achieve traffic volume estimation on the road network where GPS samples are not covered. Xu [19] used a two-dimensional linear interpolation method and a segmentation method to estimate the travel time between expressway stations by remote transportation microwave sensor data. The author of this paper has proposed a solution to the sparse RTMS data problem: Zhao et al. [20] used a two-dimensional linear interpolation method to solve the problem of sparse microwave detector data. This method has had a positive effect on improving the effectiveness of traffic congestion detection.

As a research hotspot of modern intelligent transportation systems, there are many travel time prediction methods. These methods can be divided into model-based methods (ARIMA method [21], Bayesian method [22], multivariable linear regression [10], neural network prediction algorithm [23], Kalman filter algorithm [24], support vector machine [25], etc.), and data-driven methods (k-nearest neighbor matching [5], nonparametric regression [26], etc.). In this paper, the travel time is predicted based on the RTMS data. The sampling interval is short and the sampling frequency is high, so a large amount of data can be collected in a short time. The model-based method has too many parameters when dealing with massive data, and model structure is complicated. The data-driven approach does not need establishing a model or the identification of a large number of parameters. Therefore, it is suitable for handling massive data. The K-nearest neighbor algorithm is a data-driven method, which can make use of multimodal information to cluster data from similar modes together for prediction. The basic idea behind the K-nearest neighbor algorithm is to obtain a complete historical database, extract the data characteristics, and then match the historical data pattern that is most similar to the current data pattern for future situation prediction. The author of this paper has studied travel time prediction based on the K-nearest neighbor algorithm [5]. However, in practical application, it is found that the traditional K-nearest neighbor algorithm has some limitations, such as poor ability to adapt caused by the fixed value of K and low prediction accuracy when there is a large amount of abnormal data or under congested traffic conditions. Therefore, this paper uses the K-nearest neighbor algorithm to predict travel time and further optimizes the K-nearest neighbor algorithm.

In summary, the major purpose of this paper is similar to the literature [10]: to solve the issues caused by missing data

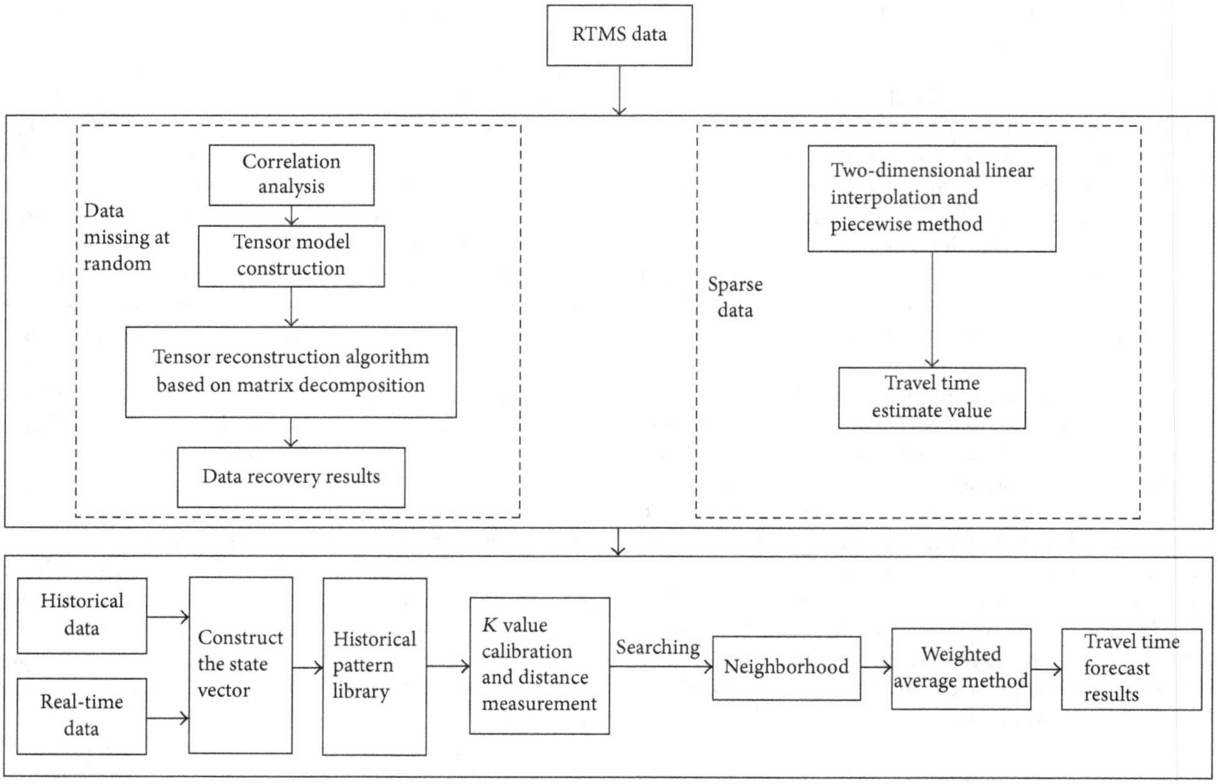

FIGURE 1: Framework of the paper.

and sparsity problem and then improve the accuracy of travel time estimation and prediction. Therefore, this paper firstly uses the data completion method based on the tensor form to complete the missing RTMS data. Then, the concept of virtual nodes is introduced, and the two-dimensional linear interpolation method is used to fill the completely missing data at the virtual node to estimate the travel time. The traditional travel time prediction method based on the K-nearest neighbor algorithm is optimized. The state vector (speed, traffic volume, and time of day) is subdivided. The degree of traffic congestion is increased with the new state vector. The K value is calibrated by the cross-validation method. Finally, the method was verified using the remote transportation microwave sensor data of Jinggangao highway. The framework of this paper is shown in Figure 1.

2. Data Preprocessing

Because RTMS has the characteristics of sublane detection, the collected data cannot be directly used for the travel time prediction. It should be sorted into section data first. Due to equipment failure, transmission failure, and other reasons, there are some abnormalities in RTMS data, which seriously affected the quality of the data. This section describes the data preprocessing process. Firstly, the sublane data is integrated into section data. Secondly, an abnormal data judgment method is based on the threshold method and traffic flow theory.

For the preprocessing of RTMS data, the author of this paper has proposed a method in the literature [19]. For the convenience of the reader's understanding, this part introduces the data preprocessing method in the literature [19].

The data in this paper was collected from the Jinggangao highway, Dujiakan station to Jingliang Road station, between January 1, 2016, and March 15, 2016. The experiment segment has three toll stations. The number of sensors is 11. The experiment segment has three lanes in each direction. The highest speed limit of the freeway is 120 km/h and the length of this highway segment (Dujiakan station to Jingliang Road station) is 9.6 km. Jinggangao Expressway is an expressway linking Beijing with major southern cities such as Guangzhou, Zhuhai, Hong Kong, and Macao. The section studied in this paper is located in the capital of China, Beijing. It is one of the main lines to get in and out of Beijing in the southwest. The traffic on the experiment segment is heavy every day and it experiences traffic congestions during peak hours.

As shown in Figure 2, the RTMS [20] is a fixed-type traffic information collection device based on digital radar technology, which can obtain real-time traffic information data. Take traffic speed detection as an example; when a vehicle enters the monitoring area in Figure 2, the RTMS will detect and record its entry time. When the vehicle leaves the detection area, the remote transportation microwave sensor will also record its departure time. The vehicle speed can be

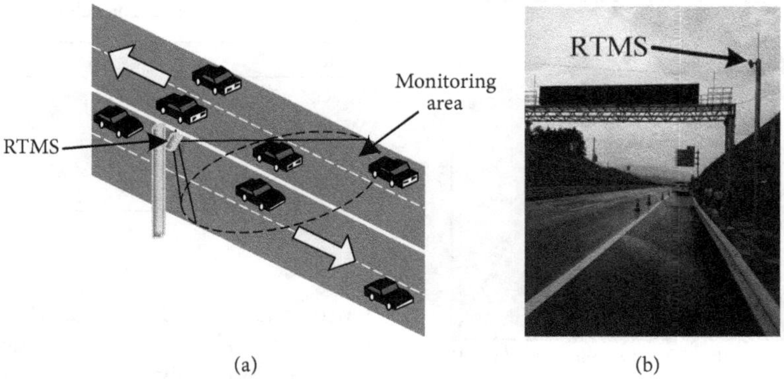

FIGURE 2: Remote transportation microwave sensor. (a) Detection principle. (b) Field application.

TABLE 1: Traffic information collected by microwave detectors.

Number	Definition	Description
(1)	*occur_time*	*Detection time*
(2)	*device_id*	*ID number of device*
(3)	*intensity*	*Traffic volume*
(4)	*speed*	*Speed of vehicle*
(5)	occupy	Time occupancy
(6)	car_class	Type of vehicle
(7)	*direction_id*	*Direction of lane*
(8)	*lane_id*	*Number of lane*
(9)	*road_id*	*Number of road*
(10)	transfered	Flag of transfer

calculated using the time difference and detection distance. Therefore, the detected traffic speed is the average speed of all the vehicles that pass through the detection zone during a certain period for a certain lane. RTMS can detect other traffic data including intensity, occupancy, and car style. Table 1 shows the traffic information that can be collected by RTMS. This paper uses 7 italic items in Table 1.

2.1. Data Integration. RTMS has the characteristic of sublane detection. RTMS detects traffic data of six lanes. However, the basis of traffic flow prediction is instantaneous section speed and intensity. Taking into account the contribution rate of traffic volume on the section speed calculation, a weighted method that can convert all the data of each lane to section data is presented [5]. The data of six lanes was integrated into available section average speed and total intensity data, as shown in (1), and the data integration period is 5 minutes.

$$v = \frac{\sum V_s \times Q_i}{q},$$
$$q = \sum Q_i, \tag{1}$$

where v represents the average speed of all vehicles passing through the detection section; V_s represents the average speed of the vehicle passing through the detection section in each lane during the detection period; Q_i represents the traffic volume of each lane passing through the detection section during the detection period; q represents the total traffic volume through the detection section during the detection period.

2.2. Abnormal Data Identification. The abnormal data judgment method is based on the threshold method and traffic flow theory. The threshold method takes into account the fact that the value of the traffic volume parameter must change within a reasonable threshold interval at a certain time interval. Traffic volume theory holds that the relationship between the three parameters of traffic volume satisfies a certain regularity. Based on the theory of traffic volume, this paper obtains the discriminant rules of threshold form and realizes the identification and judgment of abnormal data. According to the characteristics of the test section, the rules were determined as follows.

Step 1 (recognition based on the threshold method). The rules determined according to the threshold method are shown in (2). The collected traffic volume data that does not meet (2) can be recognized as abnormal data.

$$0 \le Q \le f_c \times C \times \frac{T}{60},$$
$$0 \le v \le f_v \times V_f, \tag{2}$$
$$0 \le O \le 100\%,$$

where Q is the vehicle volume; C is the road capacity; T is the data acquisition cycle; f_c is the correction factor (the value of f_c varies between 1.3 and 1.5); v is the location average speed; V_f is the limit speed of the road; O is the occupancy; f_v is the correction factor (the value of f_v varies between 1.3 and 1.5).

Step 2 (recognition based on traffic volume theory). According to the traffic volume theory and the vehicle driving conditions on the road, the abnormal data determination rule is as follows:

(1) The location average speed V is 0; the volume Q is not 0.

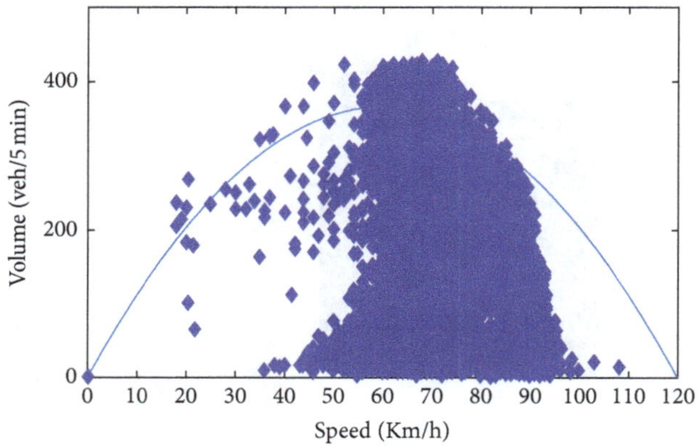

FIGURE 3: Experiment data distribution.

(2) The volume Q is 0, and the occupancy rate O and the location average speed v are not 0 at the same time.

(3) The occupancy rate is 0, and the volume rate is greater than the set value.

(4) The relationship between speed and volume is satisfied in

$$Q > \frac{4Q_m}{V_f^{\,2}} \times v \times \left(V_f - V\right) \times f_Q, \tag{3}$$

where Q_m is the maximum volume and f_Q is a correction factor.

2.3. Validation and Analysis. According to the "Technical Standard of Highway Engineering (JTG B01-2014)," it is known that the road capacity of expressways in China's plains area is 2,000 vehicles per hour and the speed limit is 120 km/h. Therefore, according to the traffic engineering theory, the relationship between speed and volume should meet

$$\begin{aligned} Q &\le \frac{4Q_m}{V_f^{\,2}} \times v \times \left(V_f - v\right) \times f_Q \\ &\le \frac{5v\,(120 - v)}{9} \times f_Q \cdot \text{veh/h} \\ &\le \frac{11v\,(120 - v)}{108} \cdot \text{veh/5 min}. \end{aligned} \tag{4}$$

Figure 3 is a speed–volume scatter plot drawn using the data from one sensor on a test road section. On the y-axis, the unit is veh/5 min in one direction. The points outside the curve are recognized as abnormal data.

3. Random Missing Traffic Data Recovery with Tensor Reconstruction

Abnormal operation, noise interference, transmission line failure, or abnormal data removal can lead to a random data-missing problem. The data-missing problem inevitably affects the quality of the traffic data collected by the RTMS. In this section, a missing data recovery method based on the static tensor theory is described. The reconstruction procedure of traffic data based on the static tensor model includes a correlation analysis of traffic data, establishment of the traffic tensor model, a low-rank analysis of traffic data, and the recovery method based on a multilinear low full-rank decomposition model.

TABLE 2: The relevance of each pattern.

Pattern	Correlation
Space	0.9763
Week	0.9910
Day	0.9011
Time	0.2656

3.1. Correlation Analysis of Traffic Data. Similarity is one of the factors impacting imputation performance. Combining the multidimensional characteristics of traffic data, mining the multimode similarities will make a great contribution to imputing missing values [11]. In this paper, the correlation analysis is used to reveal the similarity of traffic data.

The multimode characteristic is the result of people's understanding and the description of characteristics from different angles. The analysis result is different when people analyze one thing from different angles or using different methods. For the traffic data, it showed significant multimode characteristics when observed from the day, week, space, or time mode, and the researchers found that traffic data showed a strong multicorrelation in these modes. Figure 4 shows the distribution of remote transportation microwave sensor data in each mode. Figure 4(a) shows the distribution of data in day mode and time mode; Figure 4(b) is a three-dimensional diagram which shows the distribution of data in space mode, where the x-axis represents the moment, the y-axis represents the location, and the z-axis represents the volume. Figure 4(c) is a three-dimensional diagram which shows the distribution of data in week mode. The x-axis represents the moment, the y-axis represents the week, and the z-axis represents the volume. It can be seen that RTMS data has similar characteristics in different modes. It can be seen from Figure 4(a) that the volumes between 8 and 9 AM on Thursday were lower than other periods. As 8-9 AM is the peak of the weekday, traffic congestion during this time of Thursday caused a drop in traffic volume.

In this paper, the following equation in the literature [10] is used to give the correlation of the various modes of traffic data:

$$S = \frac{\sum_{n \ge i \ge j \ge 1} R\left(i, j\right)}{n\left(n - 1\right)/2}, \tag{5}$$

where n refers to the whole data points and $R(i, j)$ refers to the correlation coefficient matrix.

Table 2 shows the correlation of traffic data tensors in day mode, week mode, space mode, and time mode.

It can be seen from Table 2 that the traffic data show a high correlation in space, week, and day modes. However,

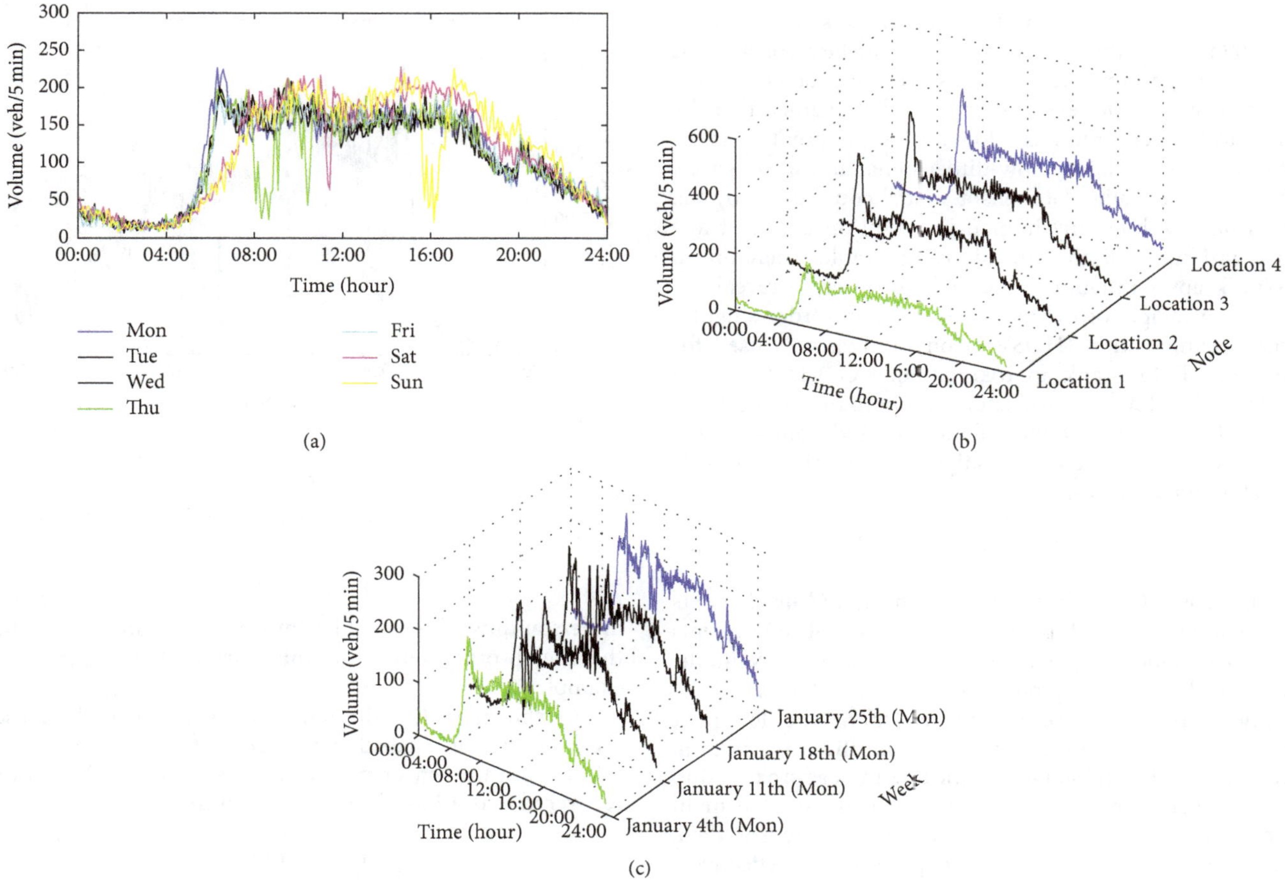

FIGURE 4: The trend of the various modes of traffic data. (a) Day mode and time mode. (b) Space mode. (c) Week mode.

in the time mode, the traffic volume shows different states at different times, which makes the traffic volume tensor exhibit a low correlation in time mode. It can also be seen that, due to the weak correlation between weekdays and weekends, compared to the "week" mode, the correlation in the "day" mode is low.

3.2. Establishment of Traffic Data Tensor Model. The tensor is a multidimensional array, which is a high-order extension of the first-order vector and the second-order matrix. The position of the element needs to be represented by three or more variables. The N-order tensor is expressed as $\mathscr{A} \in \mathbb{R}^{I_1 \times I_2 \times \cdots \times I_N}$.

Section 3.1 illustrated that the expressway traffic data show strong temporal and spatial correlation, which provides a sufficient base for the construction of a traffic data tensor model. Based on the analysis results of Section 3.1, this paper constructs the tensor form of traffic data $\mathscr{A} \in \mathbb{R}^{4\times 11\times 7\times 288}$, as shown in Figure 5, by combining the multimode traffic volume spatial-temporal characteristics.

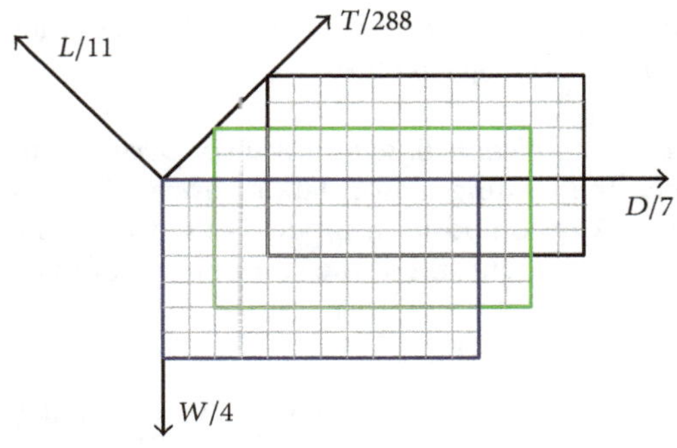

Model structure:
$L/11$ indicates that the model contains 11 sensors
$W/4$ represents a total of 4 weeks of data
$D/7$ means 7 days a week
$T/288$ represents a day of statistical traffic 288 times
Existing similarity among sensors, days, and hours

FIGURE 5: Tensor model of traffic data.

3.3. Low-Rank Analysis of Traffic Data. The reason why the tensor completion algorithm can successfully estimate the missing data in the traffic tensor is that the low-rank hypothesis can successfully approximate the multimode correlation of the traffic data. In essence, when the tensor is of low rank or has sparse characteristics, it can be said that the tensor is redundant and can be expressed in a more compact form, so that a complete tensor can be reconstructed with fewer sampling tensors. To solve the problem of lost data recovery of the tensor, it is generally assumed that the tensor to be recovered has a low-rank structure; that is, its data is distributed in a low-dimensional linear subspace. Obviously, the traffic data tensor model constructed by the actual data without any human intervention is not of low rank, and most of the time it is full rank. Therefore, a low-rank approximation is needed. In order to reveal the working principle of the low rank, the high-order singular value decomposition (HOSVD) on the tensor is used for a multimode low-rank approximation. HOSVD can use a tensor with a low-rank structure to approximate a known tensor. The HOSVD model of the tensor decomposes the tensor into a kernel tensor, multiplying a matrix of n-mode in each mode as shown in

$$\mathscr{A} \approx G \times_1 U_1 \times_2 U_2 \times \cdots \times_N U_N. \quad (6)$$

The tensor G is called a kernel tensor, and its elements represent the relationship between the principal components of various models. U_N is the expansion matrix of N-mode. $G \times_N U$ is the N-mode product of tensor G and matrix U.

Figure 6 shows the partial results of the low-rank approximation of the traffic tensor using the n-mode rank $[3, 3, 3, 3]$. As shown in Figure 6, the low-rank approximation of the tensor with rank $[3, 3, 3, 3]$ can maintain the main change of $4 \times 11 \times 7 \times 288$ traffic volume tensor data. The appropriate multi-low-rank hypothesis can effectively capture the multirelevance of traffic volume, gain the main changes of traffic volume data, and thus estimate the missing data.

In this paper, the tensor recovery algorithm based on low-rank and multilinear full-rank decomposition has the same purpose as the HOSVD model. They all make a low-rank approximation for a given tensor.

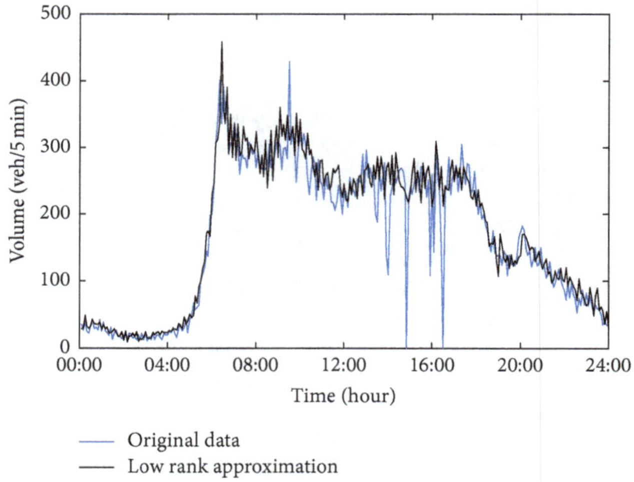

FIGURE 6: Low-rank approximation results of traffic data.

3.4. Tensor Completion Based on Matrix Decomposition. In this paper, the missing tensor data recovery via multilinear low/full-rank decomposition model is used to estimate the missing data in a tensor. For the convenience of the reader's understanding, this part first introduces the method taking the fourth-order tensor in this part of the paper as an example.

As shown in (7), the tensor completion method uses the optimization algorithm to recover the missing data by minimizing the tensor rank as low as possible. That is, the recovered tensor has a low-rank structure.

$$\begin{aligned} &\min_{\mathscr{A} \in R^{I_1 \times I_2 \times I_3 \times I_4}} \quad \operatorname{rank}(\mathscr{A}), \\ &\text{s.t.} \quad \mathscr{A}_\Omega = \mathscr{B}_\Omega, \end{aligned} \quad (7)$$

where rank($\mathscr{A}$) represents the rank of the tensor and Ω is the set of known element subscripts. $\mathscr{B}$ is the original volume (or speed) tensor and $\mathscr{A}$ is the recovered tensor. The constraint is that the value with the subscript Ω is equal to the value of the corresponding data in $\mathscr{B}$.

However, the formula is not solvable. Assuming that the given tensor has an n-mode low-rank structure, (7) can be transformed to minimize the n-ranks of the given tensor, as shown in

$$\begin{aligned} &F\left(\min_{\mathscr{A} \in R^{I_1 \times I_2 \times I_3 \times I_4}} \left(\operatorname{rank}(A_{(1)}), \operatorname{rank}(A_{(2)}), \operatorname{rank}(A_{(3)}), \operatorname{rank}(A_{(4)})\right)\right) \\ &\text{s.t.} \quad (A_{(i)})_\Omega = (B_{(i)})_\Omega, \quad i = 1, 2, 3, 4, \end{aligned} \quad (8)$$

where $A_{(i)}$ and $B_{(i)}$ are the mode-i unfolding of $\mathscr{A}$ and $\mathscr{B}$.

In order to make full use of the information of the tensor patterns, the rank of each mode of the tensor is simplified by the weighted summation strategy, as shown in

$$\begin{aligned} &\sum_{i=1}^{4} \lambda_i \left(\min_{A_{(i)}} \operatorname{rank}(A_{(i)})\right), \\ &\text{s.t.} \quad (A_{(i)})_\Omega = (B_{(i)})_\Omega \\ &\lambda_i = \frac{I_i}{I_1 + I_2 + I_3 + I_4}, \quad i = 1, 2, 3, 4. \end{aligned} \quad (9)$$

In order to simplify the tensor completion problem, it is simplified by minimizing an F norm using the full-rank decomposition of the matrix. The tensor completion problem can be transformed into

$$F\left(G_{(i)}, H_{(i)}, A_{(i)}\right) = \min_{G_{(i)}, H_{(i)}, A_{(i)}} \frac{1}{2}\left\|G_{(i)}H_{(i)} - A_{(i)}\right\|_F^2 \quad \text{s.t.} \quad \left(A_{(i)}\right)_\Omega = \left(B_{(i)}\right)_\Omega \quad \text{for } i = 1, 2, 3, 4. \tag{10}$$

$G_{(i)}$, $H_{(i)}$, and $A_{(i)}$ of the function can be obtained by solving the function $F(G_{(i)}, H_{(i)}, A_{(i)})$, as shown in

$$\begin{aligned} G_{(i)} &\longleftarrow \operatorname{orth}\left(A_{(i)}H_{(i)}{}^T\right) \longleftarrow A_{(i)}H_{(i)}{}^T \\ &\longleftarrow A_{(i)}H_{(i)}{}^{-1} \longleftarrow A_{(i)}H_{(i)}{}^T\left(H_{(i)}H_{(i)}{}^T\right)^{-1}, \\ H_{(i)} &\longleftarrow \left(\operatorname{orth}\left(A_{(i)}H_{(i)}{}^T\right)\right)^T A_{(i)} \longleftarrow G_{(i)}{}^{-1}A_{(i)} \\ &\longleftarrow \left(G_{(i)}G_{(i)}{}^T\right)^{-1} G_{(i)}{}^T A_{(i)}, \\ A_{(i)} &\longleftarrow \operatorname{orth}\left(A_{(i)}H_{(i)}{}^T\right)\left(\operatorname{orth}\left(A_{(i)}H_{(i)}{}^T\right)\right)^T \\ &\quad + \mathscr{P}_\Omega\left(B_{(i)} - G_{(i)}H_{(i)}\right) \\ &\quad \text{for } i = 1, 2, 3, 4 \text{ respectively}, \end{aligned} \tag{11}$$

$$\mathscr{P}_\Omega(B) = \begin{cases} B_{xy} & \text{if } (x, y) \in \Omega \\ 0 & \text{otherwise}, \end{cases}$$

where H^T represents the transpose of matrix H, H^{-1} represents the Moore-Penrose inverse of the matrix H, and $\operatorname{orth}(A_{(i)}H_{(i)}{}^T)$ is a set of standard orthogonal bases formed by the columns of the matrix $A_{(i)}H_{(i)}{}^T$. The core strategy of the algorithm is that when a set of variables are optimized, other group variables can be treated as constants. In the process of optimizing the tensor completion problem, three sets of variables are $G_{(i)}$, $H_{(i)}$, and $A_{(i)}$, respectively. The $G_{(i)}$, $H_{(i)}$, and $A_{(i)}$ group includes 4 variables, respectively. In order to obtain the final optimization results, the solution method used in this paper is to optimize $G_{(i)}$, $H_{(i)}$, and $A_{(i)}$ alternately, and for each optimization, the specific value of $G_{(i)}$, $H_{(i)}$, and $A_{(i)}$ will be updated once. The first optimization of the algorithm uses the initial value, and once the algorithm reaches the convergence criterion, the optimization is stopped and the result is obtained. The algorithm flow chart is shown in Figure 7.

The convergence conditions in Figure 7 are shown as

$$\begin{aligned} \mathrm{sc1} &= \frac{\left\|\mathscr{P}_\Omega\left(B_{(i)} - G_{(i)}^k H_{(i)}^k\right)\right\|_F}{\left\|\mathscr{P}_\Omega\left(B_{(i)}\right)\right\|_F} \le \mathrm{tol}, \\ \mathrm{sc2} &= \left|1 - \frac{\left\|\mathscr{P}_\Omega\left(B_{(i)} - G_{(i)}^k H_{(i)}^k\right)\right\|_F}{\left\|\mathscr{P}_\Omega\left(B_{(i)} - G_{(i)}^{k-1} H_{(i)}^{k-1}\right)\right\|_F}\right| \le \frac{\mathrm{tol}}{2}, \end{aligned} \tag{12}$$

where tol is the convergence threshold. This method converts the missing tensor recovery problem into a weighting multilinear low-rank decomposition form. Then, a differential gradient strategy and a block iterative strategy are used to solve this problem.

Figure 8 shows the traffic volume and speed data collected by a microwave detector on January 5, 2016, and the result of missing data completion.

4. Solution of the Data Sparseness Problem Based on a Two-Dimensional Linear Interpolation Method

So far, using the tensor theory, the recovery of missing data is completed. Due to the large distance between every two sensors on the highway, the collected traffic data is obviously sparse, which leads to a large calculation error when the travel time is estimated. This paper introduces virtual nodes between real nodes and solves the sparse data problem using the two-dimensional linear interpolation algorithm. Then, the velocity data after interpolation is used to estimate the travel time between two toll stations.

4.1. Virtual Node Settings. As shown in Figure 9, this paper adds some virtual microwave detection nodes on the road, which makes the distance between two neighboring nodes less than 1 km after adding the virtual node. The actual node location is where RTMS is installed.

From Figure 9, it can be seen that the sparse data is as data missing at the virtual node. In order to solve this problem, this paper uses a two-dimensional linear interpolation method [5, 20] to interpolate the velocity data at the virtual node.

4.2. Two-Dimensional Linear Interpolation Method. For the convenience of the reader's understanding, this part introduces the two-dimensional linear interpolation method in the literature [5, 20].

The principle of the two-dimensional linear interpolation method is shown in Figure 10. The first step of the method is to calculate average speeds in time of adjacent detectors utilizing the temporal linear interpolation. The equations are as follows:

$$\begin{aligned} V(d, t) &= V(d, h-1) + \frac{t - t_0}{\Delta T} \cdot \left[V(d, h) - V(d, h-1)\right], \\ V(d+1, t) &= V(d+1, h-1) + \frac{t - t_0}{\Delta T} \cdot \left[V(d+1, h) - V(d+1, h-1)\right], \end{aligned} \tag{13}$$

where $V(d, t)$ and $V(d+1, t)$ represent average speeds of detection points d and $d+1$ at time t, respectively; $V(d, h)$ represents the speed of detection point d in time period h; t_0 is the start of time period h; ΔT is the duration of h.

The second step is to calculate the average speeds of detectors by the spatial linear interpolation which can be expressed as

$$V(x, t) = V(d, t) + \frac{x(t) - x_d}{l} \cdot \left[V(d+1, t) - V(d, t)\right], \tag{14}$$

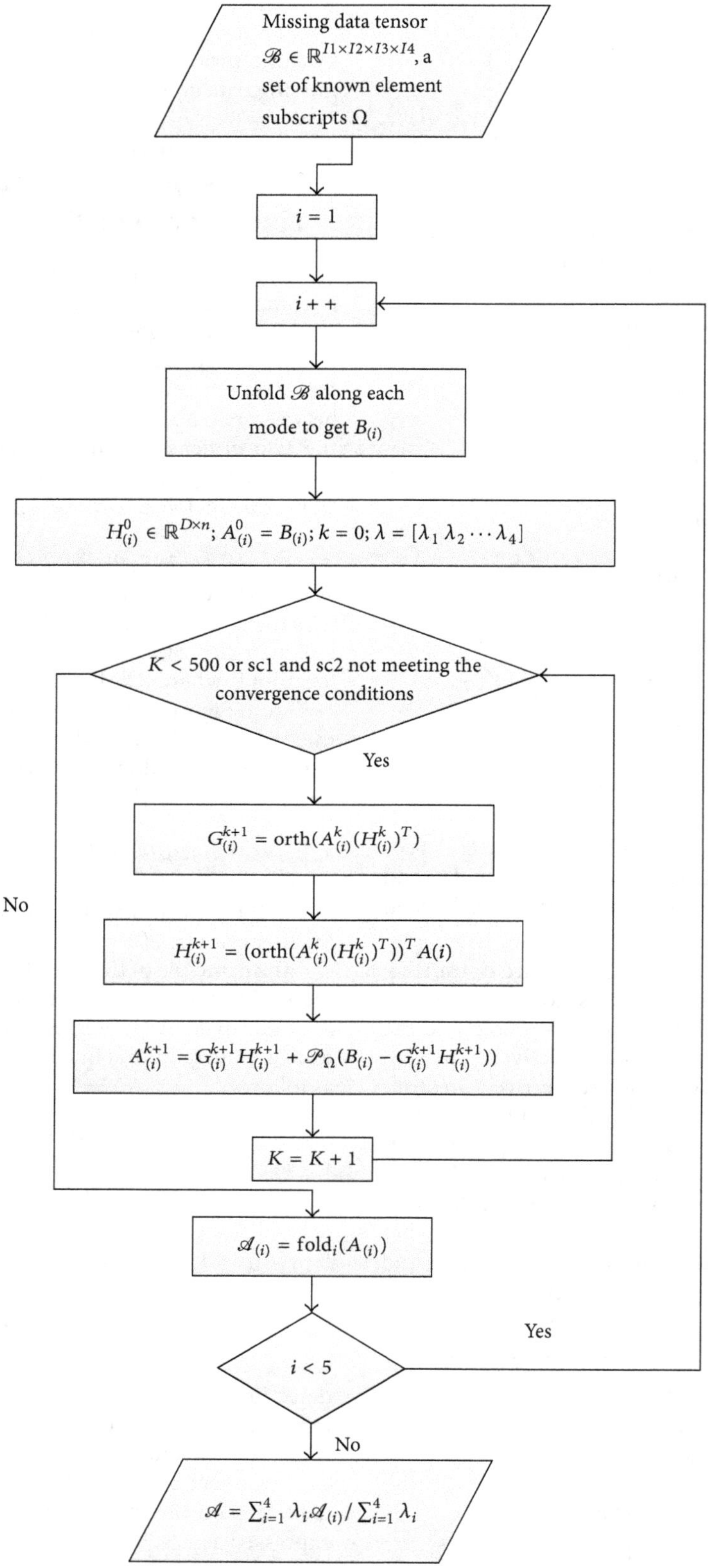

FIGURE 7: Flow chart of the algorithm.

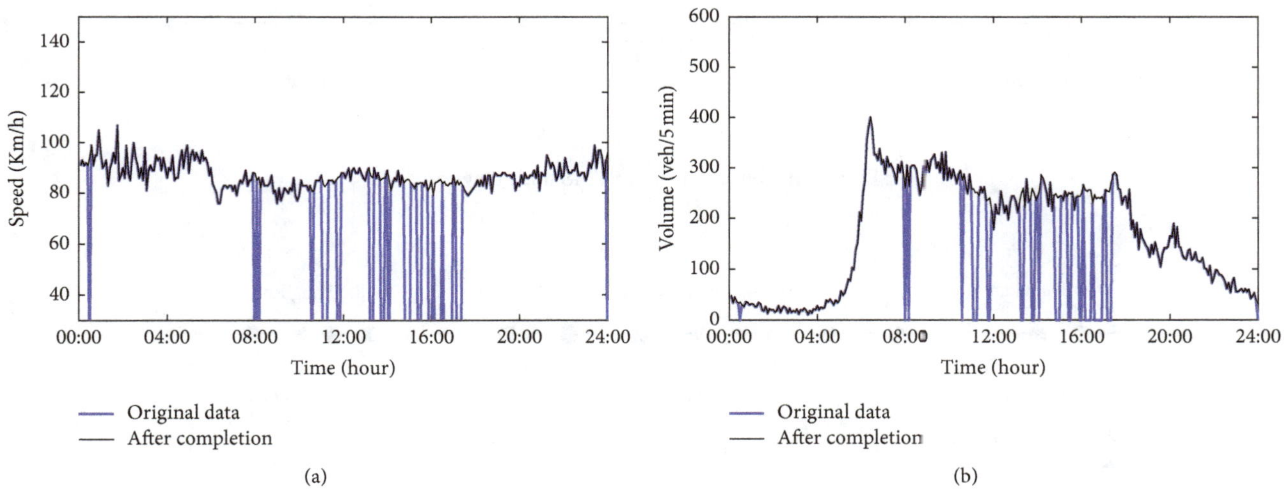

FIGURE 8: Tensor completion results (January 5, 2016). (a) Speed. (b) Volume.

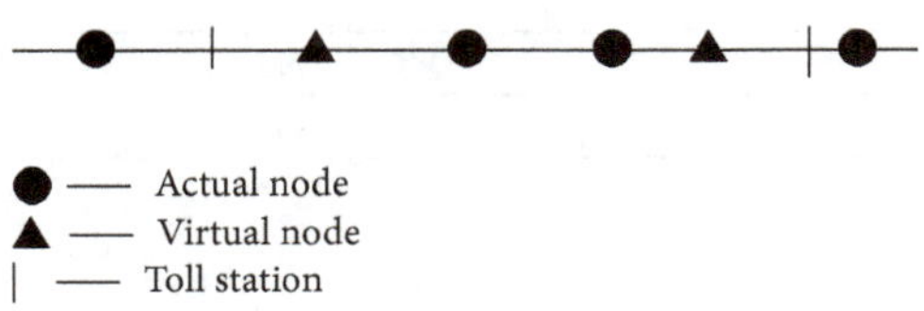

FIGURE 9: Virtual node diagram.

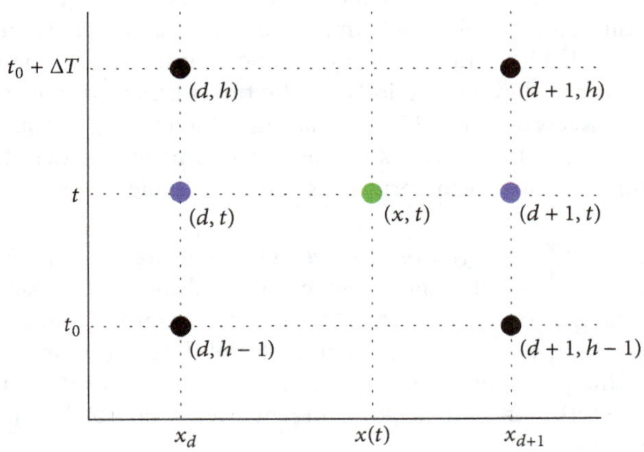

FIGURE 10: Principle of two-dimensional linear interpolation methods.

where $V(x,t)$ is the average speed of the detection point x at time t; $x(t)$ is the position where the vehicle is at time t; x_d is the position of detector d; l is the distance between the detector d and the detector $d+1$.

4.3. Travel Time Estimation. Then, the travel time is estimated based on the piecewise method as follows.

Based on Figure 11, the procedure of travel time estimation between two toll stations on the expressway is as follows.

Step 3. Using the average speed detected by RTMS before and after the toll station, the location average speed at the toll station is calculated by the two-dimensional linear interpolation method.

Step 4. The road sections between two neighboring RTMS nodes are defined as L_i. Calculate the travel time T_i on L_i.

If $L_i \leq 1000$ m, then

$$\bar{v} = \frac{(v_i + v_{i+1})}{2}, \qquad T_i = \frac{L_i}{\bar{v}}. \tag{15}$$

If $L_i > 1000$ m, then the virtual nodes are added to make the distance between the nodes less than 1 km. The travel time between two nodes can be calculated using (15).

Step 5. Calculate the travel time between toll stations A and B, $t = \sum T_i$;

Figure 12 shows the distribution of the actual detector nodes and virtual detector nodes of Jinggangao highway (Dujiakan station to Jingliang Road station).

Take the travel time estimate of Dujiakan to Jingliang Road station at 8:00 on January 4 as an example. The velocity at the virtual node and the toll station is given by the two-dimensional linear interpolation method as shown in Table 3.

Finally, the travel time calculation result is 419 s.

5. Travel Time Prediction Based on the *K*-Nearest Neighbor Algorithm

In this paper, the authors developed a travel time forecasting algorithm based on the optimal K-nearest neighbor algorithm. The traditional KNN algorithm includes five procedures: data acquisition and input, state vector construction, pattern library establishment, K value calibration, and

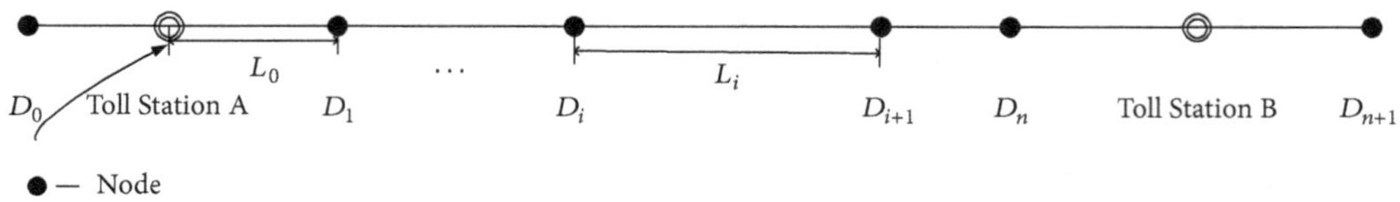

FIGURE 11: Toll station and RTMS node diagram.

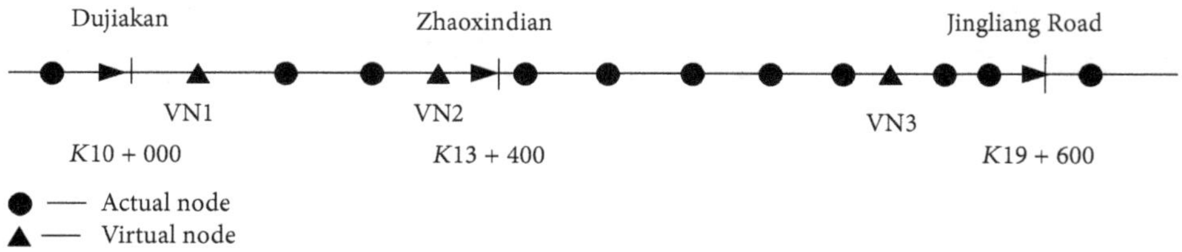

FIGURE 12: Relative position of detectors and toll stations. "▸" refers to the driving direction; "|" refers to the toll station.

TABLE 3: The section speed at the virtual node and the toll station.

Section	Dujiakan	Jingliang Road	VN1	VN2	VN3
Section speed (km/h)	46	93	64	85	90

prediction based on weighted average functions. Taking into account the deficiencies of the traditional KNN method, optimization is made in three aspects. Firstly, in addition to the three original state vectors of speed, volume, and time of day, the congestion levels are introduced as a new state vector. Secondly, by analyzing the change characteristic of traffic volume, the whole day is divided into seven periods. Thirdly, considering the low adaptability and precision caused by the fixed K value of the traditional KNN algorithm, the cross-validation method is used to calibrate the K value. The procedure of optimal K-nearest neighbor algorithm is introduced in this section.

5.1. State Vector Construction

5.1.1. Speed and Traffic Volume. The average speed and traffic volume of road sections were selected as the state vector to describe the road traffic status. As the traffic volume at the toll station cannot be measured directly, the traffic volume obtained by the first detector after the toll station consistent with vehicle heading direction was selected as a state parameter. Taking into account the continuity of the traffic state in the time dimension, the speed and volume of the first three cycles before the current state were selected as the state vector of the historical model library. A velocity vector and a traffic volume vector in the pattern library can be expressed as

$$\mathbf{M}\left(\overline{v}_{t-3T}, q_{t-3T}, \overline{v}_{t-2T}, q_{t-2T}, \overline{v}_{t-T}, q_{t-T}\right), \quad (16)$$

where t is the start time of the next travel time forecast period; T is the length of the predicted cycle; $\overline{v}_{t-nT}$ indicates the average speed of the vehicle during the period $[t - nT, t - (n-1)T]$; q_{t-nT} indicates the traffic volume during the period $[t - nT, t - (n-1)T]$.

5.1.2. Time of the Day. Time of day affects the travel time on the road; travel time is impacted by social activities and traveler habits, and the ratio of traffic volume to car type is different at different periods, so the travel times are not the same. According to [27], by analyzing the change of traffic volume at different times of the day, this paper divides the whole day into seven periods, as shown in Table 4.

5.1.3. Traffic Congestion Degree. Different weather, maintenance, accidents, and other events will affect the traffic conditions and travel times, so the road traffic condition needs to be joined into the state vector of the pattern library.

This paper uses the degree of the traffic congestion to characterize the impact of different events on road traffic conditions.

It was found that the change in road section average speed can reflect the fluctuation of the traffic volume condition. In this paper, the road section average speed was used as a standard to assess the degree of traffic congestion. Refer to the Ministry of Transport of the People's Republic of China, "Highway Network Operation Monitoring and Service Interim Technical Requirements." The degree of traffic congestion was divided into four degrees as shown in Table 5.

5.2. Pattern Library Establishment.

The state vector of the historical model library includes period, speed, traffic volume, and congestion degree. The decision attribute is travel time (estimation by method in Section 4). The state vectors and decision attributes were coded as shown in Table 6.

TABLE 4: Time of division in one day.

Time of the day	1 Flat in the morning	2 Peak in the morning	3 Trough at noon	4 Peak at noon	5 Flat at night	6 Peakat night	7 Trough at night
Start time	6:00	9:00	11:00	14:00	17:00	19:00	21:00
End time	9:00	11:00	14:00	17:00	19:00	21:00	The next day 06:00

TABLE 5: Crowded classification standard.

Traffic congestion degree	1	2	3	4
Degree	Smooth	Slightly crowded	Crowded	Seriously crowded
Speed (km/h)	>70	(50, 70]	(30, 50]	≤30

TABLE 6: Property and encoding of each state component.

Parameter type	Basic parameters (average speed and traffic volume)	Time of the day	Traffic congestion level	Decision attribute (travel time)
Attributes and coding	$\overline{v}_{t-3T}, \overline{v}_{t-2T}, \overline{v}_{t-T}$ $q_{t-3T}, q_{t-2T}, q_{t-T}$	1: flat in the morning 2: peak in the morning 3: trough at noon 4: peak at noon 5: flat at night 6: peak at night 7: trough at night	1: smooth 2: slightly crowded 3: crowded 4: seriously crowded	Travel time

5.3. *K Value Calibration.* The basic idea of KNN algorithm is to set a quantity K to search up K-nearest points to the current state in the search area, and use these points to make predictions. Use all those points if there are fewer than K points in the search area, and keep the K-nearest points if there are more than K points in the search area.

Appropriate K values play a key role in accurate predictions [5]. In order to determine the best K value of the historical data library in each period, this paper uses a cross-validation method. The specific steps are as follows.

Step 6. Assume that the minimum and maximum K values are K_{min} and K_{max}, respectively.

Step 7. The historical data of every period are randomly divided into P parts, with each being recorded as $DB_1, DB_2, \ldots, DB_P$. DB_A, $A = 1, 2, \ldots, P$ are used as test data, and the other $P - 1$ parts are combined as the new historical data library.

Step 8. Calculate the average value of the prediction performance index using the test data under different K values.

Step 9. When the performance index is optimal, the corresponding K_I is the optimal K value of this period.

The optimal K value of the test road section was calibrated as follows. The minimum and maximum K values are 1 and 50, respectively, and the historical data library in each period is divided into 10 parts on average. Taking the historical data

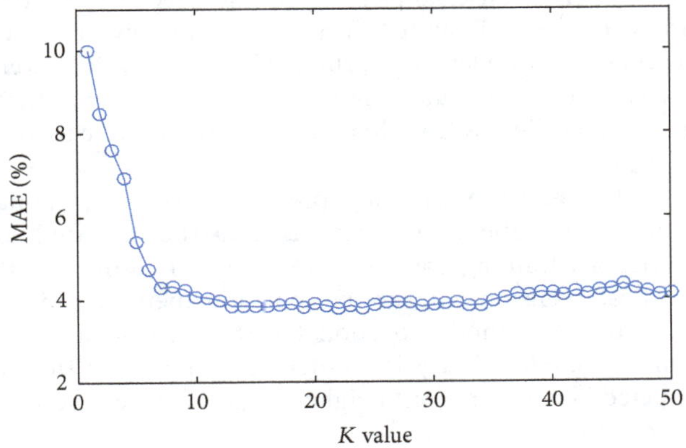

FIGURE 13: MAE under different values of K.

for the segments from Dujiakan to Jingliang Road station as an example, the results of the optimal K value calibration in different periods are shown in Table 7. Taking the peak period at night as an example, the average percentage change for absolute error in the case of different K values is shown in Figure 13. When $K = 22$, the mean absolute error is minimized.

5.4. *Travel Time Prediction Weighted Average Function.* In this paper, the weighted average method is used to make predictions based on the nearest neighbor data. Assuming

TABLE 7: Optimal values of K.

Time of the day (coding)	1	2	3	4	5	6	7
K value	20	14	13	11	16	22	7

that the searched K-nearest neighbors are $M_1 \sim M_K$, the distance between them and the current state vector is $D_1 \sim D_K$, and their decision attributes (the measured values of time) are $t_1 \sim t_K$, respectively. The predicted value of travel time (TP) is

$$\begin{aligned} \mathrm{TP} &= \sum_{i=1}^{K} \beta_i t_i, \\ \beta_i &= \frac{D_i^{-1}}{\sum_{i=1}^{K} D_i^{-1}}. \end{aligned} \tag{17}$$

D is the Euclidean distance; the Euclidean distance between two n-dimensional vectors $\mathbf{X}_1 = (x_{11}, x_{12}, \ldots, x_{1n})$ and $\mathbf{X}_2 = (x_{21}, x_{22}, \ldots, x_{2n})$ is calculated as

$$D = \sqrt{\sum_{k=1}^{n} (x_{1k} - x_{2k})^2}. \tag{18}$$

6. Test Case and Error Analysis

Taking travel time prediction for the Dujiakan to Jingliang Road station as an example, the distance between the two stations is 9.6 km. Remote transportation microwave sensor data collected from January 4, 2016, to February 29, 2016, were selected as training data. Remote transportation microwave sensor data collected from March 1 to March 15 were used as test data.

Firstly, the travel time was predicted based on original training data. The state vector is constructed based on the original training data. The travel time is estimated by the two-dimensional linear interpolation method and the segmentation method in Section 4 as the decision attribute. Then, the model library is constructed. The travel time is predicted by the K-nearest neighbor algorithm based on the test set data.

Secondly, the travel time was predicted based on data after completion. The static tensor filling method was used to fill the original data, and the state vector is constructed based on the completion data. The travel time is reestimated by two-dimensional linear interpolation and segmentation method as the decision attribute. Then, the model library is constructed. The K-nearest neighbor algorithm is used for prediction based on the test set data to obtain the travel time results.

Thirdly, the two-dimensional linear interpolation method and the segmentation method are used to estimate the travel time (ranges from March 1 to March 15) based on the test data and the result is used as the comparison data to calculate the prediction error.

In order to evaluate the travel time prediction results, the relative error (RE) and the mean absolute percentage error (MAPE) are used to measure the accuracy. The error is calculated as follows:

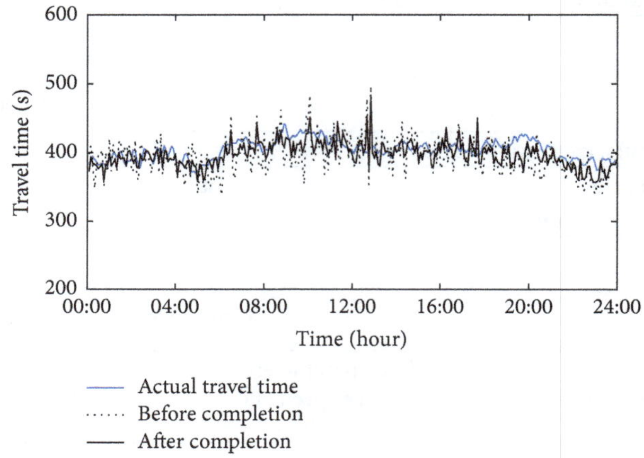

FIGURE 14: Travel time prediction result curves (03/02/2016).

$$\begin{aligned} \mathrm{RE}(t_p) &= \frac{\mathrm{TP}(t_p) - T(t_p)}{T(t_p)} \times 100\%, \\ \mathrm{MAPE} &= \frac{1}{L} \sum_{L} \frac{\left|\mathrm{TP}(t_p) - T(t_p)\right|}{T(t_p)} \times 100\%, \end{aligned} \tag{19}$$

where $\mathrm{RE}(t_p)$ is the relative error for the prediction period t_p; $\mathrm{TP}(t_p)$ is the forecast travel time for the period t_p; $T(t_p)$ is the actual travel time of the period t_p; MAPE is the average absolute relative percentage error; L is the total number of validation forecast periods.

The prediction results on March 2, 2016, March 4, 2016, and March 5, 2016, are selected as an example to analyze the prediction error of our method.

The actual travel time curve and the predicted travel time before and after data completion on March 2, 2016, March 4, 2016, and March 5, 2016, are shown in Figures 14, 16, and 18. It can be seen from Figures 14, 16, and 18 that the K-nearest neighbor algorithm can predict the travel time accurately, and the predicted travel time after the missing data is filled is closer to the actual travel time. The relative errors before and after completion are shown in Figures 15, 17, and 19. The errors before and after completion are shown in Table 8. It can be seen that the error is improved after data completion, and the mean absolute percentage error of the whole day before and after optimization is reduced.

7. Conclusion

This paper studies the travel time prediction for expressways based on RTMS data.

TABLE 8: Travel time prediction error.

Evaluation index	03/02/2016		03/04/2016		03/05/2016	
	RE	MAPE	RE	MAPE	RE	MAPE
Before completion	(−19.2, 15.3%)	7.4%	(−17.4%, 21.9%)	6.4%	(−14.9%, 12.1%)	6.1%
After completion	(−14.3, 10.8%)	3.1%	(−12.3%, 18.7%)	3.0%	(−11.1%, 9.6%)	2.8%

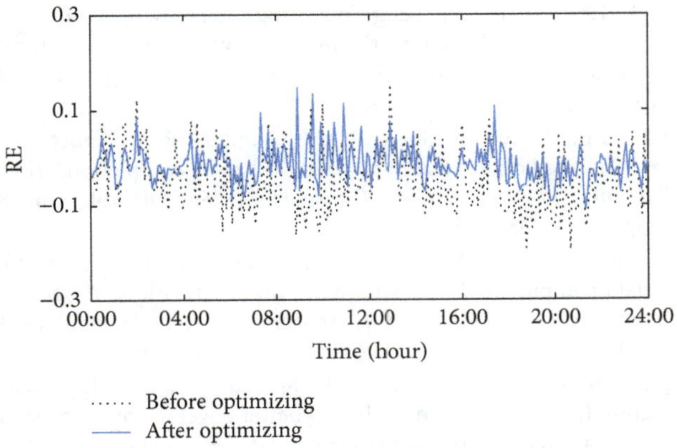

FIGURE 15: RE curves (03/02/2016).

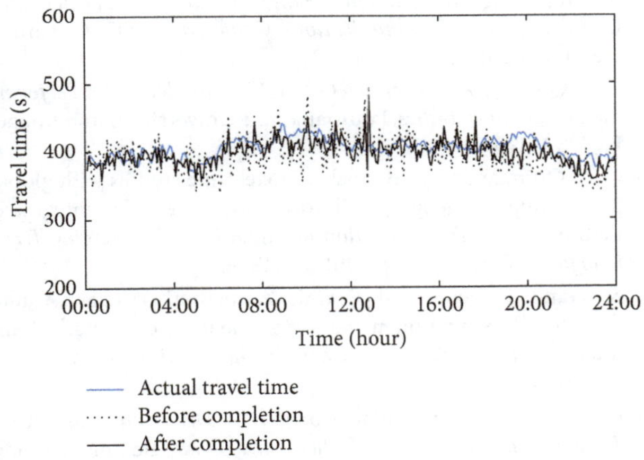

FIGURE 16: Travel time prediction result curves (03/04/2016).

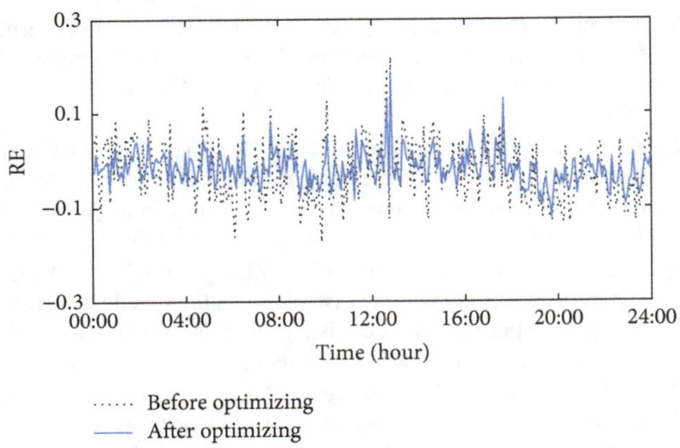

FIGURE 17: RE curves (03/04/2016).

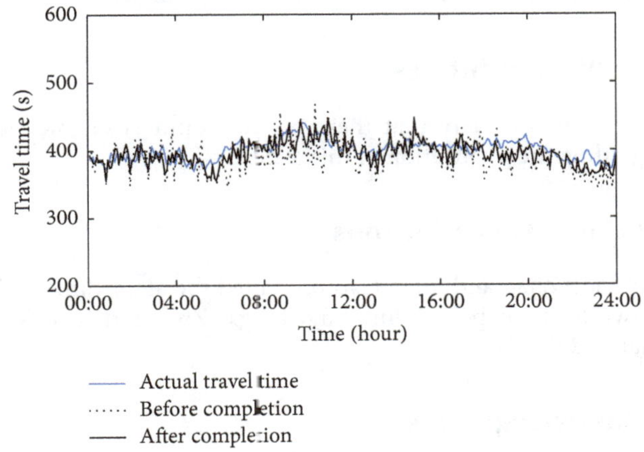

FIGURE 18: Travel time prediction result curves (03/05/2016).

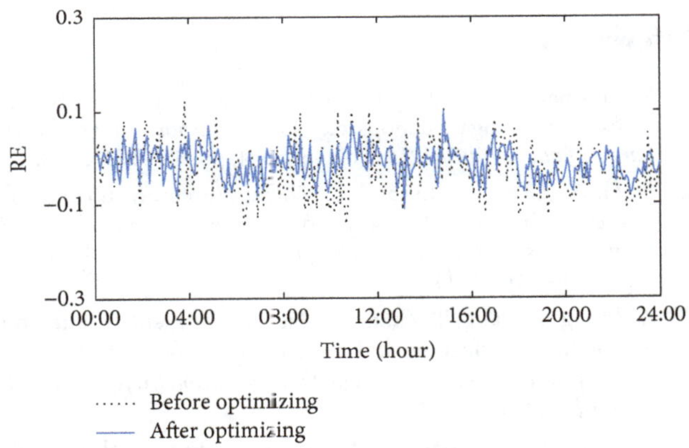

FIGURE 19: RE curves (03/05/2016).

(1) Considering the influence of the randomly missing data on prediction accuracy, the correlation and low-rank characteristics of the microwave vehicle data are analyzed, and then the traffic tensor model is established. Based on multilinear low-rank decomposition, a tensor completion algorithm is used to fill the random missing traffic data.

(2) Virtual nodes are introduced between real nodes, and the sparse data problem is solved using the two-dimensional linear interpolation algorithm. Then, the velocity data after interpolation is used to estimate the travel time between toll stations.

(3) The optimal K-nearest neighbor algorithm is proposed for travel time prediction, with the travel time as the decision attribute and the speed, traffic volume, time of the day, road section, and traffic congestion level as the state vector. The historical model library is established. Taking

the Euclidean distance as the criterion, the similarity pattern is searched using the K-nearest neighbor strategy, and the weighted average method is used to predict travel time. Practical data collected from Jinggangao highway in Beijing is used to verify the algorithm. The results show that the optimal K-nearest neighbor algorithm can accurately predict travel time, and the accuracy of prediction results using data after completion is higher than when using original data.

Conflicts of Interest

The authors declare that there are no conflicts of interest regarding the publication of this article.

Authors' Contributions

Jiandong Zhao and Yuan Gao presented the algorithms and co-wrote the paper; Jinjin Tang, Lingxi Zhu, and Jiaqi Ma analyzed the data.

Acknowledgments

This work was supported by the Fundamental Research Funds for the Central Universities (2016JBM053).

References

[1] B. Coifman, "Estimating travel times and vehicle trajectories on freeways using dual loop detectors," *Transportation Research Part A: Policy and Practice*, vol. 36, no. 4, pp. 351–364, 2002.

[2] L. Li, X. Chen, Z. Li, and L. Zhang, "Freeway travel-time estimation based on temporal-spatial queueing model," *IEEE Transactions on Intelligent Transportation Systems*, vol. 14, no. 3, pp. 1536–1541, 2013.

[3] Z. Zhang, D. Yang, T. Zhang, Q. He, and X. Lian, "A study on the method for cleaning and repairing the probe vehicle data," *IEEE Transactions on Intelligent Transportation Systems*, vol. 14, no. 1, pp. 419–427, 2013.

[4] Y. Zhang and Y. Zhang, "A comparative study of three multivariate short-term freeway traffic flow forecasting methods with missing data," *Journal of Intelligent Transportation Systems: Technology, Planning, and Operations*, pp. 1–14, 2016.

[5] J. Zhao, F. Xu, W. Liu, J. Bai, and X. Luo, "Travel time prediction based on pattern matching method," *International Journal On Smart Sensing and Intelligent Systems*, vol. 8, no. 1, pp. 658–676, 2015.

[6] L. Qu, L. Li, Y. Zhang, and J. Hu, "PPCA-based missing data imputation for traffic flow volume: a systematical approach," *IEEE Transactions on Intelligent Transportation Systems*, vol. 10, no. 3, pp. 512–522, 2009.

[7] J. Tang, G. Zhang, Y. Wang, H. Wang, and F. Liu, "A hybrid approach to integrate fuzzy C-means based imputation method with genetic algorithm for missing traffic volume data estimation," *Transportation Research Part C: Emerging Technologies*, vol. 51, pp. 29–40, 2015.

[8] Y. Duan, Y. Lv, Y.-L. Liu, and F.-Y. Wang, "An efficient realization of deep learning for traffic data imputation," *Transportation Research Part C: Emerging Technologies*, vol. 72, pp. 168–181, 2016.

[9] L. Li, Y. Li, and Z. Li, "Efficient missing data imputing for traffic flow by considering temporal and spatial dependence," *Transportation Research Part C: Emerging Technologies*, vol. 34, pp. 108–120, 2013.

[10] L. Li, X. Su, Y. Zhang, Y. Lin, and Z. Li, "Trend modeling for traffic time series analysis: an integrated study," *IEEE Transactions on Intelligent Transportation Systems*, vol. 16, no. 6, pp. 3430–3439, 2015.

[11] H. Tan, G. Feng, J. Feng, W. Wang, Y.-J. Zhang, and F. Li, "A tensor-based method for missing traffic data completion," *Transportation Research Part C: Emerging Technologies*, vol. 28, pp. 15–27, 2013.

[12] H. Tan, Y. Wu, B. Shen, P. J. Jin, and B. Ran, "Short-term traffic prediction based on dynamic tensor completion," *IEEE Transactions on Intelligent Transportation Systems*, vol. 17, no. 8, pp. 2123–2133, 2016.

[13] B. Ran, H. Tan, Y. Wu, and P. J. Jin, "Tensor based missing traffic data completion with spatial-temporal correlation," *Physica A: Statistical Mechanics and its Applications*, vol. 446, pp. 54–63, 2016.

[14] B. Ran, L. Song, J. Zhang, Y. Cheng, and H. Tan, "Using tensor completion method to achieving better coverage of traffic state estimation from sparse floating car data," *PLoS ONE*, vol. 11, no. 7, Article ID e0157420, 2016.

[15] H. Zhou, D. Zhang, K. Xie, and Y. Chen, "Spatio-temporal tensor completion for imputing missing internet traffic data," in *Proceedings of the 34th IEEE International Performance Computing and Communications Conference, IPCCC*, China, December 2015.

[16] C. A. Quiroga, *An Integrated GPS-GIS Methodology for Performing Travel Time Studies*, Louisiana state University, Baton Rouge, La, USA, 1997.

[17] C. A. Quiroga and D. Bullock, "Travel time studies with global positioning and geographic information systems: an integrated methodology," *Transportation Research Part C: Emerging Technologies*, vol. 6, no. 1-2, pp. 101–127, 1998.

[18] X. Wang, L. Peng, and T. Chi, "A method of urban traffic flow speed estimation using sparse floating car data," *Cehui Xuebao/Acta Geodaetica et Cartographica Sinica*, vol. 45, no. 7, pp. 866–873, 2016.

[19] F. F. Xu, *Highway Traffic Congestion Detection and Travel Time Prediction Based on Microwave Data*, Beijing Jiaotong University, Beijing, China, 2016.

[20] J. D. Zhao, F. F. Xu, Y. J. Guo, and Y. Gao, "Traffic congestion detection based on pattern matching and correlation analysis," *Advances in Transportation Studies*, no. 40, pp. 27–40, 2016.

[21] Y. Zou, X. Hua, Y. Zhang, and Y. Wang, "Hybrid short-term freeway speed prediction methods based on periodic analysis," *Canadian Journal of Civil Engineering*, vol. 42, no. 8, pp. 570–582, 2015.

[22] X. Fei, Y. Zhang, K. Liu, and M. Guo, "Bayesian dynamic linear model with switching for real-time short-term freeway travel time prediction with license plate recognition data," *Journal of Transportation Engineering*, vol. 139, no. 11, pp. 1058–1067, 2013.

[23] J. J. Tang, F. Liu, Y. J. Zou, W. B. Zhang, and Y. H. Wang, "An improved fuzzy neural network for traffic speed prediction considering periodic characteristic," *IEEE Transactions on Intelligent Transportation Systems*, vol. 18, pp. 2340–2350, 2017.

[24] J. Zhao, H. Wang, W. Liu, and J. Bai, "Highway travel time prediction between stations based on toll ticket data," *Tongji Daxue Xuebao/Journal of Tongji University*, vol. 41, no. 12, pp. 1849–1854, 2013.

[25] Y.-X. Sun, W. Guan, Y. Ge, G.-H. Zhang, and X. Yang, "Short-time forecasting of passenger travel time based on SVM method," *Xitong Gongcheng Lilun yu Shijian/System Engineering Theory and Practice*, vol. 34, no. 6, pp. 1587–1592, 2014.

[26] L. C. Huang, *Research on Link Travel Time Estimation and Prediction Based onData Driven*, Beijing Jiaotong University, Beijing, China, 2015.

[27] X. Wang, X.-H. Chen, and X.-M. Yang, "Short term prediction of expressway travel time based on K nearest neighbor algorithm," *Zhongguo Gonglu Xuebao/China Journal of Highway and Transport*, vol. 28, no. 1, pp. 102–111, 2015.

3

Application of Customer Segmentation for Electronic Toll Collection

Chao Qian, Meng Yang, Peiqi Li, and Shuguang Li

School of Electronic and Control Engineering, Chang'an University, Middle-Section of Nan'er Huan Road, Xi'an, Shaanxi 710064, China

Correspondence should be addressed to Chao Qian; chinaqianchao@163.com

Academic Editor: Giuseppe Musolino

Applying big data technology, this study presents a customer segmentation method of Electronic Toll Collection (ETC) based on vehicle behavioral characteristics. A segmentation index system of ETC customers comprising Recency, Frequency, and Monetary is extracted and constructed using ETC data. The whole-sample clustering analysis of ETC customers is accomplished with the Clustering LARge Applications (CLARA) algorithm while overcoming the invalidation problem of big data clustering. A decision tree on ETC customer segmentation is constructed and transformed into a set of segmentation rules. Empirical results indicate that the proposed method is better able to analyze travel characteristics and to present values and appreciation potentials for ETC customer classification. This method provides an innovative idea for implementing precision marketing and establishing hierarchical discount rates for ETC customers. Furthermore, it provides theoretical support to increase the ETC customer scale and payment ratio, thus improving the decision-making level in expressway operation and management.

1. Introduction

Electronic Toll Collection (ETC) is an essential part of the Intelligent Transportation System (ITS). ETC not only reduces travel time and energy consumption but also saves infrastructure and operation costs; thus, its advanced payment system is highly praised around the world. By the end of February 2017, 29 of 31 provinces in mainland China (except Tibet and Hainan) had realized networking of expressway ETC and cumulatively built 14,285 ETC lanes, 1,115 self-supporting service centres, and 37,502 cooperative agency centres. The number of ETC customers exceeded 47.67 million, and the daily average transaction number was over 8.1 million, accounting for 31.17% of the total traffic volume [1].

Since the 1990s, along with remarkable development in customer-oriented management, Customer Relationship Management (CRM) proposed by the Gartner Group Consulting Company has attracted extensive attention worldwide [2, 3]. CRM provides reliable, comprehensive, and complete understanding for enterprises through the application of emerging technologies to integrate customer data efficiently, helpfully maintaining and expanding a mutually beneficial relationship between customers and enterprises. Aiming to allocate service resources rationally and implement customer strategies accurately, customer segmentation classifies and evaluates types of customers, thus providing theoretical and methodological guidance for enterprises' gain of higher commercial value for customers.

Any efficient CRM needs a strong foundation of customer segmentation research. Currently, research on CRM has mainly focused on the telecom industry [4–6], the energy supply industry [7, 8], and the retail industry [9]. In the automobile dealership field, Tsai et al. (2015) considered customer transaction behavior and customer satisfaction variables, using customer segmentation to develop marketing strategies [10]. Some studies have also been conducted on the customer segmentation method in different transportation modes such as railways and aviation. For instance, Wei (2012) proposed and designed a segmentation system structure for airline customers based on ant colony clustering [11]; Teichert et al. (2008) proposed an airline customer segmentation approach by analyzing more than 5800 airline passengers' stated preference data [12]. Chiang (2017) proposed a model

TABLE 1: ETC data format.

Field	Data type	Field description
CardNo	Char(20)	ETC card number
EntryTime	Datetime	Date and time a vehicle enters an ETC lane
ExitTime	Datetime	Date and time a vehicle exits ETC lane
VehicleClass	Int	Vehicle class (passenger vehicle: 1–4)
VehicleType	Int	Vehicle type (passenger / freight vehicle)
ETCMoney	Int	Payment amount
...	...	Other fields

to discover valuable travellers for airlines and generated useful association rules to find an optimized target market for CRM systems [13]. As for railways, Cheng and Huang (2014) examined the influence of ticketing channel attributes on high-speed rail passengers' preferences and designed appropriate ticketing channel services for certain types of passengers [14]. Zhang and Peng (2017) proposed a k-means based segmentation model for railway freight customers [15]. Zhong and Guo (2008) clustered freight-customer history data and classified new customers using Bayesian classifiers [16]. Duan et al. (2016) operationalized approaches to identify market segments for rail freight services and measured the importance that customers attach to rail service attributes (i.e., transport cost, time, frequency, reliability, and safety) [17].

In highway transportation, many studies have been conducted on ETC implementation [18, 19], focusing on the analysis and evaluation of the transition phase toward this new technology, as well as cost-benefit evaluation during construction, reconstruction, or an extension period of ETC [20, 21]. Astarita et al. (2001) designed a microscopic traffic simulation model to evaluate the operational efficiency of a toll station after ETC system was progressively introduced. The results indicated that the limited capacity of manual toll gates could lead to queues spill back, interfering and reducing ETC gates capacities [22]. Zarrillo et al. (2009) emphasized the significant influence of customer satisfaction to ETC usage rate and suggested that providing an appropriate incentive for regular commuters to convert from manual usage to ETC usage would be the best way to increase the throughput of a toll plaza [23].

The current research of ETC data application mainly concentrates on traffic information extraction. Ozbay et al. (2011) demonstrated that real-time travel time could be accurately estimated using ETC data [24]. Furthermore, Yang and Ozbay (2015) illustrated the potential of ETC data mining for travel time estimation for both incident-free and incident conditions [25]. However, to date, academic literature on ETC customers is relatively rare. How to obtain consumption characteristics and tap ETC customers' payment potential by analyzing the massive ETC data, to enhance customers' value and realize precision marketing, are critical problems confronting ETC promotion and application.

This work's primary goal was to establish a customer segmentation method based on ETC consumption characteristics by applying big data analysis and mining technology. A segmentation index system was established, ETC customers were classified into categories of one to five stars, and a set of segmentation rules were extracted. In the end, travel characteristics and service strategies for each customer type were analyzed.

2. Materials and Methods

2.1. Segmentation Index. With consumption demand as a starting point, customer segmentation divides customers into similar consumer groups according to differences in their purchasing behavior. Customers on the same base have a certain degree of similarity, but customer bases show distinctions [26]. Customer segmentation models based on Recency (R), Frequency (F), and Monetary (M) or RFM behaviors, which was proposed by Hughes (1994) [27], are widely used. In this model, R represents how recently customers purchased, F how often they purchased, and M how much they spent (each time on average). Hughes believes that R, F, and M have the same degree of importance in measuring customers, and, therefore, each receives the same weight. Meanwhile, through empirical analysis of credit card data, Stone (2007) asserts that each index's weighting in customer segmentation is not the same; F should be the highest, R second, and M the lowest [28].

Expressway ETC data records various kinds of travel information, including, for example, ETC card information, travel time, vehicle information, and consumption situation. Table 1 lists ETC data's detailed format.

Each ETC datum represents an ETC customer's consumption record on a trip. An ETC customer's annual consumption can be summarized and analyzed via data aggregation. ETC customers' segmentation indexes are defined as recent consumption interval, annual frequency, and annual consumption amount (Table 2).

Hence, each ETC customer's annual consumption was aggregated according to its card number. For a particular ETC customer, assuming a frequency of F, then the indexes of R and M are calculated as follows:

$$R = T_{set} - T_{F\_out} \tag{1}$$

$$M = \sum_{i=1}^{F} S_i \tag{2}$$

where T_{set} represents a specified time, T_{F_out} means the Fth consumption time in the statistical year (driving through an

TABLE 2: Segmentation index of ETC customers.

Index	Abbreviation	Unit	Meaning
Recency	R	Hour	Time difference between the last time ETC used during the statistical year and a specified time.
Frequency	F	Times	Number of times ETC was used for a particular customer during the statistical year.
Monetary	M	Chinese yuan	How much paid for ETC for a particular customer during the statistical year.

Input:
D - ETC customer index dataset;
k - the number of clusters;
samples - number of samples to be drawn from the dataset;
sampsize - number of observations in each sample.
Output:
The clustering results of ETC customer.
Method:
(1) for $i = 1$ to *samples*, repeat (a)-(d);
(a) select *sampsize* objects randomly from ETC customer index dataset D as a sample, apply the PAM algorithm to compute the best k-medoids – $[M_1, M_2 \cdots M_k]^T$;
(b) apply k-medoids to the entire dataset D and calculate the distance from every non-medoids object in D to the closest object in the set $[M_1, M_2 \cdots M_k]^T$, reassign each ETC customer to different clusters;
(c) compute the average dissimilarity of this clustering, if the value is less than the current minimum value, then replace the current value, and form the best k-medoids and the new set of k representative objects;
(d) return to step (1), repeat the iterative process;
(2) until no change, output clustering results of ETC customer.

ALGORITHM 1: CLARA algorithm.

ETC exit lane), and S_i denotes the monetary value of the ith time paid for ETC.

2.2. Customer Clustering. A three-dimensional space of RFM indexes was obtained from the segmentation index system previously mentioned. ETC customer clustering analysis means grouping the index dataset in such a way that index data in the same group (called a cluster) are more similar (in one sense or another) to each other than to those in other clusters. This task can be summarized as making the distance between clusters as long as possible and minimizing distances from the same cluster, thus obtaining a classification method for multi-class ETC customers.

Partition-based clustering methods aim to decompose the set of objects into a set of disjoined clusters where the user predefines the resulting number of clusters. The k-means algorithm and the k-medoids algorithm are the most classical and the most commonly used partition-based clustering methods. Compared with the k-means, the k-medoids algorithm eliminates sensitivity to outliers, applicable only to small datasets because of its high computational complexity. Partitioning Around Medoids (PAM) algorithm realizes k-medoids clustering iteratively and greedily, i.e., in the iterative process, the greedy strategy is adopted to improve clustering quality by setting the maximum number of iterations. PAM works efficiently for small datasets but does not scale well for large datasets [29].

To deal with more massive datasets, Kaufman and Rousseeuw (2008) proposed a sampling-based PAM algorithm—CLARA (Clustering LARge Applications)—which solved the PAM algorithm's problem in big data processing [30]. Instead of considering the whole dataset, CLARA uses a random sample and then applies the PAM algorithm to compute the best medoids from the sample. After repeated sampling, CLARA builds clusterings from multiple random samples and returns the best clustering as output. Algorithm 1 displays the ETC customer clustering procedure using the CLARA algorithm.

The distance from every non-medoids object O_j to different medoids $M_f (f = 1, 2, \cdots, k)$, represented as $d(O_j, M_f)$ is measured by Euclidean distance in the CLARA algorithm, as shown in the following:

$$d\left(O_j, M_f\right) = \sqrt{\sum_{u=1}^{3} \left(o_{ju} - m_{fu}\right)^2} \tag{3}$$

where u represents the index dimension of ETC customer and o_{ju} and m_{fu} denote the corresponding dimension values of O_j and M_f.

Input:
D - ETC customer index dataset and their associated class labels;
minbucket - the minimum number of observations in any terminal (leaf) node.
Output:
A decision tree of ETC customer segmentation.
Method:
(1) create a node *N*;
(2) set a split point, *a*, for a specific segmentation index *A*, and split *D* into subsets D_1 and D_2. Thus, for ETC segmentation index, three set of subsets are obtained;
(3) computerize the Gini indexes of three indexes in dataset *D*, respectively. Determine an optimal splitting index;
(4) repeat steps (1)–(3) until the samples in the subset are too few or the reduction of "node impurity" cannot be below the given threshold and create a leaf node;
(5) the leaf node is labelled with the majority class in *D* to node *N*, and generate a decision tree of ETC customer segmentation;
(6) select different subtrees (branches) in the decision tree and prune it by the cross-validated error and cost complexity;
(7) output an optimal decision tree of ETC customer segmentation.

Algorithm 2: CART algorithm.

The actual distance $d(O_j, M)$ from the sample O_j to its cluster medoid is the minimum value in *k* distances:

$$d\left(O_j, M\right) = \min\left\{d\left(O_j, M_f\right), f \in (1, 2, \cdots k)\right\} \quad (4)$$

To determine whether current *k*-medoids are optimal, the average dissimilarity of this clustering, i.e., the arithmetic mean of distances from all samples in the dataset to their cluster medoid needs to be calculated, as shown in the following equation:

$$D_{average} = \frac{1}{N}\sum_{j=1}^{N} d\left(O_j, M\right) \quad (5)$$

where $D_{average}$ is the average dissimilarity and *N* is the number of samples in the ETC customer index dataset.

2.3. Segmentation Rules. After clustering analysis, each ETC customer is assigned a specific class label. Decision tree induction is the learning of decision trees from the class-labelled training dataset. The decision tree can be converted to classification of "IF-THEN" rules by tracing the path from the root node to each leaf node in the tree.

The most widely used decision tree algorithms are ID3 (Iterative Dichotomiser 3), C4.5 (a successor of ID3), and CART (Classification And Regression Trees). Compared to other decision tree algorithms, the CART algorithm simplifies the information theory based entropy model, while still retaining the entropy model's advantages using a binary tree instead of a multi-way tree and the Gini index instead of the information gain ratio [31]. This study uses the CART algorithm to induce ETC customers' segmentation decision tree, and Algorithm 2 shows the detailed procedure.

In the process of splitting, the Gini index measures the impurity of *D* or a data partition, as

$$Gini\,(D) = \sum_{i=1}^{k} p_i\left(1 - p_i\right) = 1 - \sum_{i=1}^{k} p_i^{\,2} \quad (6)$$

where p_i is the probability that the sample in *D* belongs to class C_i; *k* is the number of the class label in *D*.

If a binary split in segmentation index *A* partitions *D* into D_1 and D_2, the Gini index of *D*, given that partitioning is as follows:

$$Gini_A\,(D) = \frac{|D_1|}{D} Gini\left(D_1\right) + \frac{|D_2|}{D} Gini\left(D_2\right) \quad (7)$$

The reduction in impurity that would be incurred by a binary split on segmentation index *A* is the following:

$$\Delta Gini\,(A) = Gini\,(D) - Gini_A\,(D) \quad (8)$$

The index that maximizes reduction in impurity (or, equivalently, has the minimum $Gini_A(D)$) is selected as the splitting attribute.

To extract rules from a decision tree, one rule is created for each path from the root to a leaf node. Each splitting criterion along a given path is logically "ANDed" to form the rule antecedent ("IF" part). The leaf node holds the class prediction, forming the rule consequent ("THEN" part).

2.4. Modeling Procedure. The modeling procedure for ETC custom segmentation includes the following steps.

(1) Data Preprocessing and Index Extraction. This step includes the following: cleaning raw ETC data and extracting customer segmentation indexes; selecting data subset and forming the

Table 3: Extraction results of ETC customer segmentation index.

ETC card No.	R	F	M
61010922230010∗ ∗ ∗ ∗ ∗∗	30	1486	22055
61021101230035∗ ∗ ∗ ∗ ∗∗	1399	44	1535
61021101230035∗ ∗ ∗ ∗ ∗∗	1087	9	100
61032607230050∗ ∗ ∗ ∗ ∗∗	56	119	4400
......	......	......	......

Note. To protect privacy, the ETC card's last six digits were replaced with asterisks (∗).

ETC customer index dataset by setting a threshold value for each index.

(2) ETC Customer Clustering. This step includes the following: performing clustering analysis for the ETC customer index dataset and obtaining clustering results of the ETC customer.

(3) Segmentation Rules Extraction. This step includes the following: learning the decision tree of segmentation rules from the ETC customer index dataset (training tuple) and clustering results (class label) with the CART algorithm; extracting rules from the tree and realizing the final star-rating of the ETC customer.

Figure 1 displays the complete modeling procedure for ETC customer segmentation.

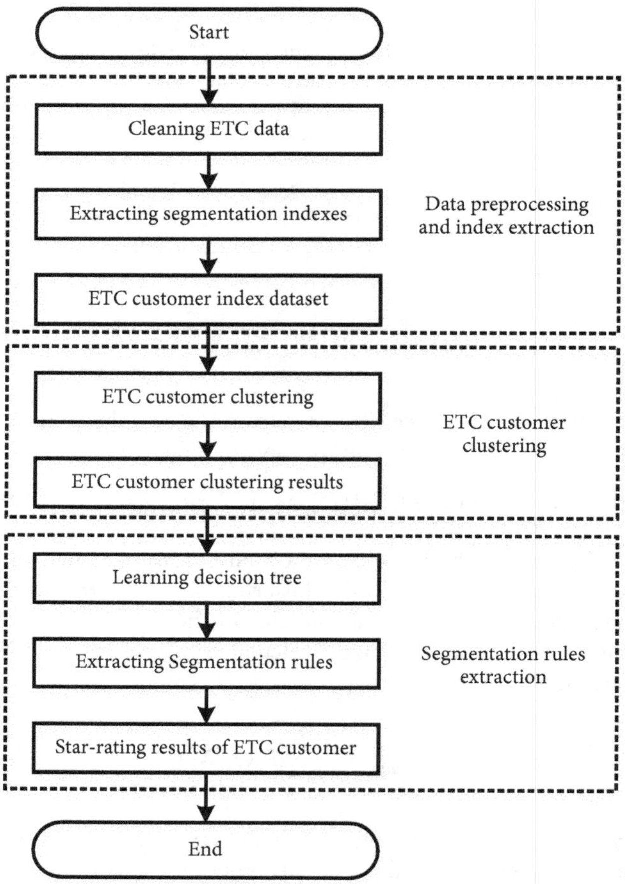

Figure 1: Modeling procedure for ETC customer segmentation.

3. Results

3.1. Data Preprocessing and Index Extraction. In this study, the 2014 annual ETC data, over 31 million, of passenger vehicles with seven seats or fewer in Shaanxi province was chosen as basic data. First, the data were cleaned. Irrelevant data (toll-free vehicles) or abnormal passing data (for instance, entrance time is later than exit time) were deleted. Then 324,585 groups of ETC customer segmentation index data were extracted with the specified time T_{set} = "2015-1-2 00:00:00." Table 3 shows the specific format.

Next, Figures 2(a)–2(c) show probability density distributions of three kinds of segmentation indexes. Further analysis indicates that when $R \leq 2160$, the percentage of ETC customers, who had consumption records within 90 days (2160 h) from the specified time, accounts for about 85% of the total. In the case of $F < 6$, that is, ETC customers with annual travel frequency of less than six times account for about 13.3%. In the case of $M < 200$, that is, ETC customers with annual monetary payments of less than 200 yuan account for about 18.6%, and those with more than 12,000 yuan account for about 0.77%.

To optimize the selected data subset and improve clustering accuracy, we filtered ETC customers who had too low frequency or extreme monetary values during data preprocessing. The filter criterion was $(F < 6) \cup (M < 200) \cup (M > 12000)$. Finally, an ETC customer index dataset containing 255,316 groups of ETC customers was formed.

Because of the massive data volume, a 2% random sampling was used to draw the scatter plot of "Frequency-Monetary", shown in Figure 3. The slope of the oblique line is 5, representing the single average toll of 5 yuan. Due to actual tolls being integral multiples of 5 yuan, the single average toll should be more than or equal to 5 yuan for normal payment vehicles (the slope is greater than or equal to 5). Figure 3 demonstrates that abnormal data generated by toll-free vehicles has been cleaned.

3.2. Clustering Results. In this study, the optimal number of clusters in ETC customer index dataset was estimated by the optimum average silhouette width [32]. The average silhouette method computes the average silhouette width of all customer samples for different values of k, and the optimal number of clusters k is the one that maximizes the average silhouette width over a range of possible values for

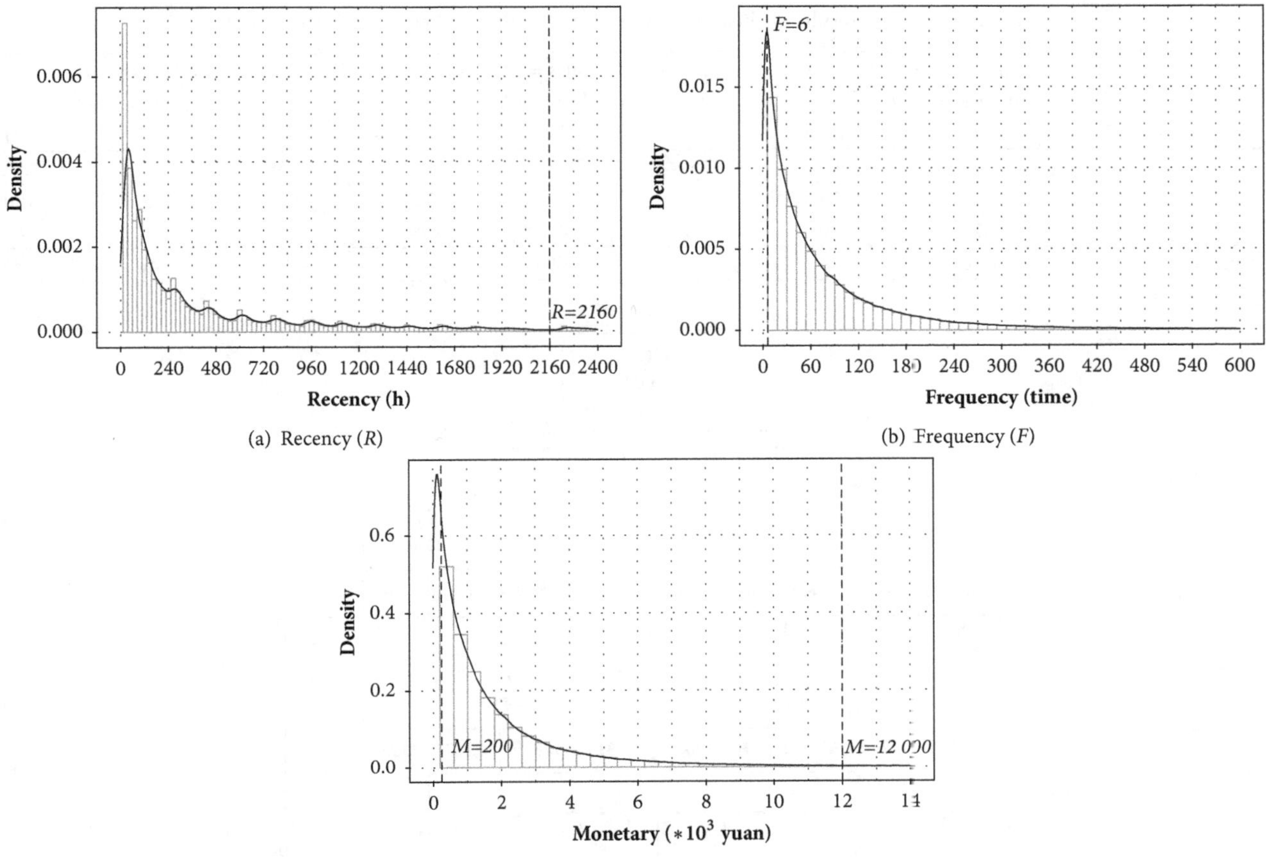

(a) Recency (R)

(b) Frequency (F)

(c) Monetary (M)

FIGURE 2: Distribution of ETC customer segmentation index.

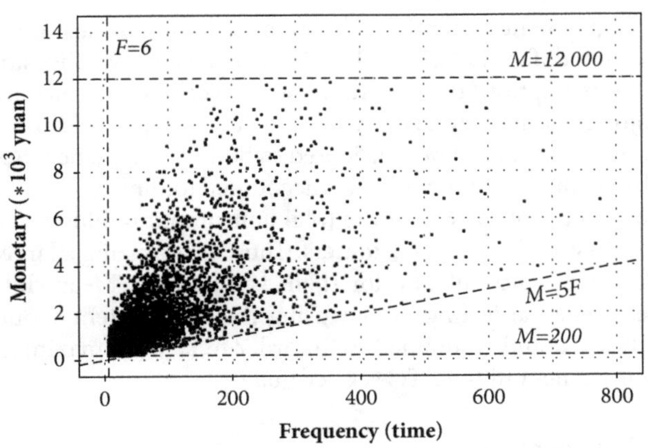

FIGURE 3: Scatter plot of "Frequency-Monetary" (sampling data).

k. The calculation indicates that $k = 3$ corresponded to the maximum width, so the optimal number of clusters is 3.

By considering filtered ETC customers during data preprocessing, the above-mentioned three types of ETC customers were expressed by C_2, C_3, and C_4. Filtered ETC customers $(F < 6) \cup (M < 200)$ and $(M > 12000)$ were, respectively, expressed as C_1 and C_5.

Due to its vast amount of data, methods like k-means, PAM, and so forth are unable to realize the whole-sample clustering of the ETC customer index dataset. In CLARA algorithm, the bigger the number of samples (*samples*) and observations (*sampsize*) is preset, the more accurate clustering results will get, but the corresponding computational expense will also increase.

By presetting different combined parameters (*samples* and *sampsize*) and executing iterative computation on ETC customer index dataset with the CLARA algorithm, the comparison results of the optimal clustering medoids and run-time (s) were obtained, as listed in Table 4.

Table 4 indicates that the clustering medoids tend to converge with the increase of *samples* and *sampsize*. Taking data volume and time effectiveness into consideration, 2% ETC customers (*sampsize* = 5000) were selected randomly at each sampling, then iteratively running ten times (*samples* = 10) with CLARA algorithm and finally getting the class label of each ETC customer.

3.3. Segmentation Results. Presetting the minimum number of observations in any leaf node at *minbucket* = 1000, a "segmentation index–customer classification (C_2, C_3, and C_4)" decision tree was built using the CART algorithm, as shown in Figure 4.

TABLE 4: Calculation results of clustering medoids under different combined parameters.

cluster		samples = 5 samplize = 500			samples = 10 samplize = 5000			samples = 20 samplize = 10000	
	R	F	M	R	F	M	R	F	M
C_2	3976	37	795	3750	27	765	3630	28	780
C_3	274	44	1025	316	43	995	302	43	1005
C_4	73	179	4340	176	160	3799	160	160	3877
run-time		0.55 s			59.69 s			846 s	

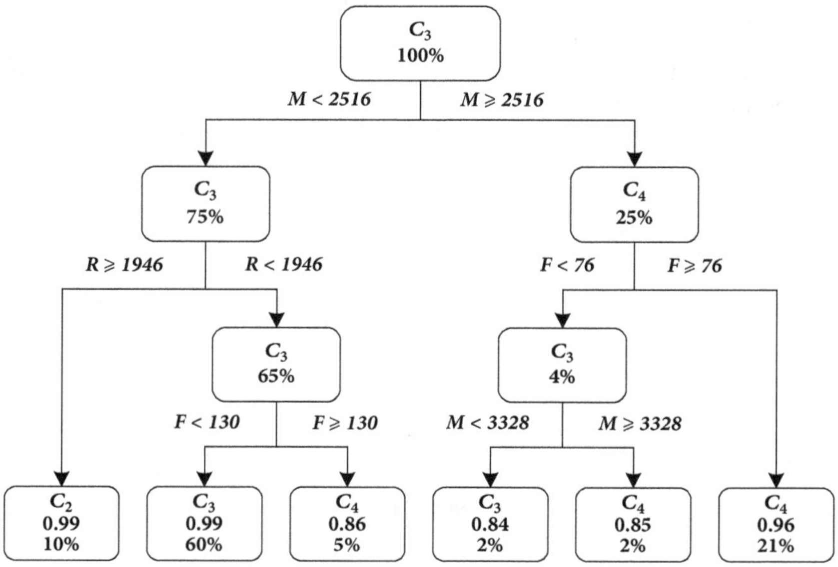

FIGURE 4: Decision tree of ETC customer segmentation.

This decision tree contains six leaf nodes. The first line in each node displays the final fitted classification of observations (ETC customers), the second line shows the probability per classification (C_2, C_3, and C_4), the third line displays the total percentage of observations (ETC customers) in this node, and the sum across all leaves is 1.

Segmentation rules of ETC customers C_2, C_3, and C_4 were extracted from Figure 4, and the filtering rules of C_1 and C_5 were also incorporated. Finally, all were transformed into a set of "IF-THEN" segmentation rules, as listed in Table 5.

All ETC customers were classified as C_1–C_5, corresponding to different stars. Different star-rating customers and their summarized details are listed in Table 6.

Table 6 indicates that 324,585 ETC customers in Shaanxi province annually paid tolls 23.13 million times, with a total consumption of 546 million yuan in 2014. According to the current 5% favorable discount rate, the actual annual ETC toll revenue was 519 million yuan.

One-star customers accounted for only about 20.57%, with a total consumption contribution of 1.33%. For such customers, a strengthened propaganda and guidance plan should be drawn to improve their ETC usage rates. Two-star customers accounted for 8.15%, with a total consumption contribution of 4.71%. According to their important characteristics, such customers should be cultivated to tap ETC payment potential. Three- and four-star customers accounted for 49.42% and 21.09%, respectively. The sum of consumption contributions was over 85%, indicating major customers in ETC service. In the future, an additional discount rate might be considered to build and enhance their self-worth. Five-star customers accounted for only 0.77%, but they contributed 7.6% to the total consumption, illustrating that they are key ETC customers. Thus, customizing a larger, additional discount rate for them is advisable. Meanwhile, their personal feelings about using the ETC system should be tracked and responded to, in order to achieve maximum improvement in the ETC service quality.

4. Discussion

An ETC customer segmentation index system was defined based on the RFM model. According to future operational requirements, new segmentation indexes can be introduced and adjusted with different weights, making star-rating results more suitable to the "Customer Pyramid" model [33].

There are differences in toll standards and usage characteristics among vehicle types. In this study, only ETC customers driving passenger vehicles with seven seats or fewer were segmented. Segmentation studies on other vehicle

TABLE 5: Segmentation rules of ETC customers.

Rule No.		Rule antecedent		Rule consequent
R1	IF	$(F < 6)$ OR $(M < 200)$	THEN	Customer = C_1
R2	IF	$M > 12000$	THEN	Customer = C_5
R3	IF	$(200 \le M < 2516)$ AND $(R \ge 1946)$	THEN	Customer = C_2
R4	IF	$(6 \le F < 76)$ AND $(2516 \le M < 3328)$	THEN	Customer = C_3
R5	IF	$(6 \le F < 76)$ AND $(3328 \le M \le 12000)$	THEN	Customer = C_4
R6	IF	$(2516 \le M \le 12000)$ AND $(F \ge 76)$	THEN	Customer = C_4
R7	IF	$(200 \le M < 2516)$ AND $(R < 1946)$ AND $(6 \le F < 130)$	THEN	Customer = C_3
R8	IF	$(200 \le M < 2516)$ AND $(R < 1946)$ AND $(F \ge 130)$	THEN	Customer = C_4

TABLE 6: Star-rating results for ETC customers.

Star	Number of customers	Total frequency	Total consumption (thousand yuan)	Ratio of customers (%)	Ratio of total consumption (%)
One-star	66765	384276	7240.2	20.57%	1.33%
Two-star	26457	963023	25727.6	8.15%	4.71%
Three-star	160405	7394715	170657	49.42%	31.25%
Four-star	68454	13187769	300927.5	21.09%	55.11%
Five-star	2504	1204661	41514.4	0.77%	7.60%

types should be conducted by following the proposed method combined with specific travel characteristics.

5. Conclusions

Applying big data technology, this study proposed an ETC customer segmentation method. Segmentation indexes were extracted from ETC data, customer clustering analysis was performed based on the CLARA algorithm, and segmentation rules were created. In this case study, ETC customer segmentation and star-rating were realized and travel characteristics and service strategies for each customer type were analyzed. The study thus provides an innovative idea for implementing precision marketing and creating hierarchical discount rates for ETC customers. Meanwhile, the study also provides theoretical support for further increase in the ETC customer scale and payment ratio, for an improved level of decision-making in expressway operation and management.

Conflicts of Interest

The authors declare that there are no conflicts of interest regarding the publication of this article.

Acknowledgments

This work was supported by the Natural Science Basic Research Plan in Shaanxi Province of China (Grant no. 2016JM5052) and the Fundamental Research Funds for the Central Universities of Ministry of Education of China (Grants nos. 310821173102 and 300102218203).

References

[1] Ministry of Transport of China, in *Proceedings of the third regular news conference in 2017*, Beijing, China, 2017, http://www.mot.gov.cn/2017wangshangzhibo/2017third/.

[2] K. A. Richards and E. Jones, "Customer relationship management: Finding value drivers," *Industrial Marketing Management*, vol. 37, no. 2, pp. 120–130, 2008.

[3] Z. Soltani and N. J. Navimipour, "Customer relationship management mechanisms: A systematic review of the state of the art literature and recommendations for future research," *Computers in Human Behavior*, vol. 61, pp. 667–688, 2016.

[4] S. H. Han, S. X. Lu and S. C. H. Leung, "Segmentation of telecom customers based on customer value by decision tree model," *Expert Systems with Applications*, vol. 39, no. 4, pp. 3964–3973, 2012.

[5] S.-Y. Kim, T.-S. Jung, E.-H. Suh, and H.-S. Hwang, "Customer segmentation and strategy development based on customer lifetime value: A case study," *Expert Systems with Applications*, vol. 31, no. 1, pp. 101–107, 2006.

[6] H. Hwang, T. Jung, and E. Suh, "An LTV model and customer segmentation based on customer value: A case study on the wireless telecommunication industry," *Expert Systems with Applications*, vol. 26, no. 2, pp. 181–188, 2004.

[7] I. Benitez, A. Quijano, J. L. Diez, and I. Delgado, "Dynamic clustering segmentation applied to load profiles of energy consumption from Spanish customers," *International Journal of Electrical Power & Energy Systems*, vol. 55, pp. 437–448, 2014.

[8] J. J. López, J. A. Aguado, F. Martín, F. Muñoz, A. Rodríguez, and J. E. Ruiz, "Hopfield-K-Means clustering algorithm: A proposal for the segmentation of electricity customers," *Electric Power Systems Research*, vol. 81, no. 2, pp. 716–724, 2011.

[9] R.-S. Wu and P.-H. Chou, "Customer segmentation of multiple category data in e-commerce using a soft-clustering approach," *Electronic Commerce Research and Applications*, vol. 10, no. 3, pp. 331–341, 2011.

[10] C.-F. Tsai, Y.-H. Hu, and Y.-H. Lu, "Customer segmentation issues and strategies for an automobile dealership with two clustering techniques," *Expert Systems with Applications*, vol. 32, no. 1, pp. 65–76, 2015.

[11] L. F. Wei, "Design and Implementation of Airline Customer Segmentation System Based on Ant Colony Clustering Algorithm," in *Proceedings of the International Conference on Materials Science and Information Technology (MSIT 2011)*, Singapore, 2011.

[12] T. Teichert, E. Shehu, and I. von Wartburg, "Customer segmentation revisited: The case of the airline industry," *Transportation Research Part A: Policy and Practice*, vol. 42, no. 1, pp. 227–242, 2008.

[13] W.-Y. Chiang, "Discovering customer value for marketing systems: an empirical case study," *International Journal of Production Research*, vol. 55, no. 17, pp. 5157–5167, 2017.

[14] Y.-H. Cheng and T.-Y. Huang, "High speed rail passenger segmentation and ticketing channel preference," *Transportation Research Part A: Policy and Practice*, vol. 66, no. 1, pp. 127–143, 2014.

[15] B. Zhang and Q.-Y. Peng, "Railway Freight Customer Segmentation Based on KFAV Model," *Jiaotong Yunshu Xitong Gongcheng Yu Xinxi/Journal of Transportation Systems Engineering and Information Technology*, vol. 17, no. 3, pp. 235–242, 2017.

[16] Y. Zhong and Y. Guo, "Research of applying data mining in segmentation of railway freight customers," *Beijing Jiaotong Daxue Xuebao/Journal of Beijing Jiaotong University*, vol. 32, no. 3, pp. 25–36, 2008.

[17] L. Duan, J. Rezaei, L. Tavasszy, and C. Chorus, "Heterogeneous valuation of quality dimensions of railway freight service by Chinese shippers choice-based conjoint analysis," *Transportation Research Record*, vol. 2546, pp. 9–16, 2016.

[18] W.-H. Lee, S.-S. Tseng, and C.-H. Wang, "Design and implementation of electronic toll collection system based on vehicle positioning system techniques," *Computer Communications*, vol. 31, no. 12, pp. 2925–2933, 2008.

[19] W.-Y. Shieh, C.-C. Hsu, S.-L. Tung, P.-W. Lu, T.-H. Wang, and S.-L. Chang, "Design of infrared electronic-toll-collection systems with extended communication areas and performance of data transmission," *IEEE Transactions on Intelligent Transportation Systems*, vol. 12, no. 1, pp. 25–35, 2011.

[20] D. Levinson and E. Chang, "A model for optimizing electronic toll collection systems," *Transportation Research Part A: Policy and Practice*, vol. 37, no. 4, pp. 293–314, 2003.

[21] H. Yamazaki, N. Uno, and F. Kurauchi, "The effect of a new intercity expressway based on travel time reliability using electronic toll collection data," *IET Intelligent Transport Systems*, vol. 6, no. 3, pp. 306–317, 2012.

[22] V. Astarita, M. Florian, and G. Musolino, "A microscopic traffic simulation model for the evaluation of toll station systems," in *Proceedings of the IEEE Intelligent Transportation Systems Conference*, pp. 692–697, Oakland, CA, USA, August 2001.

[23] M. L. Zarrillo and A. E. Radwan, "Methodology SHAKER and the capacity analysis of five toll plazas," *Journal of Transportation Engineering*, vol. 135, no. 3, pp. 83–93, 2009.

[24] K. Ozbay and M. Yildirimoglu, "Comparison of real-time travel time estimation using two distinct approaches: Universal kriging and mathematical programming," in *Proceedings of the 14th IEEE International Intelligent Transportation Systems Conference, ITSC 2011*, pp. 1083–1088, Washington, DC, USA, October 2011.

[25] H. Yang, K. Ozbay, and K. Xie, "Improved travel time estimation for reliable performance measure development for closed highways," *Transportation Research Record*, vol. 2526, pp. 29–38, 2015.

[26] A. Nairn and P. Berthon, "Creating the Customer: The Influence of Advertising on Consumer Market Segments - Evidence and Ethics," *Journal of Business Ethics*, vol. 42, no. 1, pp. 83–99, 2003.

[27] A. Hughes, *Strategic Database Marketing*, McGraw-Hill Education, New York, NY, USA, 1994.

[28] B. Stone and R. Jacobs, *Successful Direct Marketing Methods*, McGraw-Hill Education, New York, NY, USA, 8th edition, 2007.

[29] J. W. Han, M. Kamber, and J. Pei, *Data Mining: Concepts and Techniques*, Morgan Kaufmann, Waltham, MA, USA, 3rd edition, 2011.

[30] L. Kaufman and P. J. Rousseeuw, *Finding Groups in Data: An Introduction to Cluster Analysis*, John Wiley & Sons, Hoboken, NJ, USA, 2008.

[31] L. Breiman, J. Friedman, C. J. Stone, and R. A. Olshen, *Classification and Regression Trees*, Chapman and Hall/CRC, Boca Raton, FL, USA, 1984.

[32] P. J. Rousseeuw, "Silhouettes: a graphical aid to the interpretation and validation of cluster analysis," *Journal of Computational and Applied Mathematics*, vol. 20, pp. 53–65, 1987.

[33] V. A. Zeithaml, R. T. Rust, and K. N. Lemon, "The customer pyramid: Creating and serving profitable customers," *California Management Review*, no. 4, pp. 118–142, 2001.

4

Taxi Efficiency Measurements based on Motorcade-Sharing Model: Evidence from GPS-Equipped Taxi Data in Sanya

Jiawei Gui [1,2] **and Qunqi Wu** [1,2]

[1] *School of Economics and Management, Chang'an University, Xi'an, Shanxi 710064, China*
[2] *Center of Comprehensive Transportation Economic Management, Chang'an University, Xi'an, Shanxi 710064, China*

Correspondence should be addressed to Jiawei Gui; gjw@chd.edu.cn

Guest Editor: Razi Iqbal

Urban traffic congestion has become a global problem and has garnered special importance in recent years in the transportation sector, especially in taxi markets. To unlock the potential of Internet of Vehicles (IoV) in Intelligent Transportation Systems (ITS), it was vital to make efficiency measurements. In this study, Distance Formula was built to calculate distances by GPS data based on mathematical equations, and Motorcade-Sharing (MS) Model was proposed to improve the efficiency of collaborative vehicles. The experimental data of 2191 GPS-equipped taxis in Sanya of China was adopted to make comparisons between original results and modelled results. Measurement results showed that MS Model had 10.54% more leisure taxis, reduced 5 overdriving taxis, and saved 33.73% running distance in total compared to the original. This indicated that the application of MS Model could not only alleviate urban traffic congestion but also optimize urban taxi markets, and it has a bright future in the field of taxi and other collaborative vehicles. Future directions could be improving MS Model and expanding data.

1. Introduction

Nowadays, there are plenty of urgent social issues in the transportation research field. For example, urban traffic congestion has become a global problem and has garnered special importance in recent years in the transportation sector, especially in taxi markets. With the worsening situation of urban congestions and the development of autonomous vehicles, taxi markets are facing increasing opportunities as well as challenges. It is essential that appropriate counter measures must be taken to suppress or reverse or at least alleviate the worsening situation.

At the same time, there are more and more advanced technologies as well (i.e., intelligent traffic system, autonomous vehicles, etc.). To unlock the potential of Internet of Vehicles (IoV) in Intelligent Transportation Systems (ITS), it is vital to make efficiency measurements. In this study, Motorcade-Sharing Model was proposed for promoting taxi efficiency, and GPS-enabled taxi data in Sanya was adopted to support comparisons.

The rest of this paper is organized as follows: Section 2 reviews related works; Section 3 introduces formulas, models, data, and tools; Section 4 describes results from the perspective of path and efficiency; Section 5 makes a brief discussion; Section 6 summarizes main conclusions, contributions, and proposes future directions.

2. Literature Review

In this section, there are several similar problems and some related works till date, including Internet of Vehicles, collaborative vehicle routing, and taxi service improving.

2.1. Internet of Vehicles Problem. Intelligent Transportation Systems (ITS) had significant impact on our life [1], and they will consist of a crucial component of our modernized society [2, 3]. There is an amount of advanced technologies involved in ITS, for example, radio frequency identification [4], wireless sensor networks [5, 6], network coding [7, 8], network energy [9], database storage [10, 11], data mining

[12], and cloud computing security [13]. In recent years, a significant change in ITS was that much more data were collected from a variety of sources and can be processed into various forms [14]. And numerous advanced multidisciplinary journals began publishing a special issue of big data, for example, Nature in 2008 [15] and Science in 2011 [16].

Internet of Things (IoT) is a novel paradigm that is rapidly gaining ground in the scenario of modern wireless telecommunications as defined by Atzori et al. [17]. In the IoT paradigm, many of the objects that surround us will be on the network [18–20]. According to Miorandi et al. (2012) [21], it is predictable that the Internet will exist as a seamless fabric of classic networks and networked objects by 2022. And many of the hurdles, such as digitization services, of the Smart City Project can easily be solved by IoT as discussed by Ahmed and Rani [22].

Vehicular Ad Hoc Networks (VANETs) are self-organizing networks that can significantly improve traffic safety and travel comfort as defined by Chaqfeh et al. [23]. As an emerging concept transmuting the notion of VANETs, Internet of Vehicles (IoV) has a bright future. For example, Umer et al. (2018) proposed a dual ring connectivity model for VANETs under a heterogeneous traffic flow [24].

2.2. Collaborative Vehicle Routing Problem. Vehicle Routing Problem (VRP) is to find a plan to assign vehicles in such a manner that demands are satisfied and total mileage covered by the fleet is a minimum [25]. A lot of related works in this field have already been done. Some of them were with various window constraints [26], while some of them were under intelligent algorithms [27, 28].

In recent years, more and more companies tend to reduce distribution costs by creating collaborative environment on the Internet so as to share various information and resources, such as customers, warehouses, and fleets [29]. And numerous modeling methods were developed by researchers. For example, Yuan et al. (2017) proposed K–2FMM approach to improve the accuracy level of measuring travel time variability [30]. Wang et al. (2017) proposed a developed semi-nonparametric generalized multinomial logit model to extend the standard Gumbel distribution [31]. Tang et al. (2017) proposed an improved hierarchical fuzzy inference method based on map-matching algorithm [32]. Al-Mayouf et al. (2018) proposed an intersection-based segment aware algorithm for geographic routing in VANETs [33]. Laha and Putatunda (2018) proposed three incremental learning methods for next pickup location prediction problems [34]. Xu and Cai (2018) proposed a variable neighborhood search algorithm for the consistent Vehicle Routing Problem [35].

Collaborative Vehicle Routing Problem (CVRP) is to find the best solution to assign vehicles in such a manner that demands are satisfied by comprehensive efficiency being a maximum. It helps decreasing the waiting time of passengers, but its related studies were much fewer than the ones related to VRP.

2.3. Taxi Service Improving Problem. Taxi service includes varies of aspects, such as safety, convenience, efficiency, comfort, and economy. Utilizing large-scale GPS data to improve taxi services has become a popular research problem in the areas of data mining, intelligent transportation, geographical information systems, and the Internet of Things [36].

Above all, traffic safety is vital. Projects of safety have been more and more popular all over the world, such as the zero-accident goal by 2050 in EU, the Hazardous Material Cooperative Research Program in the United States, and the 5-year safety production plan in China [37]. And there are already some researches focusing on taking measures to enhance traffic safety. For example, Zou et al. (2018) proposed a developed empirical Bayes method and applied it to actual crash datasets [38]. Jain et al. (2018) proposed an algorithm to control traffic congestion and road safety for Social Internet of Vehicles [39]. According to the evaluation results of a transportation system as discussed by Lam [40], journey time was a significant factor in passenger waiting time when vehicle shortages occur. In addition, according to La et al. (2013), time pressure could impact crashes for taxi drivers [41].

In recent years, rapid technical growth of autonomous vehicles tended to accelerate the process of autonomous taxi market. For example, an autonomous taxi service has come into use on a 4.5km campus road at Seoul National University [42]. With the development of technique and policy, autonomous taxi market has a bright future. Meanwhile, understanding origin-destination distribution of taxi trips is very important for improving effects of enhancing quality of taxi services [43].

To sum up, some helpful works have already been done. However, few researches proposed MS Model and took big data into consideration. Besides, autonomous vehicles need assigning continuous directives in time while it is difficult for complex algorithms to respond immediately under the background of big data.

How to achieve collaborative vehicle routing and improve taxi service by MS Model based on Internet of Vehicles? In the next section, methods and data are involved to explore this question.

3. Methods and Materials

3.1. Measurement Steps. To solve this problem, 4 steps are necessary (see Figure 1). First is acquiring data, including the demands of passengers and the supplies of taxis. Second is measuring distances between passengers and taxis based on the real-time locations of them. Third is working out modelled results based on the MS Model. Last is making comparisons between original results and modelled results.

There were two key points during the measurement. One is Distance Formula (see Section 3.2), and the other is Motorcade-Sharing Model (see Section 3.3).

3.2. Distance Formula. There are five steps in this section and the notations for formulas are in the Appendix (see Appendix A).

The primary function of Distance Formula was to calculate distances by GPS data. First, in order to describe the

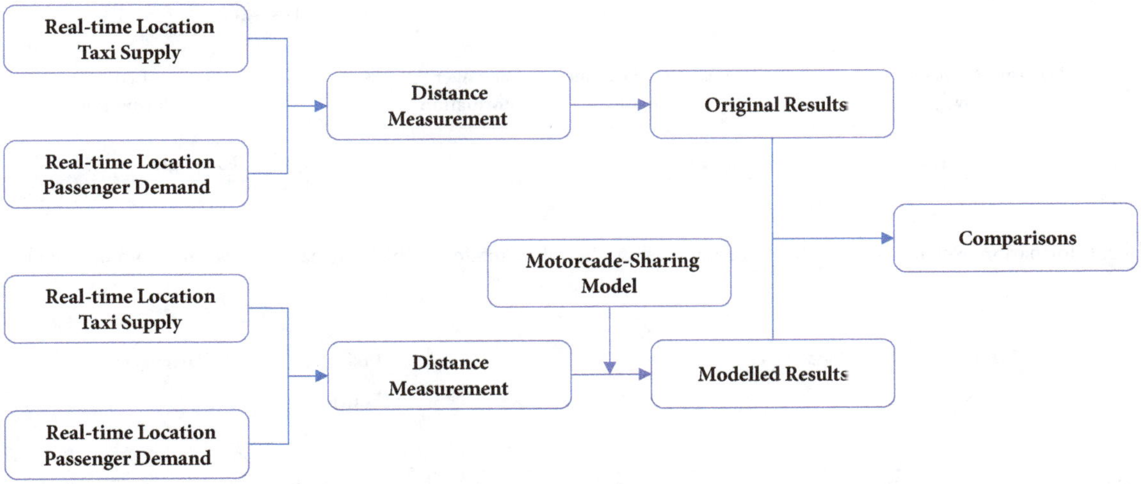

FIGURE 1: Measurement steps design.

location of taxis in a mathematical way, the polar coordinate system was established in Formula (1).

$$P(\rho, \theta, \varphi) \tag{1}$$

where $\rho \geq 0$, $0 \leq \theta \leq \pi$, and $0 \leq \varphi < 2\pi$. The variable ρ represents the radial distance. The variable θ represents the polar angle. The variable φ represents the azimuthal angle.

Second, in order to calculate the distance between two locations, the rectangular coordinate system was established in Formula (2).

$$\overrightarrow{OP} = (x, y, z) \tag{2}$$

where $x = \rho\sin\theta\cos\varphi$, $y = \rho\sin\theta\sin\varphi$, and $z = \rho\cos\theta$.

Suppose that there are two locations of taxis, called P_1 and P_2. According to Formulas (1) and (2) and $\rho = R$, their polar coordinates, $P_1(R, \theta_1, \varphi_1)$ and $P_2(R, \theta_2, \varphi_2)$, can be converted into rectangular coordinates, $\overrightarrow{OP_1} = (R\sin\theta_1\cos\varphi_1, R\sin\theta_1\sin\varphi_1, R\cos\theta_1)$ and $\overrightarrow{OP_2} = (R\sin\theta_2\cos\varphi_2, R\sin\theta_2\sin\varphi_2, R\cos\theta_2)$.

Third, in order to calculate the distance between two locations, the cosine value of the angle between vectors $\overrightarrow{OP_1}$ and $\overrightarrow{OP_2}$ was calculated by Formula (3).

$$\cos\left\langle \overrightarrow{OP_1}, \overrightarrow{OP_2} \right\rangle = \frac{\overrightarrow{OP_1} \cdot \overrightarrow{OP_2}}{\left|\overrightarrow{OP_1}\right| \left|\overrightarrow{OP_2}\right|} \tag{3}$$

Formula (3) can be simplified into Formula (4) and the derivation process is in the Appendix (see Appendix B).

$$\cos\left\langle \overrightarrow{OP_1}, \overrightarrow{OP_2} \right\rangle = \sin\theta_1\sin\theta_2\cos(\varphi_2 - \varphi_1) + \cos\theta_1\cos\theta_2 \tag{4}$$

Fourth, according to the equation of arc length calculation, the distance between vectors $\overrightarrow{OP_1}$ and $\overrightarrow{OP_2}$ was calculated by Formula (5).

$$d = R\left\langle \overrightarrow{OP_1}, \overrightarrow{OP_2} \right\rangle \tag{5}$$

where $R = 6378.1370(km)$, which represents the radius of the earth.

According to Formula (3), Formula (5) can be simplified into Formula (6).

$$d = R\arccos\left(\sin\theta_1\sin\theta_2\cos(\varphi_2 - \varphi_1) + \cos\theta_1\cos\theta_2\right) \tag{6}$$

In other words, the distance between two locations of taxis can be calculated approximately by the longitudes and latitudes of them based on GPS data, and the radius of the earth.

Suppose that the variable TD represents total distance between two locations of taxis, the variable TOD represents total occupied distance between them, and the variable TUD represents total unoccupied distance between them. And then TD can be computed by Formula (7) [44].

$$TD = TOD + TUD \tag{7}$$

where $TD = \sum_{i=1}^{I}\sum_{t=1}^{T} d_{i,t}$, $TOD = \sum_{i=1}^{I}\sum_{t=1}^{T} d_{i,t} \mid p \geq 1$, and $TUD = \sum_{i=1}^{I}\sum_{t=1}^{T} d_{i,t} \mid p = 0$.

At last, the target of this study is to minimize the total distance, and it can be indicated in Formula (8).

$$\min_{1\leq i\leq I, 1\leq t\leq T} TD = \min_{1\leq i\leq I, 1\leq t\leq T} \sum_{i=1}^{I}\sum_{t=1}^{T} d_{i,t} \tag{8}$$

Suppose that the demand of passengers is invariable, which means that TOD is constant. And then Formula (8) can be simplified into Formula (9).

$$\min_{1\leq i\leq I, 1\leq t\leq T} TD = TOD + \min_{1\leq i\leq I, 1\leq t\leq T} \sum_{i=1}^{I}\sum_{t=1}^{T} d_{i,t} \mid p = 0 \tag{9}$$

Minimizing the total distance is equivalent to minimizing the total unoccupied distance. And in the next section, MS Model is established to solve this problem.

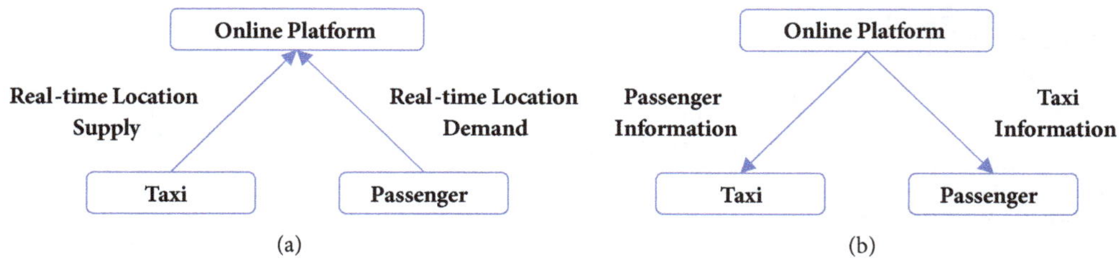

FIGURE 2: Information exchange design. (a) Information flows in online platform. (b) Information flows out from online platform.

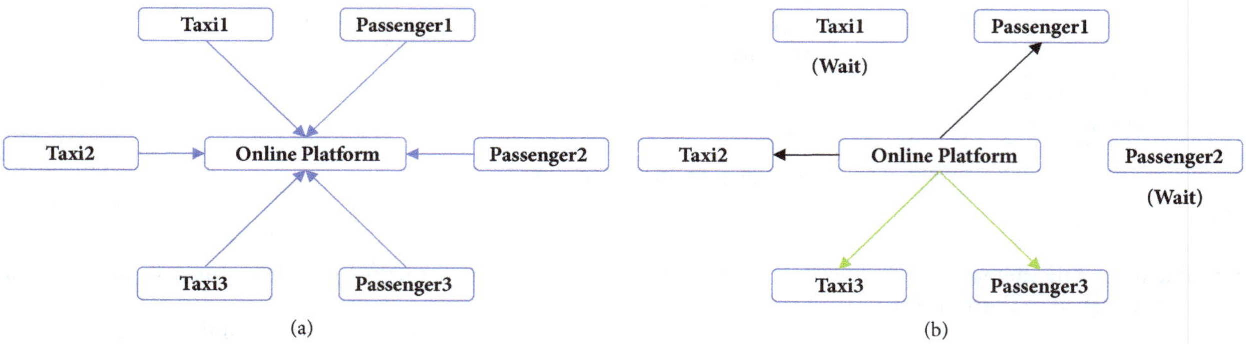

FIGURE 3: Information exchange of a hypothetical example. (a) Information flows in online platform. (b) Information flows out from online platform.

3.3. Motorcade-Sharing Model. Motorcade-Sharing (MS) Model was established to minimize the total unoccupied distance and it contained two main strategies: Proximity Strategy and Passivity Strategy.

(i) Proximity Strategy. The unoccupied taxi with closer distance has its priority to pick up passengers immediately; in other words, passengers will be picked up by the unoccupied taxi whose position is closest to them at that time.

(ii) Passivity Strategy. All the unoccupied taxis pick up passengers passively on the basis of the above-mentioned Proximity Strategy, and they keep parking after passengers arrive. That is to say, only occupied taxis keep driving while others keep parking until necessary.

MS Model needs an online platform to exchange information between taxis and passengers. Taxis and passengers send their real-time locations to the online platform (see Figure 2(a)) and get partners' information in return (see Figure 2(b)). Suppose that there are only 3 taxis and 3 passengers online (see Figure 4(a)). In the original setting, Taxi1 picks up Passenger1, Taxi2 picks up Passenger2, and Taxi3 picks up Passenger3. According to MS Model, taxis and passengers send their real-time locations to the online platform (see Figure 3(a)) and get partners' information in return (see Figure 3(b)). As a modelled result (see Figure 4(b)), Taxi3 picks up Passenger3 as usual. However, Taxi2 is going to pick up Passenger1 because Taxi2 is closer to Passenger1 than Taxi1. Besides, Taxi1 and Passenger2 keep waiting for a second. If there is still no better choice in the following seconds, Taxi1 picks Passenger2 in the end. Autonomous taxis have no conflicts of interests and need an online platform to share information. Thus, MS Model is especially suitable and useful for autonomous vehicles to improve comprehensive efficiency in the future.

When there is more than one taxi online, it is possible for those taxis to improve comprehensive efficiency by cooperating with others by MS Model. It seems that modelled routing without overlapping has higher efficiency than original setting; however, it needs evidences from real data to verify.

3.4. Data and Tool. (i) Data Source. Data in this study was collected from the big data platform, Travel Cloud, which was developed by Ministry of Transport of the People's Republic of China (see Data Availability). The GPS data of Taxis in this study was provided by Sanya Traffic and Transportation Bureau in Hainan Province in China. Major data items include vehicle ID, longitude, latitude, and time (see the file named "Data.xlsx" in the Supplementary Materials (available here)).

(ii) Data Processing. According to the origin data, the locations of 2506 taxis in Sanya were recorded every 15 seconds from 9:00 a.m. to 9:59 a.m. on Nov. 15th in 2016, adding up to 766,042 records. First, all the locations of taxis are averaged every 60 seconds so as to improve the accuracy of data; that is to say, average locations of those taxis were recorded every 60 seconds entirely. Second, the flaw data was removed so as to ensure the integrity of data. As a result, there were 2191 taxis and 131,460 records in the experimental data.

(iii) Implement Tool. Several kinds of statistics software were adopted in this study as follows. Excel was applied for original data processing and result storage. Matlab was

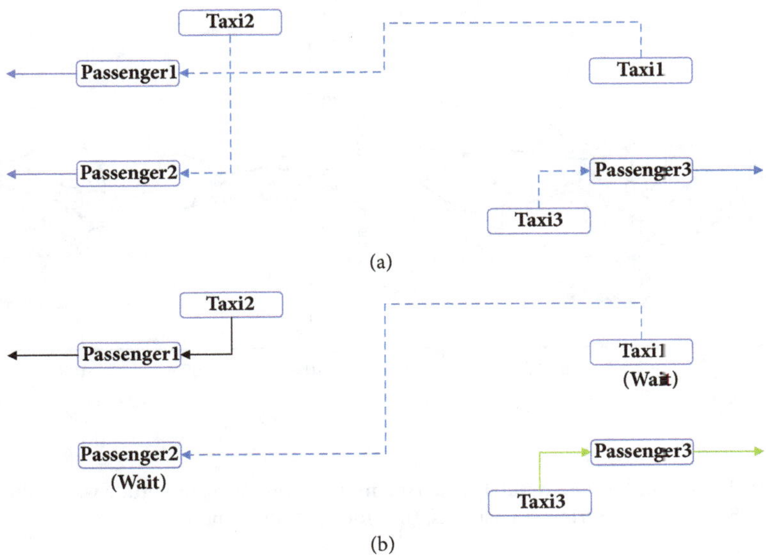

FIGURE 4: Routing comparison of a hypothetical example. (a) Original vehicle routing. (b) Improved vehicle routing.

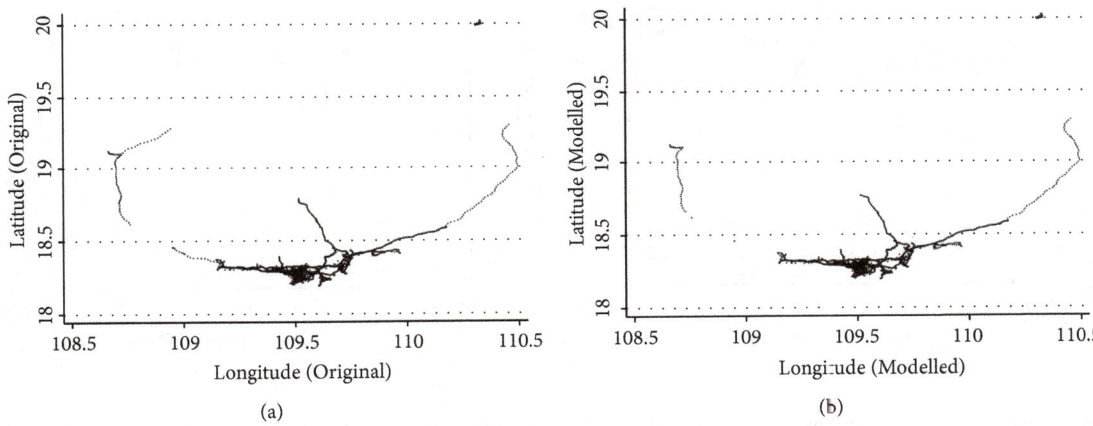

FIGURE 5: Path comparison of taxi positions located minutely at a fixed monitor where the longitude was in the range of [108°,111°] while the latitude was in the range of [18°,21°]. (a) Original positions. (b) Modelled positions.

applied for programming (see the file named "Programming.pdf" in the Supplementary Materials). Stata was applied for drawing.

4. Results

In this section, modelled results of 2191 GPS-equipped taxis in Sanya were compared with original results from the perspectives of path and efficiency.

4.1. Path Comparison. We adopted scatter diagram rather than line diagram because the former seems clearer than the latter. Taxi positions were located minutely at a fixed monitor. Thus, crowded districts lied in dark areas. To begin with, there was seldom difference between original results (see Figure 5(a)) and modelled results (see Figure 5(b)) in path comparison results of all the 2191 taxis because of compression. However, disparities gradually appeared when focusing on specific local areas. It came to a specific result by a narrow scope that the longitude was in the range of [109.4°,109.6°] while the latitude was in the range of [18.2°,18.4°] (see Figure 6). And at last, it came to another specific result by a far narrow scope that the longitude was in the range of [109.495°,109.505°] while the latitude was in the range of [18.245°,18.255°] (see Figure 7).

It can be seen from Figures 5–7 that modelled paths were obviously smoother and more unobstructed than original paths; thus, MS Model helped alleviating urban traffic congestion. However, it needs digital comparisons to quantify.

4.2. Efficiency Comparison. All the 2191 taxis were divided into different groups by their running distances. First, the efficiencies of running distances were compared (see Table 1). There were two main findings in Table 1. On the one hand,

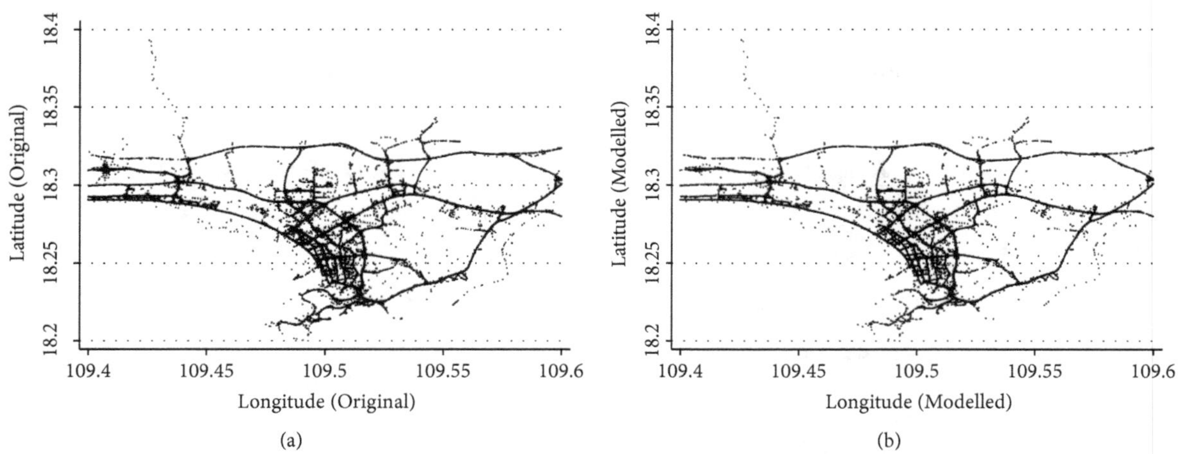

FIGURE 6: Path comparison of taxi positions located minutely at a fixed monitor where the longitude was in the range of [109.4°,109.6°] while the latitude was in the range of [18.2°,18.4°]. (a) Original positions. (b) Modelled positions.

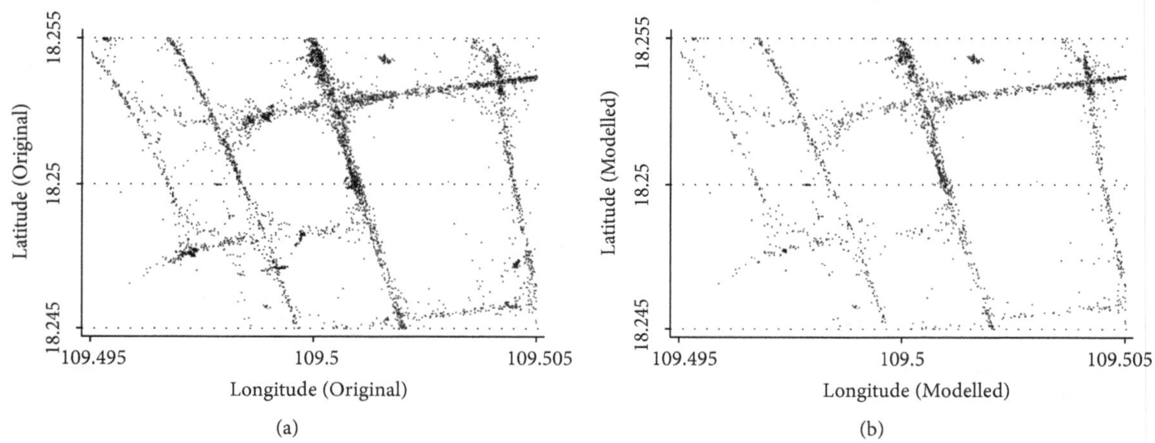

FIGURE 7: Path comparison of taxi positions located minutely at a fixed monitor where the longitude was in the range of [109.495°,109.505°] while the latitude was in the range of [18.245°,18.255°]. (a) Original positions. (b) Modelled positions.

modelled total distance (23,292.05km) was 33.73% lower than original total distance (35,147.37km), which implied that modelled scheme was more fuel-efficient and thus economical in another word. On the other hand, modelled scheme obtained reducing distance results in groups above 20 km and increasing results in groups below 20 km, which implied that modelled scheme was more efficient in general rather than unilateral.

Second, the efficiencies of running taxis were compared (see Table 2). There were three main findings in Table 2. Above all, modelled taxis with 0 km (289) were about five times the quantity of original taxis with 0 km (58), and increasing 10.54% leisure taxis implied that modelled scheme was more fuel-efficient and thus economical than the original. Furthermore, modelled taxis in groups above 60 km (5) were half the original taxis in groups above 60 km (10), and reducing 5 overdriving taxis implied that modelled scheme had fewer overdriving taxis than the original and thus was safe in another word. Last but not least, modelled scheme obtained reducing taxi results in groups above 20 km and increasing results in groups below 20 km, which implied that modelled scheme was more efficient in general rather than unilateral.

It can be seen from Tables 1 and 2 that modelled scheme had higher fuel-efficient, more economical, and safer taxis in general compared with the original. Thus, MS Model helped optimizing urban taxi markets.

5. Discussion

While making comparisons in the last section, there were still five factors that had not been contemplated within, which are described as follows.

First, Proximity Strategy has limitations. In reality, it was difficult for taxis to pick up passengers according to their relative positions even if information could be shared adequately. It required the online platform to keep good

TABLE 1: Efficiency comparison of running distances. The data in the 2nd and 4th columns are accumulation of corresponding distances; unit: km.

Distance	Original results		Modelled results	
$d \leq 10$	2,263.16	6.44%	4,354.11	18.69%
$10 < d \leq 20$	11,857.74	33.74%	8,683.47	37.28%
$20 < d \leq 30$	10,822.63	30.79%	5,076.67	21.80%
$30 < d \leq 40$	5,223.42	14.86%	2,752.53	11.82%
$40 < d \leq 50$	2,653.03	7.55%	1,303.22	5.60%
$50 < d \leq 60$	1,510.05	4.30%	761.44	3.27%
$d > 60$	817.34	2.33%	360.61	1.55%
Total	35,147.37	100.00%	23,292.05	100.00%

TABLE 2: Efficiency comparison of running taxis. The data in the 2nd and 4th columns are numbers of taxis with corresponding distances.

Distance	Original results		Modelled results	
$d = 0$	58	2.65%	289	13.19%
$0 < d \leq 10$	654	29.85%	952	43.45%
$10 < d \leq 20$	780	35.60%	612	27.93%
$20 < d \leq 30$	448	20.45%	209	9.54%
$30 < d \leq 40$	152	6.94%	80	3.65%
$40 < d \leq 50$	61	2.78%	30	1.37%
$50 < d \leq 60$	28	1.28%	14	0.64%
$d > 60$	10	0.46%	5	0.23%
Total	2191	100.00%	2191	100.00%

balance of allocation. That is to say, MS Model might have much lower efficiency because of mistakes.

Second, Passivity Strategy has limitations. In reality, it was difficult for taxis to keep parking after passengers reach their destinations because there were other requirements (i.e., traffic control, refueling, drivers' turning over, etc.). Sometimes, it was more reasonable for taxis to approach busy areas rather than staying. That is to say, MS Model might have much lower efficiency because of traffic conditions.

Third, information sharing was essential for MS Model. Several management plans for vehicular traffic (i.e., big data collections, cloud computing, platform construction, autonomous technique, etc.) should also be prepared completely. That is to say, MS Model might have much lower efficiency because of failure facilities.

Fourth, in most cities in China, morning rush hours are often centered in the period of 7:00-9:00 a.m. [45]. Thus, the experimental data in this study was not in rush hours. However, traffic conditions may vary over different time periods; more data should be analyzed and presented in the future research. That is to say, MS Model might have much higher efficiency during rush hours.

At last, this study focused on the efficiency comparison between MS Model and the original. Thus, path optimizations were not considered and taxis drove the original path in MS Model. That is to say, MS Model might have much higher efficiency by optimizing routes.

To sum up, the results could be more accurate by improving the model or expanding the sample size. Nevertheless, the trend would not change that modelled scheme would still result in higher efficiency than the original. In addition, MS Model is suitable and useful for autonomous vehicles.

6. Conclusions

In this study, the authors proposed Motorcade-Sharing (MS) Model for promoting taxi efficiency and adopted GPS-enabled taxi data in Sanya to make efficiency comparisons by Distance Formula. Given the arguments mentioned above, the major findings in this article include several contents, which are described as follows.

(i) The application of MS Model could alleviate urban traffic congestion and it resulted in less obstructed path and more leisure taxis than the original (see Section 4.1).

(ii) The application of MS Model could optimize urban taxi markets and it resulted in higher fuel-efficient, more economical, and safer taxis than the original (see Section 4.2).

(iii) The application of MS Model could improve effects according to statistic results (see Section 4.2) and it has a bright future in the field of taxi and other collaborative vehicles.

The contributions of this study could be as follows: (i) Distance Formula was built to calculate distances by GPS data based on mathematical equations. (ii) Motorcade-Sharing Model was proposed to improve the efficiency of collaborative vehicles. (iii) It was verified that modelled scheme had higher efficiency than the original.

Future directions for research can be in two ways. One is improving MS Model, and the other is expanding the sample size.

Appendix

A. Notations for Formulas

See Table 3.

B. Derivation Process of Formula (4)

$$\begin{aligned}
&\cos\left\langle \overrightarrow{OP_1}, \overrightarrow{OP_2} \right\rangle \\
&= \frac{R\sin\theta_1\cos\varphi_1 R\sin\theta_2\cos\varphi_2 + R\sin\theta_1\sin\varphi_1 R\sin\theta_2\sin\varphi_2 + R\cos\theta_1 R\cos\theta_2}{\sqrt{(R\sin\theta_1\cos\varphi_1)^2 + (R\sin\theta_1\sin\varphi_1)^2 + (R\cos\theta_1)^2}\sqrt{(R\sin\theta_2\cos\varphi_2)^2 + (R\sin\theta_2\sin\varphi_2)^2 + (R\cos\theta_2)^2}} \\
&= \frac{R^2(\sin\theta_1\sin\theta_2\cos\varphi_1\cos\varphi_2 + \sin\theta_1\sin\theta_2\sin\varphi_1\sin\varphi_2 + \cos\theta_1\cos\theta_2)}{R^2\sqrt{(\sin\theta_1\cos\varphi_1)^2 + (\sin\theta_1\sin\varphi_1)^2 + (\cos\theta_1)^2}\sqrt{(\sin\theta_2\cos\varphi_2)^2 + (\sin\theta_2\sin\varphi_2)^2 + (\cos\theta_2)^2}} \\
&= \sin\theta_1\sin\theta_2(\cos\varphi_1\cos\varphi_2 + \sin\varphi_1\sin\varphi_2) + \cos\theta_1\cos\theta_2 = \sin\theta_1\sin\theta_2\cos(\varphi_2 - \varphi_1) + \cos\theta_1\cos\theta_2
\end{aligned} \tag{B.1}$$

Table 3: Notations for Formulas.

Notation	Explanation
O	The center of the rectangular coordinate system/The center of the earth
P	The location
ρ	The radial distance
θ	The polar angle
φ	The azimuthal angle
i	The taxi ID
t	The time
I	The maximum of taxi ID
T	The maximum of time
P_i	The location of taxis
R	The radius of the earth
θ_i	The longitude of the location of taxis
φ_i	The latitude of the location of taxis
d	The distance between two locations of taxis
p	The number of passengers in a taxi
TD	Total distance between two locations of taxis
TOD	Total occupied distance between two locations of taxis
TUD	Total unoccupied distance between two locations of taxis

Conflicts of Interest

The authors declare that there are no conflicts of interest regarding the publication of this article.

Authors' Contributions

Jiawei Gui and Qunqi Wu contributed to study conception and design. Jiawei Gui contributed to data collection and software, analysis and interpretation of results, manuscript preparation-draft, and manuscript preparation-revised version. All authors reviewed the results and approved the final version of the manuscript.

Acknowledgments

The authors thank Razi Iqbal for providing objective comments twice to improve this paper and thank Aya Moez for providing editing guidance. Jiawei Gui was rewarded by the National Scholarship of China for doctoral students and appreciated that. The work described in this article was supported in part by the Strategic Planning Research Project of Ministry of Transport of China [2018-7-9] and [2018-16-9], in part by the National Social Science Fund of China [17BJY139], and in part by Chang'an University Excellent Doctoral Dissertation Cultivation Project of Chinese Universities Scientific Fund of China [2017023002].

References

[1] F.-Y. Wang, "Parallel control and management for intelligent transportation systems: concepts, architectures, and applica-

tions," *IEEE Transactions on Intelligent Transportation Systems*, vol. 11, no. 3, pp. 630–638, 2010.

[2] G. Zhang, M. Li, and J. Wang, "Application of the Advanced Public Transport System in Cities of China and the Prospect of Its Future Development," *Journal of Transportation Systems Engineering and Information Technology*, vol. 7, no. 5, pp. 24–30, 2007.

[3] X. Cheng, L. Yang, and X. Shen, "D2D for intelligent transportation systems: A feasibility study," *IEEE Transactions on Intelligent Transportation Systems*, vol. 16, no. 4, pp. 1784–1793, 2015.

[4] P. Vrba, F. Macurek, and V. Marik, "Using Radio Frequency Identification in Agent-Based Manufacturing Control Systems," in *Proceedings of the International Conference on Industrial Applications of Holonic and Multi-Agent Systems*, vol. 3593, pp. 176–187, Springer Berlin Heidelberg, Copenhagen, Denmark, 2005.

[5] V. Y. Liu and Z. Yu, "Wireless Sensor Networks for Internet of things: A systematic review and classification," *Information Technology Journal*, vol. 12, no. 16, pp. 3581–3585, 2013.

[6] S. Ali, A. Ahmad, R. Iqbal, S. Saleem, and T. Umer, "Joint RRH-Association, Sub-Channel Assignment and Power Allocation in Multi-Tier 5G C-Rans," *IEEE Access*, vol. 6, pp. 34393–34402, 2018.

[7] F. Jamil, A. Javaid, T. Umer, and M. H. Rehmani, "A comprehensive survey of network coding in vehicular ad-hoc networks," *Wireless Networks*, vol. 23, no. 8, pp. 2395–2414, 2017.

[8] S. H. Ahmed, S. H. Bouk, M. A. Yaqub, D. Kim, and H. Song, "DIFS: Distributed Interest Forwarder Selection in Vehicular Named Data Networks," *IEEE Transactions on Intelligent Transportation Systems*, vol. 19, no. 9, pp. 3076–3080, 2018.

[9] G. Auer, V. Giannini, C. Desset et al., "How much energy is needed to run a wireless network?" *IEEE Wireless Communications Magazine*, vol. 18, no. 5, pp. 40–49, 2011.

[10] F. Chang, J. Dean, S. Ghemawat et al., "Bigtable: a distributed storage system for structured data," *ACM Transactions on Computer Systems*, vol. 26, no. 2, Article ID 1365816, 2008.

[11] M. Iwazume, T. Iwase, K. Tanaka, and H. Fujii, "Big data in memory: Benchmarking in memory database using the distributed key-value store for constructing a large scale information infrastructure," in *Proceedings of the 38th Annual IEEE Computer Software and Applications Conference Workshops, COMPSACW 2014*, pp. 199–204, New York, NY, USA, 2014.

[12] C.-W. Tsai, C.-F. Lai, M.-C. Chiang, and L. T. Yang, "Data mining for internet of things: a survey," *IEEE Communications Surveys & Tutorials*, vol. 16, no. 1, pp. 77–97, 2014.

[13] K. Hashizume, D. G. Rosado, E. Fernández-Medina, and E. B. Fernandez, "An analysis of security issues for cloud computing," *Journal of Internet Services & Applications*, vol. 4, no. 1, pp. 1–13, 2013.

[14] J. Zhang, F.-Y. Wang, K. Wang, W.-H. Lin, X. Xu, and C. Chen, "Data-driven intelligent transportation systems: a survey," *IEEE Transactions on Intelligent Transportation Systems*, vol. 12, no. 4, pp. 1624–1639, 2011.

[15] F. Frankel and R. Reid, "Big data: Distilling meaning from data," *Nature*, vol. 455, no. 7209, p. 30, 2008.

[16] W. Los and J. Wood, "Dealing with data: Upgrading infrastructure," *Science*, vol. 331, no. 6024, pp. 1515-1516, 2011.

[17] L. Atzori, A. Iera, and G. Morabito, "The internet of things: a survey," *Computer Networks*, vol. 54, no. 15, pp. 2787–2805, 2010.

[18] J. Gubbi, R. Buyya, S. Marusic, and M. Palaniswami, "Internet of Things (IoT): a vision, architectural elements, and future directions," *Future Generation Computer Systems*, vol. 29, no. 7, pp. 1645–1660, 2013.

[19] M. Gohar, S. H. Ahmed, M. Khan, N. Guizani, A. Ahmed, and A. U. Rahman, "A Big Data Analytics Architecture for the Internet of Small Things," *IEEE Communications Magazine*, vol. 56, no. 2, pp. 128–133, 2018.

[20] T. A. Butt, R. Iqbal, and S. C. Shah, "Social Internet of Vehicles: Architecture and enabling technologies," *Computers & Electrical Engineering*, vol. 69, pp. 68–84, 2018.

[21] D. Miorandi, S. Sicari, F. de Pellegrini, and I. Chlamtac, "Internet of things: vision, applications and research challenges," *Ad Hoc Networks*, vol. 10, no. 7, pp. 1497–1516, 2012.

[22] S. H. Ahmed and S. Rani, "A hybrid approach, Smart Street use case and future aspects for Internet of Things in smart cities," *Future Generation Computer Systems*, vol. 79, pp. 941–951, 2018.

[23] M. Chaqfeh, A. Lakas, and I. Jawhar, "A survey on data dissemination in vehicular ad hoc networks," *Vehicular Communications*, vol. 1, no. 4, pp. 214–225, 2014.

[24] T. Umer, M. K. Afzal, E. U. Munir, and M. Alam, "A Dual Ring Connectivity Model for VANET Under Heterogeneous Traffic Flow," *Wireless Personal Communications*, vol. 98, no. 1, pp. 105–118, 2018.

[25] G. B. Dantzig and J. H. Ramser, "The truck dispatching problem," *Management Science*, vol. 6, no. 1, pp. 80–91, 1959.

[26] M. M. Solomon, "Algorithms for the vehicle routing and scheduling problems with time window constraints," *Operations Research*, vol. 35, no. 2, pp. 254–265, 1987.

[27] M. Filipec, D. Škrlec, and S. Krajcar, "Genetic algorithm approach for multiple depot capacitated vehicle routing problem solving with heuristic improvements," *International Journal of Modelling and Simulation*, vol. 20, no. 4, pp. 320–328, 2000.

[28] H. S. Mahmassani, "Dynamic Network Traffic Assignment and Simulation Methodology for Advanced System Management Applications," in *Networks and Spatial Economics*, vol. 1, pp. 267–292, 2001.

[29] B. Sun, M. Wei, and J. Yao, "Collaborative vehicle route problem based on vehicle task reliability," *Application Research of Computers*, vol. 30, no. 8, pp. 2280–2287, 2013.

[30] S. Yuan, B. Wright, Y. Zou, and Y. Wang, "Quantification of variability of valid travel times with fmms for buses, passenger cars, and taxis," *IET Intelligent Transport Systems*, vol. 11, no. 1, pp. 1–9, 2017.

[31] K. Wang, X. Ye, R. M. Pendyala, and Y. Zou, "On the development of a semi-nonparametric generalized multinomial logit model for travel-related choices," *PLoS ONE*, vol. 12, no. 10, 2017.

[32] J. Tang, S. Zhang, Y. Zou, F. Liu, and X. Ma, "An adaptive map-matching algorithm based on hierarchical fuzzy system from vehicular GPS data," *PLoS ONE*, vol. 12, no. 12, Article ID 0188796, 2017.

[33] Y. R. Al-Mayouf, N. F. Abdullah, O. A. Mahdi et al., "Real-Time Intersection-Based Segment Aware Routing Algorithm for Urban Vehicular Networks," *IEEE Transactions on Intelligent Transportation Systems*, vol. 19, no. 7, pp. 2125–2141, 2018.

[34] A. K. Laha and S. Putatunda, "Real time location prediction with taxi-GPS data streams," *Transportation Research Part C: Emerging Technologies*, vol. 92, pp. 298–322, 2018.

[35] Z. Xu and Y. Cai, "Variable neighborhood search for consistent vehicle routing problem," *Expert Systems with Applications*, vol. 113, pp. 66–76, 2018.

[36] R. Wang, C.-Y. Chow, Y. Lyu et al., "TaxiRec: Recommending road clusters to taxi drivers using ranking-based extreme learning machines," *IEEE Transactions on Knowledge and Data Engineering*, vol. 30, no. 3, pp. 585–598, 2018.

[37] C. Ren, Q. Wu, C. Zhang, and S. Zhang, "A Normal Distribution-Based Methodology for Analysis of Fatal Accidents in Land Hazardous Material Transportation," *International Journal of Environmental Research and Public Health*, vol. 15, no. 7, Article ID 15071437, 2018.

[38] Y. Zou, J. E. Ash, B.-J. Park, D. Lord, and L. Wu, "Empirical Bayes estimates of finite mixture of negative binomial regression models and its application to highway safety," *Journal of Applied Statistics*, vol. 45, no. 9, pp. 1652–1669, 2018.

[39] B. Jain, G. Brar, J. Malhotra, S. Rani, and S. H. Ahmed, "A cross layer protocol for traffic management in Social Internet of Vehicles," *Future Generation Computer Systems*, vol. 82, pp. 707–714, 2018.

[40] S. Lam, J. Taghia, and J. Katupitiya, "Evaluation of a transportation system employing autonomous vehicles," *Journal of Advanced Transportation*, vol. 50, no. 8, pp. 2266–2287, 2016.

[41] Q. N. La, A. H. Lee, L. B. Meuleners, and D. Van Duong, "Prevalence and factors associated with road traffic crash among taxi drivers in Hanoi, Vietnam," *Accident Analysis & Prevention*, vol. 50, pp. 451–455, 2013.

[42] S.-W. Kim, G.-P. Gwon, W.-S. Hur et al., "Autonomous Campus Mobility Services Using Driverless Taxi," *IEEE Transactions on Intelligent Transportation Systems*, vol. 18, no. 12, pp. 3513–3526, 2017.

[43] J. Tang, S. Zhang, X. Chen, F. Liu, and Y. Zou, "Taxi trips distribution modeling based on Entropy-Maximizing theory: A case study in Harbin city—China," *Physica A: Statistical Mechanics and its Applications*, vol. 493, pp. 430–443, 2018.

[44] H. Yang and S. C. Wong, "A network model of urban taxi services," *Transportation Research Part B: Methodological*, vol. 32, no. 4, pp. 235–246, 1998.

[45] R. Sun, M. Li, and Q. Wu, "Research on Commuting Travel Mode Choice of Car Owners Considering Return Trip Containing Activities," *Sustainability*, vol. 10, no. 10, Article ID 10103494, 2018.

5

Leader-Follower based Locally Rigid Formation Control

Krishna Raghuwaiya, Bibhya Sharma, and Jito Vanualailai

School of Computing, Information & Mathematical Sciences, The University of the South Pacific, Suva, Fiji

Correspondence should be addressed to Bibhya Sharma; bibhya.sharma@usp.ac.fj

Academic Editor: Fernando García

This paper addresses motion control of a cooperative intelligent transport system (C-ITS) of nonholonomic mobile robots navigating a dynamic environment while maintaining a locally rigid formation. We consider the design of acceleration-based control inputs that govern the motion of cooperative intelligent transport system (C-ITS) using the artificial potential fields method for the avoidance of obstacles and attraction to designated targets. The control scheme utilizes a new leader-follower strategy using Cartesian coordinates to accomplish a collision-free locally rigid formation of an autonomous and intelligent transportation system. The concepts of virtual parking bays and *minimum distance technique* (MDT) are utilized to attain prescribed orientations of the formation at the final destination. The robustness of the control scheme is established by considering the effect noise on the formation, while the effectiveness of the proposed nonlinear control laws is demonstrated through computer simulations.

1. Introduction

Connected and autonomous vehicles play a huge role in the transportation industry worldwide and will govern the industry in the coming years. The connectedness is governed through vehicle-to-vehicle (V2V) and vehicle-to-infrastructure (V2I) communication, which is facilitated using wireless and cellular technologies at current times. Such key communications of the *cooperative intelligent transport system* (C-ITS) allow respective shareholders to utilize information and coordinate task transfers in a cooperative manner. The networked V2V and V2I interactions deliver substantial social benefits in terms of safer, traffic-efficient, cooperative autonomous driving and energy-efficient transportation systems on our roads and highways [1, 2].

Cooperative tasks are more efficiently performed with desired robustness using multiple robots, which may not be possible with single robots. The benefits become apparent when considering distributed tasks, dangerous or hazardous tasks, and tasks that contain redundancies and when providing flexibility to task execution and robustness of systems [3–8]. In addition, there is the added advantage that working with multiple robots is less expensive than with just one specialized robot. To date there have been various applications of the control of C-ITS, for example, collaborative mapping and planning, search and rescue operations, flocking and schooling, transportation of large objects, undersea and space exploration, target seeking, competitive games, service robots, and military reconnaissance and surveillance [4, 8–13]. The ability of C-ITS to autonomously navigate in stable configurations while avoiding obstacles and collisions is also central to real-world applications [8, 14]. In many applications, the execution of the task requires formation control [3], and the accomplishment of the overall operation depends on each mobile robot operating in a prescribed manner. Typical examples include carrying a heavy load on roads and highways, hunting, and enclosing on an enemy to name a few.

The concept of formation control of mechanical systems, an integrated branch of motion planning and control of C-ITS, is receiving unprecedented attention from researchers all over the world for real-world applications [14]. The basic idea of formation control algorithms is to ensure that a group of mobile robots move effectively as a whole to jointly perform certain task(s). Examples of formation control tasks include delegation of feasible formations, establishing formations, maintaining and mobilizing different formation shapes, and switching between formations [15]. There are numerous approaches in literature in relation to the observance of a prescribed formation of a flock during motion [14]; however, split/rejoin maneuvers and rigid formations are the prevalent approaches. The split/rejoin maneuvers are frequently cited in flocks of birds, swarms of insects and ants, and herds of animals. The applications of split/rejoin maneuvers in the field of robotics include reconnaissance, sampling, and surveillance. In contrast, rigid formations (*globally rigid* or *locally rigid*) are required in many engineering applications, for example, parallel and simultaneous transportation of vehicles or delivery of payloads [12, 14, 16]. Alterations or distortions in formation must be detected early by a system and allow for reconfiguration for continued operation [7] in a locally rigid formation.

In recent literature, a variety of control strategies have been proposed for formation or cooperative control of mobile robots and have been roughly categorized into virtual structure, behavior-based, and leader-follower schemes. The behavior-based approach normally [3] assigns different possible behaviors (e.g., formation keeping, obstacle avoidance, collision avoidance, and target attraction) to each individual robot. The ultimate action of each robot is determined by evaluating the comparative importance of each behavior. The limitation of the behavior-based approach is that it is difficult to analyze its behavioral performance mathematically and therefore it is difficult to guarantee system stability [17]. The virtual structures [6, 18–20] consider the entire formation as a single virtual rigid body. This networked structure can then be considered similar to a physical body and thus it is easy to maintain the formation of the whole group during maneuvers; that is, the virtual rigid body progresses as a whole in a given course with some given bearing. In the leader-follower approach [14, 17, 21–24], one robot in the multirobot formation, a virtual robot, is assigned the role of the leader to pursue some team objective, while the follower robots follow their designated leader. The follower robots then place themselves relative to the leader and maintain a predefined offset with a desired relative position [17]. In [25], Sharma et al. proposed a leader-follower scheme in a Lyapunov-based decentralized formation control planner for a swarm of 2-link mobile manipulators. The acceleration-based controllers ensured a locally rigid formation for the swarm. Locally rigid formation was achieved by integrating maximum and minimum interrobot distance bounds with desired headings in the control scheme.

In this paper, we adopt the architecture of the Lyapunov-based Control Scheme (LbCS) of [14], an artificial potential field (APF) method to design attractive and repulsive potential field functions to control the formation of C-ITS. An extension to [25], this paper presents a new set of nonlinear time-invariant control laws through an amalgamation of the LbCS and a new leader-follower scheme to maintain and mobilize a locally rigid formation. The limitation of the approach is twofold; firstly it comes with the disadvantages of APFs, in particular the local minima, and secondly the dependence on the lead robot and the poor disturbance rejection properties [26] of the leader-follower scheme. However, the novelty of the new approach lies in its ability to design continuous nonlinear control laws to translate locally rigid formations of nonholonomic systems tagged with dynamic constraints. The LbCS controllers are elegant and very simple to construct compared to the mathematically and computationally intensive ones in literature, and LbCS has an in-built process of converting these constraints into artificial obstacles and incorporating them into the controllers. Finally, the virtual connectivity from the new leader-follower scheme facilitates diverse tasks compared to merely cooperative agents [26].

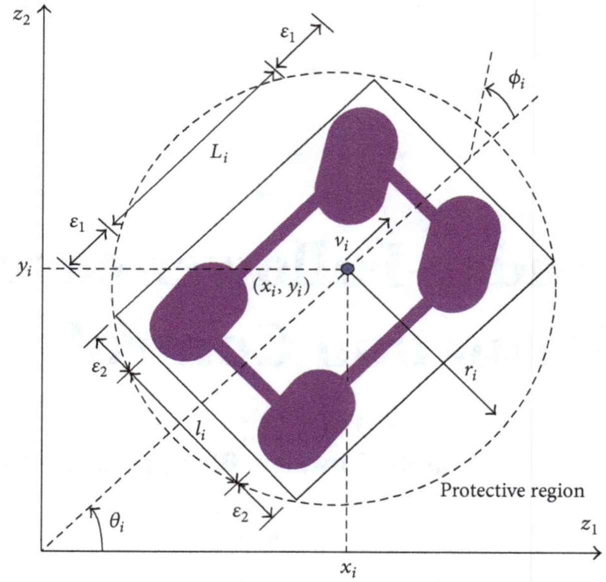

FIGURE 1: Kinematic model of the car-like mobile robot.

The advantage of the proposed scheme is that it uses Cartesian coordinate representation to avoid any singular points as encountered when using polar coordinates, even though polar coordinate representation may still be simpler to use. Another advantage of the proposed scheme in this paper is that it has a dual frame methodology, a new leader-follower approach, and the assignment of a single target for the group of vehicles. The overarching framework is a leader-follower scheme for C-ITS to establish, maintain, and translate the whole team in a locally rigid formation around the workplace performing a given task through centralized control laws. Finally, the treatment of several categories of obstacles is also included within the algorithm.

2. Vehicle Model

In this section, we derive a new kinematic model for the leader-follower based formation control of multiple vehicles. We will consider n ($n \in \mathbb{N}$) car-like vehicles in the Euclidean plane. We let $\mathscr{A}_1$ represent the leader and $\mathscr{A}_i$ for $i = 2, \ldots, n$ take the role of follower robots. With reference to Figure 1, adopted from [27], and for $i \in \{1, \ldots, n\}$, (x_i, y_i) represents the Cartesian coordinates and gives the reference point of each vehicle and θ_i gives the orientation of the ith car with respect to z_1-axis of the z_1z_2-plane. Moreover, ϕ_i gives the ith vehicles steering angle with respect to its longitudinal axis. L_i represents the distance between the centers of the front and rear axles of the ith robot, and l_i is the length of each axle.

Next, to ensure that each vehicle safely steers past an obstacle, we adopt the nomenclature of [14] and construct circular regions that protect the robot. With reference to Figure 1, given the *clearance parameters* $\epsilon_1 > 0$ and $\epsilon_2 > 0$, we enclose each vehicle by a protective circular region centered at (x_i, y_i) with radius $r_i = (1/2)\sqrt{(L_i + 2\epsilon_1)^2 + (l_i + 2\epsilon_2)^2}$.

Furthermore, we assume no slippage condition of the rear and front wheels of the mobile robots of $\mathscr{A}_i$ when in contact with a rigid surface, that is, the lateral (or tangential) velocities of the wheels of the cars are assumed to be zero. We have no slippage (i.e., $\dot{x}_i \sin\theta_i - \dot{y}_i \cos\theta_i = 0$) and pure rolling (i.e., $\dot{x}_i \cos\theta_i + \dot{y}_i \sin\theta_i = v_i$) of the car-like mobile robot which generate the *nonholonomic constraints* on the system. These assumptions of no slippage and pure rolling of the car-like mobile robots are essential to generate the nonholonomic constraints on the system. The kinematics of the system which inherently capture these nonholonomic constraints, adopted from [14], are

$$\begin{aligned} \dot{x}_i &= v_i \cos\theta_i - \frac{L_i}{2}\omega_i \sin\theta_i, \\ \dot{y}_i &= v_i \sin\theta_i + \frac{L_i}{2}\omega_i \cos\theta_i, \\ \dot{\theta}_i &= \frac{v_i}{L_i}\tan\phi_i := \omega_i, \\ \dot{v}_i &:= \sigma_{i1}, \\ \dot{\omega}_i &:= \sigma_{i2}, \end{aligned} \tag{1}$$

for $i \in \{1, \ldots, n\}$, and, without any loss of generality, we assume that $\phi_i = \theta_i$. In system (1), v_i and ω_i are, respectively, the instantaneous translational and rotational velocities, while σ_{i1} and σ_{i2} are the instantaneous translational and rotational accelerations of the ith robot.

Now, system (1) is a description of the instantaneous velocities and accelerations of $\mathscr{A}_i$. Let the vector $\mathbf{x}_i := (x_i, y_i, \theta_i, v_i, \omega_i) \in \mathbb{R}^5$ refer to the position (x_i, y_i), orientation, θ_i, and the velocities (v_i, ω_i) of $\mathscr{A}_i$ at time $t \geq 0$.

Now, let

$$\begin{aligned} \mathbf{f}_i(\mathbf{x}_i) &= (f_{i1}(\mathbf{x}_i), f_{i2}(\mathbf{x}_i), f_{i3}(\mathbf{x}_i), 0, 0) \\ &:= (\dot{x}_i, \dot{y}_i, \dot{\theta}_i, 0, 0) \in \mathbb{R}^5, \end{aligned} \tag{2}$$

and $\mathbf{u}_i(t) := (\sigma_{i1}(t), \sigma_{i2}(t)) \in \mathbb{R}^2$. Then system (1) can be written compactly as

$$\dot{\mathbf{x}}_i := \mathbf{f}_i(\mathbf{x}_i) + \mathbf{B}_i\mathbf{u}_i(t), \tag{3}$$

where $\mathbf{B}_i$ is a 5×2 matrix of the form

$$\mathbf{B}_i = \begin{bmatrix} 0 & 0 \\ 0 & 0 \\ 0 & 0 \\ 1 & 0 \\ 0 & 1 \end{bmatrix}. \tag{4}$$

Let $\mathbf{x} := (\mathbf{x}_1, \mathbf{x}_2, \ldots, \mathbf{x}_n) \in \mathbb{R}^{5n}$ refer to the positions, orientations, and the velocities of all the vehicles in the C-ITS.

Let $\mathbf{f}(\mathbf{x}) := (\mathbf{f}_1(\mathbf{x}), \mathbf{f}_2(\mathbf{x}), \ldots, \mathbf{f}_n(\mathbf{x})) \in \mathbb{R}^{5n}$ and $\mathbf{u}(t) := (\mathbf{u}_1(t), \mathbf{u}_2(t), \ldots, \mathbf{u}_n(t)) \in \mathbb{R}^{2n}$. Then we have the following initial-value problem for $\mathscr{A}_i$:

$$\begin{aligned} \dot{\mathbf{x}} &= \mathbf{f}(\mathbf{x}) + \mathbf{B}\mathbf{u}(t), \\ \mathbf{x}(t_0) &:= \mathbf{x}_0, \quad t_0 \geq 0, \end{aligned} \tag{5}$$

where if $\mathbf{0}$ is a 5×2 matrix of all zero entries,

$$\mathbf{B} = \begin{bmatrix} \mathbf{B}_1 & \mathbf{0} & \cdots & \mathbf{0} \\ \mathbf{0} & \mathbf{B}_2 & \cdots & \mathbf{0} \\ \vdots & \vdots & \ddots & \vdots \\ \mathbf{0} & \mathbf{0} & \cdots & \mathbf{B}_n \end{bmatrix}. \tag{6}$$

Now, assume that the final position of $\mathscr{A}_i$ is at the point $(x_i, y_i) = (p_{i1}, p_{i2})$ and final orientation at this point is $\theta_i = p_{i3}$. Its final instantaneous velocity vector is $(v_i, \omega_i) = (0, 0)$. Then it is clear that the points

$$\mathbf{x}_i^* := (p_{i1}, p_{i2}, p_{i3}, 0, 0) \in \mathbb{R}^5 \tag{7}$$

are the components of the equilibrium point of system (5) in which we are interested; that is,

$$\mathbf{x}_e := (\mathbf{x}_1^*, \mathbf{x}_2^*, \ldots, \mathbf{x}_n^*) \in \mathbb{R}^{5n}. \tag{8}$$

2.1. Leader-Follower Based Formation Scheme. Next we define two reference frames: the body frame which is fixed with the rotating body of the leader, $\mathscr{A}_1$, and a space frame, the inertial frame similar to one proposed in [16].

We assign a Cartesian coordinate system (X-Y) fixed on the leader body, as shown in Figure 2 adopted from [27], based on the concept of an instantaneous corotating frame of reference. Thus, when the leader $\mathscr{A}_1$ rotates, we have a rotation of the X-Y-axes.

To define the corotating frame of reference, first an origin is selected on the leader robot at (x_1, y_1). An axis of rotation is then set up, which is perpendicular to the plane of motion of the leader. Thus, at any selected moment t, the chosen rotating frame of reference rotates at an angular rate equal to the rate of rotation of the leader $\mathscr{A}_1$ about (x_1, y_1). Let r_{1k} represent the straight-line distance between the reference point of the leader and the kth follower, and α_{1k} represents the angle measured between the straight line joining the reference points of the leader and the kth follower and X-axis. Thus, given the leader's position and orientation, as long as (r_{1k}, α_{1k}), as shown in Figure 2, is fixed, the kth follower robot's position will be unique. We define the shape of the formation of the mobile robots as $\zeta = [\zeta_{12}, \zeta_{13}, \ldots, \zeta_{1n}]^T$, where $\zeta_{1k} = [r_{1k}, \alpha_{1k}]^T$ for $k \in \{2, \ldots, n\}$.

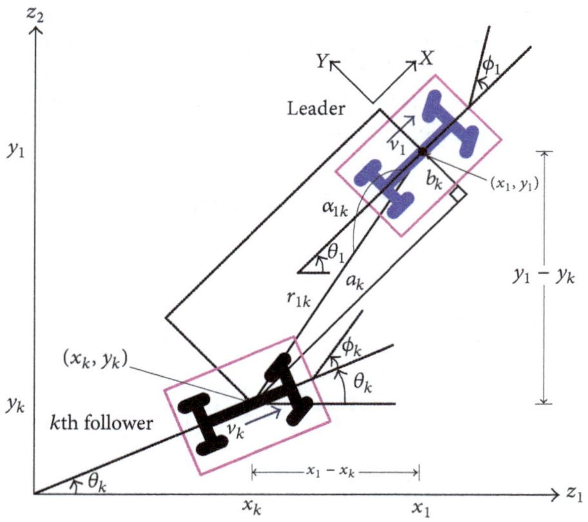

FIGURE 2: The proposed scheme utilizing a rotation of axes with the body frame fixed at the leader robot.

Definition 1. Let $2r_k < r_{1k}^d < 2r_k + \varsigma$, where, for $\varsigma > 0$ and $k \in \{2, \ldots, n\}$, the group of mobile robots make a $\zeta^d = [\zeta_{12}^d, \zeta_{13}^d, \ldots, \zeta_{1n}^d]^T$ formation, if $\exists T > 0, \forall t > T$:

$$r_{1k} = r_{1k}^d, \quad \alpha_{1k} = \alpha_{1k}^d. \tag{9}$$

This gives then the polar coordinate representation of the follower's position relative to that of the leader. However, such representations using polar coordinates lead to certain singularities in the controllers [16]. To avoid such singular points, we consider the position of the kth follower by considering the relative distances of the kth follower from the leader $\mathscr{A}_1$ along the given X and Y directions. Thus, we have

$$A_k = -(x_1 - x_k)\cos\theta_1 - (y_1 - y_k)\sin\theta_1, \quad B_k = (x_1 - x_k)\sin\theta_1 - (y_1 - y_k)\cos\theta_1, \tag{10}$$

for $k \in \{2, \ldots, n\}$, and A_k and B_k are the kth follower relative positions with respect to the X-Y coordinate system. If A_k and B_k are known and fixed, the follower's position will be distinct. Thus, to obtain a desired formation, one needs to know the design parameters a_k and b_k and the desired relative positions along the X-Y directions, such that the control objective would be to achieve $A_k \to a_k$ and $B_k \to b_k$. That is, $r_{1k} \to r_{1k}^d$ and $\alpha_{1k} \to \alpha_{1k}^d$, where $r_{1k}^d = \sqrt{a_k^2 + b_k^2}$ and $\alpha_{1k}^d = \tan(a_k/b_k)$, hence maintaining local rigidity of any formation.

Definition 2. We define a locally rigid formation as one which is maintained but allows for slight changes or distortions in the distances and angles between any two vehicles temporarily, while enroute to the target and task completion.

Remark 3 (formation initialization). Prescribed formations can be initialized by assigning appropriately the design parameters a_k and b_k. This will initiate the vehicles from their arbitrary initial positions to move into a desired formation pattern.

2.2. Formation Control Objective. Design acceleration-based control laws $(\sigma_{i1}, \sigma_{i2})$ which are functions of $\mathbf{x}_i := (x_i, y_i, \theta_i, v_i, \omega_i) \in \mathbb{R}^5$ for $i \in \{1, \ldots, n\}$, for each vehicle such that the C-ITS in a prescribed formation moves towards a predefined target of the leader in an obstacle cluttered environment. In addition, we also want the C-ITS to ensure obstacle and collision avoidance and maintain a prescribed locally rigid formation.

3. Artificial Potential Field Function

This section formulates collision-free trajectories of the vehicle system under kinodynamic constraints. We want to design the acceleration controllers, σ_{i1} and σ_{i2}, so that the team of vehicles moves safely towards the target of the leader while maintaining a prescribed locally rigid formation. We will design the attractive potential field functions which establish and translate formation and the repulsive potential field functions which ensure collision and obstacle avoidances.

In the following subsections, we will design these attractive and repulsive potential field functions which are subsequently summed to form the total potentials or artificial potential field of the system.

3.1. Attractive Potential Field Functions

3.1.1. Attraction to Target. A target is assigned to the leader. When the leader moves towards its defined target, the follower vehicles move with the leader, maintaining the formation. We want the leader $\mathscr{A}_1$ to start from an initial position, move towards a target, and finally converge at the center of the target.

Definition 4. The stationary target for the leader robot $\mathscr{A}_1$ is a circular disk with center (p_{11}, p_{12}) and radius rt.

The leader $\mathscr{A}_1$ will move towards its defined target, while the follower vehicles move with their leader maintaining a desired relative position, hence a locally rigid formation. For the attraction of $\mathscr{A}_1$ to its target, we consider an attractive potential function:

$$V_1(\mathbf{x}) = \frac{1}{2}\left[(x_1 - p_{11})^2 + (y_1 - p_{12})^2 + v_1^2 + \omega_1^2\right]. \tag{11}$$

The above function is a measure of its convergence to the target with the inclusion of the velocity components [14]. For the follower vehicle $\mathscr{A}_i$ for $i = 2, \ldots, n$ to maintain its desired relative position with respect to the leader, $\mathscr{A}_1$, we utilize the following potential function:

$$V_i(\mathbf{x}) = \frac{1}{2}\left[(A_i - a_i)^2 + (B_i - b_i)^2 + v_i^2 + \omega_i^2\right], \tag{12}$$

for $i = 2, \ldots, n$, where a_i and b_i will have unique values as dictated by the shape of the desired formation.

3.1.2. Auxiliary Function. To guarantee the convergence of the vehicles to their designated targets, we design an auxiliary function as

$$G_1(\mathbf{x}) = \frac{1}{2}\left[(x_1 - p_{11})^2 + (y_1 - p_{12})^2 + \rho_1(\theta_1 - p_{13})^2\right], \quad (13)$$

where p_{13} is the prescribed final orientation of the leader robot, $\mathscr{A}_1$, and

$$G_i(\mathbf{x}) = \frac{1}{2}\left[(A_i - a_i)^2 + (B_i - b_i)^2 + \rho_i(\theta_i - \theta_1)^2\right], \quad (14)$$

for $i = 2, \ldots, n$. The function ensures that the controllers become zero at the leader's target. The constant ρ_i is a binary constant denoted in (13) and (14) as the *angle-gain parameter* for $\theta_i, i = 1, \ldots, n$. An *angle-gain parameter* will take a value of one only if a final predefined orientation is warranted; else it takes the default value of zero. This auxiliary function is then multiplied to the repulsive potential field functions.

3.2. Repulsive Potential Field Functions. We desire each vehicle to avoid all fixed and moving obstacles intersecting its path. Hence, we construct appropriate obstacle avoidance functions that measure the Euclidean distances between mobile robots and the obstacles on our roads. To obtain the desired avoidance, we generate repulsive potential field around the obstacles by designing a repulsive potential field function for each obstacle in accordance with LbCS. The repulsive potential fields function is an inverse function that encodes an avoidance function to the denominator and a control parameter in the numerator [14].

3.2.1. Fixed Obstacles. Let us fix $n \in \mathbb{N}$ solid obstacles on the roads, such as potholes, distance marker posts, or road hazard delineators. We assume that the lth obstacle is a circular disk with center (o_{l1}, o_{l2}) and radius ro_l. For the ith vehicle to avoid the lth obstacle, we consider an avoidance function:

$$\mathrm{FO}_{il}(\mathbf{x}) = \frac{1}{2}\left[(x_i - o_{l1})^2 + (y_i - o_{l2})^2 - (ro_l + r_i)^2\right], \quad (15)$$

where $i = 1, \ldots, n$ and $l = 1, \ldots, q$.

Consider, for example, the presence of three obstacles (i.e., $q = 3$) within the workspace, with $0 < z_1 < 70$ and $0 < z_2 < 70$. The total potentials, where $\alpha_{1l} > 0$ is a control or tuning parameter, which govern the motion of the leader $\mathscr{A}_1$ are

$$V_1(\mathbf{x}) + \sum_{l=1}^{3} \frac{\alpha_{1l}}{\mathrm{FO}_{1l}(\mathbf{x})}. \quad (16)$$

Figure 3 presents a three-dimensional view of the total potentials and Figure 4 presents the corresponding contour plot produced by (16).

The total potentials are generated for target attraction and avoidance of three stationary disk-shaped obstacles. For better visualization, the target of the leader is located at $(p_1, p_2) = (35, 35)$, and the disks are fixed at

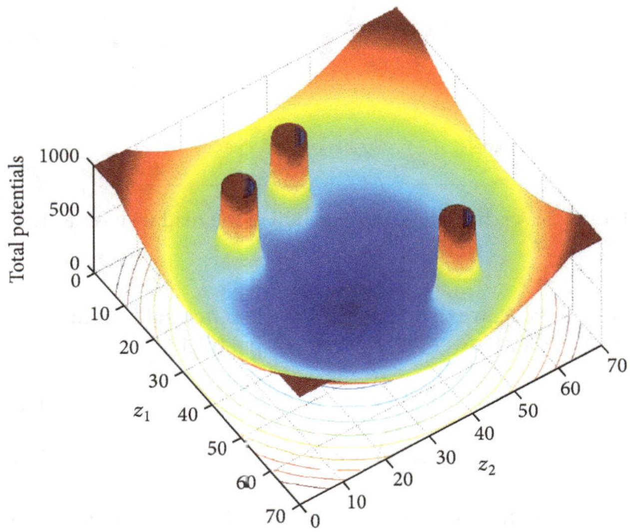

FIGURE 3: A three-dimensional view of the total potentials.

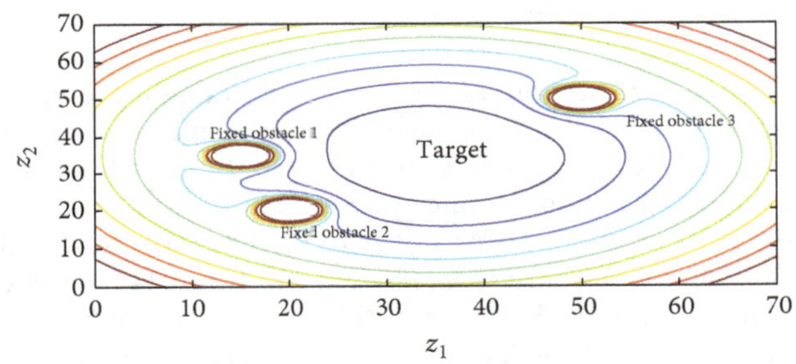

FIGURE 4: The contour plot of the total potential.

$(o_{11}, o_{12}) = (15, 35)$, $(o_{21}, o_{22}) = (20, 20)$, and $(o_{31}, o_{32}) = (50, 50)$, with radii of $ro_l = 3$, while $\alpha_{1l} = 1000$, for $l = 1, \ldots, 3$. Also, the velocity and angular components of the lead vehicle have been treated as constants ($v_1 = 0.5$, $\omega_1 = 0$, and $\theta_1 = 0$).

3.3. Boulevard Limitations

Definition 5. Consider a section of the road defined, for some $\eta_1 > 2r_i$, for $i = 1, \ldots, n$, as

$$WS = \left\{(z_1, z_2) \in \mathbb{R}^2 : 0 \le z_2 \le \eta_1\right\}. \quad (17)$$

The boundaries of the road section, illustrated in Figure 5, are defined as follows:

(a) Lower boundary: $B_1 = \{(z_1, z_2) \in \mathbb{R}^2 : z_2 = 0\}$

(b) Upper boundary: $B_2 = \{(z_1, z_2) \in \mathbb{R}^2 : z_2 = \eta_1\}$

We require the prescribed formation to stay within the boundaries of the road at all time $t \ge 0$. In our LbCS, these boundaries are considered as *fixed obstacles*. For the ith robot

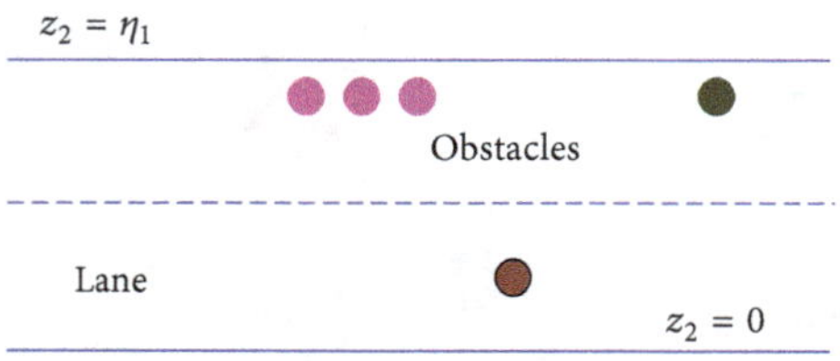

FIGURE 5: A schematic representation of the road with obstacles.

to avoid these, we define the following potential functions for the upper and lower boundaries, respectively:

$$W_{i1}(\mathbf{x}) = y_i - r_i, \tag{18a}$$

$$W_{i2}(\mathbf{x}) = \eta_1 - (y_i + r_i). \tag{18b}$$

Now, since $\eta_1 > 2r_i$, each of the functions is positive in WS, for $i = 1, \ldots, n$. As discussed before, these obstacle avoidance functions will be combined with appropriate tuning parameters to generate repulsive potential field functions for the boundaries of the workspace.

3.4. Final Orientation of Formation. Although the final position is reachable, it is virtually impossible to simultaneously harvest exact orientations via continuous feedback controllers of nonholonomic systems [14], a direct result of Brockett's theorem [28]. We utilize the concepts of ghost wall and *minimum distance technique* (MDT) from [14] to force final orientations of the vehicles. This is needed to confine the vehicles to within the lane markings and later for parking a vehicle into a parking bay. To begin, we construct ghost walls along the three sides of the final positions of the vehicles, treated as a parking bay, with the orientation depending on the desired final orientation of the vehicles. To avoid the ghost walls of each vehicle's parking bay, we utilize MDT, which gives the perpendicular distance between the closest point on each kth ghost wall and (x_i, y_i). The avoidance of these closet points on a given ghost wall at any time $t \geq 0$ essentially results in the avoidance of the entire wall by a vehicle.

Now let us consider the kth ghost wall in the z_1z_2-plane from the point (a_{k1}, b_{k1}) to the point (a_{k2}, b_{k2}). We assume that the point (x_i, y_i) is closest to it at the tangent line which passes through the point. From geometry, it is known that if (Lx_{ik}, Ly_{ik}) is the point of intersection of this tangent, then

$$\begin{aligned} Lx_{ik} &= a_{k1} + \lambda_{ik}(a_{k2} - a_{k1}), \\ Ly_{ik} &= b_{k1} + \lambda_{ik}(b_{k2} - b_{k1}), \end{aligned} \tag{19}$$

where

$$\begin{aligned} \lambda_{ik} &= (x_i - a_{k1}) d_k + (y_i - b_{k1}) r_k, \\ d_k &= \frac{a_{k2} - a_{k1}}{(a_{k2} - a_{k1})^2 + (b_{k2} - b_{k1})^2}, \\ r_k &= \frac{b_{k2} - b_{k1}}{(a_{k2} - a_{k1})^2 + (b_{k2} - b_{k1})^2}. \end{aligned} \tag{20}$$

If $\lambda_{ik} \geq 1$, then we let $\lambda_{ik} = 1$; if $\lambda_{ik} \leq 0$, then we let $\lambda_{ik} = 0$; otherwise we accept the value of λ_{ik} between 0 and 1, in which case there is a perpendicular line to the point (Lx_{ik}, Ly_{ik}) on the ghost wall from the center (x_i, y_i) of the ith vehicle at every time $t \geq 0$.

The leader will be avoiding the 1st, 2nd, and 3rd ghost walls, while the ith follower will be avoiding the $(3i-2)$, $(3i-1)$, and $(3i)$ ghost walls for $i = 2, \ldots, n$. Now, for the ith vehicle to avoid the closest point of each of the kth line segments, we consider an avoidance function

$$LS_{ik}(\mathbf{x}) = \frac{1}{2}\left[(x_i - Lx_{ik})^2 + (y_i - Ly_{ik})^2 - r_i^2\right], \tag{21}$$

for $k \in \{3i-2, 3i-1, 3i\}$ and $i = 1, \ldots, n$.

3.5. Moving Obstacles. To generate feasible trajectories, we consider moving obstacles in the workspace, in which the C-ITS has prior knowledge. Here, each vehicle has to be treated as a moving obstacle for all other vehicles on the road. The vehicles will have to travel towards their targets while avoiding another vehicle in their path. For the vehicle $\mathcal{A}_i$ to avoid the vehicle $\mathcal{A}_j$, via vehicle-to-vehicle (V-V) communication, we adopt an avoidance function:

$$\mathrm{MO}_{ij}(\mathbf{x}) = \frac{1}{2}\left[(x_i - x_j)^2 + (y_i - y_j)^2 - (r_i)^2\right], \quad \text{for } i, j = 1, \ldots, n \text{ with } i \neq j. \tag{22}$$

3.5.1. Dynamic Constraints. Practically, the steering angle of the ith autonomous vehicle is limited due to mechanical singularities, while the translational speed is restricted due to safety reasons [14]. Subsequently, we have the following;

(i) $|v_i| < v_{\max}$, where $v_{\max}$ is the *maximal achievable speed* of the ith vehicle.

(ii) $|\omega_i| < v_{\max}/|\rho_{\min}|$, where $\rho_{\min} := L_i/\tan(\phi_{\max})$. This condition arises due to the boundedness of the steering angle ϕ_i. That is, $|\phi_i| \leq \phi_{\max} < \pi/2$, where $\phi_{\max}$ is the *maximal steering angle.*

Remark 6. For simplicity, the values of $v_{\max}$ and $\phi_{\max}$ will be kept the same for each vehicle.

As per the LbCS, for each dynamic constraint, we design a corresponding *artificial obstacle.* For example, we consider the artificial obstacle $\mathrm{AO}_i = \{v_i \in \mathbb{R} : v_i \leq -v_{\max} \text{ or } v_i \geq v_{\max}\}$ for the constraint tagged to v_i. We can create similar artificial obstacles for the other limitations as well. Hence, we consider the following avoidance functions:

$$\begin{aligned} U_{i1}(\mathbf{x}) &= \frac{1}{2}(v_{\max} - v_i)(v_{\max} + v_i), \\ U_{i2}(\mathbf{x}) &= \frac{1}{2}\left(\frac{v_{\max}}{|\rho_{\min}|} - \omega_i\right)\left(\frac{v_{\max}}{|\rho_{\min}|} + \omega_i\right), \end{aligned} \tag{23}$$

for $i = 1, \ldots, n$.

4. Design of the Acceleration-Based Controllers

4.1. Lyapunov Function. We now construct the total potentials, that is, a Lyapunov function for system (1). First, for $i = 1, \ldots, n$, we introduce the following *control parameters* that we will use in the repulsive potential functions:

(i) $\alpha_{il} > 0$, $l = 1, \ldots, q$, for the collision avoidance of q disk-shaped obstacles.

(ii) $\beta_{is} > 0$, $s = 1, 2$, for the avoidance of the artificial obstacles from dynamic constraints.

(iii) $\eta_{ij} > 0$, $j = 1, \ldots, n$, $i \neq j$, for the collision avoidance between any two vehicles.

(iv) $\kappa_{ip} > 0$, $p = 1, \ldots, 2$, for the avoidance of the lane boundaries.

(v) $\gamma_{ik} > 0$, $k = 1, \ldots, 3n$, for the avoidance of the parking bays.

The above parameters are determined via heuristics, but there is also an opportunity to obtain optimal values through optimization techniques. Using these, we now construct the following Lyapunov function for system (1) with two components, namely, the attractive and repulsive potential field functions:

$$L_{(1)}(\mathbf{x}) = \sum_{i=1}^{n} \left[V_i(\mathbf{x}) + G_i(\mathbf{x}) \cdot \left(\sum_{l=1}^{q} \frac{\alpha_{il}}{\mathrm{FO}_{il}(\mathbf{x})} + \sum_{s=1}^{2} \frac{\beta_{is}}{U_{is}(\mathbf{x})} + \sum_{p=1}^{2} \frac{\kappa_{ip}}{W_{ip}(\mathbf{x})} \right) \right] + \sum_{i=1}^{n} G_i(\mathbf{x}) \left(\sum_{\substack{j=1 \\ j \neq i}}^{n} \frac{\eta_{ij}}{\mathrm{MO}_{ij}(\mathbf{x})} + \sum_{k=1}^{3n} \frac{\gamma_{ik}}{LS_{ik}(\mathbf{x})} \right). \tag{24}$$

4.2. Nonlinear Acceleration Controllers. The process of designing the feedback controllers begins by noting that the functions f_{ik} to g_{ij} for $i = 1, \ldots, n$, $j = 1, 2$, and $k = 1, \ldots, 3$ are defined as in the Appendix (on suppressing $\mathbf{x}$).

Remark 7. The choice of total potential $L(\mathbf{x})$ given in (24) with its terms specified in (11)–(16), (18a), (18b), and (21)–(23) has the following properties:

(i) It attains a minimum value when the robots are at their desired locations.

(ii) It goes to infinity whenever one or more robots come in contact with an obstacle.

Remark 8. With the interrobot bounds (see (12) and (22)) in place, it is guaranteed that the robots reestablish the predetermined formation if the robot positions are slightly distorted with the encounter of obstacle(s) soon after the avoidance and before reaching the target.

So, we design the following theorem.

Theorem 9. *Consider n car-like mobile robots in the C-ITS, whose motion is governed by the ODEs described in system (1). The principal goal is to establish and control a prescribed formation, facilitate maneuvers of the vehicles within a constrained environment, and reach the target configuration while maintaining a desired formation. The subtasks include restrictions placed on the workspace, convergence to predefined targets, and consideration of kinodynamic constraints. Utilizing the attractive and repulsive potential field functions, the following continuous time-invariant acceleration control laws can be generated in accordance with the LbCS of system (1):*

$$\sigma_{i1} = -\frac{\left[\delta_{i1} v_i + f_{i1} \cos\theta_i + f_{i2} \sin\theta_i\right]}{g_{i1}},$$
$$\sigma_{i2} = -\frac{\left[\delta_{i2}\omega_i + (L_i/2)\left(f_{i2}\cos\theta_i - f_{i1}\sin\theta_i\right) + f_{i3}\right]}{g_{i2}}, \tag{25}$$

for $i = 1, \ldots, n$, where $\delta_{i1} > 0$ and $\delta_{i2} > 0$ are constants commonly known as convergence parameters.

5. Stability Analysis

We utilize Lyapunov's direct method to provide a mathematical proof of stability of system (1).

Theorem 10. *Let (p_{11}, p_{12}) be the position of the target of the leader and let p_{i3}, for $i = 1, \ldots, n$, be the prescribed final orientations of the robots. Let p_{i1} and p_{i2} satisfy*

$$a_i = -(p_{11} - p_{i1})\cos\theta_1 - (p_{12} - p_{i2})\sin\theta_1,$$
$$b_i = (p_{11} - p_{i1})\sin\theta_1 - (p_{12} - p_{i2})\cos\theta_1, \tag{26}$$

for any given a_i and b_i, for $i = 2, \ldots, n$. If $\mathbf{x}_e \in \mathbb{R}^{5n}$ as defined in (8) is an equilibrium point for (1), then $\mathbf{x}_e \in D(L_{(1)}(\mathbf{x}))$ is a stable equilibrium point of system (1).

Proof. One can easily verify the following, for $i \in \{1, \ldots, n\}$:

(1) $L_{(1)}(\mathbf{x})$ is defined as continuous and positive over the domain $D(L_{(1)}(\mathbf{x})) = \{\mathbf{x} \in \mathbb{R}^{5n} : \mathrm{FO}_{il}(\mathbf{x}) > 0,\ l = 1, \ldots, q;\ \mathrm{MO}_{ij}(\mathbf{x}) > 0,\ j = 1, \ldots, n,\ j \neq i;\ W_{ip}(\mathbf{x}) > 0,\ p = 1, \ldots, 4;\ LS_{ik}(\mathbf{x}) > 0,\ k = 1, \ldots, 3n;\ U_{is}(\mathbf{x}) > 0,\ s = 1, 2\}$.

(2) $L_{(1)}(\mathbf{x}_e) = 0$; $\dot{L}_{(1)}(\mathbf{x}_e) = 0$.

(3) $L_{(1)}(\mathbf{x}) > 0\ \forall \mathbf{x} \in D(L_{(1)}(\mathbf{x}))/\mathbf{x}_e$.

(4) $\dot{L}_{(1)}(\mathbf{x}) = -\sum_{i=1}^{n}(\delta_{i1} v_i^2 + \delta_{i2}\omega_i^2) \leq 0,\ \forall \mathbf{x} \in D(L_{(1)}(\mathbf{x}))$.

(5) $L_{(1)}(\mathbf{x}) \in C^1(D(L_{(1)}(\mathbf{x})))$.

Hence, $L_{(1)}(\mathbf{x})$ is classified as a Lyapunov function for system (1) and $\mathbf{x}_e$ is a stable equilibrium point in the sense of Lyapunov. □

6. Simulation Results

In this section, we illustrate the effectiveness of the proposed continuous controllers by simulating a number of virtual scenarios for the C-ITS on our roads.

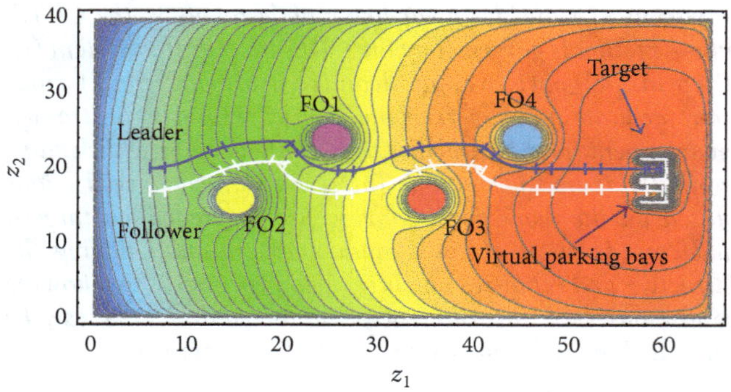

FO: Fixed obstacle

FIGURE 6: The evolution of C-ITS trajectories and the contour plot in the presence of obstacles in Scenario 1.

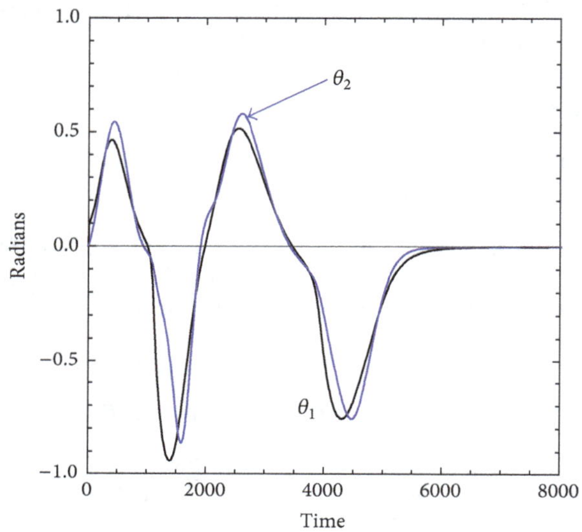

FIGURE 7: Orientations of $\mathscr{A}_1$ and $\mathscr{A}_2$ for Scenario 1.

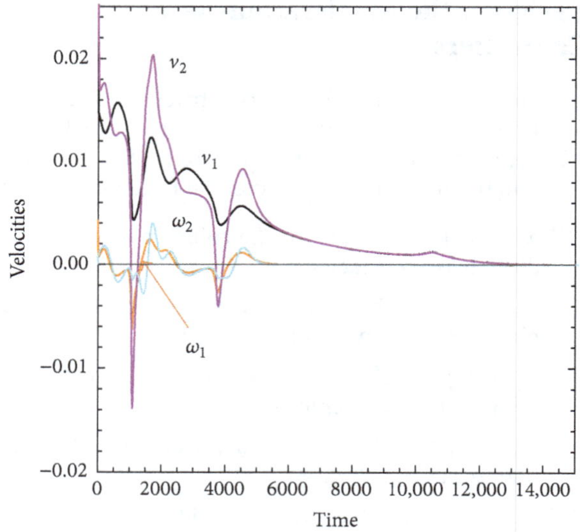

FIGURE 8: Translational and rotational velocities for Scenario 1.

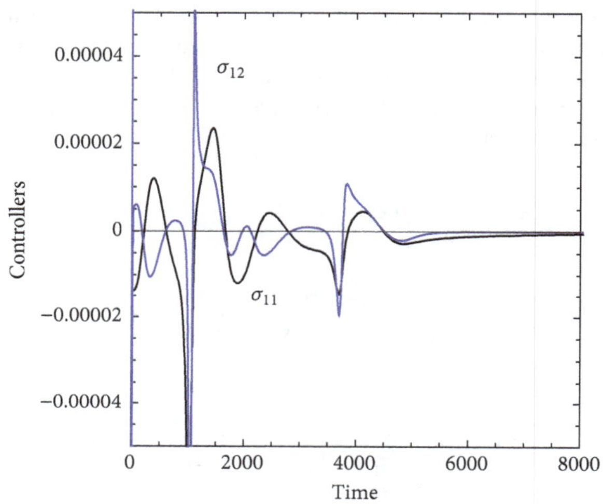

FIGURE 9: Accelerations σ_{11} and σ_{12} of $\mathscr{A}_1$ for Scenario 1.

6.1. Scenario 1: Line Formation in the Presence of Obstacles. We consider the motion of two automated vehicles in a *line formation* in the presence of obstacles. The follower vehicle is prescribed a position relative to the leader $\mathscr{A}_1$ as seen in Figure 6. While the leader moves towards its target, the follower maintains a desired distance and orientation relative to the leader, therefore maintaining a locally rigid formation throughout the journey.

Assuming that the appropriate units have been accounted for, Table 1 provides the corresponding initial and final configurations of the two vehicles and other parameters required to simulate Scenario 1.

Figure 6 also depicts the contour plot of the potential fields of the system enroute to the final destination. Figures 7 and 8 show the orientations and velocities of $\mathscr{A}_1$ and $\mathscr{A}_2$, respectively. Figure 9 depicts the time evolution of the nonlinear controllers of the leader $\mathscr{A}_1$, and Figure 10 compares the relative distance r_{12} to the desired relative distance r_{12}^d. It is evident that the formation is slightly distorted when the pair of cooperative cars encounter an obstacle; however, the desired formation is reestablished before the pair reach the final destination.

6.2. Scenario 2: Effect of Noise on Line Formation. To evaluate the robustness of the proposed scheme, we look at the effect of noise on the formation of the C-ITS. It is sufficient to include the noise parameters in the components A_k and B_k which define the follower vehicles' relative position to the leader vehicle, with respect to the X-Y coordinate system, similar to the one proposed in [29]. Thus we have

$$\begin{aligned} A_k &= -(x_1 - x_k)\cos\theta_1 - (y_1 - y_k)\sin\theta_1 + \xi\gamma_k(t), \\ B_k &= (x_1 - x_k)\sin\theta_1 - (y_1 - y_k)\cos\theta_1 + \xi\nu_k(t). \end{aligned} \tag{27}$$

The terms $\xi\gamma_k(t)$ and $\xi\nu_k(t)$ are the small disturbances, where $\xi \in [0,1]$ is the noise level, while $\gamma_k(t)$ and $\nu_k(t)$ are randomized time-dependent variables such that $\gamma_k(t) \in [-1,1]$

TABLE 1: Numerical values of initial and final states, constraints, and parameters for Scenario 1.

	Initial configuration
Rectangular positions	$(x_1, y_1) = (7, 20)$ and $(x_2, y_2) = (7, 17)$
Velocities	$v_i = 0.5$ and $\omega_i = 0$, for $i = 1, 2$
Angular positions	$\theta_i = 0$, for $i = 1, 2$
Parking bays	$(a_{11}, b_{11}) = (57.5, 21.3)$ and $(a_{12}, b_{12}) = (60.1, 21.3)$ $(a_{31}, b_{31}) = (57.5, 18.7)$ and $(a_{32}, b_{32}) = (60.1, 18.7)$ $(a_{41}, b_{41}) = (57.5, 18.2)$ and $(a_{42}, b_{42}) = (60.1, 18.2)$ $(a_{61}, b_{61}) = (57.5, 15.6)$ and $(a_{62}, b_{62}) = (60.1, 15.6)$
	Constraints and parameters
Dimensions of robots	$L_i = 1.6$ and $l_i = 1.2$ for $i = 1, 2$
Leader's target	$(p_{11}, p_{12}) = (57, 20)$; $rt = 0.5$
Final orientations	$p_{13} = p_{23} = 0$
Max. translational velocity	$v_{\max} = 5$
Max. steering angle	$\phi_{\max} = \frac{\pi}{2}$
Clearance parameters	$\epsilon_1 = 0.1$ and $\epsilon_2 = 0.05$
Fixed obstacles	$(o_{11}, o_{12}) = (15, 16)$, $(o_{21}, o_{22}) = (25, 24)$, and $(o_{31}, o_{32}) = (35, 16)$ $(o_{41}, o_{42}) = (45, 24)$ and $ro_l = 2$ for $l = 2, \ldots, 4$
	Control and convergence parameters
Collision avoidance	$\eta_{ij} = 0.001$, for $i, j = 1, 2$, $j \neq i$ $\kappa_{ip} = 0.1$, for $i = 1, 2$, $p = 1, \ldots, 2$ $\gamma_{ik} = 0.1$, for $i = 1, 2$, $k = 1, \ldots, 3n$ $\alpha_{11} = 1$, $\alpha_{12} = \alpha_{14} = 0.1$, and $\alpha_{13} = 2$ $\alpha_{2l} = 0.01$ for $l = 1, \ldots, 4$
Dynamics constraints	$\beta_{is} = 0.01$, for $i, s = 1, 2$
Convergence	$\delta_{11} = 3000$, $\delta_{12} = 100$, $\delta_{21} = 10$, and $\delta_{22} = 100$

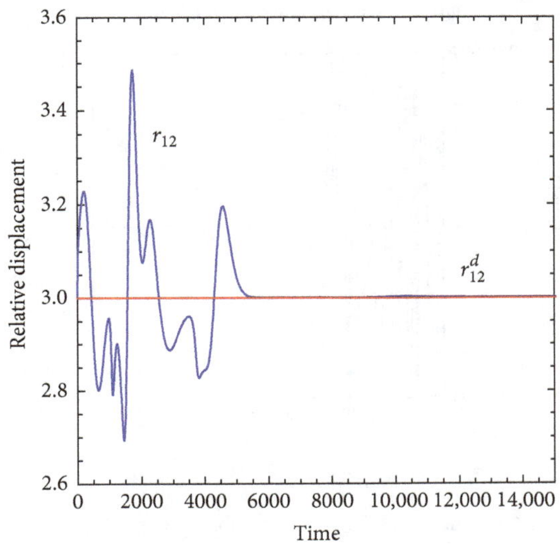

FIGURE 10: The relative distance of the follower to the leader vehicle compared to the desired relative distance in Scenario 1.

and $v_k(t) \in [-1, 1]$. Figure 11 shows the trajectories and the control signals under the influence of small disturbances, $\xi \in [0, 1]$. It is observed that the pair of cooperative cars maintain the line formation throughout their road journey even under the influence of the noise terms. There are slight distortions in their formation when the pair of cooperative vehicles encounter obstacles but these distortions are temporary. Also, the disturbances in the controllers at $t = 11{,}000$ units are a result of the parking bays. The two cars align themselves to achieve the desired orientations inside the parking bays. The final orientations are forced as a result of the repulsive potentials created by the walls of the parking bays.

6.3. Scenario 3: Diamond Formation. In this scenario, we have considered the leader vehicle at the center of a Diamond Formation with the followers positioned at each vertex (see Figure 12). The figure shows the formation maneuvered from an initial state to a predefined final state, with collision and obstacle avoidance.

Figure 12 also depicts the contour plot of the potential fields and the corresponding collision-free path over the defined workspace, $0 < z_1 < 40$, containing the obstacles on the road. Figures 13 and 14 depict the time evolution of the nonlinear controllers of the leader and its follower vehicles. Assuming that the appropriate units have been accounted for, Table 2 (if different from Table 1) provides the corresponding initial and final configurations of the 5-car C-ITS and other parameters required to simulate Scenario 3. The coordinates of the parking bays can be obtained from Figure 12.

Clearly the translational and rotational accelerations of the vehicles decrease as the formation approaches a fixed obstacle and increase once it is able to evade it. Moreover, Figures 15 and 16 compare the relative distance A_k to the

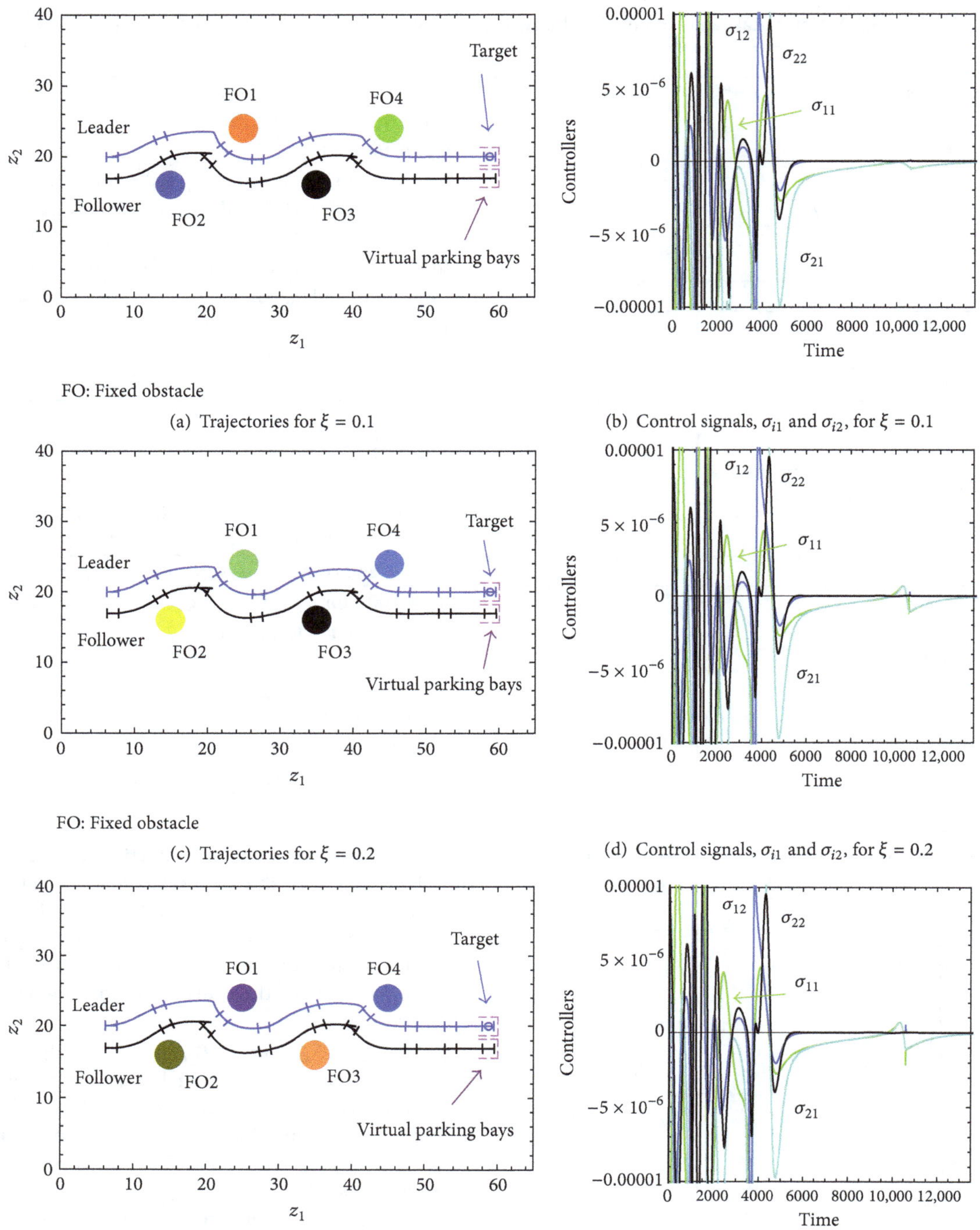

(a) Trajectories for $\xi = 0.1$
(b) Control signals, σ_{i1} and σ_{i2}, for $\xi = 0.1$
(c) Trajectories for $\xi = 0.2$
(d) Control signals, σ_{i1} and σ_{i2}, for $\xi = 0.2$
(e) Trajectories for $\xi = 0.5$
(f) Control signals, σ_{i1} and σ_{i2}, for $\xi = 0.5$

FIGURE 11: Trajectories of $\mathscr{A}_1$ and $\mathscr{A}_2$ and the evolution of control signals under various influences of noise.

desired relative distance a_k and compares B_k to the desired relative distance b_k, respectively. It is evident from the behavior shown in the figures that the formation gets slightly distorted when the C-ITS encounters an obstacle; however, the desired formation is reestablished before the leader $\mathscr{A}_1$ reaches its designated target, thus ensuring a locally rigid formation along the trajectory of the C-ITS.

7. Conclusion

This paper presents a set of nonlinear control laws using the LbCS to achieve a novel technique of extracting a locally rigid formation of a C-ITS made up of multiple car-like autonomous robots. A leader-following strategy is proposed to operate within the control scheme for the maintenance

TABLE 2: Numerical values of initial and final states, constraints, and parameters for Scenario 3.

	Initial configuration
Rectangular positions	$(x_1, y_1) = (7, 20)$, $(x_2, y_2) = (4, 20)$, $(x_3, y_3) = (7, 23)$
	$(x_4, y_4) = (10, 20)$, and $(x_5, y_5) = (7, 17)$
	$(a_2, b_2) = (3, 0)$, $(a_3, b_3) = (0, -3)$
	$(a_4, b_4) = (-3, 0)$, and $(a_5, b_5) = (0, 3)$
Velocity and angular position	$v_i = 0.5$, $\omega_i = 0$, and $\theta_i = 0$ for $i = 1, \ldots, 5$
	Constraints and parameters
Leader's target	$(p_{11}, p_{12}) = (57, 37)$; $r_l^* = 0.5$
Final orientations	$p_{i3} = 0$ for $i = 1, \ldots, 5$
	Control and convergence parameters
	$\alpha_{il} = 10$, for $i = 1, \ldots, 5$, $l = 1, \ldots, 3$
Obstacle avoidance	$\kappa_{ip} = 0.1$, for $i = 1, \ldots, 5$, $p = 1, \ldots, 2$
	$\gamma_{ik} = 1.2 \times 10^{-4}$, for $i = 1, \ldots, 5$, $k = 1, \ldots, 3i$, $k = 1, \ldots, 3n$
Collision avoidance	$\eta_{ij} = 0.01$, for $i, j = 1, \ldots, 5$, $j \neq i$
Dynamics constraints	$\beta_{is} = 0.001$, for $i = 1, \ldots, 5$, $s = 1, 2$
Convergence	$\delta_{11} = 20{,}500$, $\delta_{12} = 10{,}500$, $\delta_{i1} = 60$, and $\delta_{i2} = 60$ for $i = 1, \ldots, 5$

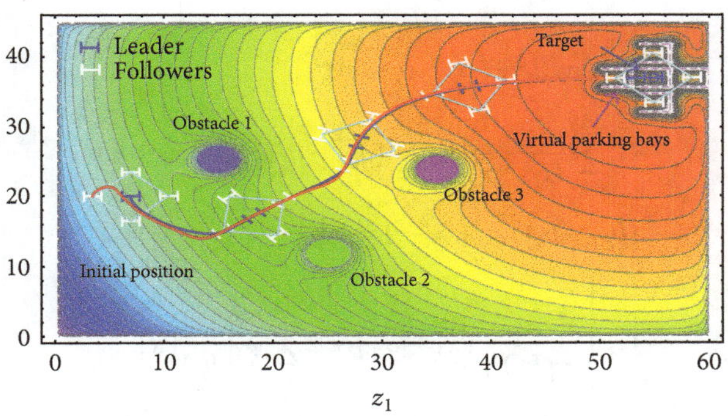

FIGURE 12: The evolution of the 5-robot C-ITS and the contour plot in the presence of obstacles in Scenario 3.

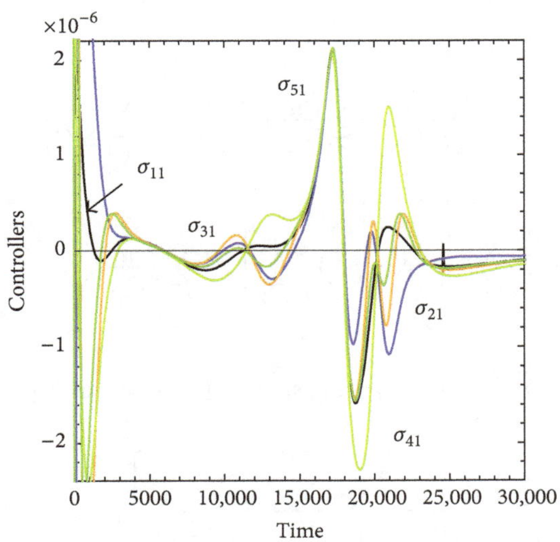

FIGURE 13: Translational accelerations, σ_{1i} for Scenario 3.

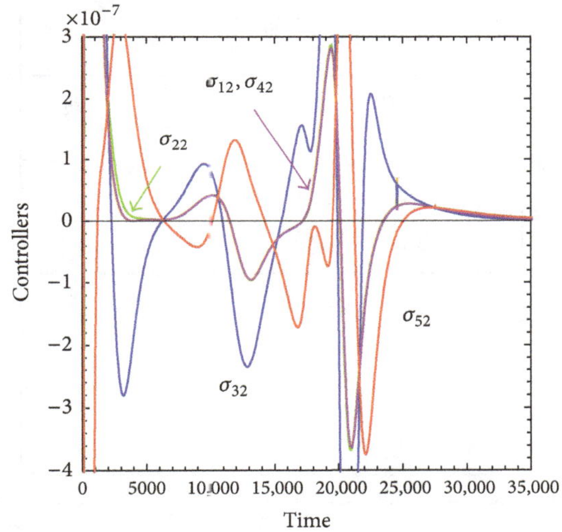

FIGURE 14: Rotational accelerations, σ_{2i} for Scenario 3.

of the locally rigid formation for a C-ITS which navigates in a constrained environment. This leader-follower scheme uses a Cartesian coordinate system fixed on the leader's body based on the concept of an instantaneous corotating frame of reference to uniquely assign a position to each follower. The advantage of such an approach is the complete avoidance of singularities inherent to the polar coordinate representations.

The new time-invariant acceleration-based controllers produce feasible trajectories and ensured a nice convergence of the system to its equilibrium state while satisfying the necessary kinematic and dynamic constraints. The scheme enables the vehicles to follow a predetermined leader while maintaining a locally rigid formation. The robustness of the proposed scheme is established by considering the effect of noise on the formation. while stability of the system is guaranteed using the direct method of Lyapunov. The assumptions of no slippage and pure rolling of the car-like mobile robots are also considered to generate the nonholonomic constraints

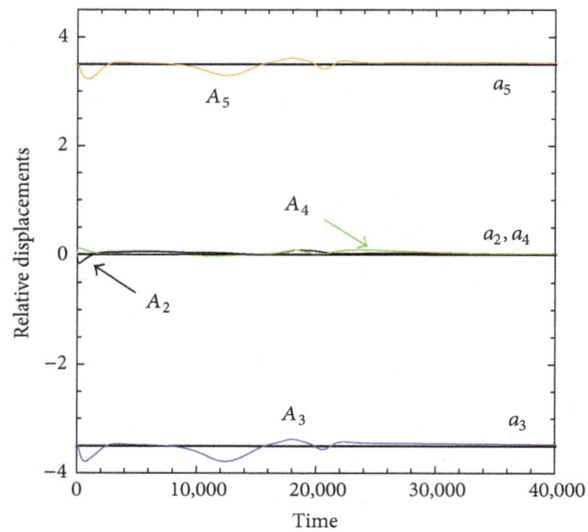

FIGURE 15: Relative distance error, A_k to a_k for Scenario 3.

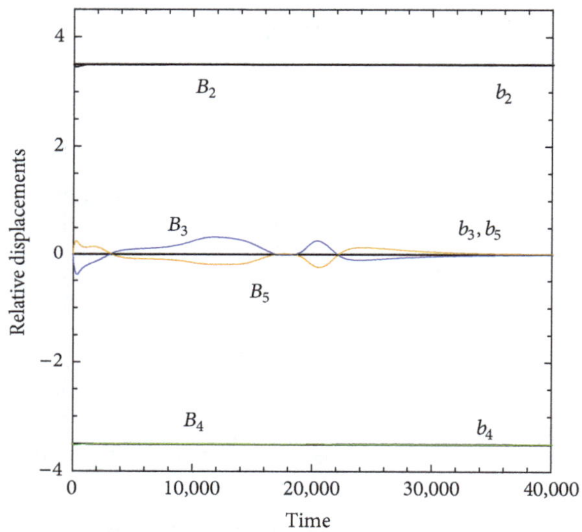

FIGURE 16: Relative distance error, B_k to b_k for Scenario 3.

on the given system. Further, the LbCS utilizes the *minimum distance technique* and parking bays to guarantee desired parking maneuvers and establish feasible prescribed posture of each vehicle in its designated parking bay.

This paper is a theoretical exposition into the applicability of an amalgamation of the new artificial potential field method, LbCS, and the leader-follower scheme. We restrict ourselves to showing the effectiveness of the control laws using computer-based simulations of interesting scenarios and numerical proofs. We shall be undertaking experiential design of the above system to see its effectiveness in the future. Future research will also address irregular shaped multimodal obstacles in partially known or completely known environments with the adaptive usage of sensing zones and relevant parameters. The optimization process of parameters within the LbCS still remains an open problem that warrants a careful thought.

Appendix

$$
\begin{aligned}
f_{11} = {} & \left[1 + \sum_{l=1}^{q} \frac{\alpha_{1l}}{\mathrm{FO}_{1l}} + \sum_{s=1}^{2} \frac{\beta_{1s}}{U_{1s}} + \sum_{p=1}^{2} \frac{\kappa_{1p}}{W_{1p}} + \sum_{\substack{j=1\\ j\neq i}}^{n} \frac{\eta_{1j}}{\mathrm{MO}_{1j}} \right. \\
& \left. + \sum_{k=1}^{3n} \frac{\gamma_{1k}}{LS_{1k}} \right] (x_1 - p_{11}) + \sum_{r=2}^{n} \left[1 + \sum_{l=1}^{q} \frac{\alpha_{1l}}{\mathrm{FO}_{1l}} \right. \\
& \left. + \sum_{s=1}^{2} \frac{\beta_{1s}}{U_{1s}} \right] \left[-(A_r - a_r) \cos\theta_1 + (B_r - b_r) \sin\theta_1 \right] \\
& + \sum_{r=2}^{n} \left[\sum_{p=1}^{2} \frac{\kappa_{1p}}{W_{1p}} + \sum_{\substack{j=1\\ j\neq i}}^{n} \frac{\eta_{1j}}{\mathrm{MO}_{1j}} + \sum_{k=1}^{3n} \frac{\gamma_{1k}}{LS_{1k}} \right] [-(A_r \\
& - a_r) \cos\theta_1 + (B_r - b_r) \sin\theta_1] \\
& - G_1 \sum_{k=1}^{3} \frac{\gamma_{1k}}{LS_{1k}^2(\mathbf{x})} \{ [1 - (a_{k2} - a_{k1}) d_k] (x_1 - Lx_{1k}) \\
& - (b_{k2} - b_{k1}) d_k (y_1 - Ly_{1k}) \} \\
& - G_1 \left[\sum_{l=1}^{q} \frac{\alpha_{1l}}{\mathrm{FO}_{1l}^2} (x_1 - o_{l1}) + 2 \sum_{\substack{j=1\\ j\neq i}}^{n} \frac{\eta_{1j}}{\mathrm{MO}_{1j}^2} (x_1 \right. \\
& \left. - x_j) \right],
\end{aligned}
$$

$$
\begin{aligned}
f_{12} = {} & \left[1 + \sum_{l=1}^{q} \frac{\alpha_{1l}}{\mathrm{FO}_{1l}} + \sum_{s=1}^{2} \frac{\beta_{1s}}{U_{1s}} + \sum_{p=1}^{2} \frac{\kappa_{1p}}{W_{1p}} + \sum_{\substack{j=1\\ j\neq i}}^{n} \frac{\eta_{1j}}{\mathrm{MO}_{1j}} \right. \\
& \left. + \sum_{k=1}^{3n} \frac{\gamma_{1k}}{LS_{1k}} \right] (y_1 - p_{12}) + \sum_{h=2}^{n} \left[1 + \sum_{l=1}^{q} \frac{\alpha_{1l}}{\mathrm{FO}_{1l}} \right. \\
& \left. + \sum_{s=1}^{2} \frac{\beta_{1s}}{U_{1s}} \right] \left[-(A_h - a_h) \sin\theta_1 - (B_h - b_h) \cos\theta_1 \right] \\
& + \sum_{h=2}^{n} \left[\sum_{p=1}^{2} \frac{\kappa_{1p}}{W_{1p}} + \sum_{\substack{j=1\\ j\neq i}}^{n} \frac{\eta_{1j}}{\mathrm{MO}_{1j}} + \sum_{k=1}^{3n} \frac{\gamma_{1k}}{LS_{1k}} \right] [-(A_h \\
& - a_h) \sin\theta_1 - (B_h - b_h) \cos\theta_1] - G_1 \left[\sum_{l=1}^{q} \frac{\alpha_{1l}}{\mathrm{FO}_{1l}^2} (y_1 \right.
\end{aligned}
$$

$$
\begin{aligned}
&\left.- o_{l2}) + 2\sum_{\substack{j=1\\ j\neq i}}^{n} \frac{\eta_{1j}}{\mathrm{MO}_{1j}^2}\left(y_1 - y_j\right)\right] - \frac{\kappa_{12}}{W_{12}^2} + \frac{\kappa_{14}}{W_{14}^2} \\
&- G_1 \sum_{k=1}^{3} \frac{\gamma_{1k}}{LS_{1k}^2(\mathbf{x})} \{[1 - (b_{k2} - b_{k1}) r_k](y_1 - Ly_{1k}) \\
&- (a_{k2} - a_{k1}) r_k (x_1 - Lx_{1k})\}, \\
f_{13} &= \left[\sum_{l=1}^{q} \frac{\alpha_{1l}}{\mathrm{FO}_{1l}} + \sum_{s=1}^{2} \frac{\beta_{1s}}{U_{1s}} + \sum_{p=1}^{2} \frac{\kappa_{1p}}{W_{1p}} + \sum_{\substack{j=1\\ j\neq i}}^{n} \frac{\eta_{1j}}{\mathrm{MO}_{1j}}\right. \\
&\left.+ \sum_{k=1}^{3n} \frac{\gamma_{1k}}{LS_{1k}}\right] \rho_1 (\theta_1 - p_{13}) - \sum_{i=2}^{n} \left[\sum_{l=1}^{q} \frac{\alpha_{1l}}{\mathrm{FO}_{1l}} + \sum_{s=1}^{2} \frac{\beta_{1s}}{U_{1s}}\right. \\
&\left.+ \sum_{p=1}^{2} \frac{\kappa_{1p}}{W_{1p}} + \sum_{\substack{j=1\\ j\neq i}}^{n} \frac{\eta_{1j}}{\mathrm{MO}_{1j}} + \sum_{k=1}^{3n} \frac{\gamma_{1k}}{LS_{1k}}\right] \rho_i (\theta_i - \theta_1), \\
g_{11} &= 1 + G_1 \frac{\beta_{11}}{U_{11}^2}, \\
g_{12} &= 1 + G_1 \frac{\beta_{12}}{U_{12}^2},
\end{aligned}
\tag{A.1}
$$

and, for $i = 2, \ldots, n$,

$$
\begin{aligned}
f_{i1} &= \left[1 + \sum_{l=1}^{q} \frac{\alpha_{il}}{\mathrm{FO}_{il}} + \sum_{s=1}^{2} \frac{\beta_{is}}{U_{is}} + \sum_{p=1}^{2} \frac{\kappa_{ip}}{W_{ip}} + \sum_{\substack{j=1\\ j\neq i}}^{n} \frac{\eta_{ij}}{\mathrm{MO}_{ij}}\right. \\
&\left.+ \sum_{k=3i-2}^{3i} \frac{\gamma_{ik}}{LS_{ik}}\right] (A_i - a_i) \cos\theta_1 - \left[1 + \sum_{l=1}^{q} \frac{\alpha_{il}}{\mathrm{FO}_{il}}\right. \\
&\left.+ \sum_{s=1}^{2} \frac{\beta_{is}}{U_{is}} + \sum_{p=1}^{2} \frac{\kappa_{ip}}{W_{ip}} + \sum_{\substack{j=1\\ j\neq i}}^{n} \frac{\eta_{ij}}{\mathrm{MO}_{ij}} + \sum_{k=3i-2}^{3i} \frac{\gamma_{ik}}{LS_{ik}}\right] (B_i \\
&- b_i) \sin\theta_1 - G_i \sum_{k=3i-2}^{3i} \frac{\gamma_{ik}}{LS_{ik}^2(\mathbf{x})} \{[1 \\
&- (a_{k2} - a_{k1}) d_k] (x_i - Lx_{ik}) - (b_{k2} \\
&- b_{k1}) d_k (y_i - Ly_{ik})\} \\
&- G_i \left[\sum_{l=1}^{q} \frac{\alpha_{il}}{\mathrm{FO}_{il}^2} (x_i - o_{l1}) + 2\sum_{\substack{j=1\\ j\neq i}}^{n} \frac{\eta_{ij}}{\mathrm{MO}_{ij}^2} (x_i - x_j)\right], \\
f_{i2} &= \left[1 + \sum_{l=1}^{q} \frac{\alpha_{il}}{\mathrm{FO}_{il}} + \sum_{s=1}^{2} \frac{\beta_{is}}{U_{is}} + \sum_{p=1}^{2} \frac{\kappa_{ip}}{W_{ip}} + \sum_{\substack{j=1\\ j\neq i}}^{n} \frac{\eta_{ij}}{\mathrm{MO}_{ij}}\right. \\
&\left.+ \sum_{k=1}^{3n} \frac{\gamma_{ik}}{LS_{ik}}\right] (A_i - a_i) \sin\theta_1 + \left[1 + \sum_{l=1}^{q} \frac{\alpha_{il}}{\mathrm{FO}_{il}}\right. \\
&\left.+ \sum_{s=1}^{2} \frac{\beta_{is}}{U_{is}} + \sum_{p=1}^{2} \frac{\kappa_{ip}}{W_{ip}} + \sum_{\substack{j=1\\ j\neq i}}^{n} \frac{\eta_{ij}}{\mathrm{MO}_{ij}} + \sum_{k=1}^{3n} \frac{\gamma_{ik}}{LS_{ik}}\right] (B_i \\
&- b_i) \cos\theta_1 - G_i \sum_{k=3i-2}^{3i} \frac{\gamma_{ik}}{LS_{ik}^2(\mathbf{x})} \{[1 \\
&- (b_{k2} - b_{k1}) r_k] (y_i - Ly_{ik}) - (a_{k2} - a_{k1}) r_k (x_i \\
&- Lx_{ik})\} - G_i \left[\sum_{l=1}^{q} \frac{\alpha_{il}}{\mathrm{FO}_{il}^2} (y_i - o_{l1})\right. \\
&\left.+ 2\sum_{\substack{j=1\\ j\neq i}}^{n} \frac{\eta_{ij}}{\mathrm{MO}_{ij}^2} (y_i - y_j)\right] - \frac{\kappa_{i2}}{W_{i2}^2} + \frac{\kappa_{i4}}{W_{i4}^2}, \\
f_{i3} &= \left[\sum_{l=1}^{q} \frac{\alpha_{il}}{\mathrm{FO}_{il}} + \sum_{s=1}^{2} \frac{\beta_{is}}{U_{is}} + \sum_{p=1}^{2} \frac{\kappa_{ip}}{W_{ip}} + \sum_{\substack{j=1\\ j\neq i}}^{n} \frac{\eta_{ij}}{\mathrm{MO}_{ij}}\right. \\
&\left.+ \sum_{k=1}^{3n} \frac{\gamma_{ik}}{LS_{ik}}\right] \rho_i (\theta_i - \theta_1), \\
g_{i1} &= 1 + G_i \frac{\beta_{i1}}{U_{i1}^2}, \\
g_{i2} &= 1 + G_i \frac{\beta_{i2}}{U_{i2}^2}.
\end{aligned}
\tag{A.2}
$$

Conflicts of Interest

The authors declare that they have no conflicts of interest.

References

[1] A. Festag, "Cooperative intelligent transport systems standards in Europe," *IEEE Communications Magazine*, vol. 52, no. 12, pp. 166–172, 2014.

[2] E. Ben Hamida, H. Noura, and W. Znaidi, "Security of cooperative intelligent transport systems: Standards, threats analysis

and cryptographic countermeasures," *Electronics*, vol. 4, no. 3, pp. 380–423, 2015.

[3] T. Balch and M. Hybinette, "Social potentials for scalable multi-robot formations," *Proceedings-IEEE International Conference on Robotics and Automation*, vol. 1, pp. 73–80, 2000.

[4] C. De La Cruz and R. Carelli, "Dynamic model based formation control and obstacle avoidance of multi-robot systems," *Robotica*, vol. 26, no. 3, pp. 345–356, 2008.

[5] F. Arrichiello, *Coordination Control of Multiple Mobile Robots [Ph.D. thesis]*, Universita Degli Studi Di Cassino, Cassino, Italy, November 2006.

[6] T. H. A. Van Den Broek, N. Van De Wouw, and H. Nijmeijer, "Formation control of unicycle mobile robots: a virtual structure approach," in *Proceedings of the 48th IEEE Conference on Decision and Control held jointly with 2009 28th Chinese Control Conference, CDC/CCC 2009*, pp. 8328–8333, Shanghai, China, December 2009.

[7] H. Yamaguchi, "A distributed motion coordination strategy for multiple nonholonomic mobile robots in cooperative hunting operations," *Robotics and Autonomous Systems*, vol. 43, no. 4, pp. 257–282, 2003.

[8] Y. H. Esin and M. Unel, "Formation control if nonholonomic mobile robots using implicit polynomials and elliptic fourier descriptors," *Turkish Journal of Electrical Engineering & Computer Sciences*, vol. 18, no. 5, pp. 765-80, 2010.

[9] R. Fierro, A. K. Das, V. Kumar, and J. P. Ostrowski, "Hybrid control of formations of robots," in *Proceedings of the IEEE Conference Robotics and Automation*, vol. 1, pp. 374–379, Seoul, Korea, May 2001.

[10] W. L. Seng, J. C. Barca, and Y. A. Şekercioğlu, "Distributed formation control of networked mobile robots in environments with obstacles," *Robotica*, vol. 12, pp. 1–13, 2014.

[11] F. Fahimi, "Full formation control for autonomous helicopter groups," *Robotica*, vol. 26, no. 2, pp. 143–156, 2008.

[12] R. Olfati-Saber, "Flocking for multi-agent dynamic systems: algorithms and theory," *Institute of Electrical and Electronics Engineers Transactions on Automatic Control*, vol. 51, no. 3, pp. 401–420, 2006.

[13] R. Falconi, L. Sabattini, C. Secchi, C. Fantuzzi, and C. Melchiorri, "Edge-weighted consensus-based formation control strategy with collision avoidance," *Robotica*, vol. 33, no. 2, pp. 332–347, 2015.

[14] B. Sharma, *New Directions in the Applications of the Lyapunov-based Control Scheme to the Findpath Problem [Ph.D. thesis]*, University of the South Pacific, Suva, Fiji Islands, July 2008.

[15] E. Yang, D. Gu, and H. Hu, "Nonsingular formation control of cooperative mobile robots via feedback linearization," in *Proceedings of the IEEE IRS/RSJ International Conference on Intelligent Robots and Systems (IROS '05)*, pp. 3652–3657, August 2005.

[16] W. Kang, N. Xi, J. Tan, and Y. Wang, "Formation control of multiple autonomous robots:theory and experimentation," *Intelligent Automation & Soft Computing*, vol. 10, no. 2, pp. 1–17, 2004.

[17] A. Bazoula, M. S. Djouadi, and H. Maaref, "Formation control of multi-robots via fuzzy logic technique," *International Journal of Computers, Communications & Control*, vol. 3, pp. 179–184, 2008.

[18] N. H. M. Li and H. H. T. Liu, "Formation UAV flight control using virtual structure and motion synchronization," in *Proceedings of the 2008 American Control Conference (ACC '08)*, pp. 1782–1787, Washington, Wash, USA, June 2008.

[19] W. Ren and R. W. Beard, "Virtual structure based spacecraft formation control with formation feedback," in *Proceedings of the AIAA Guidance, Navigation, and Control Conference and Exhibit*, California, Calif, USA, August 2002.

[20] M. Egerstedt and X. Hu, "Formation constrained multi-agent control," in *Proceedings of the 2001 IEEE International Conference on Robotics and Automation (ICRA '01)*, pp. 3961–3966, Seoul, Republic of Korea, May 2001.

[21] N. Sorensen and W. Ren, "A unified formation control scheme with a single or multiple leaders," in *Proceedings of the 2007 American Control Conference (ACC '07)*, pp. 5412–5418, New York, NY, USA, July 2007.

[22] L. Consolini, F. Morbidi, D. Prattichizzo, and M. Tosques, "Leader-follower formation control of nonholonomic mobile robots with input constraints," *Automatica*, vol. 44, no. 5, pp. 1343–1349, 2008.

[23] W. J. Dong, Y. Guo, and J. A. Farrell, "Formation control of nonholonomic mobile robots," in *Proceedings of the 2006 American Control Conference*, pp. 5902-5607, Minneapolis, MN, USA, June 2006.

[24] K. Raghuwaiya and S. Singh, "Formation types of multiple steerable 1-trailer mobile robots via split/rejoin maneuvers," *New Zealand Journal of Mathematics*, vol. 43, pp. 7–21, 2013.

[25] B. Sharma, J. Vanualailai, and A. Prasad, "Formation control of a swarm of mobile manipulators," *Rocky Mountain Journal of Mathematics*, vol. 41, no. 3, pp. 909–940, 2011.

[26] B. Sharma, J. Vanualailai, and A. Prasad, "A $d\phi$ strategy: facilitating dual-formation control of a virtually connected team," *Journal of Advanced Transportation*, vol. 2017, Article ID 9213805, 17 pages, 2017.

[27] K. Raghuwaiya, B. Sharma, and J. Vanualailai, *Advanced Computer and Communication Engineering Technology: Lecture Notes in Electrical Engineering*, Cooperative Control of Multi Robot Systems with a low-degree formation, Springer, 2015.

[28] R. W. Brockett, *Differential Geometric Control Theory*, Asymptotic Stability and Feedback Stabilisation, Springer, 1983.

[29] A. Prasad, B. Sharma, and J. Vanualailai, "A new stabilizing solution for motion planning and control of multiple robots," *Robotica*, vol. 34, no. 5, pp. 1071–1089, 2016.

6

High-Level Interpretation of Urban Road Maps Fusing Deep Learning-Based Pixelwise Scene Segmentation and Digital Navigation Maps

Carlos Fernández,[1] Jesús Muñoz-Bulnes,[2] David Fernández-Llorca ,[3] Ignacio Parra,[3] Iván García-Daza,[3] Rubén Izquierdo,[3] and Miguel Á. Sotelo[3]

[1] *Karlsruhe Institute of Technology, 76131 Karlsruhe, Germany*
[2] *ALTRAN Innovation, Madrid, Spain*
[3] *Computer Engineering Department, University of Alcalá, Madrid, Spain*

Correspondence should be addressed to David Fernández-Llorca; david.fernandezl@uah.es

Academic Editor: Jaeyoung Lee

This paper addresses the problem of high-level road modeling for urban environments. Current approaches are based on geometric models that fit well to the road shape for narrow roads. However, urban environments are more complex and those models are not suitable for inner city intersections or other urban situations. The approach presented in this paper generates a model based on the information provided by a digital navigation map and a vision-based sensing module. On the one hand, the digital map includes data about the road type (residential, highway, intersection, etc.), road shape, number of lanes, and other context information such as vegetation areas, parking slots, and railways. On the other hand, the sensing module provides a pixelwise segmentation of the road using a ResNet-101 CNN with random data augmentation, as well as other hand-crafted features such as curbs, road markings, and vegetation. The high-level interpretation module is designed to learn the best set of parameters of a function that maps all the available features to the actual parametric model of the urban road, using a weighted F-score as a cost function to be optimized. We show that the presented approach eases the maintenance of digital maps using crowd-sourcing, due to the small number of data to send, and adds important context information to traditional road detection systems.

1. Introduction

The high number of casualties on the road can be explained by many factors. As reported in the analytical report on road safety [1], more than 12.000 lives can be saved per year on European roads if everybody fasten their seat belt, respect speed limits, and do not drive under the influence of alcohol. Distraction is another factor since drivers need to keep their attention focused on surrounding traffic continuously, not just for their own safety but for the sake of their passengers and other road users too. In addition to the aforementioned driver factors, all road elements (vehicles, drivers, infrastructure, etc.) play an important role in the probability of crash or the final outcome. In order to reduce human errors and improve traffic efficiency and safety, Assistance Intelligent Transportation Systems (AITS) [2], Advanced Driver Assistance System (ADAS), and Autonomous Driving (AD) were developed.

Self-driving cars require a precise and robust perception of the urban environment. It is a crucial point in the development of autonomous vehicles because the perception layer is the base for higher level systems, such as path planning and control algorithms. Most of the major car makers aim to produce fully autonomous vehicles by 2020. This has made the research on ADAS and AD a high priority research issue for both private and public agents.

Different levels of autonomy have been demonstrated in highways, urban scenarios, and cooperative environments [3–6]. However, in all cases a high definition 2D, or even 3D, map is required. Enhanced maps integrate precise information of the environment such as road shape, lane markings, curbs, intersections, and buildings. The main drawbacks of this type of maps are their size ($\sim 2GB/km$), the complexity of measurements integration, and their maintenance.

On the other hand, the advent of deep learning techniques, namely, convolutional neural networks (CNNs), has involved a breakthrough in the field of artificial intelligence, with strong implications in a large variety of application domains. Thus, research on self-driving cars is experiencing a significant thrust due to the enhanced perception capabilities that the deployment of CNNs are making possible today. Powerful CNN architectures, such as AlexNet, VGG-16, or ResNet, are endowing self-driving cars with advanced capabilities to robustly and accurately interpret road scenes, even in complex urban scenarios with a great deal of clutter and complex road shapes. On top of that, the sensor costs of these road detection approaches are considerably low since they only involve the use of digital cameras. Thus, it eases the maintenance and it is more robust to environmental changes. Other approaches based on high cost LIDARs ($\sim 75K\$$) are not competitive for the car industry.

In this paper we present a high-level interpretation approach capable of estimating a parametric model of the road representing the real scenario appearing in front of the vehicle even in complex and cluttered environments. Based on our previous works [8, 9], a hybrid vision-map method is proposed. However, instead of using the best estimate of the road shape from an enhanced digital map as a feature of a hand-crafted road segmentation classifier [8], we propose a deep learning framework using ResNet-101 network with a fully convolutional architecture and multiple upscaling steps for image interpolation, to obtain an accurate estimation of the road, outperforming our previous results. We demonstrate that significant generalization gains in the learning process are attained by randomly generating augmented training data using several geometric transformations and pixelwise changes, such as affine and perspective transformations, image cropping, mirroring, blur, noise, distortions, and color changes. In addition, we show that the use of a 4-step upscaling strategy provides better learning results as compared to other similar techniques that perform data upscaling based on shallow layers with scarce representation of the scene data. This pixelwise segmentation process is combined with previously designed hand-crafted features extraction methods ([8, 10–12]) to provide a multiclass segmentation of the road and nearby regions. Finally, a high level interpretation module is trained to learn the best set of parameters of a mapping function that is capable of transforming the multiclass segmentation output to the actual parametric model of the urban road. The presented approach makes the maintenance of digital maps using crowd-sourcing easy, due to the small number of data to send, adding important context information to traditional road detection systems and increasing their robustness.

2. Related Work

2.1. Road Segmentation. Different vision-based methodologies to detect the road can be distinguished. Some of them are based on road appearance learning, where the main hand-crafted features are texture and color information. The second approach focuses on road limits detection, assuming that the space between limits is the road surface. Finally, model-based approaches try to extract a compact high level representation of the road. In order to have a better overview of the different sensing technologies, road appearance and limits modeling techniques, geometrical models, and features integration we refer to [8].

It has been proved that CNNs can improve state-of-the-art results on image classification [13–16], and they have also been successfully applied to object detection [17–19] as well as monocular color image segmentation. There are several approaches to obtain a CNN-based pixelwise classification of an input image. First, the widely adopted fully convolutional networks (FCN) [7] adapt classifier networks, such as AlexNet [13] or VGG [14], to the segmentation task by replacing fully connected layers with convolutional ones and using a progressive interpolation approach. Other approaches follow this trend using other base networks, such as the ResNet [15]. In [20] dilated convolutions are introduced to reduce the downsampling performed by the convolutional layers, removing the need of the progressive interpolation stages. These kinds of dilated convolutions are further explored in [21] along with another upsampling method called *dense upsampling convolution*.

A more complex approach such as DeconvNet [22] learns a deep deconvolutional network on top of the convolutional one. SegNet [23] uses an encoder-decoder architecture. PSPNet [24] exploits pyramidal pooling to introduce global contextual priors in a dilated fully convolutional network. FRRN [25] is based on a novel architecture that keeps a stream with full-resolution features.

Other models do not directly produce a full image classification but work patchwise instead. In [26] the inputs to the classifier layers are enhanced with spatial information of the patch, and in [27] they use a deconvolution scheme. However, the current trend is to use fully convolutional, end-to-end models, from images to pixelwise classification masks [19].

In addition, there are specialized methods for road detection. One example is [28], where the goal is to optimize the models to speed-up inference and make them capable of being used in a real road detection scenario. In [29], MultiNet system is presented, which performs simultaneous street classification, vehicle detection, and road segmentation, all with the same CNN encoder and three different decoders.

In order to learn the huge number of parameters of CNNs architectures, thousands of images labeled with the categories to learn are required. There are different datasets available for autonomous vehicles. One of the most popular datasets is KITTI benchmark [30]. This dataset has labeled 289 images with road and nonroad labels. However, CityScapes dataset [31] is increasing the number of submissions due to the number of labeled categories (30) and the number of images (25K). The results of CNNs in the field of road detection have outperformed all the algorithms that use other approaches.

(a) LIDAR-based HD map (b) Our approach

FIGURE 1: Map-based models: comparison between heavy 3D cloud-based models and light high-level interpretation maps.

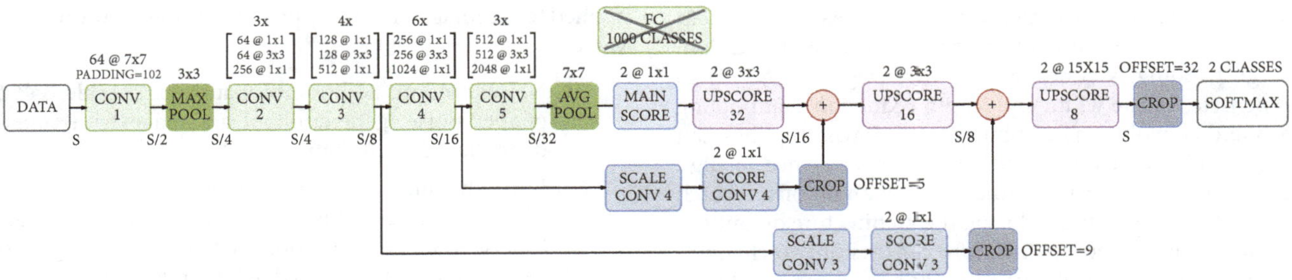

FIGURE 2: High-level schematic of the CNN-based road detector, implementing the FCN-8s architecture [7] on a ResNet-50.

2.2. Map Road Model Fitting. The simplest geometric models used for road boundaries are straight lines. Due to the pin hole camera model, straight parallel lines converge in a vanishing point. This principle is exploited in the state of the art to model the road using an edge descriptor extracted from the images [32]. More complex models are used to model curved roads, such as parabolic curves [33], clothoids [34], B-splines [35], or active contours (snakes) [36]. These parametric models improve the noisy bottom-up detections due to their constraints of width and curvature. However, urban environments are more difficult to model due to the presence of intersections and the variety of curvatures and width changes.

Nonparametric models are less common because they demand only that the line should be continuous. It provokes the model to be less robust than parametric models but more flexible to adapt to the irregular shapes present in urban environments [37] or rural paths [38].

Map-based models (see Figure 1(a)) are dominated by high definition maps. They have demonstrated to be a robust way to navigate [3–5] and they are usually built integrating several measurements of a multibeam LIDAR [39, 40] or multiple single beam LIDARs [41]. The drawback is their size ($\sim 2GB/km$), which is difficult to manage in a long trip or in a city. The ultimate goal is to drive everywhere with full functionality, but there are two main points of view on how to get to it. The first approach tries to drive in some places with full functionality using 3D detailed map ($\sim GB/km$) and low resolution sensing. On the other side, the goal is to drive everywhere with partial functionality using low resolution maps and high resolution sensing. The problem of using a 3D highly detailed map is the scalability of the map and the updates. The problem of using low resolution images is that more intelligent and realtime perception algorithms are needed. The ultimate goal is to get cognitive perception as humans do.

3. CNN-Based Road Classifier

On the one hand, and given the generalized use of CNNs on the road detection problem, this paper proposes a road detector based on the ResNet network model [15] adapting its last stages to the fully convolutional architecture [7]. First, a ResNet-50 model already trained on the ImageNet dataset (1000 labels at image level) has been used. In contrast, our system is designed on the KITTI road detection dataset [30], which only defines 2 labels (road/nonroad) at pixel level. Accordingly, the original ResNet-50 architecture has to be transformed into a fully convolutional network to admit an input of an arbitrary size and to produce an output of the same size with pixelwise segmentation. In order to do that, the last inner-product fully connected layer (1000 outputs) is replaced with a new convolutional layer (two outputs) that will be learned from scratch. In addition, due to the fact that ResNet has an overall downsampling factor of 32, some upsampling stages are needed. As can be observed in Figure 2 the upsampling is performed in three interpolation stages:

(1) UPSCORE 32: the main output scores are upsampled by a factor of 2. The output from the previous block

CONV 4 is added, since both scores have the same accumulated downsampling factor (16).

(2) UPSCORE 16: the previous result is upsampled by a factor of 2. The output from the previous block CONV 3 (downsampling factor of 8) is then added.

(3) UPSCORE 8: the result is upsampled by a factor of 8 to recover the scores in the original input size.

This process allows to recover pixelwise scores smoothly, with a high level of detail. The final output combines the coarser global features (MAIN SCORE) with some finer local features (SCORE CONV 4 and SCORE CONV 3).

Upsampling layers are initialized with bilinear interpolation kernels that do not need to be trained. The scores from the shallower layers are obtained with a two-output convolutional layer in the same manner as the main score. Also, a learnable scaling layer is placed before each one to help the network to adapt the different features to their addition. A large padding is added on the first stage (CONV 1) to compensate for the width and length reduction that pooling layers and convolutions combined with downsamplings can cause. Finally, some croppings have to be performed to align the score maps and match dimensions, with an offset which is calculated automatically during the architecture definition. The final output of our architecture consists of two channels that represent the probability maps for background and road respectively, obtained from the final SOFTMAX layer.

3.1. Data Augmentation. One of the main weaknesses of CNNs is their dependence on the training data. With data augmentation a better generalization can be achieved and different road conditions can be simulated, increasing the robustness of the network against illumination, color or texture changes, or variations in the orientation of the cameras. We adopt an online augmentation approach where modifications are performed at random each time. This way, the CNN never sees the same augmented image twice and this virtually infinite dataset does not require extra storage space on disk.

It can be distinguished between geometric transformations and pixel-value modifications. It is also possible to apply several augmentations on the same image or to apply different augmentations for each label (road or background) or to patches.

3.1.1. Geometric Transformations. These transformations are applied to both the image and the ground truth mask. Zero padding is added when needed to keep the original image size and the padding pixels are assigned to the "ignore label" of the classifier. The applied transformations include the following:

(i) Random affine transformations: translations, rotations, scalings, and shearings are performed in order to change the positions of the points, while keeping lines parallel.

(ii) Mirroring: apart from the affine transformations, horizontal flipping is performed independently to easily double the size of the training set.

(iii) Cropping and scaling: the original image is cropped and scaled to the original size. Crops are defined by a random top left corner and also random size, within image limits.

(iv) Distortion: random distortion parameters are applied to the image.

(v) Perspective transformations: the original positions are selected empirically on road limits. The final positions are calculated adding Gaussian noise to the original ones with two restrictions. First, the shift of top points is the opposite of that of the bottom ones. Second, top points should not cross each other to prevent reflected images.

3.1.2. Pixel-Value Changes. These transformations are only applied to the image, since they produce changes only on pixel values.

(i) Noise: random addition of Gaussian, speckle, salt & pepper noise, generation of an image with signal-dependent Poisson noise.

(ii) Blur: the filters are applied independently to the image, creating a blur effect. The selected filters are: Gaussian, diagonal motion (left-to-right or right-to-left, at random), box, median and bilateral filtering.

(iii) Color changes: three types of transformations are applied. The first one is casting, which consists in adding a random constant to each RGB channel [42]. The second one is an additive jitter, which is generated at random by means of exponentiation, multiplication and addition of random values. The last one is a PCA-based shift [13] which consists in adding to each pixel a linear combination of the found three principal components with magnitudes proportional to their corresponding eigenvalue times a random Gaussian variable (zero mean and standard deviation of 0.01).

3.2. Network Components and Training Variants. Regarding fully convolutional CNNs, there are several elements that can be optimized, such as the initialization of the score layers (with zeros, noise, etc.) and the initialization and training of the upsampling layers. We can also use more complex activation functions rather than the simple ReLU, such as parametric ReLUs (PReLUs), which are recommended in combination with MSRA initialization [43]. Although it is not possible to change the original pretrained ResNet-50 structure, we can add PReLUs to the new score layers. Training alternatives involve trying different learning rates and learning rate policies, such as decreasing the learning rate when the training stalls in previous trials or doing a *warmup* stage [15] at a reduced learning rate until error goes under 20%. Another alternative is to have a higher learning rate for score layers, which are learned from scratch and a lower rate for inherited layers. Moreover, in [7], several training schemes are defined: the standard accumulated learning (batch size of 20 and standard momentum of 0.9) or the *heavy learning* scheme, which uses a single image per gradient

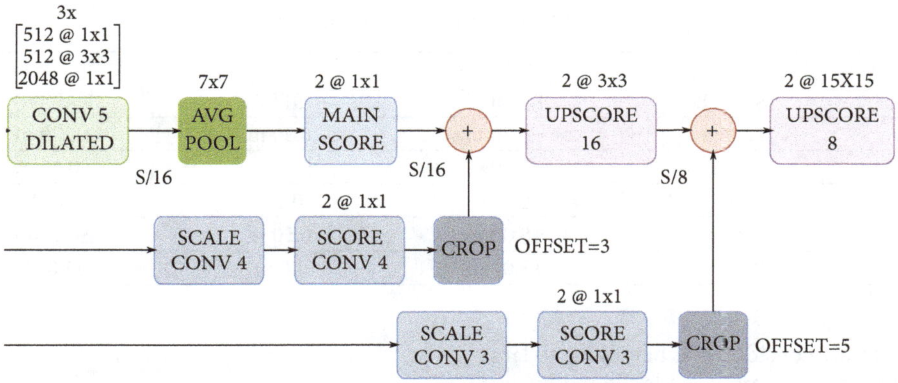

FIGURE 3: Detail of the scoring stage with dilated convolution in CONV 5. It removes the necessity of using the UPSCORE 32 layer.

actualization and a high momentum of 0.99, that simulates the gradient accumulation effect of the batch size. In [21] they use a variant of the accumulated learning (batch size of 12) with a polynomially decreasing learning rate which we try in the form $2.5 \cdot 10^{-4} \cdot (1 - iter/max\_iter)^{0.9}$.

3.3. Training in Bird's Eye View. The traditional training approach uses images in perspective view and obtains detections in this space. However, since KITTI benchmark evaluates its results with the F1-measure in Bird's Eye View (BEV) [44], we also train the network model directly in BEV. In this case, a less aggressive data augmentation strategy is used since geometric transformations in BEV creates important distortions.

3.4. Deeper Models. A ResNet-101 [15] model has been adapted using the same procedure applied to the ResNet-50, to test a deeper model in this problem. On the one hand, this model has an increased learning capacity. On the other hand, the risk of overfitting becomes more relevant.

3.5. Upsampling Variants. Apart from the schematics presented in Figure 2, the number of connections from shallower layers is modified. In order to obtain a more fine grained classification, both the full step-by-step upsampling (with additional connections from CONV 2 and CONV 1) and the four-step one (only additional connection from CONV 2) have been tested. Likewise, a two-step approach has also been evaluated to cover all possible cases, as well as the basic approach with no skip connections and just one large interpolation step.

There are other methods to increase the field of view of the deeper layers without downsampling the input features. The dilated convolution [20] and its improved version [21], which is claimed to avoid grid effects, are evaluated. This approach replaces the downsampling performed in one or more blocks with dilated convolutions in all of the subsequent layers. However, downsampling not only is necessary to enlarge the field of view but also plays an important role reducing the size of the input features to reduce the GPU memory consumption. If downsampling is completely removed, the model would not fit in memory. For this reason, our tested method combines a dilated convolution in the deeper block of the ResNet-50, with two upsampling steps to achieve a tradeoff (see Figure 3).

4. High Level Interpretation Module

4.1. Digital Navigation Maps. There are two main types of maps. The first ones are navigation maps, which provide information about the steps to reach our destination. The second ones are high definition maps, which provide 3D information of the environment with centimeter precision. Most of the autonomous navigation vehicles are based on these types of maps [3]. In contrast to that, our approach is closer to the human way of driving. Human drivers do not need high definition maps. They drive using visual perception and local navigation methods. The only information they need are the indications and steps on the navigation map to reach the destination.

The digital navigation map used for our approach is Open Street Map. This collaborative map is created by a large community around the world and all the information stored in the map is editable and it is freely accessible. The map consists of a list of streets called ways. Every way is composed of a list of nodes with a location and its relations with the other nodes and ways. Thanks to the location and relation between the nodes, the shape of the current street and its surroundings can be estimated. In addition, digital maps include the number of lanes and road type.

4.2. Hand-Crafted Features. In order to make the paper self-contained, we provide a brief description of the main features used in the high-level interpretation module. See [8] for more details.

(i) Vegetation areas: in order to segment the green areas of the image, multithresholding is applied over an area of the Hue channel of the HSV color space. A basic filtering is then applied on the resulted image.

(ii) Obstacles: 3D points from the stereo cameras are processed to estimate normal (n_i) and curvature vectors (γ_i). Points having components $n_x \geq 0.5$,

TABLE 1: Model parameter range of values.

PARAM	START	COARSE RANGE	COARSE STEP	FINE RANGE	FINE STEP
Lane Width	3.20 m	±0.40 m	0.20 m	±0.20 m	0.10 m
Lateral Displacement	0.0 m	±8.00 m	0.50 m	±0.50 m	0.10 m
Angular Offset	0.0°	±5.00°	0.25°	±0.20°	0.10°
Longitudinal Offset	0.0 m	±8.00 m	0.50 m	±0.50 m	0.10 m
Curvature Radius	7.00 m	±3.00 m	1.0 m	±1.0 m	0.50 m

$n_y \geq 0.5, \gamma_z \geq 0.5$ are considered as belonging to large obstacles. Every vector component is filtered by area and merged together to obtain the final result.

(iii) Curbs: the curvature vectors variation of the 3D points is a good feature for the detection of curbs. However, depending on the curb height, curvature values differ in each scene. Five types of curvatures are segmented using five pairs of thresholds for different curb heights. The resulting clusters are filtered independently using morphological operators and contour analysis and finally merged.

(iv) Road markings: a median filter is applied to the input image to remove white objects with horizontal size lower than the maximum horizontal size of markings. An adaptive window size of the median filter is needed due to perspective constraints. On the other hand, an adaptive thresholding with variable window size is applied to the input image to obtain white objects corresponding to markings. Both images are then subtracted to get the final result. This approach can be also carried out using the BEV image, with constant window sizes.

4.3. Road Map Modeling. The features extracted from the cameras (including the features obtained in [8]) are fused with digital navigation maps to obtain a high level interpretation of the urban scene. The map includes relevant information about the presence of railways, parking areas, buildings, or intersections, which are key points for a correct scene interpretation. Most of the elements in the map model are fixed (buildings, gardens, etc.). However, the road width should be adapted depending on the road type (residential, highway, intersection, etc.). Our proposed model has 6 degrees of freedom: number of lanes (l), lane width (w), lateral offset (y), longitudinal offset (x), angular offset (a), and curvature radius in intersections (r); see Figure 4.

It is assumed that the number of lanes is determined by the map. Nowadays, most of the maps indicate the number of lanes and the lane you should drive to reach your destination. The other parameters are evaluated in two steps, the first one for a coarse adjustment and the second one for a fine adaptation. Table 1 shows the range of every parameter, obtaining more than ~ $800K$ and ~ $15K$ combinations in the coarse and fine adjustment, respectively. The integration of the vehicle ego-motion with the previous models along the time creates a prior knowledge where the model should be. This prior knowledge removes the coarse adjustment and

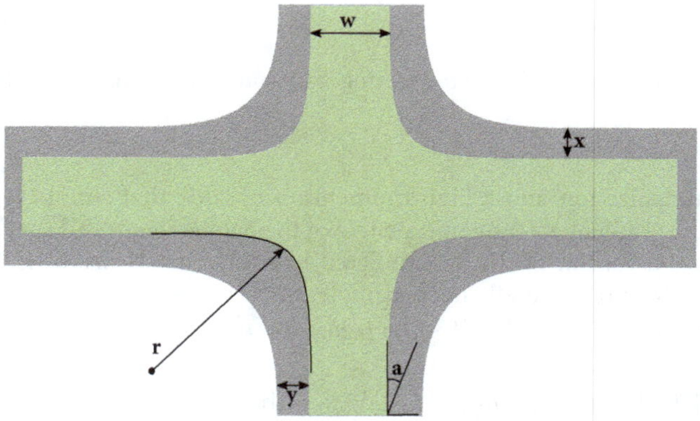

FIGURE 4: Parameters to adjust in the proposed model of the urban road. The real scene is painted in grey and the model to adjust is shown in green.

the fine adjustment could be extended. The selected option to reduce the number of combinations divides the process in three steps: the first step combines lane width and lateral displacement, the second one adjusts the angular offset, and finally the third step combines the longitudinal offset and the curvature radius. This process reduces the number of combinations from more than ~ $815K$ to only 512.

The metric to evaluate the best adjustment is calculated using equations (1) and (2), where the precision and recall are computed by matching the model and the sensing of the environment. The matching is evaluated in 4 different groups (G). The first one compares vegetation areas (garden, grass, and forest) and obstacles (barriers, buildings and walls) [8]. The second one is the road provided by the CNN model. The third group is composed of curbs and road markings [8]. The last one only compares curbs [8] in order to reinforce the correct adjustment of the road boundaries. The weights of every group in the final score (S) are set after a training stage to optimize the correct adjustment. The combination of the parameters with the highest score (S) generates a map-based model which is the output of the algorithm. This map is then projected at the same space of the images obtained from the cameras.

$$F_{score} = \frac{2 \cdot precision \cdot recall}{precision + recall} \quad (1)$$

$$S = \sum_i w_i \cdot F_{score}(G_i) \quad (2)$$

(a) Input image

(b) Sensing result with road in grey, road markings in white, curbs in blue, obstacles in red, and vegetation areas in green

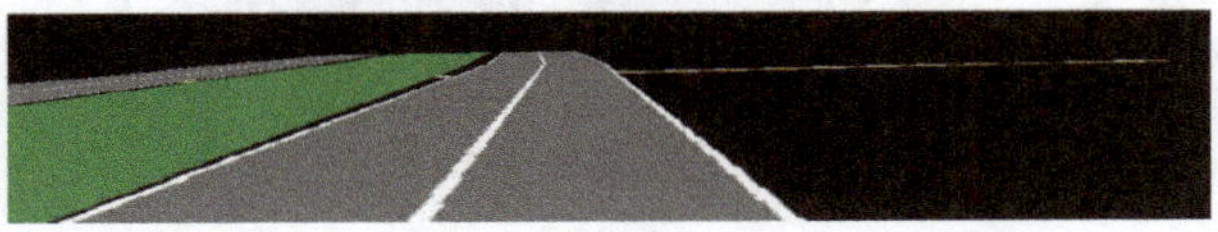

(c) Road model generated from the sensing results and the information from the digital navigation map. Road is painted in grey, lanes are delimited in white, and vegetation and obstacles are painted in green and red, respectively

FIGURE 5: Results of the high-level interpretation module.

As an example, we depict in Figure 5 the results of the sensing stage. The map model is adapted to the current scenario taking into account every detected feature and their correspondence in the map model.

Note that the resulting model is based on static elements such as the road, vegetation areas, and buildings. This is a clear advantage in order to maintain an updated and enhanced version of the high-level structure of the digital navigation maps. When performing autonomous navigation, the drivable area is directly provided by the CNN-based road classifier which will not include dynamic objects such as pedestrians, bicycles, and cars. Previous hand-crafted [45–48] and more recent deep learning-based approaches [19] can be here adopted to detect dynamic obstacles.

5. Results

5.1. Experimental Setup. Our CNN-based road segmentation model was trained on the KITTI dataset, which is composed of 289 images manually labeled with two classes: road and nonroad. 50% of the images were used for training the net, and the remaining 50% are kept aside for validation.

More specifically, the ResNet-50 model previously trained on ImageNet was used for weight initialization, and then the full net is fine-tuned on the road detection task. This is performed with stochastic gradient descent at a constant learning rate of $5 \cdot 10^{-5}$ (except for the bilinear filters, which are fixed), weight decay of $5 \cdot 10^{-4}$, one image per iteration and high (0.99) momentum. This scheme is referred as *heavy learning* in [7]. The training is run for 24K iterations, with validation checkpoints every 4K iterations.

The Caffe framework [49] was used for the network prototype definition and the control of training and testing processes. In addition, a Python input data layer, adapted for KITTI images, is used for loading images and labels into the net: each training image is randomly picked along with its corresponding label, and some minimal preprocessing must be done. The ImageNet per-channel pixel mean is subtracted, and the label images are converted into a 1 × height × width integer array of label indexes to be compatible with the loss function. As stated before, instead of passing the original image to the network, some data augmentation operations were applied to extend the training set, prevent overfitting, and make the net more robust to image changes. The data augmentation layer runs on CPU, and the rest of the processing can be done on GPU. It takes between 2 and 3 hours to complete the standard training on a single NVIDIA® Titan X GPU.

The parameters of the high-level interpretation module were estimated using the same images used to train the ResNet segmentation module from KITTI dataset. This module was implemented in C/C++. Training stage takes less than 1 minute and online estimation is performed in real time (< 5ms).

5.2. Road Segmentation Results. In this subsection we present the results obtained from the different variations on the network previously mentioned. As proposed in [44], quantitative results are calculated in terms of F1-measure, computed over the validation subset on Bird's Eye View (converting from perspective view when the road detector is trained in this space). Namely, the MaxF is computed using the working point (confidence threshold) in the precision-recall curve that maximizes F1-measure.

5.2.1. Data Augmentation. We can see, in the training and validation losses curves (Figure 6), that the use of data augmentation prevents the network from overfitting, since the gap between training and validation losses disappears: training losses rise slightly whereas validation losses decrease. We have observed that geometric transformations introduce higher variability than pixel value changes, and using both kinds we obtain the best results. We transform the full image with a single random operation each time. This way, we get a high variability, as we note from the width of the losses curve in Figure 6(b), but with an acceptable level of noise.

With this method we can achieve an improvement of approximately 1% in F-measure when training in perspective space (from 94.59% to 95.76%), and 2% when training in Bird's Eye View, which will be discussed later on.

Moreover, the trained model was tested on some sequences at the campus of the University of Alcala, Madrid (Spain), to test the network in a different environment from that used in the training. Figure 7 demonstrates that data augmentation makes the model more robust against illumination, texture, perspective, and orientation changes.

5.2.2. Network Components and Training Variants. The upscore described in Figure 2 is composed of fixed bilinear kernels and score layers are initialized using the MSRA method because it is considered robust against symmetries in gradient propagation. Neither finetuning the bilinear filters (slower convergence and we get the same kernels in the end)

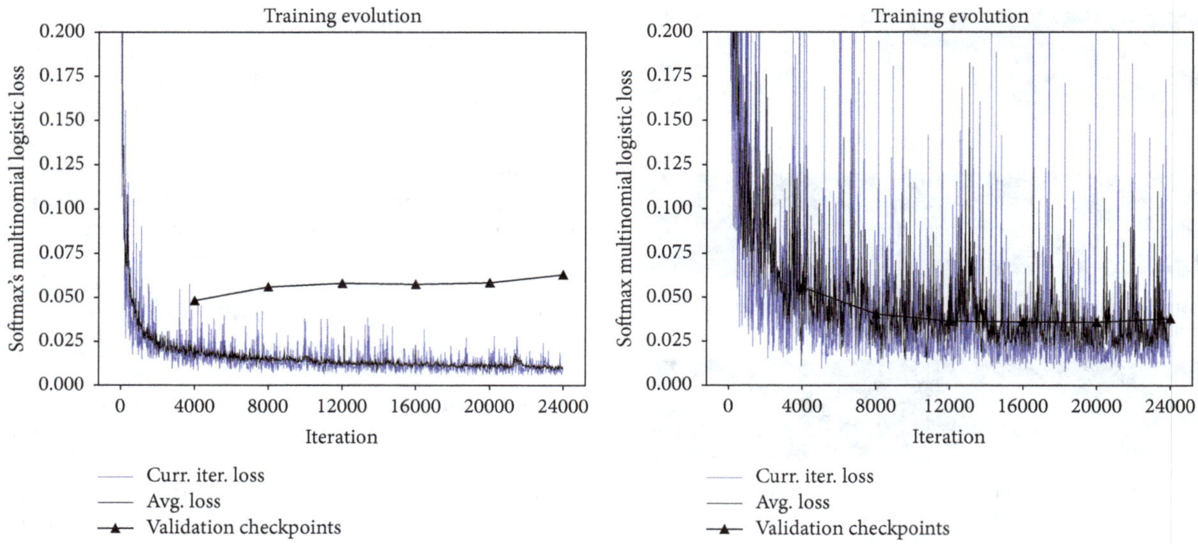

(a) Training without data augmentation

(b) Training with data augmentation

FIGURE 6: Comparison of the training and validation losses without data augmentation and with data augmentation.

(a) Training without data augmentation

(b) Training with data augmentation

FIGURE 7: The qualitative results of the proposed model are coloured as follows: TP in green, FP in blue, and FN in red. The scenarios with strong illumination changes and challenging road textures are better detected in the model trained with data augmentation.

nor learning them from scratch (smoother convergence and we get different interpolation kernels that use information from both classes' scores and scattered road limits) improves our previous performance. Regarding the learning rate, three different rates are compared ($1 \cdot 10^{-6}$, $5 \cdot 10^{-5}$, and $1 \cdot 10^{-4}$). The slower one ($1 \cdot 10^{-6}$) does not converge even with 40K iterations, the faster one ($1 \cdot 10^{-4}$) adds instability to the process, and the best results are obtained with the trade off between both approaches ($5 \cdot 10^{-5}$).

Since we are performing a fine-tuning with few iterations, changing the learning rate does not seem to have positive effects: decreasing policies can yield a monotonically decreasing validation losses curve, but the final losses and the F-measure are not better, and if the decrease is too abrupt, the training will become unstable, probably due to the high momentum; with the *warmup* scheme, there are no improvements either. It also appears to be better to let the whole net adapt to the new task of per-pixel road detection, instead of using a reduced (0.1×) learning rate for inherited layers. Finally, accumulated learning policies lead to worse results and are also much slower (proportionally to the batch size). The polynomially decreasing learning rate variant from [21] is better but still not superior to *heavy learning* in our trials.

5.2.3. Training in Bird's Eye View. In general, the model is able to learn better (less training losses) and also to generalize better (smaller gap with validation losses) during the training in perspective view because perspective images have more information about the scene, and more aggressive data augmentation recipes can be applied while maintaining the meaning of the image. Thus, without data augmentation, the model trained in BEV (94.08%) is worse than the one in perspective view (94.59%).

Data augmentation can significantly reduce the gap between training and validation losses and makes it worthwhile to train in BEV. Although the BEV approach with data augmentation is still worse at learning than the perspective one, the fact of learning in the same space as the evaluation obtains a better performance (96.06%). Analyzing in detail the performance, the model trained in BEV performs similarly at near and further pixels, whereas the perspective model

TABLE 2: Comparative results of deeper models performance.

Model	F-measure	Training Time	Iterations	Memory
ResNet-50	95.76	2h00	24K	7GB
ResNet-101	95.88	2h30	20K	10GB

TABLE 3: Quantitative results in F-measure on KITTI dataset for the main network variants evaluated.

Data aug.	BEV train	ResNet-101	4-step up.	F-measure
	√			94.08%
				94.59%
√				95.76%
√			√	95.80%
√		√		95.88%
√	√			96.06%
√		√	√	96.09%
√	√	√		96.13%
√	√	√	√	**96.31%**

has more problems with further pixels. Some problems of the BEV approach is that, in some cases, buildings at the end of the road or incoming tunnels can be confused with a continuation of the road.

5.2.4. Deeper Models. The training tests in perspective space over deeper models establish that the ResNet-101 achieves slightly better results over ResNet-50, which are obtained consistently with fewer iterations. As a drawback, the training takes slightly more time to complete than with ResNet-50 and more GPU memory is needed; see Table 2.

In the experiments evaluated in BEV, it is observed overfitting in the learning curve (training losses decrease while validation ones do not) because the deeper model has more learning capacity and needs a larger training set to generalize. Therefore, the training is stopped at 20K iterations to avoid the problem and the obtained F-measure is (96.13%). In conclusion, ResNet-101 offers a small but consistent improvement in detection performance, at the expense of needing more computing resources and time.

5.2.5. Upsampling Variants. Different upsampling variants are evaluated in a ResNet-50 trained in perspective view. As expected, the detections with the full step-by-step upsampling scheme have the highest resolution, but they are noisier and the F-measure is worse (95.49%), probably because the extracted features come from too shallow layers with little knowledge of the full scene. In the four-step case, the resolution is higher than in the original setup and the F-measure is slightly upraised (95.80%). The four-step approach has also been tried with a ResNet-50 trained in BEV and a ResNet-101 trained in both perspective and BEV spaces. Whereas the ResNet-50 gives similar results (95.97%), the ResNet-101 yields the best detections so far, with a F-measure of 96.09% and 96.31% in perspective and BEV spaces, respectively. The dilated convolution approaches yield also similar results. In particular, the method from [20] combined with two upsampling steps seems to be as good as the four-step approach in a ResNet-50 and less (20K) iterations, but it does not improve the results with the ResNet-101.

(a) Single cyclist well segmentated

(b) Group of cyclists considered a single obstacle

FIGURE 8: Results from the perspective-trained model in a scene with cyclists.

5.2.6. Global Road Segmentation Results. Table 3 summarizes the quantitative results in F-measure over our KITTI validation subset for the most interesting network variants. The baseline algorithm is the ResNet-50 model with the fully convolutional architecture and three-step upsampling shown in Figure 2.

The two best-performing methods, namely, the ResNet-101 with data augmentation and four-step interpolation, are trained with perspective and BEV images. Small obstacles such as pedestrians, cyclists, or cars are well differentiated from the road areas (Figure 8(a)), although two cyclists riding together are considered as a single obstacle (Figure 8(b)) since the space between them is not well segmented. This problem is also present when training in BEV and may be solved with higher resolution approaches.

(a) FN gaps inside road area

(b) FN areas due to incorrect road limits detection

(c) Missdetection of a distant intersection

(d) FN of an incoming lane

(e) Correct detection of the building

FIGURE 9: Results from the perspective-trained model.

(a) FP in parking areas in the closer meters

(b) Improved road detection

(c) Improved far intersection detection

(d) Improved near intersection detection

(e) FP detection of a building

FIGURE 10: Results from the BEV-trained model.

Both models sometimes leave FN gaps (Figures 9(a) and 10(a) on top-right corner), as well as FP patches outside road limits (Figure 9(d)) that could be filtered with some postprocessing methods. However, the model trained in BEV seems to be better delimiting road limits in the same image (Figure 10(b)) because in this representation they are straighter.

It can be seen that the BEV-trained model is better at detecting irregular road limits (Figures 10(c) and 10(d)) than the perspective-trained one (Figures 9(c) and 9(d)). However, the main problem of the BEV-approach is that, in some particular cases, the resulting image is so distorted and the net confuses buildings with the continuation of the road (Figure 10(e)). This would be very unlikely to happen if the image was analyzed in perspective space.

5.3. High-Level Interpretation Results. Considering the handcrafted features of our scene interpretation module, we remark the following statements. The curb detection method was analyzed in detail in [10]. The algorithm was compared using different sources for the 3D cloud data (LIDAR and stereo), obtaining a lateral RMSE of 12cm in a range from 6m to 20m. The proposed algorithm can deal with curbs of different curvature and heights, from as low as 3cm, in a range up to 20m whenever that curbs are connected in the curvature image. Finally, the use of fixed or empirical thresholds is avoided given that the proposed function is adapted automatically for different road scenes depending on the predominant curvature value. The boosting classifier was described and tested in [8, 11]. The weight of each feature in the final road classification reveals that 3D features (Y and Z coordinates) and its 2D representation (column and row) are the most discriminant features. However, some of the other 2D features are still important for the boosting classifier such as the grey value of (HSV) or the vegetation. Some of the road misclassifications are produced in sidewalks, where the only difference between the road and sidewalk is a small curb. Furthermore, in challenging urban scenarios the limit of a drivable area is difficult to distinguish from the nondrivable area, such as a cyclist lane. In some cases the limit is just a road marking or a variation in the pavement texture. It is remarkable that road markings have a low weight in the final response.

According to the results obtained in [8, 11], the weight of curbs and road markings in the road classification is very small because that type of features describes road limits instead of the road surface. The proposed algorithm follows a human reasoning. Therefore road markings and curbs should have a relevant role in the interpretation of the environment.

Table 4: Quantitative results on F-score in KITTI dataset.

	Image plane				Bird Eye View			
Classifier	**UM**	**UMM**	**UU**	**All**	**UM**	**UMM**	**UU**	**All**
Boosting	85.24	93.09	82.08	87.06	85.15	89.37	62.60	79.05
Map Model	89.35	88.56	90.20	89.37	85.17	84.76	86.66	85.53
CNN	96.64	95.36	94.89	95.50	95.09	94.94	93.39	94.59

Table 5: F-measure score for urban marked road, urban multiple marked lane, and urban unmarked road respectively of the KITTI test benchmark.

Type of road	**UM**	**UMM**	**UU**	**All**
F-measure	94.16%	95.45%	91.17%	93.98%

That is the reason to match the map model with 4 different groups of features: vegetation + obstacles, road, curbs + road markings, and curbs. After a weight training stage for each group, the qualitative results demonstrate the effectiveness of the proposed method to infer complex urban environments.

When evaluating our previous boosting-based classifier [8] with the new CNN-based approach, we can state that our CNN model with data augmentation clearly outperforms the classic hand-crafted features + classifier approach.

Table 4 shows the F-measure using the training set of the KITTI dataset (50%/50% training/validation) in both perspective view and BEV of urban marked road (UM), urban multiple marked lane road (UMM), and urban unmarked road (UU) and for all types of road (All). As can be observed, the CNN-based road segmentation provides an overall F-measure a 8.44% and a 15.54% better than the value provided by our previous approach in the perspective image and BEV, respectively. In addition, we have obtained the F-measure using the KITTI test dataset for the CNN-based approach (see Table 5). As can be observed, the performance decreased by 0.61% compared with the validation results, which clearly demonstrates that overfitting has been avoided.

We have also obtained the F-measure of the high-level interpretation module (see Table 4) for our validation dataset, yielding overall values of 89.37% and 85.53% in perspective view and BEV, respectively (note that we have not been able to obtain results from the KITTI test dataset since GPS positions are not available). These values are a 6.13% and 9.06% worse than the CNN classifier in the image plane and BEV, respectively. This is obviously an unfair comparison since the high-level road estimation module is not a pixelwise approach trained with the ground truth location of the road. It does not include dynamic obstacles, and, in some cases, it provides some regions of the road that are not even labeled in the ground truth (opposite lanes or not visible intersections). In general, pixelwise classifiers outperform model-based approaches because there is not any model that fits as close to the ground truth as the pixelwise classifiers. Figure 11 shows an example of a very complex intersection where the pixelwise classifier outperforms the map model

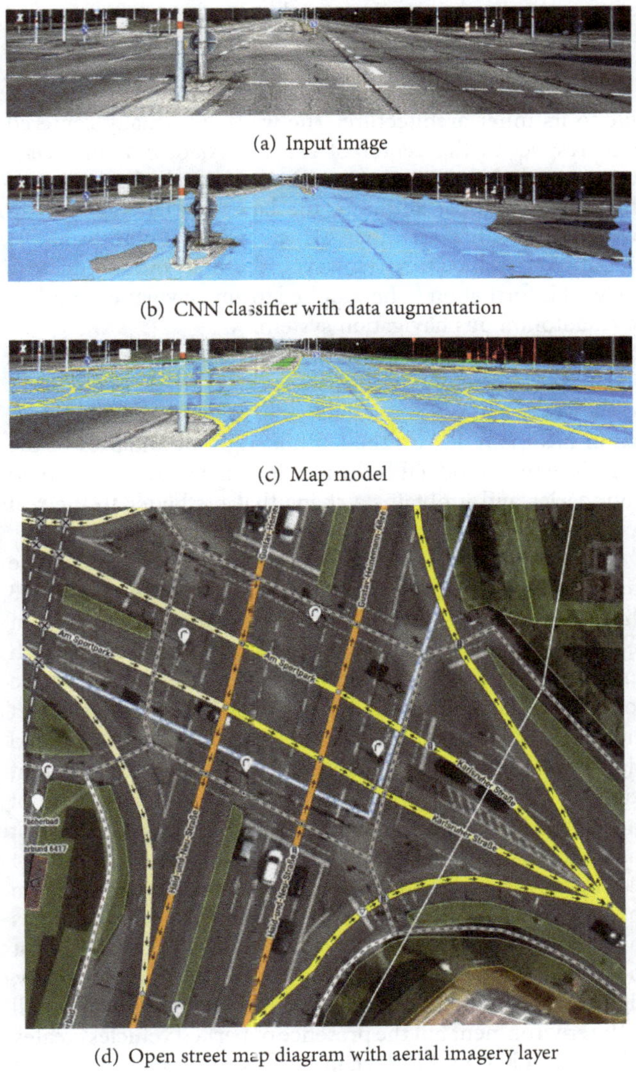

(a) Input image

(b) CNN classifier with data augmentation

(c) Map model

(d) Open street map diagram with aerial imagery layer

Figure 11: Example of a very complex intersection where the map-based model is not well fitted to the road limits. However, it provides richer high-level information to be used in autonomous navigation systems.

(a) Input image

(b) CNN segmentation without data augmentation

(c) Map model

FIGURE 12: Example of a scene where the pixelwise classifier has a large number of false negatives and the use of a map-based model improves the result.

due to its inner architecture. The ways are usually centered with respect to the center of the real road, but this scene has ways that have different number of lanes before and after the intersection. In addition, the lanes for left turning are overlapped with each other, which is impossible to adjust to the map architecture. However, the map model includes relevant information to be supplied to any navigation module of an autonomous navigation system.

In some other scenarios, the map-based model improves pixelwise classifiers. Figure 12 shows an example of an urban scene where the CNN road classifier without data augmentation has many FN on the left and right boundaries of the road. The road model fills the missing pixels close to the road boundaries and it obtains a shape that fits better to the real scenario.

The qualitative results show the added value of the interpretative approach presented in this paper. Figure 13 shows an urban street with one lane for each direction separated by a fence. Furthermore, on the right side of each lane, there are slots for parking and buildings. Even with a coarse detection of the road (Figure 13(b)) or neither detect the road, the map model fits well to the scene and add high level information to the system. The use of different features for the map matching increases the robustness of the method because the absence of one of them does not make the system to fail. The scene represented in Figure 14 highlights the use of high level information extracted from the map. The use of map information reduces the possibilities to infer the cycle lane as a road lane and because of that the railways and the cycle lane are correctly labeled. Finally, Figure 15 shows an urban street where the map model fits well to the environment but the presence of parked vehicles creates a high number of FP. Dynamic obstacles, such as cyclist and vehicles, are effectively removed and not considered as drivable by the CNN-based road segmentation system. The high-interpretation module maintains the actual structure of the road besides the dynamic obstacles.

6. Conclusions and Future Work

A novel high-level interpretation approach that integrates pixelwise road segmentation, a set of hand-crafted features that describe road limits (vegetation, road markings and curbs), and enhanced data provided by digital navigation maps were presented. A ResNet-50 CNN model with a fully convolutional architecture and three interpolation steps, which are finetuned in perspective KITTI images, were used to obtain an accurate detection of the road, which represents the drivable area. Several variations were introduced to improve the training such as data augmentation, training in BEV space, training parameters tuning, using deeper models, and other upsampling architectures. Data augmentation offers a significant improvement within 2% in F-measure. The use of a ResNet-101 model with a four-step upsampling scheme, trained directly in BEV with data augmentation, improves the results up to a 96.31% in the validation subset. The proposed approach clearly outperforms our previous Boosting-based road detection approach [8].

The presented approach can be applied to update digital navigation maps using floating vehicles equipped with cameras. In parallel, an accurate estimation of the drivable area is supplied by our CNN-based road segmentation module.

Future works will be devoted to obtain smoother road detection results by adding a postprocessing layer into the system. More degrees of freedom can be considered to be capable of handling multilane roads with different lane widths. In addition, due to the new advances in semantic segmentation [19], the data provided by digital navigation maps and the output given by the high-level interpretation module will be enriched with new variables, including traffic lights, traffic signs, or even urban furniture (benches, bins, bus stops, etc.). Finally, different sensor ensembles will be tested to facilitate a quantitative evaluation of the high level interpretation module.

Conflicts of Interest

The authors declare that they have no conflicts of interest.

Acknowledgments

This work was funded by Research Grants SEGVAUTO S2013/MIT-2713 (CAM), DPI2017-90035-R (Spanish Min. of Economy), BRAVE Project (H2020, Contract No. 723021), and UAH (CCGP2017-EXP/055). In addition, this project has received funding from the Electronic Component Systems for European Leadership Joint Undertaking under Grant Agreement no. 737469 (AutoDrive Project). This Joint Undertaking receives support from the European Unions Horizon 2020 Research and Innovation programme and Germany, Austria,

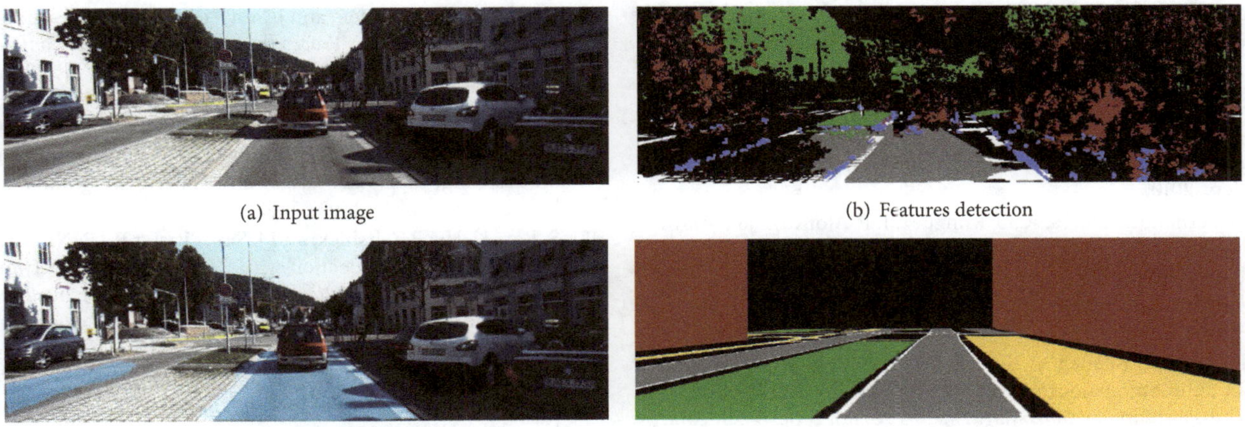

(a) Input image

(b) Features detection

(c) CNN road classifier

(d) Map model

FIGURE 13: Results in a scene with parking slots on the right and vegetation on the left. Even with a coarse detection of the road (Figure 13(b)), the map model fits well to the scene and add high-level information to the system.

(a) Input image

(b) Features detection

(c) CNN road classifier

(d) Map model

FIGURE 14: The use of a map model increases the robustness of a road classifier because the other boundary features compensate some road missclassifications.

(a) Input image

(b) Features detection

(c) CNN road classifier

(d) Map model

FIGURE 15: Dynamic obstacles (cyclist and vehicles) are effectively not considered as drivable by the CNN-based road segmentation system. The high-interpretation module maintains the actual structure of the road besides the dynamic obstacles.

Spain, Italy, Latvia, Belgium, Netherlands, Sweden, Finland, Lithuania, Czech Republic, Romania, and Norway.

References

[1] Analytical report on road safety, "European Commission, Tech. Rep., 2010".

[2] D. Fernandez-Llorca, R. Q. Minguez, I. P. Alonso et al., "Assistive Intelligent Transportation Systems: The Need for User Localization and Anonymous Disability Identification," *IEEE Intelligent Transportation Systems Magazine*, vol. 9, no. 2, pp. 25–40, 2017.

[3] J. Ziegler, P. Bender, M. Schreiber et al., "Making bertha drive-an autonomous journey on a historic route," *IEEE Intelligent Transportation Systems Magazine*, vol. 6, no. 2, pp. 8–20, 2014.

[4] P. Czerwionka, *A Three Dimensional Map Format for Autonomous Vehicles*, Freie Universität, Berlin, Germany, 2014.

[5] E. Guizzo, "How google's self-driving car works," *IEEE Spectrum*, 2011, http://spectrum.ieee.org/automaton/robotics/artificial-intelligence/how-google-self-driving-car-works.

[6] I. Parra Alonso, R. Izquierdo Gonzalo, J. Alonso, A. Garcia-Morcillo, D. Fernandez-Llorca, and M. A. Sotelo, "The experience of drivertive-driverless cooperative vehicle-team in the 2016 gcdc," *IEEE Transactions on Intelligent Transportation Systems*, 13 pages, 2017.

[7] E. Shelhamer, J. Long, and T. Darrell, "Fully Convolutional Networks for Semantic Segmentation," *IEEE Transactions on Pattern Analysis and Machine Intelligence*, vol. 39, no. 4, pp. 640–651, 2017.

[8] C. Fernández, D. Fernández-Llorca, and M. A. Sotelo, "A Hybrid Vision-Map Method for Urban Road Detection," *Journal of Advanced Transportation*, vol. 2017, Article ID 7090549, 21 pages, 2017.

[9] J. Munoz-Bulnes, C. Fernandez, I. Parra, D. Fernandez-Llorca, and M. A. Sotelo, "Deep fully convolutional networks with random data augmentation for enhanced generalization in road detection," in *Proceedings of the 2017 IEEE 20th International Conference on Intelligent Transportation Systems (ITSC)*, pp. 366–371, Yokohama, October 2017.

[10] C. Fernandez, D. F. Llorca, C. Stiller, and M. A. Sotelo, "Curvature-based curb detection method in urban environments using stereo and laser," in *Proceedings of the IEEE Intelligent Vehicles Symposium, IV 2015*, pp. 579–584, July 2015.

[11] C. Fernandez, R. Izquierdo, D. F. Llorca, and M. A. Sotelo, "A Comparative Analysis of Decision Trees Based Classifiers for Road Detection in Urban Environments," in *Proceedings of the 18th IEEE International Conference on Intelligent Transportation Systems, ITSC 2015*, pp. 719–724, September 2015.

[12] C. Fernandez, R. Izquierdo, D. F. Llorca, and M. A. Sotelo, "Road curb and lanes detection for autonomous driving on urban scenarios," in *Proceedings of the 2014 IEEE 17th International Conference on Intelligent Transportation Systems (ITSC)*, pp. 1964–1969, Qingdao, China, October 2014.

[13] A. Krizhevsky, I. Sutskever, and G. E. Hinton, "Imagenet classification with deep convolutional neural networks," in *Proceedings of the 26th Annual Conference on Neural Information Processing Systems (NIPS '12)*, pp. 1097–1105, Lake Tahoe, Nev, USA, December 2012.

[14] K. Simonyan and A. Zisserman, "Very Deep Convolutional Networks for Large-Scale Image Recognition," *Computer Vision and Pattern Recognition*, 2014, https://arxiv.org/abs/1409.1556.

[15] K. He, X. Zhang, S. Ren, and J. Sun, "Deep residual learning for image recognition," in *Proceedings of the 2016 IEEE Conference on Computer Vision and Pattern Recognition, CVPR 2016*, pp. 770–778, July 2016.

[16] C. Szegedy, S. Ioffe, V. Vanhoucke, and A. Alemi, "Inception-v4, Inception-ResNet and the Impact of Residual Connections on Learning," *Computer Vision and Pattern Recognition*, 2016, https://arxiv.org/abs/1602.07261.

[17] S. Ren, K. He, R. Girshick, and J. Sun, "Faster R-CNN: Towards Real-Time Object Detection with Region Proposal Networks," *IEEE Transactions on Pattern Analysis and Machine Intelligence*, vol. 39, no. 6, pp. 1137–1149, 2017.

[18] J. Redmon and A. Farhadi, "YOLO9000: Better, Faster, Stronger," in *Proceedings of the 2017 IEEE Conference on Computer Vision and Pattern Recognition (CVPR)*, pp. 6517–6525, Honolulu, HI, July 2017.

[19] K. He, G. Gkioxari, P. Dollár, and R. B. Girshick, "Mask R-CNN," *CoRR*, vol. abs/1703, 2017, http://arxiv.org/abs/1703.06870.

[20] L. Chen, G. Papandreou, I. Kokkinos, K. Murphy, and A. L. Yuille, "DeepLab: Semantic Image Segmentation with Deep Convolutional Nets, Atrous Convolution, and Fully Connected CRFs," *IEEE Transactions on Pattern Analysis and Machine Intelligence*, vol. 40, no. 4, pp. 834–848, 2018.

[21] P. Wang, P. Chen, Y. Yuan et al., "Understanding Convolution for Semantic Segmentation," in *Proceedings of the 2018 IEEE Winter Conference on Applications of Computer Vision (WACV)*, pp. 1451–1460, Lake Tahoe, NV, March 2018.

[22] H. Noh, S. Hong, and B. Han, "Learning deconvolution network for semantic segmentation," in *Proceedings of the 15th IEEE International Conference on Computer Vision, ICCV 2015*, pp. 1520–1528, Santiago, Chile, December 2015.

[23] V. Badrinarayanan, A. Kendall, and R. Cipolla, "SegNet: A Deep Convolutional Encoder-Decoder Architecture for Image Segmentation," *IEEE Transactions on Pattern Analysis and Machine Intelligence*, vol. 39, no. 12, pp. 2481–2495, 2017.

[24] H. Zhao, J. Shi, X. Qi, X. Wang, and J. Jia, "Pyramid Scene Parsing Network," in *Proceedings of the 2017 IEEE Conference on Computer Vision and Pattern Recognition (CVPR)*, pp. 6230–6239, Honolulu, HI, July 2017.

[25] T. Pohlen, A. Hermans, M. Mathias, and B. Leibe, "Full-Resolution Residual Networks for Semantic Segmentation in Street Scenes," in *Proceedings of the 2017 IEEE Conference on Computer Vision and Pattern Recognition (CVPR)*, pp. 3309–3318, Honolulu, HI, July 2017.

[26] C. Brust, S. Sickert, M. Simon, E. Rodner, and J. Denzler, "Convolutional Patch Networks with Spatial Prior for Road Detection and Urban Scene Understanding," in *Proceedings of the International Conference on Computer Vision Theory and Applications*, pp. 510–517, Berlin, Germany, March 2015.

[27] R. Mohan, "Deep deconvolutional networks for scene parsing," *Machine Learning*, 2014, https://arxiv.org/abs/1411.4101.

[28] G. L. Oliveira, W. Burgard, and T. Brox, "Efficient deep models for monocular road segmentation," in *Proceedings of the 2016 IEEE/RSJ International Conference on Intelligent Robots and Systems (IROS)*, pp. 4885–4891, Daejeon, South Korea, October 2016.

[29] M. Teichmann, M. Weber, M. Zoellner, R. Cipolla, and R. Urtasun, "MultiNet: Real-time Joint Semantic Reasoning for Autonomous Driving," *Computer Vision and Pattern Recognition*, 2016, https://arxiv.org/abs/1612.07695.

[30] A. Geiger, P. Lenz, and R. Urtasun, "Are we ready for autonomous driving? the KITTI vision benchmark suite," in

Proceedings of the IEEE Conference on Computer Vision and Pattern Recognition (CVPR '12), pp. 3354–3361, Providence, RI, USA, June 2012.

[31] M. Cordts, M. Omran, S. Ramos et al., "The Cityscapes dataset for semantic urban scene understanding," in *Proceedings of the 2016 IEEE Conference on Computer Vision and Pattern Recognition, CVPR 2016*, pp. 3213–3223, USA, July 2016.

[32] H. Kong, J.-Y. Audibert, and J. Ponce, "General road detection from a single image," *IEEE Transactions on Image Processing*, vol. 19, no. 8, pp. 2211–2220, 2010.

[33] J. C. McCall and M. M. Trivedi, "Video-based lane estimation and tracking for driver assistance: survey, system, and evaluation," *IEEE Transactions on Intelligent Transportation Systems*, vol. 7, no. 1, pp. 20–37, 2006.

[34] S. Nedevschi, R. Schmidt, T. Graf et al., "3D lane detection system based on stereovision," in *Proceedings of the Proceedings - 7th International IEEE Conference on Intelligent Transportation Systems, ITSC 2004*, pp. 161–166, October 2004.

[35] A. Wedel, H. Badino, C. Rabe, H. Loose, U. Franke, and D. Cremers, "B-spline modeling of road surfaces with an application to free-space estimation," *IEEE Transactions on Intelligent Transportation Systems*, vol. 10, no. 4, pp. 572–583, 2009.

[36] L. Bentabet, S. Jodouin, D. Ziou, and J. Vaillancourt, "Road vectors update using SAR imagery: a snake-based method," *IEEE Transactions on Geoscience and Remote Sensing*, vol. 41, no. 8, pp. 1785–1803, 2003.

[37] J. Beck and C. Stiller, "Non-parametric lane estimation in urban environments," in *Proceedings of the 25th IEEE Intelligent Vehicles Symposium, IV 2014*, pp. 43–48, June 2014.

[38] A. V. Nefian and G. R. Bradski, "Detection of drivable corridors for off-road autonomous navigation," in *Proceedings of the 2006 IEEE International Conference on Image Processing, ICIP 2006*, pp. 3025–3028, October 2006.

[39] F. Moosmann and C. Stiller, "Velodyne SLAM," in *Proceedings of the 2011 IEEE Intelligent Vehicles Symposium, IV'11*, pp. 393–398, June 2011.

[40] R. Valencia, E. H. Teniente, E. Trulls, and J. Andrade-Cetto, "3D mapping for urban service robots," in *Proceedings of the IEEE/RSJ International Conference on Intelligent Robots and Systems (IROS '09)*, pp. 3076–3081, IEEE, St. Louis, Mo, USA, October 2009.

[41] P. Pfaff, R. Triebel, C. Stachniss, P. Lamon, W. Burgard, and R. Siegwart, "Towards mapping of cities," in *Proceedings of the 2007 IEEE International Conference on Robotics and Automation, ICRA'07*, pp. 4807–4813, April 2007.

[42] J. Gao, T. Zhang, X. Yang, and C. Xu, "Deep relative tracking," *IEEE Transactions on Image Processing*, vol. 26, no. 4, pp. 1845–1858, 2017.

[43] K. He, X. Zhang, S. Ren, and J. Sun, "Delving deep into rectifiers: surpassing human-level performance on imagenet classification," in *Proceedings of the 15th IEEE International Conference on Computer Vision (ICCV '15)*, pp. 1026–1034, IEEE, Santiago, Chile, December 2015.

[44] J. Fritsch, T. Kuhnl, and A. Geiger, "A new performance measure and evaluation benchmark for road detection algorithms," in *Proceedings of the 16th International IEEE Conference on Intelligent Transportation Systems (ITSC '13)*, pp. 1693–1700, IEEE, The Hague, The Netherlands, October 2013.

[45] M. Sotelo, J. Nuevo, L. Bergasa, M. Ocana, I. Parra, and D. Fernandez, "Road Vehicle Recognition in Monocular Images," in *Proceedings of the IEEE International Symposium on Industrial Electronics*, pp. 1471–1476, Dubrovnik, Croatia, June 2005.

[46] S. Álvarez, M. A. Sotelo, I. Parra, D. F. Llorca, and M. Gavilán, "Vehicle and pedestrian detection in esafety applications," in *Proceedings of the World Congress on Engineering and Computer Science*, 2009.

[47] D. F. Llorca, V. Milanes, I. Parra Alonso et al., "Autonomous pedestrian collision avoidance using a fuzzy steering controller," *IEEE Transactions on Intelligent Transportation Systems*, vol. 12, no. 2, pp. 390–401, 2011.

[48] V. Milanés, D. F. Llorca, J. Villagrá et al., "Intelligent automatic overtaking system using vision for vehicle detection," *Expert Systems with Applications*, vol. 39, no. 3, pp. 3362–3373, 2012.

[49] Y. Jia, E. Shelhamer, J. Donahue et al., "Caffe: convolutional architecture for fast feature embedding," in *Proceedings of the ACM Conference on Multimedia (MM '14)*, pp. 675–678, ACM, Orlando, Fla, USA, November 2014.

7

Accident Management System based on Vehicular Network for an Intelligent Transportation System in Urban Environments

Yusor Rafid Bahar Al-Mayouf,[1] Omar Adil Mahdi,[1] Namar A. Taha,[1] Nor Fadzilah Abdullah,[2] Suleman Khan,[3] and Muhammad Alam[4]

[1]*Department of Computer Science, College of Education for Pure Science (Ibn Al-Haitham), University of Baghdad, Iraq*
[2]*Centre of Advanced Electronic and Communication Engineering, Faculty of Engineering and Built Environment, Universiti Kebangsaan Malaysia, Malaysia*
[3]*School of Information Technology, Monash University, Bandar Sunway 47500, Malaysia*
[4]*Department of Computer Science and Software Engineering, Xi'an Jiaotong-Liverpool University, Suzhou, China*

Correspondence should be addressed to Yusor Rafid Bahar Al-Mayouf; engineer_mu2007@yahoo.com and Suleman Khan; suleman.khan@monash.edu

Academic Editor: Syed Hassan Ahmed

As cities across the world grow and the mobility of populations increases, there has also been a corresponding increase in the number of vehicles on roads. The result of this has been a proliferation of challenges for authorities with regard to road traffic management. A consequence of this has been congestion of traffic, more accidents, and pollution. Accidents are a still major cause of death, despite the development of sophisticated systems for traffic management and other technologies linked with vehicles. Hence, it is necessary that a common system for accident management is developed. For instance, traffic congestion in most urban areas can be alleviated by the real-time planning of routes. However, the designing of an efficient route planning algorithm to attain a globally optimal vehicle control is still a challenge that needs to be solved, especially when the unique preferences of drivers are considered. The aim of this paper is to establish an accident management system that makes use of vehicular ad hoc networks coupled with systems that employ cellular technology in public transport. This system ensures the possibility of real-time communication among vehicles, ambulances, hospitals, roadside units, and central servers. In addition, the accident management system is able to lessen the amount of time required to alert an ambulance that it is required at an accident scene by using a multihop optimal forwarding algorithm. Moreover, an optimal route planning algorithm (ORPA) is proposed in this system to improve the aggregate spatial use of a road network, at the same time bringing down the travel cost of operating a vehicle. This can reduce the incidence of vehicles being stuck on congested roads. Simulations are performed to evaluate ORPA, and the results are compared with existing algorithms. The evaluation results provided evidence that ORPA outperformed others in terms of average ambulance speed and travelling time. Finally, our system makes it easier for ambulance to quickly make their way through traffic congestion so that the chance of saving lives is increased.

1. Introduction

The concept of a smart city is linked to a paradigm shift where interest is being directed towards proposals and technological initiatives, which seek to make cities smart so that the quality of the lives of the people living in these cities can be improved [1, 2]. For instance, in Europe, the European Commission has established the European Initiative on Smart Cities in 2010. This initiative seeks to address the four important elements in the life of a city: climate management systems, building, transport, and electricity.

One of the main elements that smart cities seek to contain is the issue of increasing levels of road accidents, which have resulted from increasing numbers of vehicles, leading to congestion. It has been proven that vehicle accidents are one of the main causes of fatalities with more than 100 deaths being reported in the United States on a daily basis [3]. The same report indicates that the year 2007 experienced more than 43,000 deaths as a result of over 10 million reported accidents. Moreover, in most of the countries in the world, a major cause of accidents is poor infrastructure for managing traffic. The concept of traffic management systems refers to intelligent public transport mechanisms founded on information obtained in real-time, which helps drivers to avoid congested areas and increase safety while also ensuring that vehicles operate in a manner that is more environmentally friendly. Within the last few years, researchers, both from academic institutions and from industry have taken advantage of the progress made in wireless sensing technology to make the current traffic management systems more effective and ensure their ability to cope with the issues within the smart city discussed above.

Traffic congestion, as one of the leading problems affecting modern societies, has been given increasing attention. The delay in the arrival of emergency services at accident scenes like ambulances, the police, and fire, and rescue operations is one of the critical results of traffic congestion. This is an undesirable situation as the safety of the general public depends on these services being able to arrive at the scene of an accident in the efficient and shortest time possible. Intelligent transportation systems (ITSs) have not yet been able to resolve the challenge of traffic congestion. For example, Google Maps still uses communication networks that exist such as Wi-Fi, global positioning system (GPS), and cellular networks to plan journeys and mitigate the challenge of being stuck in traffic. However, apart from being expensive, these technologies still do not have the capacity to respond with speed when there is an emergency resulting from a road accident. These systems have limited access to real-time traffic data. Hence, it is important that these emergency services should be equipped with a system that can enhance the adaptability of route planning so information can be collected more efficiently and real-time traffic data can be further exploited to avoid traffic congestion.

Firstly, emerging vehicular ad hoc networks (VANETs) can be used to deliver on ITS system with improved communications capabilities for obtaining real-time traffic information more efficiently and in a cost-effective manner [4]. VANETs support vehicle-to-roadside unit (V2R) and vehicle-to-vehicle communications [5, 6], so that real-time updates can be transmitted from and to vehicles and roadside units (RSUs) [7, 8]. This real-time information which will be collected can then be consequently used for route planning in individual vehicles [9], freeway-traffic-flow management [10, 11], and vehicle localization [12–14]. Secondly, once the real-time traffic information has been obtained, various algorithms can be designed with the aim of discovering the most efficient routes that individual vehicles can take [15, 16]. However, if the route planning is deployed in a manner which is not coordinated, this could lead to even more congestion. While most of the available routes planning global algorithms pay attention to the improvements linked to networks, they generally overlook driver preferences, such as travel distances and times. This is important, particularly if one considers that replanning decisions are a consequence of the need to avoid traffic congestion and balancing of traffic as opposed to the discovery of optimal routes. Hence, some drivers may actually prefer to take longer routes even though this may entail higher costs. Hence, the design of algorithms should be done in a way that considers both the reduction of the average cost of running the vehicle and network traffic.

To this end, we propose an accident management system based on VANET for ITS in urban environments. The major contributions of this paper are listed as follows:

(1) We provide an accident management system that makes use of VANETs coupled with systems that employ cellular technology in public transport. The provided system ensures the possibility of real-time communication among vehicles, ambulances, hospitals, roadside units, and central servers. A multihop optimal forwarding algorithm is used in this system, which provides the optimal route between the sources (accident vehicle) and the destination (ambulance). This can lower the amount of time required to alert an ambulance that it is required at an accident scene.

(2) We propose an optimal route planning algorithm which provides the optimal route algorithm from ambulance to accident vehicle and then to hospital. The proposed algorithm employs VANET communication proficiencies so that vehicles can mitigate the challenge of congestion. With the aim of balancing the preferences of the drivers and the need for overall smoothness in the traffic flow, we consider vehicle travel costs and network spatial utilization.

The remainder of this paper is organized as follows. Section 2 of this paper looks at related works on route planning, while Section 3 discusses the system model. Section 4 of the paper presents the traffic routing algorithms, while Section 5 delivers a demonstration of the performance evaluation. The paper is finally concluded in Section 6.

2. Related Works

There are many consequences of an unbalanced flow of traffic and the subsequent accidents and incidents it causes: arriving late and suffering added cost are some of the results drivers have to contend with. Route navigation and planning can, however, mitigate the challenges linked to traffic congestion.

Some of the methods which can be used to manage and plan vehicle routes include the shortest-route based GPS navigation, advance route reservation, and accident duration prediction [17–19]. However, all these methods do not have the capacity to deliver immediate information when an accident or congestion happens suddenly, considering the fact that they lack a timely update capacity to deliver information on traffic conditions. Hence, the availability of real-time traffic information becomes vital for supporting vehicular real-time route planning algorithms.

Loop directors and cellular systems are the most commonly used methods in standard ITS when collecting time-varying traffic condition data. The use of mobile sensors connected through cellular networks was investigated, for the collection of real-time traffic information for the forecasting of traffic or experimental research reconstruction [20–22]. A traffic management system which employs loop detectors for continuously measuring traffic and also monitoring of arterial routes is introduced in [15]. However, both loop detectors and cellular-based systems have their own challenges. With regard to cellular services, they are not designed for the purpose of collecting traffic data and the transmission of data through them can also be expensive. When too much data is transmitted through such networks, they may also crash due to added congestion. Loop detectors can also be quite expensive to deploy. Also, when it comes to short distance transmission, inaccurate position measurements can also present problems, especially in networks that are dense, resulting in the degrading of the performance of the route management planning [23].

VANETs V2R and V2V methods of communication have the capacity of making the delivery of real-time messages much quicker, more efficient, and cost-effective when compared to prevailing methods, even in shorter distances and networks that are thick [24, 25]. More importantly, the collection and distribution of data can be enhanced by the RSUs, in VANETs making it possible for coordinated route planning to be achieved for clusters of vehicles [26]. With the aim of improving the quality of experience, the multimedia delivery application can be supported by a vehicular network based on the multimedia system, which could still be a victim of extensive delays in transmission [27, 28].

A variety of studies have concentrated on real-time route planning assisted by VANETs. A method for distributed route planning was advanced to mitigate the challenge of congestion through the employment of real-time data obtained from VANETs, with the increase in the flow of traffic [15]. With regard to fuel saving in vehicles, [29] has designed a navigation system which assists drivers to steer away from congested roads. However, the lack of coordination and selfish behavior of drivers could lead to more congestion when the individual-user manual schemes are deployed. Hence, the requirement for the routes of groups of vehicles to be planned jointly with the aim of balancing the traffic network exists. Multivehicle route planning is considered in [30, 31]. However, these works do not pay attention to the preferences of drivers, nor the average cost of travelling. It also remains unexplained how communications in VANETs can affect the route planning algorithm.

Hence, in this study, a route planning algorithm which is globally optimal is proposed to assist vehicles to avoid traffic congestion as a consequence of accidents within an urban setting. It is expected that the collected real-time traffic data and the efficient route decision-making will help facilitate better use of the resources within the road network, resulting in the average costs associated with travelling in a vehicle being reduced.

3. System Model and Architecture

In this section, we aim to provide an accident management system. We initially introduce the system model and then provide its architecture.

3.1. System Model. Within the urban environment, we consider a vehicular network that incorporates intersections and their roads. Within every road, there are two lanes in which the vehicles move in opposite directions as is illustrated in Figure 1.

The circle, in which intersection ID is found, represents the intersection, while the arrow is an indication of the direction in which the vehicles in the lane are travelling. Hospitals 1 and 2 denote the destinations that are assumed to be placed within two of the intersections. In each intersection, an RSU is located. There is one ambulance located near each hospital. All parts of the system are connected to the central server. When vehicles move through the road, they do so in either a similar or opposite direction. We make the assumption that all ambulances and vehicles have an on-board navigation mechanism and the capacity to use wireless communication similar to assumption in [32]. Within each ambulance and vehicle, there is a digital map that also contains knowledge of the locations of neighboring intersections. In addition, all vehicles are able to obtain necessary information regarding the location, speed, and direction of neighboring vehicles, which is maintained in tables accessible to neighbors by exchanging of periodic messages. While still generalizing, we also make the assumption that all vehicles have a range of transmission. We also make the assumption that the source node is armed with location services that allow it to obtain destination position in the event that it is needed [33, 34].

3.2. System Architecture. The accident management system consists of five units as is illustrated in Figure 2: vehicular, central server, RSUs, ambulance, and hospital. In this section, we will discuss each unit in detail as follows.

3.2.1. Vehicular Unit. The role of the vehicular unit is to manage the overall activities that the vehicle carries out, including the detection of accidents, while also maintaining communication with other units. In each vehicular unit, there are three modules.

(1) Sensor module: it is used to monitor and control different sensors attached to the vehicle. Our system uses two kinds of sensors for the purpose of detecting accidents: biomedical sensors which detect the individual's heart rate, temperature, and pressure and vehicle sensor which monitor acceleration, vibration, and the deployment of the airbag. Both sensors have their strengths and weaknesses. For instance, the vehicle sensor is easier to use but will not provide a precise picture of the accident. While it is able to detect that an accident has occurred, it is unable to give details regarding the gravity of the accident and how badly the passengers in the accident require medical attention. On the other hand, biomedical sensors are prone to raising false alarms. In order to mitigate against this challenge, we propose that both sensors should be used simultaneously. All readings provided by the sensors are always being measured and delivered to the local processor in the vehicle. The

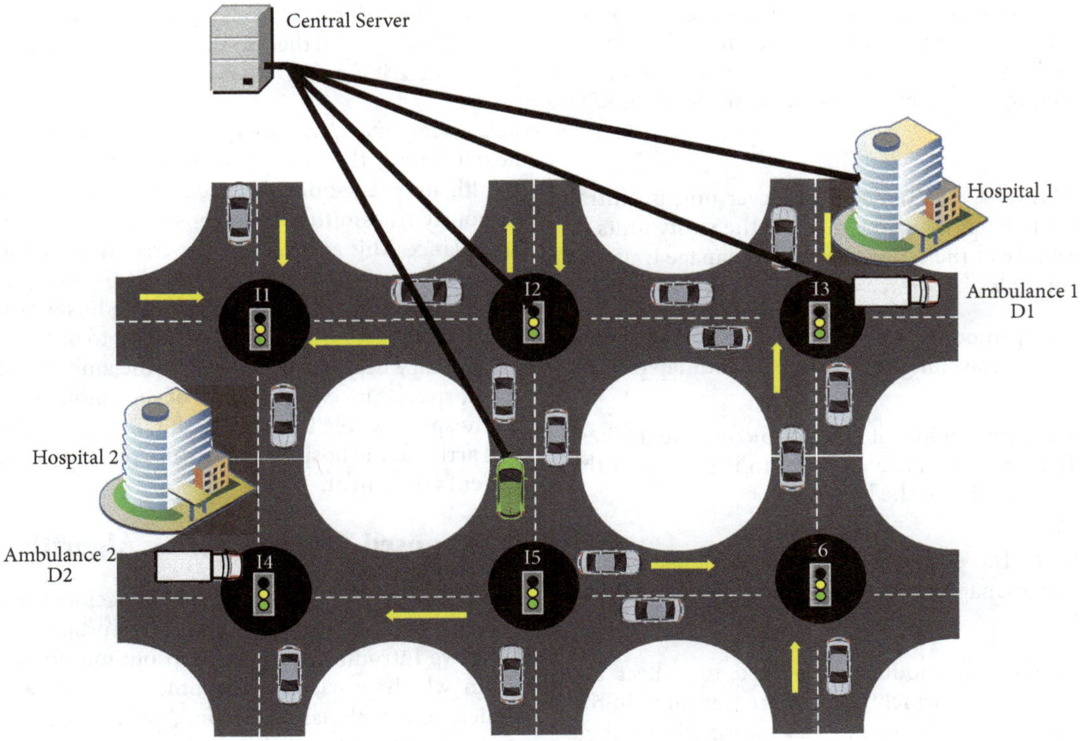

FIGURE 1: Model of accident management system.

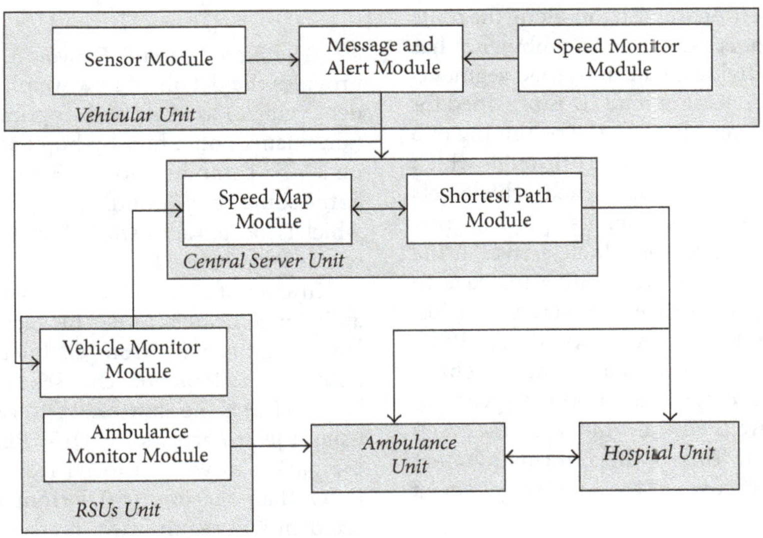

FIGURE 2: Architecture of accident management system.

processor is used for setting the threshold level for each reading of the sensor. The role of the processor is to continuously check the reading of the sensor against the threshold value. Once any abnormal values, compared to the threshold are detected, in both the vehicular and biomedical sensors, then an accident is confirmed by the processor. Information about the accident is them sent to the alert module by the local processor. Also, the local processor obtains information regarding the location of the on-board GPS. This ensures its ability to continuously locate the vehicle.

(2) Speed monitor module: it is used to maintain a record of the speed at which the vehicle is travelling and records the average speed at which the vehicle travels at each road. This message is then transmitted to the message and alert module.

(3) Message and alert module: it is used to maintain communication with the central server, RSU units,

and the ambulance. It also obtains the average speed, from the speed monitor module, at which the vehicle is travelling at each section of the road. This information is then transmitted to the corresponding RSU unit.

3.2.2. Central Server Unit. The central server unit is central to our model which has connections to all the entity units. It is the responsibility of the central server to manage traffic. It contains two modules:

(1) Speed map module: it uses the average speeds that vehicles are travelling at to create a roadmap of each road.

(2) Optimal path module: it is used to compute the best path that an ambulance can take to the scene of the accident and also to the hospital.

3.2.3. RSU Unit. The RSU unit has the responsibility of monitoring and managing traffic and also consists of two modules.

(1) Vehicle monitor module: it is used to collect the average speeds at which vehicles are travelling in the adjacent section of the road. This information is then transmitted to the central server unit.

(2) Ambulance monitor module: it is used for tracking the ambulance and setting traffic lights along the route in a manner that reduces the time an ambulance has to wait. Our model divides roads into road segments between two RSUs. A distinct road ID is provided for each road segment. Every RSU, at the beginning of a road segment, transmits the road ID through which the vehicle is going to pass. Once each vehicle gets the message, it begins to monitor its speed on that segment of the road. Once the vehicle arrives at the end of the road segment, it transmits a message to the RSUs. Within each message, there are two fields: the matching road ID and average speed. The RSUs then gather the vehicle's speed details from the vehicle and then compute the aggregate average speeds of all vehicles, which are then transmitted by the RSUs to the central server. This is the information that the central server then uses to create the city's speed roadmap.

3.2.4. Ambulance Unit. The vehicular unit and the central server unit provide the ambulance with accident information and the optimal route information, respectively. With the assistance of the GPS, the ambulance then uses this information to identify the best route. As it travels, the ambulance will be transmitting an alert message which contains two fields: the current and next road IDs. At the beginning of each road segment, upon the RSU receiving the alert message containing the current road ID, it transmits the message to the RSU at the end of the road segment. Once the transmitted message has been received, the RSU will extract the following segment of the road that the ambulance is travelling towards and sets the traffic light to allow the ambulance quick passage. This will be done all the way until the ambulance arrives at the place where it is going.

3.2.5. Hospital Unit. There is a connection between the hospital unit, the ambulance units, and the central server. Health updates about the situation of the patient are continuously transmitted to the hospital unit, directly from the ambulance. This ensures that the treatment is started as soon as possible once the patient arrives at the hospital. Owing to the fact that the ambulance keeps the hospital informed of its current location, the hospital is able to obtain the patient's preliminary health status ahead of the ambulance arriving at the hospital. Since the location of the ambulance is updated, the hospital is able to estimate the time at which the patient will arrive at the hospital and ensure preliminary steps for the patient's treatment.

4. Proposed Traffic Routing Algorithms

In this section, we discuss the postaccident action in our system. Traffic routing consists of two proposed algorithms. First, we introduce the multihop optimal forwarding algorithm which provides the optimal route between accident vehicle and ambulance. Second, we introduce the optimal route planning algorithm which provides the optimal route algorithm from ambulance to accident vehicle and then to hospital.

4.1. Multihop Optimal Forwarding Algorithm. This section provides the details of how we plan to use the sources (accident vehicle) to identify the optimal route to the destination (ambulance) on a hop-by-hop basis. The algorithm is based on a V2V communications. Figure 3 shows the green vehicle as the source and Ambulance 1 is shown as destination 1 (D1), which is located at I3, while Ambulance 2 is destination 2 (D2) whose location is I4.

In case of an accident, the source sends a packet to the ambulance nearest to it. The accident vehicle computes the length of the path from all intersections that link it with both destinations and chooses the destination whose path length (PL) is the shortest. With regard to D1, from Figure 3, using the red arrows PL2D1 < PL5D1, it can be shown that the packet forwarded to D1 from intersection I2-I3 will be faster than the one to D1 from I5-I6-I3. Accordingly, the accident will require two intersections to get to Ambulance 1. Meanwhile, with regard to D2, as shown in Figure 3 by green arrows, PL5D2 < PL2D2, the packet to intersection I5-I4 to D2 will be quicker when compared to the packet transmitted from I2-I1-I4. Based on this, the accident vehicle will choose the path that has more vehicles to ensure continuous connectivity. Consequently, two intersections are required by the accident vehicle in order for it to reach Ambulance 2. What this implies is that we have two paths whose length is similar: I5-I4-D2 and I2-13-D1. For this case, it would be more prudent for the accident vehicle to choose a path that has more vehicles so that maximum connectivity can be maintained. According to Figure 3, the path chosen by the accident vehicle is PL2.

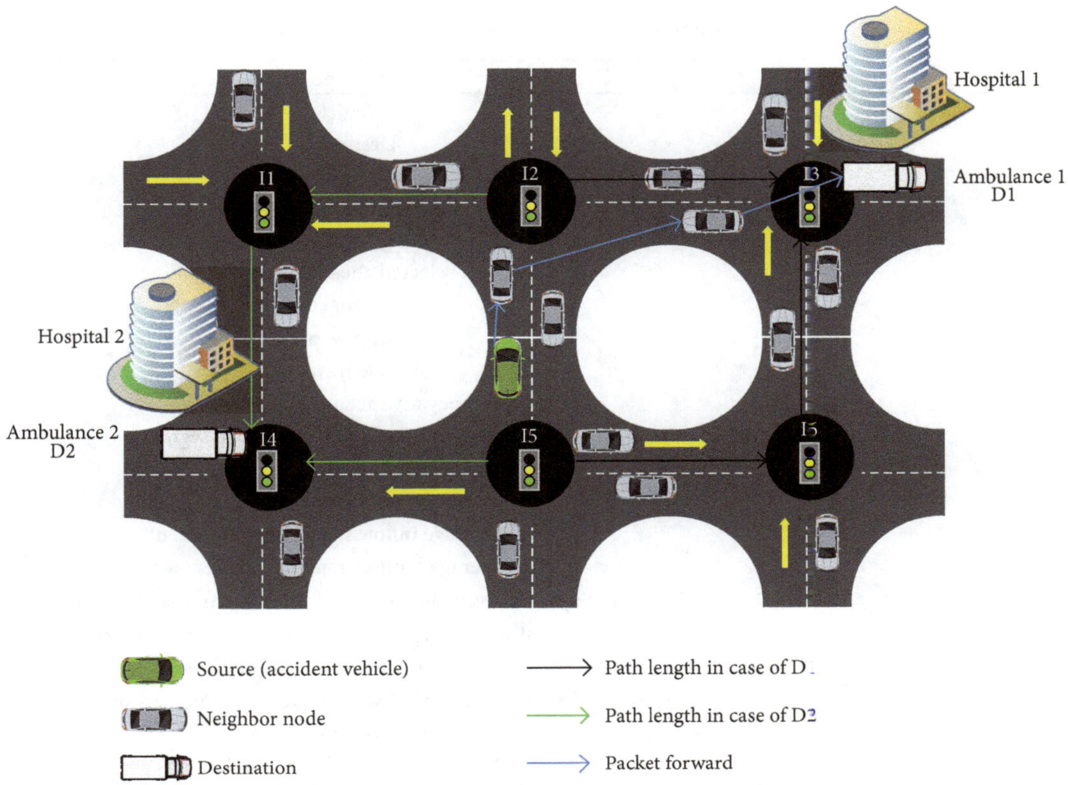

FIGURE 3: Source vehicle calculates the path length to destination.

The blue arrows denote the accident vehicle forwarding the packet to the next candidate vehicle travelling towards the intersection that has the shortest path length to D1, depending on [35, 36]. This is a process that continues all the way until the packet arrives at its estimation. This not only does ensure fast progress, but also makes sure that there is minimal traffic on the network.

4.2. Optimal Route Planning Algorithm. Our aim in this section is to identify the optimal route planning algorithm (ORPA) from the ambulance to the accident vehicle and then to the hospital. The reasoning behind this algorithm is the avoidance of traffic congestion and the reduction of the average cost of travelling.

It is expected that each vehicle will trial a planned route right from the point where it starts its journey to the point where the journey terminates. In this instance, the planned route will be referred to, in GPS, as a route preset, depending on what the choices of the driver are and also depending on where the starting and termination points are located. The driver is also expected to follow the route which has been preset until they receive an alert that there is congestion along that route. The central server, through managing the optimal route planning algorithm, will be responsible for identifying the optimal substitute routing for a particular ambulance once congestion or an accident happens. To facilitate the understanding of technical aspects, a table of notations would be useful as illustrated in Table 1.

The optimal route planning algorithm is planned in a manner that balances the avoidance of congestion while also ensuring that the cost of travel is reduced. Hence, for this purpose, we suggest two metrics. The first one is the weighted average cost metric which determines the average cost for vehicles whose routes have been replanned, while also ensuring that the preferences of the driver are also considered. Some of the mathematical notions used in this algorithm are v, d, (i, j), and T representing vehicles, destination, road segment, and sample time duration, respectively. In this case, we consider PL_{ci}^{vd} as the prevailing length of the path (first based on what the driver prefers such as the one they are familiar with or the shortest one), where ci is the path selected ahead of the replanning. PL_{pij}^{vd} represents the changed path, where, pij represents the new path which has been planned and by going via (i, j) the path of v is changed in the direction of d. $C(T)$ is vehicle travel cost and will be formulated as

$$C(T) = \begin{cases} 0, & \text{if } i \notin \text{current path of } vd \\ \rho\left(\left|\mathrm{PL}_{pij}^{vd}\right| - \left|\mathrm{PL}_{ci}^{vd}\right|\right), & \text{if } i \in \text{current path of } vd \end{cases} \quad (1)$$

where ρ represents a nonnegative accumulative function used to measure the effect of the increase in path length [37].

Table 1: Summary of the important mathematical notations.

Symbol	Description
v	Vehicles
d	Destination
(i, j)	Road segment
T	Sample time duration
$\mathbf{PL_{ci}^{vd}}$	Prevailing length of the path
ci	The path selected ahead of the re-planning
$\mathbf{PL_{pij}^{vd}}$	The changed path
pij	The new path
C(T)	Vehicle travel cost
ρ	A nonnegative accumulative function
$\mathbf{C_{ij}(T)}$	The average cost of vehicles on road segment
$\mathbf{C_i(T)}$	The average cost of vehicles on intersection i
$\mathbf{J_i}$	The set of adjacent crossing of intersection i
$\boldsymbol{\lambda}_{\mathbf{ij}}\mathbf{(T)}$	The average inflow rates of road segment
$\boldsymbol{\mu}_{\mathbf{ij}}\mathbf{(T)}$	The average outflow rates of road segment
$\boldsymbol{\lambda}_{\mathbf{ij}}\mathbf{(t)}$	The upstream vehicle arrivals respectively on road segment
$\boldsymbol{\mu}_{\mathbf{ij}}\mathbf{(t)}$	The upstream vehicle departure rates on road segment
Δ	Sample time duration
$\mathbf{Q_i\ (T)}$	The total length of all virtual queues of intersection i
WI	Weighted intersection
WI(T)	Intersection i's weight in sample time
$\mathbf{Q_i^d(T)}$	Virtual queues representing the number of the shielded vehicles travelling through intersection i
$\mathbf{Q_j^d(T)}$	Virtual queues representing the number of the shielded vehicles travelling through intersection j
$\mathbf{j*}$	The huge differential row build-up for intersection i and the other adjacent intersection j
$\mathbf{q_{j*}^d(T)}$	Re-planning to intersection $j*$ of a number of vehicles

Computing the average cost of vehicles on road segment (i, j) can be done as follows:

$$C_{ij}(T) = \frac{\sum C(T)}{V} \tag{2}$$

where V represents the number of vehicles travelling on a road segment (i, j). For an intersection, considering that there could be a number of adjacent intersections which are candidates of the approaching intersections, determining the average cost of vehicles on intersection i is defined as

$$C_i(T) = \frac{\sum_{j\in J_i} C_{ij}(T)}{J_i} \tag{3}$$

where J_i represents the set of adjacent crossing of intersection i. Secondly, in the weighted network stability, we used the inflow/outflow system (Figure 4) as a method of modelling vehicle traffic [38].

With the Tth time interval based on the rates of the traffic flow involved, as gathered by RSUs, the average inflow and outflow rates of road segment (i, j) are marked as $\lambda_{ij}(T)$ and $\mu_{ij}(T)$, respectively, and formulated as

$$\lambda_{ij}(T) = \frac{1}{\Delta}\sum_{t=(T-1)\Delta}^{T\Delta} \lambda_{ij}(t)$$

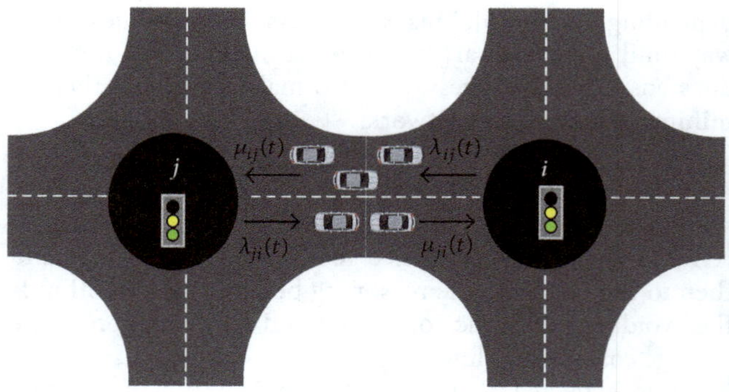

Figure 4: Traffic flow model.

$$\mu_{ij}(T) = \frac{1}{\Delta}\sum_{t=(T-1)\Delta}^{T\Delta} \mu_{ij}(t) \tag{4}$$

where $\lambda_{ij}(t)$ and $\mu_{ij}(t)$ denote the upstream vehicle arrivals and departure rates, respectively, on road segment (i, j) whose destination on the timeslot t is the same. Δ is a sample time duration that includes a series of time slots as distinguished by sampling theorem with the aim of mitigating the risk of loss of information in the compressive sensing for estimation of traffic in [39]. It is vital to be conscious of the reality that all variables for the road segment (i, j) on the opposite direction identified as road segment (j, i) could be defined as $\lambda_{ji}(T)$ and $\mu_{ji}(T)$, respectively. The information on a virtual queue of length $Q_i(T)$ is needed in order to determine if an intersection is stable. Defining the leftover number of vehicles on road segment (i, j) will be done as follows:

$$Q_{ij}(T) = \lambda_{ij}(T-1) + \mu_{ij}(T-1) \quad (5)$$

From (5) we can get the total length $Q_i(T)$ of all virtual queues of intersection i for all destinations as

$$Q_i(T) = \sum_{j \in Ji} Q_{ij}(T) \quad (6)$$

Provided that there is stability on all queues within the network, the traffic flow across the whole network will equally be stable. In the event that the central server gets an incident warning message gathered by vehicles through VANETs and RSUs, the optimal route planning algorithm will be computed to update and also determine $\lambda_{ij}(T)$ with the aim of reducing the cost of travelling and improving spatial utility. With regard to the time on the sample, T, based on the optimal route planning algorithm, it is possible to dispatch a vehicle whose destination is d from one intersection towards another. The required vehicle will be navigated through the updated path as delivered to the GPS device. To meet the goal of balancing between reduce travel cost and spatial utility, we have designed an innovative metric referred to as weighted intersection (WI). Allow $\mathrm{WI}(T)$ to represent intersection i's weight in sample time T, formulated as shown below:

$$\mathrm{WI}_i(T) = \left[Q_i^d(T) - Q_j^d(T)\right] - KC_i(T) \quad (7)$$

where $Q_i^d(T)$ and $Q_j^d(T)$ are virtual queues representing the number of the shielded vehicles travelling through intersections i and j in the direction of destination d in sample time T. K is a nonnegative constant defined by the central server used for every vehicle [40]. In the beginning, vehicles travelling through an intersection that has a heavy weight are replanned. Eventually, vehicles with destination d, kept at i, have to be dispatched to queue $Q_{j*}^d(T)$ of intersection $j*$, where $j*$ represents the huge differential row build-up for intersection i and the other adjacent intersection j and destination d as

$$j* = Q_i^d(T) - Q_j^d(T) \quad (8)$$

According to short increased path lengths, there will be a replanning to intersection $j*$ of a number of vehicles whose destination is d, using

$$q_{j*}^d(T) = Q_i^d(T) - Q_{j*}^d(T) \quad (9)$$

Subsequently, queues at all remaining intersections are correspondingly updated as follows:

$$\begin{aligned} Q_i^d(T) &= Q_i^d(T) - q_{j*}^d(T) \\ Q_{j*}^d(T) &= Q_{j*}^d(T) + q_{j*}^d(T) \end{aligned} \quad (10)$$

A similar process continues all the way till all linked intersections have been processed. Algorithm 1 shows the detail of OPPA.

5. Performance Evaluation

For purposes of conducting the case study, a general Manhattan grid model has been selected. In order to design the scenario and extract the intersection images, MATLAB and Google Maps are used, respectively. Simulations that deploy assistance mobility models with the capacity to make vehicles accelerate, change direction, and decelerate, can enable realistic scenarios to be achieved. For this part of the study, a familiar mobility model which is car following model [4] is used. Using the average speed information obtained from the RSUs, the map road segment is separated into two speed zones. If the average vehicle speed on a certain road segment is low, then that segment is considered congested and those with high average speeds are seen as free from congestion. However, the method of incessantly averaging speeds of vehicles is prone to errors as a result of accumulated historical data. The result of this is that the data represents a historical characteristic as opposed to the prevailing average speed at a given time. To mitigate against this challenge, we create divisions based on 5-minute time slots with the aggregate average being computed in any slot using the data obtained from only that slot. Another challenge is the high overhead resulting from regular updates of speed between central servers and RSUs. To mitigate this problem, the speed will be separated into time zones, with the RSU only updating the average speed if there are any changes in the average speed from one time zone to the next. Based on the RSU, the central server update will dynamically provide updates of the time zone to reflect the road segment's congestion.

We do a comparison of the assessment of four algorithms within our ambulance setting using their average speed and travelling time as a basis. The Dijkstra algorithm deploys the characteristic for planning paths without any changes. In this case, each road segment speed assigns a distinct average value for all segments. When using this algorithm, congestion is not considered, and speed cannot be dynamically altered. There are similarities between that the A$*$ algorithm and the Dijkstra. However, in the A$*$ algorithm the experimental function $H(n)$ is presented. The $H(n)$ denotes the angle between the adjacent node and the straight line in the direction of the endpoint node. With regard to the VBA$*$ algorithm, the prevailing speed on a road segment can be varied, dynamically, and a distribution manner can be used to operate it. We finally do a comparison of the proposed ORPA, A$*$, Dijkstra, and VBA$*$ algorithms. The implemented simulation parameters are presented in Table 2.

Figure 5 shows the travelling time of Dijkstra, A$*$, VBA$*$, and ORPA in the low- and high-density scenario. The ORPA obtains a lower travelling time than Dijkstra, A$*$ and VBA$*$. Specifically, in low-density the travelling time of ORPA, Dijkstra, A$*$, and VBA$*$ reaches 1400, 2400, 2200, and 2000 seconds, respectively. Hence, the travelling time of ORPA is 71.4%, 57.14%, and 42.85% lower than that of Dijkstra, A$*$, and VBA$*$, respectively, whereas in high-density the ORPA

(1) **Begin Algorithm 1**
(2) A candidate set of intersections $Ic = \emptyset$
(3) **For** intersection $i \in$ set of all intersections **Do**
(4) **If** $i \in$ current path of vd **Then**
(5) Compute the vehicle travel cost $\mathbf{C(T)}$ on (i, j);
(6) Determining the average $\mathbf{C_{ij}(T)}$;
(7) Determining the average $\mathbf{C_i(T)}$ on i;
(8) Compute the average inflow $\boldsymbol{\lambda}_{\mathbf{ij}}\mathbf{(T)}$ and outflow rate $\boldsymbol{\mu}_{\mathbf{ij}}\mathbf{(T)}$s on (i, j);
(9) Compute the virtual queue of length $\mathbf{Q_{ij}(T)}$ on (i, j);
(10) Compute the total length $\mathbf{Q_i(T)}$ of all virtual queues on i;
(11) Compute the weighted intersection $\mathbf{WI_i(T)}$ on i;
(12) **Else**
(13) Go to (3);
(14) **End If**
(15) **If** $\mathbf{WI_i(T)} \neq 0$ **Then**
(16) Update the set $Ic \leftarrow Ic \cup \{i\}$;
(17) **End If**
(18) **End For**
(19) **While** intersection $Ic \neq \emptyset$ **Do**;
(20) Intersection $i = \mathbf{WI_i(T)}$;
(21) **For** each destination **d Do**
(22) Find huge differential row build-up $\mathbf{j*}$ for intersections i and j;
(23) Find re-planning path length $\mathbf{q^d_{j*}(T)}$ to intersection $\mathbf{j*}$;
(24) Update $\mathbf{Q^d_i(T)}$ and $\mathbf{Q^d_{j*}(T)}$;
(25) **End For**
(26) $Ic \leftarrow Ic \setminus \{i\}$;
(27) **End While**
(28) **End Algorithm 1**

Algorithm 1: ORPA.

Table 2: Simulation parameters.

Parameter	Value
Simulation area	$5 \times 5\,\text{km}^2$
Number of lanes	2 bidirectional
Intersections	24
Segments	38
Speed	(40–80) km/hr.
Average vehicles density	High density 40–50 veh/lane/km & low density 6–8 veh/lane/km
Communication radius	300 m
MAC protocol	IEEE 802.11p
Total simulation time	3600 s

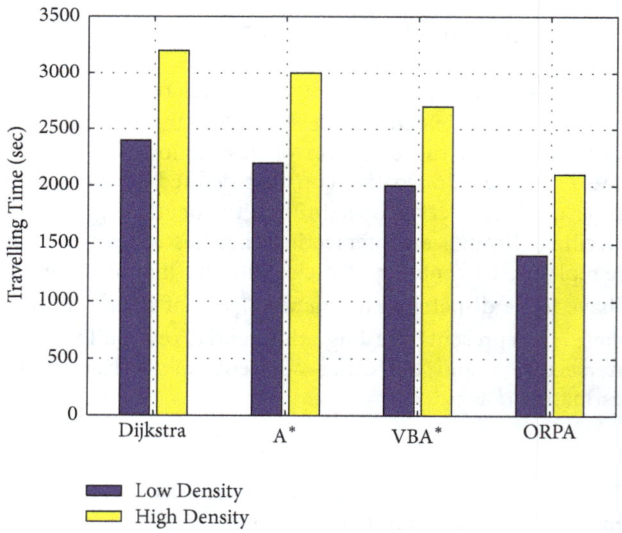

Figure 5: Travelling time in low- and high-density.

obtains a travelling time of 2100 seconds, while Dijkstra, A∗, and VBA obtain a travelling time of 3200, 3000, and 2700 seconds. Thus, ORPA decreases the total travelling time by 52.38%, 42.3%, and 28.57% for Dijkstra, A∗, and VBA, respectively. Figure 5 illustrates that the travelling time of high-density is greater than low-density due to congestion.

Figure 6 depicts the average ambulance speed of four algorithms in low- and high-density scenario. The ORPA obtains a higher average ambulance speed than Dijkstra, A∗, and VBA∗. Specifically, in low-density the average ambulance speed of ORPA, Dijkstra, A∗, and VBA∗ reaches 45, 23, 26, and 32 km/hr, respectively. Hence, the average ambulance speed of ORPA is 48.8%, 42.2%, and 28.8% higher than that of Dijkstra, A∗, and VBA∗, respectively, whereas in high-density the average ambulance speed of ORPA, Dijkstra, A∗, and VBA∗ reaches 30, 12, 17, and 20 km/hr, respectively.

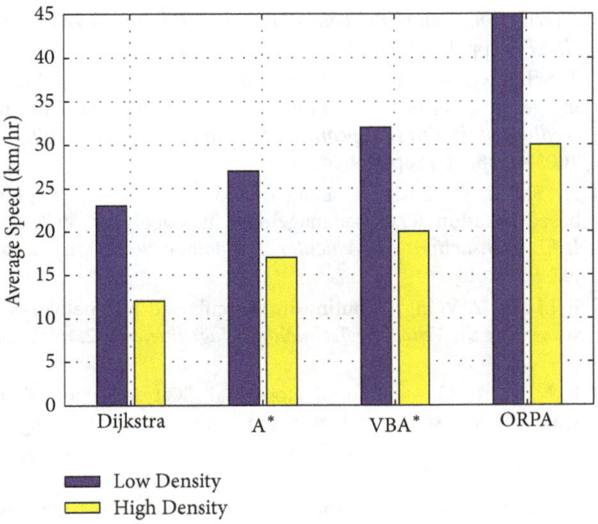

FIGURE 6: Average ambulance speed in low- and high-density.

Thus, the average ambulance speed of ORPA is 60%, 43.3%, and 33.3% higher than that of Dijkstra, A∗, and VBA∗, respectively. This figure further depicts that the average ambulance speed of low-density environments is greater than that of high-density environments.

6. Conclusion

Through this paper, we have developed a common system to manage accidents so that vehicles are able to avoid congested areas within an ITS. Initially, we established an accident management system which employs cellular systems of the public transportation systems and VANETs to make efficient real-time communication between vehicles possible, including ambulances, hospitals, RSUs, and central servers. We subsequently propose a real-time algorithm for planning routes with the aim of improving the overall use of space while at the same time reducing the cost of travelling, through vehicles' ability to avoid congested road segments. Finally, we have shown that the path planning algorithm we propose will reduce the time taken by ambulances to be alerted and dispatched to a scene of accident through being able to avoid road segments that are congested and will increase the chance of saving the lives of accident victims.

Conflicts of Interest

The authors declare that there are no conflicts of interest regarding the publication of this paper.

Acknowledgments

This work is fully funded by Universiti Kebangsaan Malaysia under Grant no. GUP-2016-005.

References

[1] M. Y. I. Idris, A. M. A. Abu Znaid, A. W. A. Wahab, L. K. Qabajeh, and O. A. Mahdi, "Low communication cost (LCC) scheme for localizing mobile wireless sensor networks," *Wireless Networks*, vol. 23, no. 3, pp. 737–747, 2017.

[2] A. S. Sadiq, S. Khan, K. Z. Ghafoor, M. Guizani, and S. Mirjalili, "Transmission power adaption scheme for improving IoV awareness exploiting: evaluation weighted matrix based on piggybacked information," *Computer Networks*, vol. 137, pp. 147–159, 2018.

[3] G. Egilmez and D. McAvoy, "Benchmarking road safety of U.S. states: A DEA-based Malmquist productivity index approach," *Accident Analysis & Prevention*, vol. 53, pp. 55–64, 2013.

[4] H. Hartenstein and K. Laberteaux, *VANET: Vehicular Applications and Inter-Networking Technologies*, vol. 1, John Wiley & Sons, 2009.

[5] Y. R. B. Al-Mayouf, N. F. Abdullah, M. Ismail, A. W. A. Wahab, and O. A. Mahdi, "Efficient routing algorithm for VANETs based on distance factor," in *Proceedings of the 2016 International Conference on Advances in Electrical, Electronic and Systems Engineering, ICAEES 2016*, pp. 567–571, mys, November 2016.

[6] Y. Sun, R. Lu, X. Lin, X. Shen, and J. Su, "An efficient pseudonymous authentication scheme with strong privacy preservation for vehicular communications," *IEEE Transactions on Vehicular Technology*, vol. 59, no. 7, pp. 3589–3603, 2010.

[7] Y. R. B. Al-Mayouf, M. Ismail, N. F. Abdullah, S. M. Al-Qaraawi, and O. A. Mahdi, "Survey on VANET technologies and simulation models," *ARPN Journal of Engineering and Applied Sciences*, vol. 11, no. 15, pp. 9414–9427, 2016.

[8] M. Wang, H. Liang, R. Zhang, R. Deng, and X. Shen, "Mobility-aware coordinated charging for electric vehicles in VANET-enhanced smart grid," *IEEE Journal on Selected Areas in Communications*, vol. 32, no. 7, pp. 1344–1360, 2014.

[9] I. Leontiadis, G. Marfia, D. Mack, G. Pau, C. Mascolo, and M. Gerla, "On the effectiveness of an opportunistic traffic management system for vehicular networks," *IEEE Transactions on Intelligent Transportation Systems*, vol. 12, no. 4, pp. 1537–1548, 2011.

[10] O. A. Mahdi, A. W. Abdul Wahab, M. Y. I. Idris et al., "A comparison study on node clustering techniques used in target tracking WSNs for efficient data aggregation," *Wireless Communications and Mobile Computing*, vol. 16, no. 16, pp. 2663–2676, 2016.

[11] A. Skabardonis and N. Geroliminis, "Real-time monitoring and control on signalized arterials," *Journal of Intelligent Transportation Systems: Technology, Planning, and Operations*, vol. 12, no. 2, pp. 64–74, 2008.

[12] N. M. Drawil and O. Basir, "Intervehicle-communication-assisted localization," *IEEE Transactions on Intelligent Transportation Systems*, vol. 11, no. 3, pp. 678–691, 2010.

[13] O. A. Mahdi, A. W. Abdul Wahab, M. Y. I. Idris, A. Abu Znaid, Y. R. B. Al-Mayouf, and S. Khan, "WDARS: a weighted data aggregation routing strategy with minimum link cost in event-driven WSNs," *Journal of Sensors*, vol. 2016, Article ID 3428730, 12 pages, 2016.

[14] A. M. A. Abu Znaid, M. Y. I. Idris, A. W. Abdul Wahab, L. Khamis Qabajeh, and O. Adil Mahdi, "Sequential Monte Carlo Localization Methods in Mobile Wireless Sensor Networks: A Review," *Journal of Sensors*, vol. 2017, Article ID 1430145, 2017.

[15] A. H. Khosroshahi, P. Keshavarzi, Z. D. Koozehkanani, and J. Sobhi, "Acquiring real time traffic information using VANET

and dynamic route guidance," in *Proceedings of the IEEE 2nd International Conference on Computing, Control and Industrial Engineering, CCIE 2011*, pp. 9–13, August 2011.

[16] O. A. Mahdi, A. W. Abdul Wahab, M. Y. I. Idris, A. Abu Znaid, S. Khan, and Y. R. B. Al-Mayouf, "ESAM: endocrine inspired sensor activation mechanism for multi-target tracking in WSNs," in *Proceedings of the 4th International Conference on Wireless and Optical Communications*, vol. 9902 of *Proceedings of SPIE*, International Society for Optics and Photonics, Beijing, China, June 2016.

[17] M. Abboud, L. Jaoude, and Z. Kerbage, "Real time GPS navigation system," in *Proceedings of the 3rd FEA Student Conference*, Department of Electrical and Computer Engineering, American University of Beirut, Citeseer, 2004.

[18] Y. Chung, "Development of an accident duration prediction model on the Korean Freeway Systems," *Accident Analysis & Prevention*, vol. 42, no. 1, pp. 282–289, 2010.

[19] Y. Zhao, K. Triantis, D. Teodorović, and P. Edara, "A travel demand management strategy: The downtown space reservation system," *European Journal of Operational Research*, vol. 205, no. 3, pp. 584–594, 2010.

[20] J. C. Herrera, D. B. Work, R. Herring, X. Ban, Q. Jacobson, and A. M. Bayen, "Evaluation of traffic data obtained via GPS-enabled mobile phones: the Mobile Century field experiment," *Transportation Research, Part C: Emerging Technologies*, vol. 18, no. 4, pp. 568–583, 2010.

[21] R. Herring et al., "Using mobile phones to forecast arterial traffic through statistical learning," in *Proceedings of the 89th Transportation Research Board Annual Meeting*, 2010.

[22] J. Jariyasunant et al., *Mobile transit trip planning with real-time data*, 2011, Mobile transit trip planning with real-time data.

[23] B. Hoh, M. Gruteser, R. Herring et al., "Virtual trip lines for distributed privacy-preserving traffic monitoring," in *Proceedings of the 6th international conference on Mobile systems, applications, and services*, p. 15, Breckenridge, CO, USA, June 2008.

[24] H. T. Cheng, H. Shan, and W. Zhuang, "Infotainment and road safety service support in vehicular networking: from a communication perspective," *Mechanical Systems and Signal Processing*, vol. 25, no. 6, pp. 2020–2038, 2011.

[25] J. Zhao and G. Cao, "VADD: vehicle-assisted data delivery in vehicular Ad hoc networks," *IEEE Transactions on Vehicular Technology*, vol. 57, no. 3, pp. 1910–1922, 2008.

[26] T. H. Luan, X. Ling, and X. Shen, "Provisioning QoS controlled media access in vehicular to infrastructure communications," *Ad Hoc Networks*, vol. 100, no. 2, pp. 231–242, 2012.

[27] C. Xu, F. Zhao, J. Guan, H. Zhang, and G.-M. Muntean, "QoE-driven user-centric vod services in urban multihomed P2P-based vehicular networks," *IEEE Transactions on Vehicular Technology*, vol. 62, no. 5, pp. 2273–2289, 2013.

[28] L. Zhou, Y. Zhang, K. Song, W. Jing, and A. V. Vasilakos, "Distributed media services in P2P-based vehicular networks," *IEEE Transactions on Vehicular Technology*, vol. 60, no. 2, pp. 692–703, 2011.

[29] P. Chen, Y. Guo, and W. Chen, "Fuel-Saving Navigation System in VANETs," in *Proceedings of the 2010 IEEE Vehicular Technology Conference (VTC 2010-Fall)*, pp. 1–5, Ottawa, ON, Canada, September 2010.

[30] M. Kimura, Y. Taoda, Y. Kakuda, S. Inoue, and T. Dohi, "A novel method based on VANET for alleviating traffic congestion in urban transportations," in *Proceedings of the 2013 IEEE Eleventh International Symposium on Autonomous Decentralized Systems (ISADS)*, pp. 1–7, Mexico City, Mexico, March 2013.

[31] T. Schouwenaars, B. De Moor, E. Feron, and J. How, "Mixed integer programming for multi-vehicle path planning," in *Proceedings of the 6th European Control Conference, ECC 2001*, pp. 2603–2608, prt, September 2001.

[32] H. Saleet, O. Basir, R. Langar, and R. Boutaba, "Region-based location-service-management protocol for VANETs," *IEEE Transactions on Vehicular Technology*, vol. 59, no. 2, pp. 917–931, 2010.

[33] F. Li and Y. Wang, "Routing in vehicular ad hoc networks: a survey," *IEEE Vehicular Technology Magazine*, vol. 2, no. 2, pp. 12–22, 2007.

[34] R. N. Murty, G. Mainland, I. Rose et al., "Citysense: an urban-scale wireless sensor network and testbed," in *Proceedings of the IEEE Conference on Technologies for Homeland Security*, pp. 583–588, IEEE, 2008.

[35] Y. R. B. Al-Mayouf, N. F. Abdullah, M. Ismail, S. M. Al-Qaraawi, O. A. Mahdi, and S. Khan, "Evaluation of efficient vehicular ad hoc networks based on a maximum distance routing algorithm," *EURASIP Journal on Wireless Communications and Networking*, vol. 2016, no. 1, article no. 265, 2016.

[36] Y. R. Al-Mayouf, M. Ismail, N. F. Abdullah et al., "Efficient and stable routing algorithm based on user mobility and node density in urban vehicular network," *PLoS ONE*, vol. 11, no. 11, Article ID e0165966, 2016.

[37] H. Liang, B. J. Choi, W. Zhuang, and X. Shen, "Optimizing the energy delivery via V2G systems based on stochastic inventory theory," *IEEE Transactions on Smart Grid*, vol. 4, no. 4, pp. 2230–2243, 2013.

[38] J. Chen, W. Xu, S. He, Y. Sun, P. Thulasiraman, and X. Shen, "Utility-based asynchronous flow control algorithm for wireless sensor networks," *IEEE Journal on Selected Areas in Communications*, vol. 28, no. 7, pp. 1116–1126, 2010.

[39] Z. Li, Y. Zhu, H. Zhu, and M. Li, "Compressive sensing approach to urban traffic sensing," in *Proceedings of the 31st International Conference on Distributed Computing Systems (ICDCS '11)*, pp. 889–898, IEEE, July 2011.

[40] M. J. Neely, "Stochastic network optimization with application to communication and queueing systems," *Synthesis Lectures on Communication Networks*, vol. 3, no. 1, 211 pages, 2010.

8

Neuromorphic Vision based Multivehicle Detection and Tracking for Intelligent Transportation System

Guang Chen ,[1,2] Hu Cao,[3] Muhammad Aafaque,[2] Jieneng Chen,[4] Canbo Ye,[1] Florian Röhrbein,[2] Jörg Conradt,[5] Kai Chen,[1] Zhenshan Bing,[2] Xingbo Liu,[1] Gereon Hinz,[2] Walter Stechele ,[6] and Alois Knoll[2]

[1]*College of Automotive Engineering, Tongji University, China*
[2]*Robotics and Embedded Systems, Technische Universität München, Germany*
[3]*State Key Laboratory of Advanced Design and Manufacturing for Vehicle Body, Hunan University, China*
[4]*College of Electronics and Information Engineering, Tongji University, China*
[5]*Department of Computational Science and Technology, KTH Royal Institute of Technology, Sweden*
[6]*Integrated Systems, Technische Universität München, Germany*

Correspondence should be addressed to Guang Chen; guang@in.tum.de

Academic Editor: Krzysztof Okarma

Neuromorphic vision sensor is a new passive sensing modality and a frameless sensor with a number of advantages over traditional cameras. Instead of wastefully sending entire images at fixed frame rate, neuromorphic vision sensor only transmits the local pixel-level changes caused by the movement in a scene *at the time they occur*. This results in advantageous characteristics, in terms of low energy consumption, high dynamic range, sparse event stream, and low response latency, which can be very useful in intelligent perception systems for modern intelligent transportation system (ITS) that requires efficient wireless data communication and low power embedded computing resources. In this paper, we propose the first neuromorphic vision based multivehicle detection and tracking system in ITS. The performance of the system is evaluated with a dataset recorded by a neuromorphic vision sensor mounted on a highway bridge. We performed a preliminary multivehicle tracking-by-clustering study using three classical clustering approaches and four tracking approaches. Our experiment results indicate that, by making full use of the low latency and sparse event stream, we could easily integrate an online tracking-by-clustering system running at a high frame rate, which far exceeds the real-time capabilities of traditional frame-based cameras. If the accuracy is prioritized, the tracking task can also be performed robustly at a relatively high rate with different combinations of algorithms. We also provide our dataset and evaluation approaches serving as the first neuromorphic benchmark in ITS and hopefully can motivate further research on neuromorphic vision sensors for ITS solutions.

1. Introduction

Neuromorphic vision sensors, inspired by biological vision, use an event-driven frameless approach to capture transients in visual scenes. In contrast to conventional cameras, neuromorphic vision sensors only transmit local pixel-level changes (called "events") caused by movement in a scene *at the time of occurrence* and provide an information rich stream of events with a latency within tens of microseconds. To be specific, a single event is a tuple (t, x, y, p), where x, y are the pixel coordinates of the event in 2D space, t is the time-stamp of the event, and p is the polarity of the event, which is the sign of the brightness change (increasing or decreasing). Furthermore, the requirements for data storage and computational resources are drastically reduced due to the sparse nature of the event stream. Apart from the low latency and high storage efficiency, neuromorphic vision sensors also enjoy a high dynamic range of 120 dB. In combination, these properties of neuromorphic vision sensors inspire entirely new designs of intelligent transportation systems. In order to elucidate the mechanism of neuromorphic sensors more clearly, a comparison between standard frame-based cameras and neuromorphic vision sensors is shown in Figure 1.

Traditionally, frame-based vision sensors serve as the main information sources for vision perception tasks of ITS, which results in well-known challenges, such as the limited real-time performance and substantial computational costs. The key problem lies in the fact that conventional cameras sample their environment with a fixed frequency and produce a series of frames, which actually contain enormous amounts of redundant information but lost all the information between two adjacent frames. Hence, traditional vision sensors waste memory access, energy, computational power, and time on the one hand and also discard significant information between continuous frames on the other hand. These properties bring about great limitations on its applications. For an intelligent transportation system equipped with conventional cameras, appearance feature extraction based on learning methods is the major strategy of environment perception tasks, which is acknowledged to be computationally demanding [3]. Moreover, in order to get good detection and tracking performance, large amounts of labeled data as well as dedicated and expensive hardware such as GPU are indispensable for the training and learning process.

In this paper, a novel approach for the tracking system of the intelligent transportation systems (ITS) is proposed based on the neuromorphic vision sensor. And we will publish our dataset and evaluation approaches as well, aiming to provide the first neuromorphic benchmark in ITS and motivate further research on neuromorphic vision sensors for ITS solutions. To fully demonstrate the feasibility and potential of the approach, different detection and tracking algorithms are presented and compared in this paper. In detection stage, we utilize and evaluate three classical clustering approaches: mean-shift clustering (MeanShift) [4], density based spatial clustering of applications with noise (DBSCAN) [5], and WaveCluster [6]. In terms of tracking stage, we carry out online multitarget tracking via four different algorithms: simple online and real-time tracking (SORT) [7], the Gaussian mixture probability hypothesis density filter (GM-PHD) [8], the cardinalized probability hypothesis density filter (GM-CPHD) [9], and probabilistic data association filter (PDAF) [10].

In combination, we propose the first neuromorphic vision based multivehicle detection and tracking system in ITS, with the unique properties of neuromorphic vision sensors mentioned above. The performance of the system is evaluated with a dataset recorded by a neuromorphic vision sensor mounted on a highway bridge. According to the experiment results, the tracking-by-clustering system can run at a rate of more than 110Hz, which far exceeds the real-time performance of traditional frame-based cameras. With priority given to accuracy, the tracking task can also be performed more robustly and precisely using different algorithm combinations. This work is extended from a conference paper which is published on the Joint German/Austrian Conference on Artificial Intelligence, 2017 [11]. We extended it from 4 aspects. First, we extend the testing data sequences to 3 sequences for the experiment section. Second, we evaluate 3 detection-by-clustering approaches instead of 2 in [11]. Third, we extend to evaluate 4 tracking approaches instead of 1 in [11]. Finally, based on these differences, we analyze the results in different views with new outcomes [11].

The rest of this paper is organized as follows. In Section 2, we list the related work in the context of previous multivehicle detection and tracking methods. In Section 3, we introduce the variety of neuromorphic vision sensors and dataset. The algorithms utilized for detection and tracking are illustrated, respectively, in Section 4. The experiment results are analyzed and discussed in Section 5. In Section 6, we draw the conclusion and point out the possible further work.

2. Related Work

In the past decade, detecting and tracking multiple vehicles in traffic scenes for traffic surveillance, traffic control, and road traffic information systems is an emerging research area for intelligent transport systems [12–15]. Most of the existing vehicle tracking systems are based on the video cameras [16]. Previous approaches of vision based multiple vehicles detection and tracking could be subdivided into four categories: frames difference and motion based methods [17–19], background subtraction methods [15, 20], and feature based methods [21, 22]. Meanwhile, a few camera-based datasets for vehicle detection and tracking come to light in recent years [23–25], which promote the research for ITS.

All previous multivehicle detection and tracking methods leverage images acquired by traditional frame-based cameras. Conventional cameras may suffer from various motion-related issues (motion blur, rolling shutter, etc.) which may impact performance for high-speed vehicles detection and tracking. Neuromorphic vision sensors are widely applied to robotics [26–29] and vehicles [30–32]. A few relevant neuromorphic vision datasets [33, 34] have been released in recent years, which facilitate the neuromorphic vision application for object detection and tracking. Recent years also witness the various applications for detection and tracking tasks with neuromorphic vision sensor such as feature tracking [35, 36], line tracking [37], and microparticle tracking [38].

However, there is still a lack of neuromorphic datasets and relevant applications with neuromorphic vision sensors in intelligent transport system, albeit such sensors instinctively enjoy superiority in high-speed motion recording, which can correspondingly facilitates the high-speed multiple vehicle detection and tracking in ITS systems. Thus, it is meaningful to apply neuromorphic vision techniques to ITS systems.

3. Neuromorphic Vision Sensor and Dataset

3.1. Neuromorphic Vision Sensor. A short description of different versions of neuromorphic vision sensors is provided in this section which is also mentioned in [11]. The purpose is to encourage researchers who are not familiar with neuromorphic vision sensors to explore the potential applications

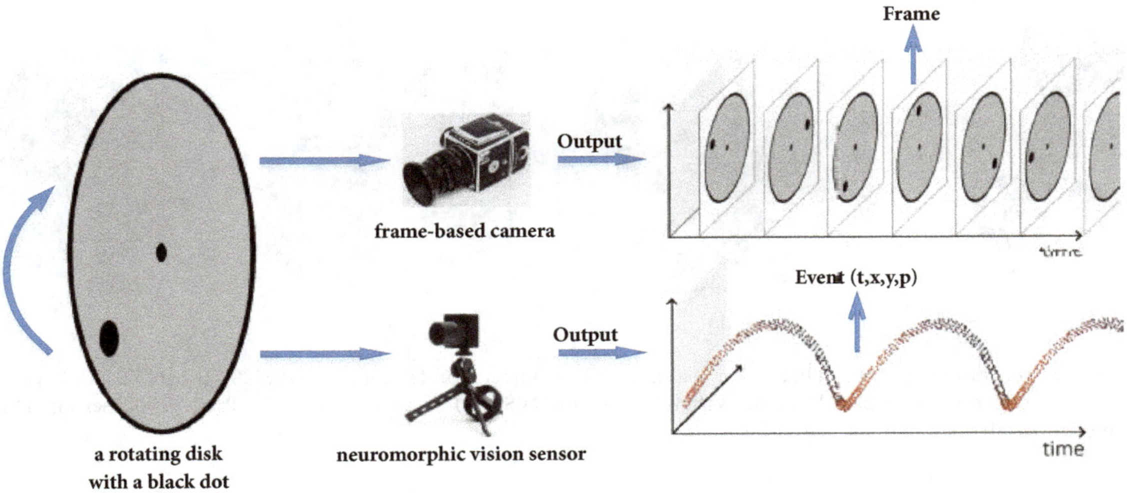

FIGURE 1: Visualization of the output from a neuromorphic vision sensor and a standard frame-based camera when facing a rotating disk with a black dot. Comparing to conventional frame-based camera which transmitted complete images at fixed latency, the neuromorphic vision sensor [1] emitted events individually and asynchronously at the time they occur. This figure is adopted from [2].

in the intelligent system. Figure 2 shows different versions of neuromorphic vision sensors.

Dynamic Vision Sensor (DVS). Dynamic Vision Sensors (DVS) are a new generation of cameras that are sensitive to intensity change, more specifically, to intensity logarithmic change. A DVS pixel typically generates one to four events (spikes) when an edge crosses it. DVS output consists of a continuous flow of events (spikes) in time, each with submicrosecond time resolution, representing the observed moving reality as it changes, without waiting to assemble or scan artificial time-constrained frames (images).

Embedded Dynamic Vision Sensor (eDVS). For embedded systems in mobile robotics such as unmanned aerial vehicle, an USB interface to transmit raw events is not desirable, nor is a desktop PC for event processing acceptable. For this purpose, a small embedded DVS (eDVS) is developed consisted of a DVS chip and a compact 64MHz 32bit microcontroller directly connected to the DVS chip.

Miniature Embedded Dynamic Vision Sensor (meDVS). The miniaturized version of the eDVS(meDVS) has minimum size (18cm×18cm) and lightest weight (2.2g) of DVS so far. The typical power consumption is 300mW. The strengths of meDVS make it desirable to any applications on the limited storage, bandwidth, and low latency of the on-board embedded system of the intelligent system.

Dynamic and Active Pixel Vision Sensor (DAVIS). In this paper we use a new neuromorphic vision sensor which is named the Dynamic and Active Pixel Vision Sensor (DAVIS) [39]. The model DAVIS240 camera has a higher resolution of 240x180, higher dynamic range, and lower power consumption and allows a concurrent readout of global shutter image frames, which are captured using the same photodiodes as for the DVS event generation. In this work, we only use the event data.

3.2. Dataset and Benchmark. We present a labeled dataset for the evaluation of an online multivehicle detection and tracking system in ITS domain. The raw event data are collected by a neuromorphic vision sensor which is mounted on the bridge in a highway scenario. The neuromorphic vision sensor used in this paper is called dynamic and active pixel sensor (DAVIS) with a model No. DAVIS240C. We have labeled three event sequences in this work. The first event sequence (named EventSeq-Vehicle1) is of length $45.4s$ having $110.7Mevents$ and on average contains $2,438Keps$ (Kilo events per second). The second event sequences (named EventSeq-Vehicle2) is of length $32.4s$ with $79.4Mevents$ and on average contains $2,450Keps$. The third event sequence (named EventSeq-Vehicle3) is of length $21.8s$ having $53.4Mevents$ and on average contains $2,450Keps$. The vehicles are moving in both the directions, i.e., towards and away from the camera in multiple lanes. The vehicles in the dataset range from small cars to the trailers and trucks, which makes the dataset diverse and challenging in nature.

We manually annotated all the vehicles' positions and unique identity in three event sequences using the openly available video annotation tool called ViTBAT [40]. For annotation, the video was created from the binary events data file. We accumulated events data into video frames at three different time intervals: $10ms$, $20ms$, and $30ms$. The description and data format of our dataset can be seen from Table 1.

4. Online Multitarget Detection and Tracking

We illustrate our multiobject tracking-by-clustering system in this section. In contrast to traditional object detection

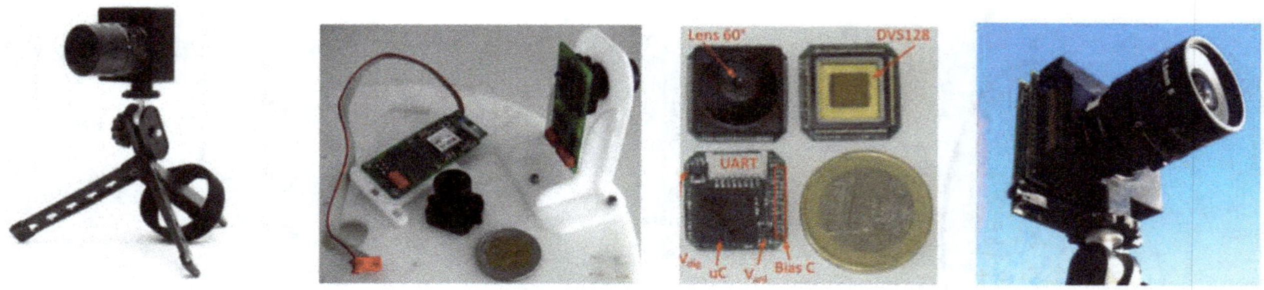

(a) (b) (c) (d)

FIGURE 2: Different versions of neuromorphic vision sensors: (a) Dynamic Vision Sensor 128 (DVS128). (b) Embedded Dynamic Vision Sensor (eDVS). (c) Miniature Embedded Dynamic Vision Sensor (meDVS). (d) Dynamic and Active Pixel Vision Sensor (DAVIS). This figure is adopted from [11].

TABLE 1: The standard file format in our benchmark. In event.txt, timestamp recodes the timestamp of each raw event, x and y are the coordinates of the event's origin, and p is the event's polarity. In det.txt, box_x and box_y are x and y coordinates of the bounding box and box_width and box_height are the width and height of each bounding box. The events which fall into a bounding box from start timestamp to end timestamp are accumulated to a cluster. Each cluster has an ID. The items' definitions in gt.txt and track.txt are similar to det.txt file, except that the gt.txt shows the ground truth and track.txt shows the result of tracking.

File name	Description	Format
event.txt	One event per line	(timestamp, x, y, p)
det.txt	One detection measurement per line	(starttimestamp, endtimestamp, box_x, box_y, box_width, box_height)
gt.txt	One ground-truth measurement per line	(starttimestamp, endtimestamp, ID, box_x, box_y, box_width, box_height)
track.txt	One tracking measurement per line	(starttimestamp, endtimestamp, ID, box_x, box_y, box_width, box_height)

approaches, we generate our object hypothesis directly from the measurements with a classic clustering method. The advantage is that we can skip the background modeling step (dynamic foreground segmentation), as most events transmitted by the dynamic vision sensor are generated by dynamic objects. In order to estimate the states of the actual objects, we integrate an online multitarget tracking method into our system. It is our opinion that only highly effective and online tracking methodology can take full advantage of neuromorphic vision cameras.

4.1. Vehicle Detection by Clustering. As neuromorphic sensors only transmit relative light intensity changes for each pixel, methods using appearance features, such as color and texture as input, cannot be utilized. Clustering methods, on the contrary, are very suitable for this situation. Hence we present three different clustering algorithms in this section, which do not depend upon the prior knowledge of the number and shape of the clusters. In addition, only dynamic information in the form of sparse streams of asynchronous time-stamped events can be gained from neuromorphic vision sensors. In order to arrive at a meaningful interpretation and make the most of neuromorphic vision sensors' advantages, it is necessary to accumulate event streams before applying clustering algorithms. We accumulate event data for different time intervals (10ms, 20ms, and 30ms), making it synchronized and more informative, after which three classic clustering approaches, MeanShift [4], DBSCAN [5] and WaveCluster [6], are carried out and compared. The following subsections illustrate these clustering approaches briefly.

Detection by MeanShift [11]. Estimation of the gradient of a density function via MeanShift and the iterative mode seeking procedure was developed by Fukunaga and Hostetler in [41]. The mean-shift algorithm has been exploited in low level computer vision tasks, including image segmentation, color space analysis, face tracking, etc., by reason of its properties of data compaction and dimensionality reduction [4]. Specifically, the mean-shift algorithm considers the input as a probability density function and the objective of the algorithm is to find the modes of this function. These modes represent the centers of the discovered clusters. The input points are fed to the kernel density estimation and then the gradient ascent method is applied to the density estimate. The density estimation kernel uses two inputs: the total amount of points and the bandwidth or the size of the window. The main disadvantage of the mean-shift algorithm lies in its iterative nature and difficulty of filtering out noise.

Detection by DBSCAN [11]. DBSCAN uses density based spatial clustering for applications with noise. For each point *p*, the associated density is calculated by counting the number of points in a search area of specified radius, *Eps* (the maximum radius of the neighbourhood from point), around the point. The points with density higher than the specified threshold value, MinPts (the minimum number of points required to form a dense region), are classified as core points while the rest are classified as noncore points. A cluster is yielded if p is a core point; otherwise, if p is a border point, then no point is density-reachable from p and DBSCAN takes the next point from the database [5]. The main advantage of DBSCAN is that it can find the clusters of arbitrary shapes.

Detection by WaveCluster. The basic idea of WaveCluster is to quantize the feature space of the image firstly and then apply discrete wavelet transform on it, after which we can find the connected components (clusters) in the subbands of transformed feature space [6]. For best clustering result, the quantization scale as well as the component connection algorithm should be applied according to the raw data. In the context of this paper, the accumulated event data can be regarded as 2-dimensional data. With selected interval m in each dimension, we can now divide the event data into m^2 grids, and each grid i contains N_i data point. Considering the multiresolution property of wavelet transform, different grid sizes can be adopted at different scales of transform. In the second step of WaveCluster algorithm, discrete wavelet transform will be applied on the quantized feature space [6]. Afterwards, a new feature space T_k is acquired. We can also filter out the noise in T_k with a selected threshold. With the new set of units T_k, connected components in the transformed feature space can be detected as clusters. Details of the algorithm can refer to [6].

4.2. Online Multitarget Tracking. In order to make full use of the advantages of event data, we have chosen four classic tracking algorithms, which are relatively small in computation and highly effective. Our online multitarget tracking is a simple and standard method which is widely explored in traditional camera-based multiobject tracking [42]. As the event data have no texture information, we use the bounding box overlap as a simple association metric for the data association problem. All these tracking algorithms are briefly described in the following sections.

Tracking by SORT [11]. We utilize a single hypothesis tracking methodology with standard Kalman filter and data association using Hungarian method [7]. In order to assign detected clusters to existing targets, each target's geometry and image coordinates are estimated by predicting its new state in the current frame. The cost matrix for each detected cluster and each existing target is calculated as the intersection over union distance (IOU). The Hungarian algorithm is used to optimally solve the assignment problem. We also define a minimum IOU to reject assignments where the detected cluster to target cluster overlap is less than the threshold. When a new cluster enters into the camera field of view or when an existing target leaves the camera view, target identities get updated, either by adding new IDs or by according deletion. The same methodology for tracking has been used in this work as presented in [7]. Instead of solving for detection for tracking in a global assignment problem, we choose an early deletion of lost targets policy, which prevents unbounded growth of the number of trackers.

Tracking by GM-PHD. GM-PHD filter is a recursive algorithm which jointly estimates the time-varying number of targets and their states from the observation sets in the presence of data association uncertainty, noise, false alarms, and detection uncertainty. The algorithm models the respective collection of targets and measurements as random finite sets and applies the probability hypothesis density (PHD) recursively for posterior intensity propagation, which is basically the first order-statistic of the random finite set in time. With linear and Gaussian assumptions, the target dynamics and birth process and the posterior intensity at any time step are considered to be Gaussian mixture. The recursions with number of Gaussian components management increase the efficiency. In tracking world, the intensity is also known as probability hypothesis density [8]. The further mathematical insights into the algorithm and its recursive linear Gaussian version can be studied in [8]. As stated in the previous section, the birth model for the targets is chosen to be linear in this work, which also stands for this and upcoming approaches for tracking.

Tracking by GM-CPHD. In probability hypothesis density (PHD) filter, the posterior intensity of the random finite set of targets is propagated, recursively. In cardinalized PHD (CPHD) filter, both the posterior intensity and posterior cardinality distribution are propagated jointly, hence making it a generalization of PHD recursion. The accuracy and stability are increased by incorporating the cardinality information [9]. This work is basically the implementation of closed-form solution to CPHD recursion under the assumption of linear Gaussian target dynamics and birth model. The algorithm can also be extended to nonlinear models using linearization and unscented transformation techniques. While comparing with standard PHD filter, CPHD filter not only side steps the need of data association task in conventional tracking methods but also improves the accuracy of the individual target state estimates and the variance of the estimated number of targets [9].

Tracking by PDAF. The probabilistic data association filter (PDAF) computes the probabilities for target being tracked for each valid measurement. This measurement origin uncertainty is accounted by this probabilistic or Bayesian information. As the linear models for the targets birth dynamics and measurement equations are assumed, therefore, the developed PDAF algorithm is based on Kalman filter. PDAF works on the validated measurements at the current time and for each measurement, an association probability is calculated for computing the weight of current measurement in a combined innovation. This combined innovation helps in updating the estimation of the state. And finally, the state covariances are updated for computing the measurement origin uncertainty [10]. The detailed mathematical insights into the PDAF algorithm with its extensions can be studied from [10].

TABLE 2: Evaluation metrics for vehicle detection, TP is the total number of true positives, FP is the total number of false positives, and GT indicates ground truth.

Metrics	Better	Perfect	Description
Precision	higher	100%	Ratio of TP / (TP+FP)
Recall	higher	100%	Ratio of correct detections to total number of GT boxes

5. Experiments and Results

We evaluate the performance of various tracking-by-clustering implementations on our dataset. The evaluation results are provided by following the standard MOT challenge metrics [43]. We analyze the performance and runtime of the three classical clustering algorithms, as well as the four tracking algorithms for multivehicle tracking-by-clustering task, where stream inputs are accumulated at different intervals (10ms, 20ms, and 30ms time intervals).

5.1. Metrics. For performance evaluation, we follow the current evaluation protocols for visual object detection and multiobject tracking. Although these protocols are designed for frame-based vision sensors, they are still suitable for quantitative evaluation of our tracking method. In this work, we accumulate events to frames in different time intervals. In this work we have two evaluation metrics (see Table 2) which are defined in [44].

Since our detection results from clustering methods have no probability score, we are not able to provide the mean precision to summarize the shape of the precision/recall (ROC) curve which is widely adopted in object detection evaluation in computer vision. The evaluation metrics for multivehicle tracking used in this work is defined in [43], well-known as the MOT challenge metrics. Evaluation scripts are available on MOT Challenge official website (https://motchallenge.net). More details are as follows:

(i) MOTA(↑): Multiple Object Tracking Accuracy. This measure combines three error sources: false positives, missed targets, and identity switches.

(ii) MOTP(↑): Multiple Object Tracking Precision. The misalignment between the annotated and the predicted bounding boxes.

(iii) MT(↑): mostly tracked targets. The ratio of ground-truth trajectories that are covered by a track hypothesis for at least 80% of their respective lifespan.

(iv) PT(↑): number of partially tracked trajectory.

(v) ML(↓): mostly lost targets. The ratio of ground-truth trajectories that are covered by a track hypothesis for at most 20% of their respective lifespan.

(vi) FP(↓): the total number of false positives.

(vii) FN(↓): the total number of false negatives (missed targets).

(viii) IDs(↓): the total number of identity switches.

(ix) FM(↓): the total number of times a trajectory is fragmented (i.e., interrupted during tracking).

5.2. Performance Evaluation. In this section, we report the performance and runtime of the selected approaches for multivehicle detection and tracking. Firstly, we compare the detection performance of the three clustering methods (DBSCAN, MeanShift, and WaveCluster). Then the impacts of different sampling time intervals on detection results are studied. Finally, the tracking performance and runtime for different tracking methods are evaluated.

5.2.1. Online Multivehicle Detection. In this work, event data are considered as pure 2D point data. The clustering technique is applied to generate object proposals. The event data for different time intervals (10ms, 20ms, and 30ms) are accumulated and can be seen in Figure 3. It is straightforward to see that clusters of event data reflect moving vehicles. The noise events surrounding each cluster are mainly generated by the environmental changes and sensor noise. Therefore, prior to generating object hypotheses, a background activity filtering step is performed to filter out the noise from the events. For each event, background activity filter checks whether one of the 8 (vertical and horizontal) neighbouring pixels has had an event within the last "us_Time" microseconds. If not, the event being checked will be considered as noise and removed. In other words, whether a new event is considered as "signal" or "noise" is determined by whether there is a neighbouring event generated within a set interval (us_Time). Figure 4 shows the accumulated events frame before and after the application of activity filter.

Figure 5(a) shows DBSCAN clustering results. For DBSCAN, the search radius, *Eps*, is chosen as 5 and the density, MinPts, is chosen as 10. The points with density higher than the specified threshold value, MinPts, are classified as core points while the rest are classified as noncore points. Those noncore points are also classified as noise points. Seven clusters including noise events have been detected. Figure 5(b) shows the MeanShift clustering results, with a chosen bandwidth of 20. The MeanShift algorithm successfully detected six clusters. And we can see many clusters were detected using WaveCluster from Figure 5(c). MeanShift divides many noises and objects into one cluster, and WaveCluster treats many noises as a single cluster. Their common shortcoming is that they cannot distinguish between object (here is car) and noise well.

The detection performance is assessed by clustering approach in terms of the metrics of recall and precision. The evaluation of DBSCAN, MeanShift, and WaveCluster on our neuromorphic data with different time intervals is shown in Table 3. We can see that the performance of clustering algorithms increases significantly from 10ms time interval to 20ms time interval, which shows that detection-by-clustering methods, used in this work, perform better with more events per time interval. But, from the performance of 30 ms interval, we can also know that, with the accumulation of events, more and more noise points appear, and the accuracy of the detection algorithm decreases. The results indicate

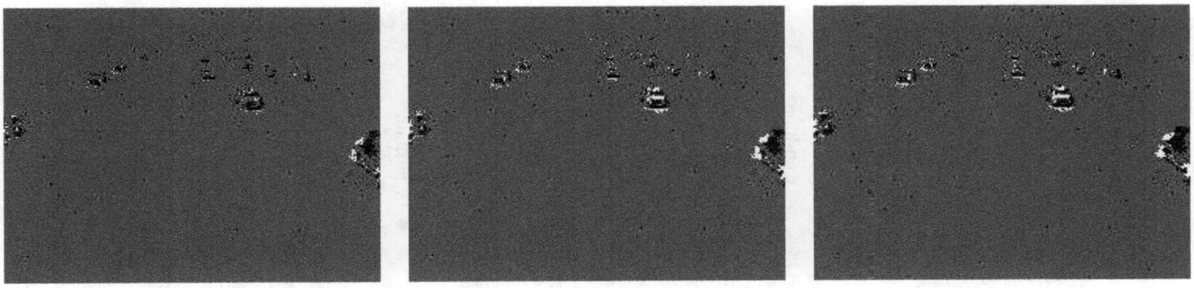

(a) (b) (c)

FIGURE 3: (a) Events accumulated in 10ms time interval. (b) Events accumulated in 20ms time interval. (c) Events accumulated in 30ms time interval.

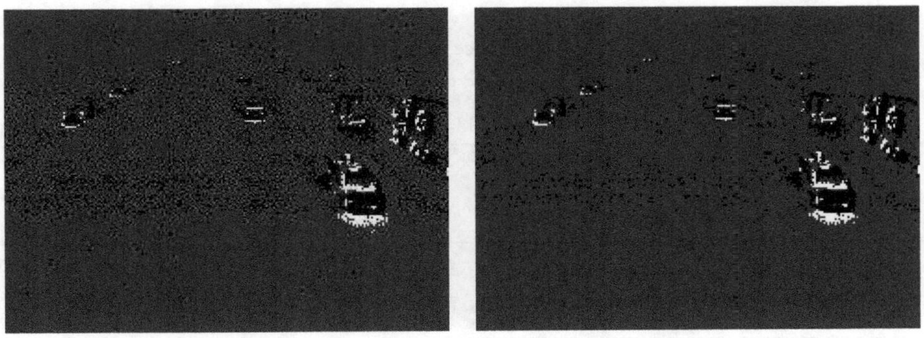

(a) (b)

FIGURE 4: Background activity filter. (a) Events accumulated in 20ms time interval prior to filtering. (b) Events accumulated in 20ms time interval after filtering. This figure is adopted from [11].

that the detection performance is highly dependent on the number of events during the accumulated time. This points out an alternative way of accumulating a constant number of events instead of constant time intervals may increase the robustness of our detection-by-clustering approach. Among the three algorithms, MeanShift performs the worst. The reason behind it is that the density estimation of MeanShift is affected by the random noise from DAVIS. Secondly, since the MeanShift is aiming at globular clustering, it may merge some small targets when detecting as illustrated in Figure 5. Lastly, the kernel bandwidth and window size remain the same in detection, resulting in bad performance when detecting fast moving and size-changing vehicles in our scenario. From Table 3, the detection accuracy of WaveCluster is higher overall. But, the detection effect of WaveCluster at 10ms time interval is relatively poor, noise cannot be eliminated, and the detection performance is greatly affected by the number of events. In order to make the tracking algorithms get a better performance in different time intervals of the three datasets. We choose DBSCAN as detection algorithm used for comparing the tracking results.

5.2.2. Online Multivehicle Tracking. In this part, the four tracking algorithms have been implemented, i.e., simple online and real-time tracking (SORT), GM-PHD filter, GM-CPHD filter, and the PDA filter. The tracking performance for the four trackers applied to the three vehicle sequences datasets is presented below.

Figure 6 shows the tracking results of SORT, GM-PHD filter, GM-CPHD filter, and the PDA filter using a series of input events for 20ms time interval. It can be seen from the continuous figures, such as Figures 6(a), 6(b), and 6(c), that our tracking algorithms perform better with moving vehicles when a new vehicle enters into the camera field of view or when an existing target leaves the camera view, target identities get updated, either by adding new IDs or by according deletion.

If any detected target in the current event frame had an overlap with an untracked detected target in previous frame, it would be registered with a new ID. As can be seen from Figure 6, most of the targets are well tracked. Especially, in the same continuous time interval, SORT tracks 29 targets, which is the largest number of targets tracked in the four algorithms. And the GM-PHD tracks 19 targets, followed by GM-CPHD with 15. However ID switching or target missing errors can also be witnessed from Figures 6(d)–6(l), PDAF performs the worst in terms of the problems of the number of targets, ID assignment and missed targets. And it can be

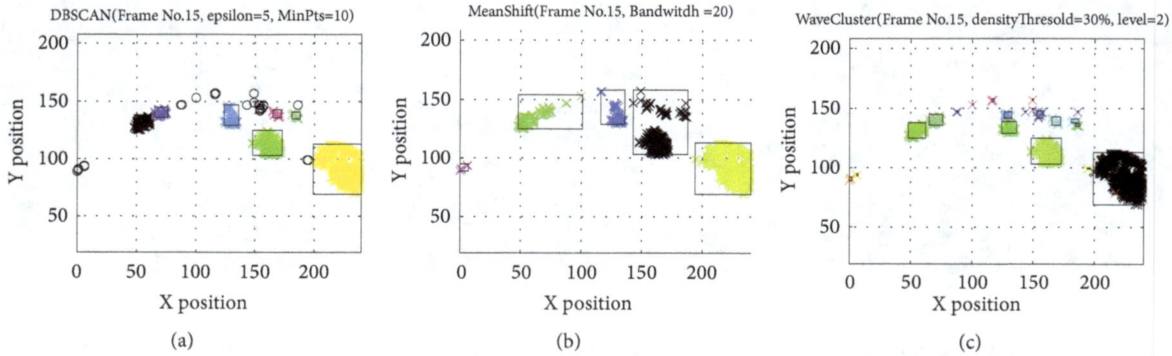

FIGURE 5: Detection-by-clustering results of (a) DBSCAN clustering applied to the events data accumulated at 20ms time interval. (b) MeanShift clustering applied to the events data accumulated at 20ms time interval. (c) WaveCluster clustering applied to the events data accumulated at 20ms time interval (best viewed in color).

TABLE 3: The performance of clustering algorithm.

Dataset	Tis	DBSCAN		MeanShift		WaveCluster	
		Recall	Precision	Recall	Precision	Recall	Precision
EventSeq-Vehicle1	10ms	53.1%	60.6%	44.5%	41.9%	44.4%	49.2%
EventSeq-Vehicle1	20ms	62.8%	64.5%	46.6%	40.7%	63.1%	64.4%
EventSeq-Vehicle1	30ms	61.9%	61.9%	44.3%	40.7%	61.8%	64.9%
EventSeq-Vehicle2	10ms	46.4%	51.9%	39.8%	38.0%	38.7%	41.4%
EventSeq-Vehicle2	20ms	52.3%	53.3%	38.3%	35.0%	51.9%	53.0%
EventSeq-Vehicle2	30ms	46.2%	47.2%	33.2%	33.5%	47.2%	52.1%
EventSeq-Vehicle3	10ms	41.1%	54.4%	35.1%	40.1%	32.1%	41.8%
EventSeq-Vehicle3	20ms	49.7%	59.4%	35.5%	36.8%	49.6%	58.0%
EventSeq-Vehicle3	30ms	47.7%	55.6%	33.5%	35.3%	49.0%	57.5%

clearly seen from Figures 6(k) and 6(l) that the same target is given different ID at different times, indicating that target lost occurs. Hence, the performance of our algorithms shows the limitations of the tracking-by-clustering system to some extent.

Table 4 shows the tracking performance metrics, i.e., MOTA, MOTP, MT, PT, ML, FP, FN, IDs, and FM for all the four trackers, i.e., SORT, GM-PHD filter, GM-CPHD filter, and the PDA filter for each 10ms, 20ms, and 30ms time intervals fed with EventSeq-Vehicle1. As the tracking component is highly dependent on the detection results, the number of times an ID-switched (IDs) is pretty large due to the inconsistent detection results. From the overall tracking performance evaluation results in Table 4, the value of MOTA and MOTP for four tracking algorithms is relatively higher. After applying these frame-by-frame-based tracking approaches, it is not surprising that we get large number of false detection, missed detection, ID switch, and fragmentation (FM). One possible way to decrease the number of missed detection, ID switch, and fragmentation (FM) is replacing the simple association metric in this paper to a more informed metric including motion information; it is able to track objects through longer periods of occlusions and disappearances. Tables 5 and 6 present the tracking performance metrics for EventSeq-Vehicle2 and EventSeq-Vehicle3, which are not as good as that of EventSeq-Vehicle1. It is especially obvious in EventSeq-Vehicle2 with 30ms time interval, where the evaluation metrics of MOTA for tracking algorithms are very low. The main reason behind it lies in the occasional flash of huge amount of noise as shown in Figure 7, which would seriously obscure the tracking targets, resulting in periodic fluctuation in the performance of the algorithm. This "noise flash" phenomenon can attributed to the unstable working state of the sensor and variable environmental conditions. It also indicates that our three datasets are very representative and challenging. Such limitation of the neuromorphic vision sensor will also be discussed in Section 6.2.

As the first work of multitarget tracking based on neuromorphic vision sensor, we are not able to compare to state-of-the-art tracking algorithms. Instead, we provide our evaluation results as a baseline tracker for future neuromorphic vision based multiobject tracking methods.

Runtime. The experiment is carried out on a laptop with Intel Core™i7-6700HQ CPU with 2.60GHz quad core processor and 8.00 GB of RAM. Table 7 shows that the average FPS of the DBSCAN algorithm is 36, 17, and 8 for 10ms, 20ms, and 30ms time intervals, respectively. The decreasing frame rate is due to the increased number of events in the density search area, resulting in a more iterative process. Of course, the runtime performance is related to the selection of algorithms; for example, WaveCluster has almost the same frame rate at

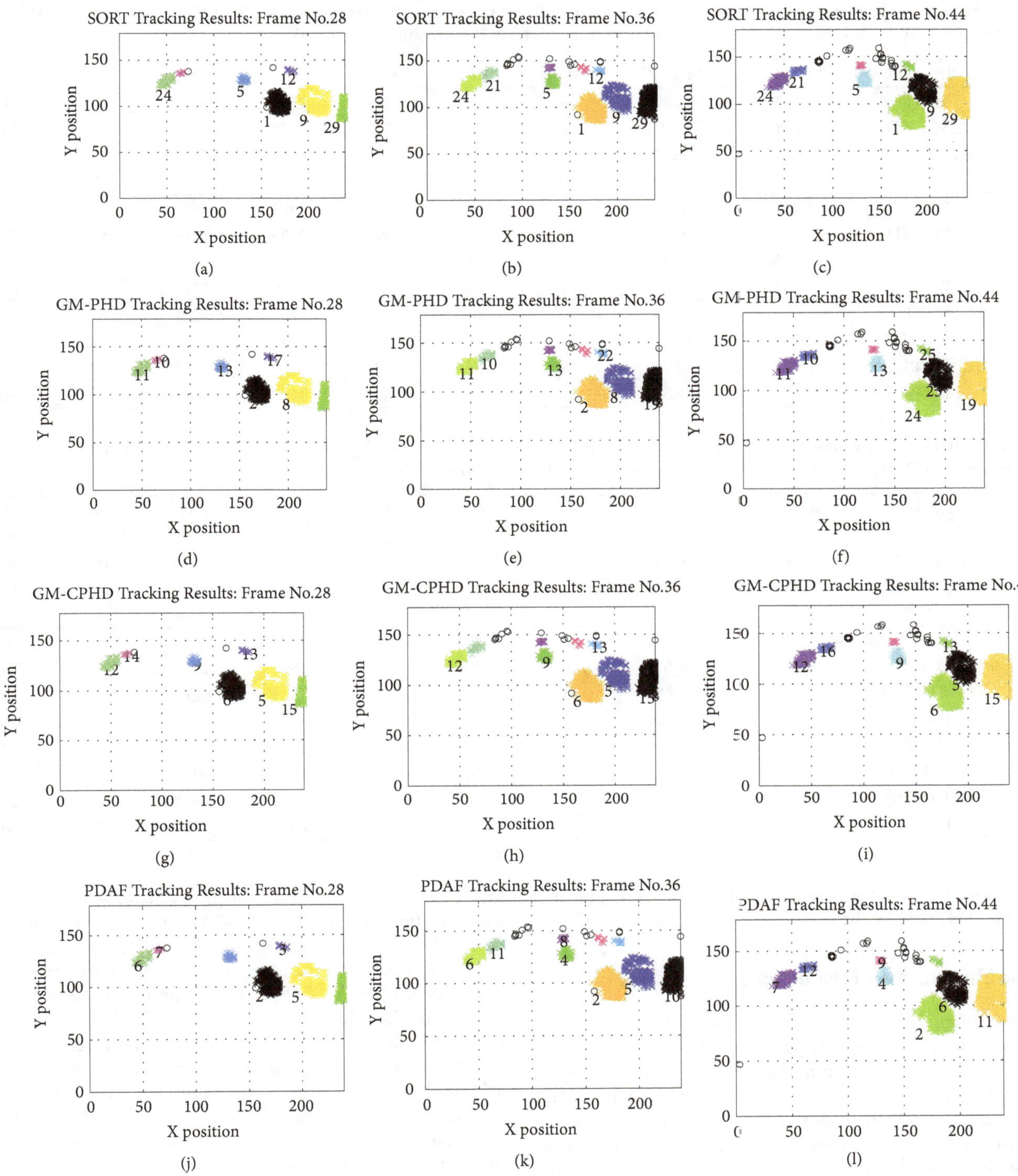

FIGURE 6: Tracking results from Frame 28 to Frame 44 for SORT, GM-PHD, GM-CPHD, and PDAF tracker. The numbers in black font near to cluster show the tracking ID (each object hypothesis is associated a tracker with a unique ID). The ID switching and lost IDs can also be witnessed from those sequences of input data (best viewed in color).

different time intervals. Additionally, in spite of the ordinary computer resources, MeanShift has a high running efficiency. For the tracking component, SORT is able to reach 552 FPS as shown in Table 8. Such a high frame rate indicates a promising application of the sensors. According to the experimental results, our tracking-by-clustering system can run at a rate of more than 110 Hz when our tracking algorithms is combined with efficient detection algorithms, such as MeanShift. In

Table 4: The tracking performance for EventSeq-Vehicle1 using different tracking methods (SORT, GM-PHD, GM-CPHD, and PDAF) with detection through DBSCAN clustering. The impact of time intervals from 10ms to 30ms was studied.

Tis	Tracker	MOTA↑	MOTP↑	MT↑	PT↑	ML↓	FP ↓	FN↓	IDs↓	FM↓
10ms	SORT	36.2%	69.2%	8	79	20	2891	16369	146	1302
10ms	GM-PHD	24.0%	69.1%	1	85	21	4924	16649	1541	4097
10ms	GM-CPHD	21.1%	69.2%	3	89	15	7480	15621	900	3616
10ms	PDAF	20.9%	69.1%	0	86	21	5653	18228	158	4678
20ms	SORT	35.0%	70.2%	18	71	18	2905	6893	92	444
20ms	GM-PHD	35.1%	70.6%	18	70	19	2523	7019	323	770
20ms	GM-CPHD	25.7%	70.5%	12	75	20	3974	7180	152	716
20ms	PDAF	24.5%	70.4%	4	80	23	3576	7815	95	1371
30ms	SORT	28.5%	70.4%	12	69	26	1950	5190	94	265
30ms	GM-PHD	23.6%	70.8%	14	67	26	2323	5224	190	478
30ms	GM-CPHD	18.3%	70.7%	8	76	23	2870	5259	135	481
30ms	PDAF	19.3%	70.5%	1	76	30	2402	5701	66	900

Table 5: The tracking performance for EventSeq-Vehicle2 using different tracking methods (SORT, GM-PHD, GM-CPHD, and PDAF) with detection through DBSCAN clustering. The impact of time intervals from 10ms to 30ms was studied.

Tis	Tracker	MOTA↑	MOTP↑	MT↑	PT↑	ML↓	FP ↓	FN↓	IDs↓	FM↓
10ms	SORT	24.4%	70.2%	3	53	29	2170	14929	183	942
10ms	GM-PHD	13.4%	69.4%	0	60	25	3807	14994	1000	2524
10ms	GM-CPHD	7.8%	69.7%	2	69	14	7452	13084	528	2524
10ms	PDAF	13.8%	69.8%	0	57	28	4403	15206	100	3050
20ms	SORT	5.7%	68.1%	7	49	28	3304	7393	81	331
20ms	GM-PHD	15.6%	70.6%	11	52	21	2839	6524	290	729
20ms	GM-CPHD	11.3%	70.6%	10	57	17	3824	6196	118	655
20ms	PDAF	11.5%	70.5%	4	58	22	3091	6948	76	995
30ms	SORT	0%	67.3%	3	45	37	2069	5496	53	183
30ms	GM-PHD	7.6%	70.3%	5	50	30	1888	5001	149	328
30ms	GM-CPHD	-0.7%	70.1%	4	57	24	2872	4694	100	389
30ms	PDAF	5%	69.9%	3	55	27	1969	5216	49	542

comparison, DeepSort method [45] only reaches a runtime speed of 40 Hz despite the use of high performance GPU.

6. Conclusion and Discussion

6.1. Conclusion. In this paper, the first neuromorphic vision based multivehicle detection and tracking system in ITS is proposed. We provide our datasets and approaches as a baseline tracker for future neuromorphic vision based multiobject tracking methods. A variety of algorithms to perform the tracking task are presented, of which different combinations can be chosen for different accuracy and rate requirements. Hopefully, our preliminary study can motivate further research in this field, considering that the sparse stream of event data from the sensor captures only motion and salient information, which is perfect for the intelligent infrastructure systems. The proposed event-based online multiple target tracking-by-clustering system utilizes strikingly simple algorithms while it achieves good detection and tracking performance with respect to runtime requirement.

Specifically, three clustering algorithms, i.e., DBSCAN, MeanShift, and WaveCluster, were explored to deal with the sparse data from neuromorphic sensor. After studying the detection results, the DBSCAN was selected for further detection stage due to its more robust and accurate outcome. Based on the detection results from DBSCAN, four different trackers were studied and their results were compared. The selected trackers were SORT, GM-PHD filter, GM-CPHD filter, and the PDA filter. From the experimental results, the tracking algorithm combined with DBSCAN can achieve higher accuracy, while combined with MeanShift can achieve higher frame rate of more than 110Hz. Different combinations of algorithms can be applied depending on different requirements of accuracy and real-time performance.

6.2. Discussion. To the best of our knowledge, the presented system is the first application of neuromorphic vision sensor on ITS which makes it well suited as a baseline, allowing for new researcher to work on intersection of the neuroscience and intelligent system. In our future work, different event encoding methods will be tried, adaptive algorithms will be explored, and the benchmark will be extended to pedestrian detection and tracking. Different filters, other than the basic

TABLE 6: The tracking performance for EventSeq-Vehicle3 using different tracking methods (SORT, GM-PHD, GM-CPHD, and PDAF) with detection through DBSCAN clustering. The impact of time intervals from 10ms to 30ms was studied.

Tis	Tracker	MOTA↑	MOTP↑	MT↑	PT↑	ML↓	FP ↓	FN↓	IDs↓	FM↓
10ms	SORT	24.6%	69.5%	1	34	24	1258	10838	109	611
10ms	GM-PHD	12.9%	68.9%	1	42	16	2484	10956	666	1903
10ms	GM-CPHD	12.5%	69.1%	2	41	16	3844	9957	363	1646
10ms	PDAF	13.8%	69.0%	0	38	21	2710	11173	70	2166
20ms	SORT	10.1%	69.3%	5	33	21	1990	5225	64	235
20ms	GM-PHD	21.4%	70.1%	5	38	16	1475	4699	188	490
20ms	GM-CPHD	13.4%	70.4%	5	36	18	2218	4698	94	402
20ms	PDAF	17.4%	70.2%	1	40	18	1582	4948	59	707
30ms	SORT	4.0%	69.0%	1	32	26	1329	3815	41	141
30ms	GM-PHD	14.3%	70.3%	7	35	17	1199	3311	120	285
30ms	GM-CPHD	6.1%	70.4%	4	34	21	1712	3300	58	240
30ms	PDAF	12.9%	70.6%	3	34	22	1112	3548	46	406

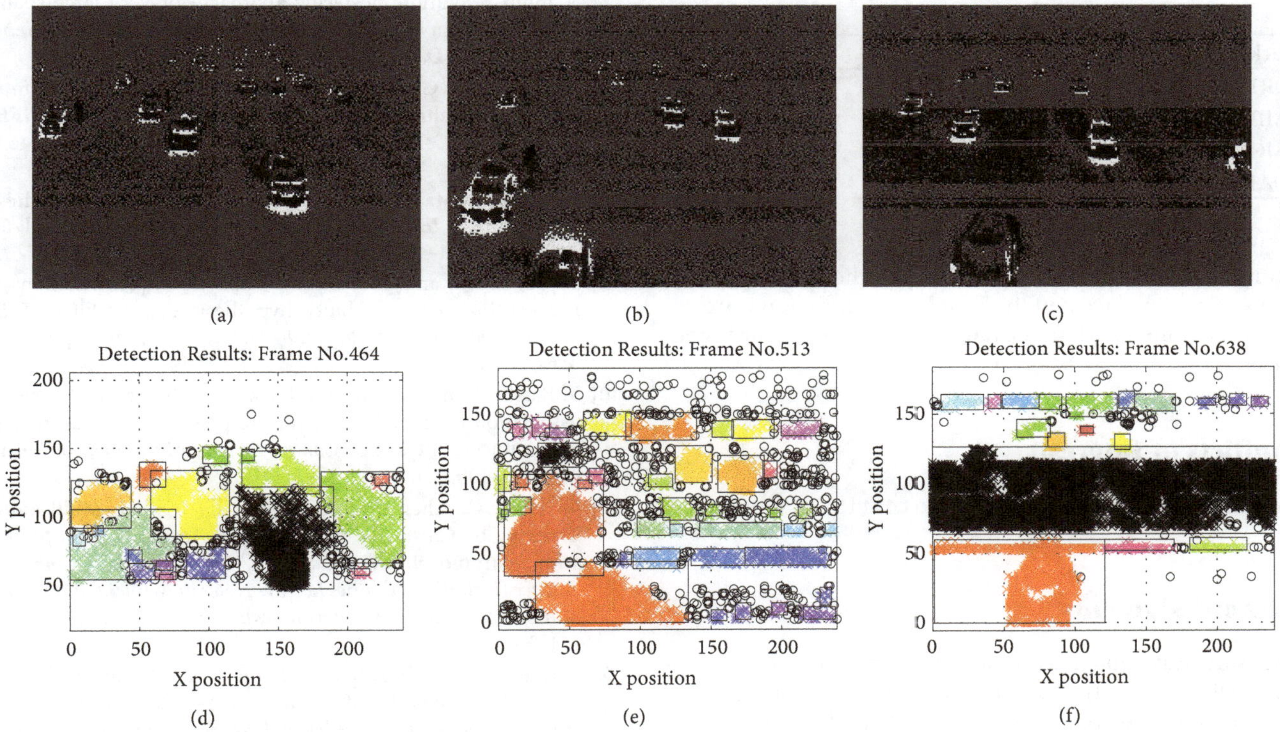

FIGURE 7: (a), (b), and (c) are the original images with the noise; (d), (e), and (f) are the results of the detection (best viewed in color).

activity filter, can be exploited to filter out the noise from the input data received from neuromorphic vision sensor. As a baseline, new approaches including recent deep learning based methods are supposed to improve the detection and tracking performance, especially the ability to identify vehicle types, such as trucks and cars and different pedestrians such as the elderly, children, etc.

Limitation. Admittedly, our algorithm still has some shortcomings. As can be seen in Figure 7, with noise becoming severer, the tracking system will make errors, such as missed detection, multiple targets detection as one, the false detection of noise points as the target, and so on. The reason mainly lies in the immaturity of neuromorphic sensor technology. To be specific, the inherent defects of the current neuromorphic sensor lead to instability in collecting event information, which affects the quality of data and thus degrades the performance of the algorithms. Hence, a development of the sensor is indispensable before wide application of it in intelligent transportation system (ITS). It is also significant to note that, in order to take full advantage of event data, completely new neuromorphic vision algorithms are

TABLE 7: The FPS for EventSeq-Vehicle1 using different detection methods (DBSCAN, MeanShift, and WaveCluster). The impact of time intervals from 10ms to 30ms was studied.

Detector	Tis	FPS
DBSCAN	10ms	36
MeanShift	10ms	160
WaveCluster	10ms	17
DBSCAN	20ms	18
MeanShift	20ms	107
WaveCluster	20ms	19
DBSCAN	30ms	8
MeanShift	30ms	71
WaveCluster	30ms	15

TABLE 8: The FPS for EventSeq-Vehicle1 using different tracking methods (SORT, GM-PHD, GM-CPHD, and PDAF).

Tracker	FPS
SORT	552
GMPHD	3
GMCPHD	4
PDAF	46

required instead of extending existing methods of computer vision, taking into account the brand new information stream and the extremely high frame rate of the neuromorphic vision sensor.

Conflicts of Interest

The authors declare that they have no conflicts of interest.

Acknowledgments

This work was supported by the German Research Foundation (DFG) and the Technical University of Munich. Part of the research has received funding from the European Unions Horizon 2020 Research and Innovation Program under Grant Agreement no. 785907(HBP SGA2) and from the Shanghai Automotive Industry Sci-Tech Development Program under Grant Agreement no. 1838. The authors would like to thank Zhongnan Qu for technical assistance and the help of data acquisition.

References

[1] P. Lichtsteiner, C. Posch, and T. Delbruck, "A 128 × 128 120 dB 15 μs latency asynchronous temporal contrast vision sensor," *IEEE Journal of Solid-State Circuits*, vol. 43, no. 2, pp. 566–576, 2008.

[2] E. Mueggler, C. Bartolozzi, and D. Scaramuzza, "Fast event-based corner detection," in *Proceedings of the British Machine Vision Conference (BMVC)*, 2017.

[3] X. Lagorce, C. Meyer, S.-H. Ieng, D. Filliat, and R. Benosman, "Asynchronous event-based multikernel algorithm for high-speed visual features tracking," *IEEE Transactions on Neural Networks and Learning Systems*, vol. 26, no. 8, pp. 1710–1720, 2015.

[4] D. Comaniciu and P. Meer, "Mean shift: a robust approach toward feature space analysis," *IEEE Transactions on Pattern Analysis and Machine Intelligence*, vol. 24, no. 5, pp. 603–619, 2002.

[5] M. Ester, H.-P. Kriegel, J. Sander, and X. Xu, "A density-based algorithm for discovering clusters a density-based algorithm for discovering clusters in large spatial databases with noise," in *Proceedings of the Second International Conference on Knowledge Discovery and Data Mining*, pp. 226–231, AAAI Press, 1996, http://dl.acm.org/citation.cfm.

[6] G. Sheikholeslami, S. Chatterjee, and A. Zhang, "Wavecluster: A multi-resolution clustering approach for very large spatial databases in," in *Proceedings of the 24rd International Conference on Very Large Data Bases*, vol. 98, pp. 428–439, 1998.

[7] A. Bewley, Z. Ge, L. Ott, F. Ramos, and B. Upcroft, "Simple online and realtime tracking," in *Proceedings of the 2016 IEEE International Conference on Image Processing (ICIP)*, pp. 3464–3468, Phoenix, AZ, USA, September 2016.

[8] B. Vo and W. Ma, "The Gaussian mixture probability hypothesis density filter," *IEEE Transactions on Signal Processing*, vol. 54, no. 11, pp. 4091–4104, 2006.

[9] B. Vo, B. Vo, and A. Cantoni, "Analytic implementations of the cardinalized probability hypothesis density filter," *IEEE Transactions on Signal Processing*, vol. 55, no. 7, part 2, pp. 3553–3567, 2007.

[10] Y. Bar-Shalom, F. Daum, and J. Huang, "The probabilistic data association filter: estimation in the presence of measurement origin uncertainty," *IEEE Control Systems Magazine*, vol. 29, no. 6, pp. 82–100, 2009.

[11] G. Hinz, G. Chen, M. Aafaque et al., "Online Multi-object Tracking-by-Clustering for Intelligent Transportation System with Neuromorphic Vision Sensor," *Lecture Notes in Computer Science (including subseries Lecture Notes in Artificial Intelligence and Lecture Notes in Bioinformatics): Preface*, vol. 10505, pp. 142–154, 2017.

[12] S. R. E. Datondji, Y. Dupuis, P. Subirats, and P. Vasseur, "A Survey of Vision-Based Traffic Monitoring of Road Intersections," *IEEE Transactions on Intelligent Transportation Systems*, vol. 17, no. 10, pp. 2681–2698, 2016.

[13] J. C. Rubio, J. Serrat, A. M. López, and D. Ponsa, "Multiple-target tracking for intelligent headlights control," *IEEE Transactions on Intelligent Transportation Systems*, vol. 13, no. 2, pp. 594–605, 2012.

[14] X. Zhang, S. Hu, H. Zhang, and X. Hu, "A real-time multiple vehicle tracking method for traffic congestion identification," *KSII Transactions on Internet and Information Systems*, vol. 10, no. 6, pp. 2483–2503, 2016.

[15] J. Zhou, D. Gao, and D. Zhang, "Moving vehicle detection for automatic traffic monitoring," *IEEE Transactions on Vehicular Technology*, vol. 56, no. 1, pp. 51–59, 2007.

[16] S. Sivaraman and M. M. Trivedi, "Looking at vehicles on the road: a survey of vision-based vehicle detection, tracking, and behavior analysis," *IEEE Transactions on Intelligent Transportation Systems*, vol. 14, no. 4, pp. 1773–1795, 2013.

[17] R. Cucchiara, M. Piccardi, and P. Mello, "Image analysis and rule-based reasoning for a traffic monitoring system," in *Proceedings of the 1999 IEEE/IEEJ/JSAI International Conference on Intelligent Transportation Systems*, pp. 758–763, October 1999.

[18] M.-C. Huang and S.-H. Yen, "A real-time and color-based computer vision for traffic monitoring system," in *Proceedings of the 2004 IEEE International Conference on Multimedia and Expo (ICME)*, pp. 2119–2122, Taipei, Taiwan.

[19] W. Zhang, Q. M. J. Wu, and H. B. Yin, "Moving vehicles detection based on adaptive motion histogram," *Digital Signal Processing*, vol. 20, no. 3, pp. 793–805, 2010.

[20] B.-F. Wu, S.-P. Lin, and Y.-H. Chen, "A real-time multiple-vehicle detection and tracking system with prior occlusion detection and resolution," in *Proceedings of the 5th IEEE International Symposium on Signal Processing and Information Technology*, pp. 311–316, Greece, December 2005.

[21] B. Aytekin and E. Altuğ, "Increasing driving safety with a multiple vehicle detection and tracking system using ongoing vehicle shadow information," in *Proceedings of the 2010 IEEE International Conference on Systems, Man and Cybernetics, SMC 2010*, pp. 3650–3656, Turkey, October 2010.

[22] M. Betke, E. Haritaoglu, and L. S. Davis, "Real-time multiple vehicle detection and tracking from a moving vehicle," *Machine Vision and Applications*, vol. 12, no. 2, pp. 69–83, 2000.

[23] A. Geiger, P. Lenz, and R. Urtasun, "Are we ready for autonomous driving? the KITTI vision benchmark suite," in *Proceedings of the IEEE Conference on Computer Vision and Pattern Recognition (CVPR '12)*, pp. 3354–3361, Providence, RI, USA, June 2012.

[24] L. Wen, D. Du, Z. Cai et al., "UA-DETRAC: A new benchmark and protocol for multi-object detection and tracking," *Computer Vision and Pattern Recognition*, 2015.

[25] L. Leal-Taixé, A. Milan, I. Reid, S. Roth, and K. Schindler, "MOTChallenge 2015: Towards a benchmark for multi-target tracking," *Computer Vision and Pattern Recognition*, 2015.

[26] D. P. Moeys, F. Corradi, E. Kerr et al., "Steering a predator robot using a mixed frame/event-driven convolutional neural network," in *Proceedings of the 2nd International Conference on Event-Based Control, Communication, and Signal Processing, EBCCSP 2016*, Poland, June 2016.

[27] Z. Jiang, Z. Bing, K. Huang, G. Chen, L. Cheng, and A. Knoll, "Event-Based Target Tracking Control for a Snake Robot Using a Dynamic Vision Sensor," in *Neural Information Processing*, vol. 10639 of *Lecture Notes in Computer Science*, pp. 111–121, Springer International Publishing, Cham, Switzerland, 2017.

[28] G. Chen, Z. Bing, F. Roehrbein et al., "Efficient Brain-inspired Learning with the Neuromorphic Snake-like Robot and the Neurorobotic Platform," in *Proceedings of the IEEE Transactions on Cognitive and Developmental Systems*, p. 1, 2018.

[29] H. Blum, A. Dietmüller, M. Milde, J. Conradt, G. Indiveri, and Y. Sandamirskaya, "A neuromorphic controller for a robotic vehicle equipped with a dynamic vision sensor," in *Proceedings of the Robotics: Science and Systems*, Cambridge, MA, USA, July 2017.

[30] M. Litzenberger, A. N. Belbachir, and N. Donath, "Estimation of vehicle speed based on asynchronous data from a silicon retina optical sensor," in *Proceedings of the IEEE Intelligent Transportation Systems Conference (ITSC '06)*, pp. 653–658, Toronto, Canada, September 2006.

[31] M. Litzenberger, C. Posch, D. Bauer et al., "Embedded vision system for real-time object tracking using an asynchronous transient vision sensor," in *Proceedings of the 2006 IEEE 12th Digital Signal Processing Workshop and 4th IEEE Signal Processing Education Workshop, DSPWS*, pp. 173–178, USA, September 2006.

[32] A. I. Maqueda, A. Loquercio, G. Gallego, N. Garca, and D. Scaramuzza, "Event-based vision meets deep learning on steering prediction for self-driving cars in," in *Proceedings of the IEEE Conference on Computer Vision and Pattern Recognition*, pp. 5419–5427, 2018.

[33] J. Binas, D. Neil, S.-C. Liu, and T. Delbruck, "Ddd17: End-to-end davis driving dataset, arXiv preprint," *Computer Vision and Pattern Recognition*, 2017.

[34] Y. Hu, H. Liu, M. Pfeiffer, and T. Delbruck, "DVS benchmark datasets for object tracking, action recognition, and object recognition," *Frontiers in Neuroscience*, vol. 10, 2016.

[35] D. Tedaldi, G. Gallego, E. Mueggler, and D. Scaramuzza, "Feature detection and tracking with the dynamic and active-pixel vision sensor (DAVIS)," in *Proceedings of the 2nd International Conference on Event-Based Control, Communication, and Signal Processing, EBCCSP 2016*, Poland, June 2016.

[36] B. Kueng, E. Mueggler, G. Gallego, and D. Scaramuzza, "Low-latency visual odometry using event-based feature tracks," in *Proceedings of the 2016 IEEE/RSJ International Conference on Intelligent Robots and Systems, IROS 2016*, pp. 16–23, Republic of Korea, October 2016.

[37] L. Everding and J. Conradt, "Low-latency line tracking using event-based dynamic vision sensors," *Frontiers in Neurorobotics*, vol. 12, 2018.

[38] Z. Ni, C. Pacoret, R. Benosman, S. Ieng, and S. Régnier, "Asynchronous event-based high speed vision for microparticle tracking," *Journal of Microscopy*, vol. 245, no. 3, pp. 236–244, 2012.

[39] C. Brandli, R. Berner, M. Yang, S.-C. Liu, and T. Delbruck, "A 240 × 180 130 dB 3 μs latency global shutter spatiotemporal vision sensor," *IEEE Journal of Solid-State Circuits*, vol. 49, no. 10, pp. 2333–2341, 2014.

[40] T. A. Biresaw, T. Nawaz, J. Ferryman, and A. I. Dell, "ViTBAT: Video tracking and behavior annotation tool," in *Proceedings of the 13th IEEE International Conference on Advanced Video and Signal Based Surveillance, AVSS 2016*, pp. 295–301, USA, August 2016.

[41] K. Fukunaga and L. D. Hostetler, "The estimation of the gradient of a density function, with applications in pattern recognition," *IEEE Transactions on Information Theory*, vol. 21, no. 1, pp. 32–40, 1975.

[42] G. Chen, F. Zhang, D. Clarke, and A. Knoll, "Learning to track multi-target online by boosting and scene layout," in *Proceedings of the 2013 12th International Conference on Machine Learning and Applications, ICMLA 2013*, pp. 197–202, USA, December 2013.

[43] L. Leal-Taixé, A. Milan, I. D. Reid, S. Roth, and K. Schindler, "Motchallenge 2015: Towards a benchmark for multi-target tracking," *Computer Vision and Pattern Recognition*, 2015.

[44] M. Everingham, L. van Gool, C. K. I. Williams, J. Winn, and A. Zisserman, "The pascal visual object classes (VOC) challenge," *International Journal of Computer Vision*, vol. 88, no. 2, pp. 303–338, 2010.

[45] N. Wojke, A. Bewley and D. Paulus, "Simple online and realtime tracking with a deep association metric," in *Proceedings of the 2017 IEEE International Conference on Image Processing (ICIP)*, pp. 3645–3649, Beijing, China, September 2017.

9

Microscopic Congestion Detection Protocol in VANETs

Mushtaq Ahmad,[1] **Qingchun Chen**,[2] **and Zahid Khan**[1]

[1]*Southwest Jiaotong University, Chengdu, Sichuan 611756, China*
[2]*Guangzhou University, Guangzhou, Guangdong 510006, China*

Correspondence should be addressed to Qingchun Chen; qcchen@swjtu.edu.cn

Academic Editor: María Calderon

Effective transportation status surveillance imposes critical challenges for the Intelligent Transportation System (ITS) design. In this paper, the microscopic congestion detection protocol (MCDP) is proposed to make the vehicle-to-vehicle (V2V) communication capable of monitoring vehicle density and identifying traffic jam. By introducing transportation control domain in the existing network protocol header, each vehicle can count its neighbors and estimate the time spacing among vehicles. MCDP provides an infrastructure-less solution to the estimate of vehicle density, flow, and average velocity in a microscopically manner. Moreover, the safety speed limit is introduced to make each vehicle calculate its time to cover the intervehicle distance, such that every vehicle is able to assess the transportation congestion by comparing with some predefined safety time threshold. Monte Carlo simulations of the MCDP over four typical Chinese highways are presented to compare the MCDP scheme with the existing Green-Shield congestion detection scheme. In addition, real road traces generated by SUMO over NS2 are utilized to show the achieved performance in terms of throughput, end-to-end delay, and packet delivery rate (PDR) in comparison to DSR and AOMDV in IEEE 802.11p and IEEE 802.11ac scenarios. On the basis of all the results, we conclude that MCDP is an inexpensive transport congestion detection technique for Vehicular Ad Hoc Networks (VANETs).

1. Introduction

The ever-increasing traffic density requires more effective traffic control technique to avoid the serious traffic jam [1]. Intelligent transportation system (ITS) provides innovative transport and traffic management technique. So far, a number of research efforts have been devoted to traffic congestion detection in both infrastructure mode and infrastructure-free mode. And these protocols aim to monitor road traffic and to estimate vehicle speed, density, and the arrival time [2–4]. Basically, sensing devices such as induction loop detectors [5], infrared detectors [6], microwave radars [7], and video recording devices [8] can be utilized to monitor vehicles on the highway. However, the coverage of long highway with the aforementioned dedicated devices becomes too expensive to afford the huge installations of these sensor devices. Compared to the aforementioned infrastructure method, the vehicular ad hoc network (VANET) provides us a cost-effective infrastructure-free technique to support a variety of ITS applications, such as safety surveillance, road monitoring, traffic flow management, and vehicle density estimation [9, 10]. In the framework of the VANET, each vehicle acts as a sensor node to collect information for transportation congestion estimate. However, there are several technical problems that need to be addressed.

Basically, a reasonable information exchange mechanism among moving vehicles will be the basis to realize the traffic monitoring in the VANET framework. The direct adaptation of VANETs for traffic detection looks attractive, but it has to cope with challenges like bandwidth flooding and duplication [11], delay and inaccurate traffic evaluation [12], and the reliability problems [13]. In [2], the contents oriented communication (COC) protocol was proposed for traffic congestion and accident detection. Whenever a vehicle receives a packet, it calculates the congestion and estimates speed by itself. By exchanging the calculated results, the COC offers a feasible scheme to gather real-time position for congestion detection and speed estimation. Nonetheless, too many content exchanges in the COC protocol will consume high bandwidth. Efficient congestion detection (ECODE) protocol was proposed for VANET in [14] to evaluate traffic characteristics as well. However, ECODE uses multihop communications and geocast principles to collect and analyze vehicle data for each road section, which will lead to high bandwidth consumption, overhead, and complexity issues. How to design a cost-effective and efficient signaling mechanism among different vehicles to realize traffic monitoring is the first motivation of our work in this paper.

Intervehicle spacing on the highway is a critical parameter to assess traffic congestion status. Given the intervehicle spacing, the human driver can make an accurate decision to cope with the emergency situation. As an alternative, the vehicle density estimation is to know the number of vehicles per unit length of road or cell; here the cell can be any area defined by the specific clustering algorithm [15–17]. The density estimation of heterogeneous traffic was performed microscopically in [18] by assuming a fixed spacing between neighboring vehicles to simplify the model and to assess the impact of density on vehicle speed. By assuming fixed spacing, the congestion detection issues were addressed in [19–22]. Obviously, the fixed spacing assumption is not true and unrealistic. The Green-Shield car-following model was adopted in [23] for traffic congestion avoidance in single lane road. Basically, Green-Shield model is based on the macroscopic parameters like traffic density, flow, and average velocity. In this paper, the limitations of the fixed spacing among vehicles become another motivation of our work in this paper. In the proposed microscopic congestion detection protocol (MCDP) for VANETs, we considered more realistic intervehicle spacing, which is modeled as an exponential distribution.

In this paper, we propose the MCDP to reduce the computational cost by eliminating those unnecessary beacon messages exchange among vehicles to enable a cost-effective congestion detection scheme. In fact, the MCDP provides an infrastructure-less V2V approach for traffic congestion surveillance and driver assistance without installation of road side unit (RSU) or distributed sensors. The proposed MCDP will disseminate only the microscopic parameters associated with congestion detection through beacon messages and process the parameters microscopically (processing at each individual vehicle) to derive the intervehicle spacing and time headway for congestion detection, as well as speed estimation [24]. On the basis of the MCDP, the vehicles on the highway are capable of calculating the number of neighbors N_c and estimating the spacing coverage time/headway T_g. If T_g is less than the predefined safety threshold τ_0, the congestion is detected, and the message will be broadcast. It should be addressed that most of the existing congestion related ITS research works are some kind of physical layer or MAC layer mechanism, while the proposed MCDP protocol is a network layer technique. As a network layer mechanism, the MCDP logically divides VANET into subsections without installation of any physical component. Hence, the network layer approach is more cost-effective than the MAC/PHY layer techniques.

In order to assess the effectiveness of the proposed MCDP protocol, we consider three typical scenarios. Firstly, theoretical analysis of MCDP over selected single and multilane Chinese highways was presented to show the achieved performance. Secondly, the MCDP characteristics were assessed through temporal congestion detection and speed estimation analysis. Thirdly, selected communication capability assessment was conducted to show the achieved network performance of the MCDP. In this paper, we have conducted a comprehensive study that combines transportation and communications, and we aim to address interdependent issues. In practice, vehicles with congestion detection and speed estimation capabilities can facilitate a safer and more comfortable driving experience. Our initial experimental analysis reveals that V2V communication among vehicles helps to refine traffic flow throughput, because it reduces the driver's perception-reaction time (PRT), which allows higher speed and safer movement of vehicles on road. All the analysis in this paper confirms that the proposed MCDP provides us with an effective network layer technique to detect and manage traffic congestion, even in the multiple lanes scenarios.

The remainder of this paper is organized as follows. In Section 2, previous works related to congestion detection are briefly reviewed to highlight the challenges. In Section 3, the MCDP and its working philosophy are presented. Analysis and simulation results are presented in Section 4 and the paper is concluded in Section 5.

2. The State of the Art

Traffic evaluation always remains a concentrative point for the ITS. Many existing protocols in literature are available for traffic evaluation. In this paper, we focus on the VANET based congestion detection technology. Some VANET based congestion detection and traffic evaluation technologies are summarized in Table 1.

In addition to the protocols of ECODE [14] and COC [2], the voting protocol was proposed in [25] to estimate congestion level from neighboring vehicle characteristics. Each vehicle within transmission range disseminates its own information to neighbor vehicles, and the neighbor vehicles estimate congestion by comparing the current moving speed with the maximal allowed speed in that specific zone. The receiver vehicle cumulates the neighbor speed and votes for or against the conclusion of congestion. The congestion decision depends on the majority of votes, because it is possible that on the highway some vehicles may voluntarily move slowly. More specifically, the majority voting for slow-moving implies that highway is congested at that moment. If the other surrounding vehicles are moving at a relatively higher speed, we may arrive at the conclusion that this vehicle is voluntarily driving slowly. In contrast, if a vehicle is moving slowly over a particular road segment and the surrounding vehicles are traveling at approximately the same speed, one can conclude that congestion has occurred. Each region requires relatively high traffic density to confirm congestion decision; otherwise, congestion assessment will be less accurate.

The clustered area approach was proposed in [26], wherein the geographic area on the highway is subdivided into small managerial zones. The passing time of each zone in normal traffic condition can be defined in advance. Every vehicle measures its passing time with respect to every zone and compares it with the normal coverage time. Each

TABLE 1: Highway traffic monitoring protocols.

Protocol	Architecture	Transportation	Bandwidth	Accuracy
ECODE [14]	V2V	Directed density detection	High	Optimal
COC [2]	V2V	Density estimation	High	Good
Voting [25]	V2V	Speed detection	High	Low
Clustered Area[26]	V2V	Arrival time estimation	Medium	Low
SOTIS[27]	V2V	Vehicle Speed	High	Good
IFTIS[28]	V2V	Density detection	Low	Good
StreetSmart[3]	V2I	Traffic Speed	Low	Low

vehicle on highway exchanges its entering and leaving time with each other, which would become a prediction notice of congestion for ongoing vehicles. The frequent and large exchange of entering/exit messages with high transmission power is the primary issue in clustered area approach. Self-organizing the intervehicle network (SOTIS) [27] is another specialized approach, wherein every vehicle is assumed to be provisioned with an internal database of geographic information of highway and digital map. And all ongoing vehicles are assumed to exchange their locations, speeds, and road conditions with each other periodically. In this way, the forehead vehicle is able to inform the following vehicles about the road conditions. The main issues of the SOTIS approach are that now every vehicle is assumed to have proactive information about the highway, which will be utilized for comparison later on with the reactive information.

An infrastructure-free traffic information system (IFTIS) was proposed for vehicular networks in [28], which was the first approach in the literature to monitor road condition in a segment-wise manner. The primary advantage of the IFTIS approach is that it provides a protocol to assist the driver in selecting a noncongested lane. In the proposed IFTIS protocol, the investigated road will be segmented into overlapping location-based groups. Each group has a vehicle centralized to the cell called group leader, which has information about the whole group. The traffic density of each group is evaluated by the group leader and disseminate to the intersections. IFTIS is able to evaluate traffic density in both directions. Sometimes IFTIS experiences high density from one direction but low from other direction, and this may cause long end-to-end delays and decrease the packet delivery ratio, especially if the vehicles on the road are in the opposite direction to the desired destination. Secondly, the overlapping in clusters also reduces the accuracy of traffic evaluation across the road segment.

On the basis of clustering and epidemic communication strategy, the StreetSmart aims to find dynamic patterns and report to adjacent clusters [3]. The dynamic patterns are filtered to find out an unexpected status that is forwarded to the rest of vehicles. Each vehicle summarizes gathered statistics locally and concludes the road situation. Unfortunately, the decentralized nature of StreetSmart does not prevent the flash crowding effect nor contributes to the optimized efficiency at a global scale. After a sudden change in traffic conditions, it does not update the delay information in a timely manner. The low level of accuracy is the main limitation of the StreetSmart because a number of vehicles can get involved in multiple clusters due to the overlapping cluster areas. Other congestion detection schemes for VANETs include Virtual Sink [29], V2X [11], and Lattice [4].

A comparative study of two approaches for road traffic density estimation from traffic video scenes was presented in [30]. Both the extracted microscopic parameters (i.e., individual vehicle motion parameters) and the macroscopic parameters (i.e., global motion parameters) are applied in classifiers to enable classification of light, medium, and heavy road traffic status. It is shown that a very high accuracy can be achieved by using the traffic video classification. However, this kind of traffic monitoring is dependent on the widely deployed traffic surveillance camera, which may become too expensive to afford if we consider a very large number of cameras. Moreover, an infrared camera may be needed for the surveillance at night. Therefore, infrastructure-less solution, for instance, VANETs with reasonable traffic monitoring and congestion detection capabilities, may become highly desirable for ITS.

On the basis of the VANETs, a strategy was proposed in [31] to reduce traffic congestion, wherein periodically emitted beacons of V2V communication were proposed to enable traffic flow estimate and to warn drivers about the possible traffic breakdown. The research efforts in [31] were dedicated to the VANETS-assisted traffic jam reduction mechanism. In our work, we primarily focus on how to introduce transportation control domain in the existing network protocol header, such that each vehicle can count its neighbors and estimate the time spacing among vehicles. Because every vehicle with VANET capability is able to estimate vehicle density, flow, and average velocity in a microscopical manner, and the similar driver behavior recommendation strategy in [31] can also be employed in order to eliminate the forthcoming congestion.

Unlike all the aforementioned works, in this paper, we propose a novel approach called microscopic congestion detection protocol (MCDP), which detects the highway lane congestion and disseminate the detection results in both interlane and intralane manner. As will be shown in the following discussion, MCDP provides us with an inexpensive but effective transport congestion detection technique in VANETs.

3. The MCDP Mechanism

By integrating microscopic properties of vehicles [33] and external parameters, such as highway type and safety distance between vehicles, we propose a novel MCDP

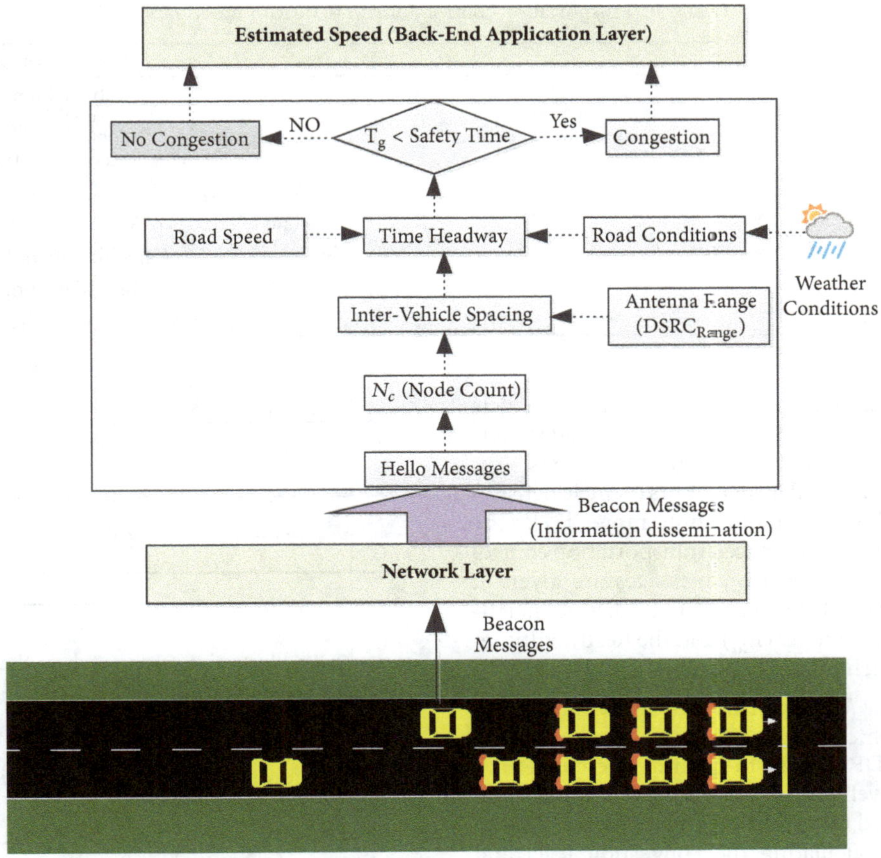

FIGURE 1: The work flow illustration of the MCDP.

to support transportation congestion detection over the existing MANETs. MCDP operates in a fully distributed infrastructure-less mode that is independent of any additional information, such as traffic data from local authorities. And MCDP is independent of the highway length. In [18], the authors considered microscopic properties for density estimation and road monitoring by assuming a fixed length intervehicle distance d, which is unrealistic because the fixed spacing between vehicles will affect safety regulations. Safety is an important feature of VANET, which relies on intervehicle spacing. A driver can make a reasonable decision to cope with the emergency situation only on the basis of accurate estimation of leading vehicle distance. In order to design a feasible transportation congestion detection scheme for VANETs, we should take into account the relationship between vehicle speed and interspacing. More specifically, we consider more realistic intervehicle spacing modeled as an exponential distribution. In addition, we consider *macroscopic* vehicle parameters that are collected *microscopically* for congestion detection. In MCDP, vehicles within communication range exchange macroscopic information locally, and each vehicle can thus estimate its surrounding vehicle density. The density estimation is enabled by measuring distinct neighbors and time headway (spacing coverage time) to predict upcoming congestion, as will be elaborated in detail in Section 3.1. Furthermore, every vehicle calculates its *headway* (spacing coverage time) and predicts upcoming congestion. The work flow of the MCDP scheme can be depicted in Figure 1, which is supposed to operating at every individual vehicle.

3.1. Information Dissemination. Information dissemination specifies the periodic exchange of beacons messages in ad hoc vehicular networks without any complicated negotiation among vehicles, which is necessary for protocol maintenance. In order to realize the transportation surveillance, it is expected to transmit the aforementioned macroscopic information to some neighboring vehicles in very short time. Many existing works, for instance [2, 14, 23, 25–28, 34–36], assume a control beacon based information dissemination mechanism as well. The information dissemination in MCDP can be subdivided into three types, namely, basic parameter exchange, speed assistance message, and interlane vehicle density.

(i) *Basic Parameter Exchange*: In this study, we assume that every vehicle has a GPS navigation system. On this basis, each vehicle is assumed to be able to collect vehicle current position, driving direction, speed, session time, and the topology of road networks from GIS. In the proposed MCDP framework, vehicle on the highway will periodically exchange beacon

TABLE 2: Transportation control domain in beacon message.

Content	Description
V_ID	Vehicle identifier
Session Time T_s	Message generation time
Speed S	Vehicle's Speed (Read from odometer)
Direction d	Vehicle's Direction (Obtained from Navigator system)
Acceleration a	Rate of change of Vehicle speed (Read from odometer)
Lane ID L_id	Lane identity of Highway
Position p	Vehicle's Position (GPS)
Road Type H	Obtained from Transportation department

messages to disseminate vehicle information (vehicle ID, position, speed, lane ID, session time, and vehicle length) to all its one-hop neighbors. In this way, every vehicle can count its neighbors (vehicle density) within its DSRC communication range [37]. The details of the newly introduced transportation control domain for periodic beacon messages are given in Table 2. Each vehicle will calculate its distinct neighbor in the same Lane ID (L_{id}) and the headway/time-gap between each other.

(ii) *Speed Assistance Message:* This message is multicast to all intralane vehicles to control the moving speed. The proposed MCDP will try to quantify the congestion level of each vehicle. As long as T_g is lower than some predefined threshold τ_0 (for instance, $\tau_0 = 2$), MCDP starts calculating the congestion level and estimating the moving speed V_{est}.

(iii) *Interlane Vehicle Density:* This message is exchanged among different highway lanes to share density level. Interlane density exchange is important for better utilization of highway lanes.

3.2. Transport Congestion Detection. Let us consider a road of length L, which can be subdivided into the following n small road segments:

$$L = \{s_1, s_2, s_3, \dots \dots \dots \dots s_n\} \tag{1}$$

At the same time, the vehicle density on the road is assumed to follow a Poisson distribution; namely, the likelihood that k vehicles are found in a space of s meters can be expressed as

$$P(k, s) = \frac{(\rho s)^k e^{-\rho s}}{k!}, \quad k \geq 0 \tag{2}$$

where ρ represents the vehicle density (unit: [veh/Km]). And the probability that there is no vehicle in the segment with length s on the road is given by

$$g(s) = P(0, s) = e^{-\rho s} \tag{3}$$

The intervehicle distance is random with mean $1/\rho_s$, where ρ_s is the vehicles spatial density. The probability that there is at least one single vehicle in the segment with length s can be given by

$$P_r(\text{at least one vehicle in s}) = 1 - e^{-\rho s} \tag{4}$$

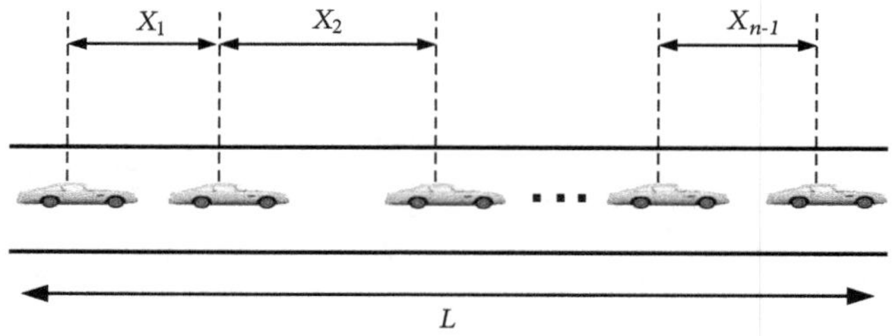

FIGURE 2: An illustrative scenario where the vehicles are randomly distributed along the road.

Let X_i in Figure 2 represent the intervehicle distance between vehicle i and vehicle $i + 1$. And X_i is independent and identically distributed (*i.i.d.*) with the probability density function (*pdf*) denoted by $P_X(x)$ as

$$P_X(x) = \lambda e^{-\lambda x}, \quad x \geq 0 \tag{5}$$

where λ is the vehicle arrival rate that can be modeled as Poisson distribution and the speed V assigned to the vehicles typically follows the normal distribution. If V is normally distributed and intervehicle distance is exponentially distributed, the time headway T_g is also exponentially distributed, namely,

$$T_g(t) = \lambda e^{-\lambda t}, \quad t \geq 0 \tag{6}$$

It can be readily derived that the time headway t between two vehicles must be greater than or equal to zero since

$$T_g(t \geq 0) = \int_0^{\infty} \lambda e^{-\lambda t} dt = -e^{-\lambda \infty} + e^{-\lambda 0} = 1 \tag{7}$$

Congestion detection is based on the time headway T_g among the vehicles. If the time headway t is greater than the safety transportation limit τ_0, the congestion probability will be

$$T_g(t \geq \tau_0) = 1 - P_T(t < \tau_0) = 1 - \int_0^{\tau_0} \lambda e^{-\lambda t} dt = e^{-\lambda \tau_0} \tag{8}$$

If the time headway t lies between intervals of $(\tau_0, (1 + \delta)\tau_0)$, the congestion probability will be

$$\begin{aligned} &T_g[\tau_0 \leq t \leq (\tau_0 + \delta\tau_0)] \\ &= P_T[t \geq \tau_0] - P_T[t \geq (\tau_0 + \delta\tau_0)] \\ &= e^{-\lambda\tau_0} - e^{\lambda(\tau_0 + \delta\tau_0)} \end{aligned} \tag{9}$$

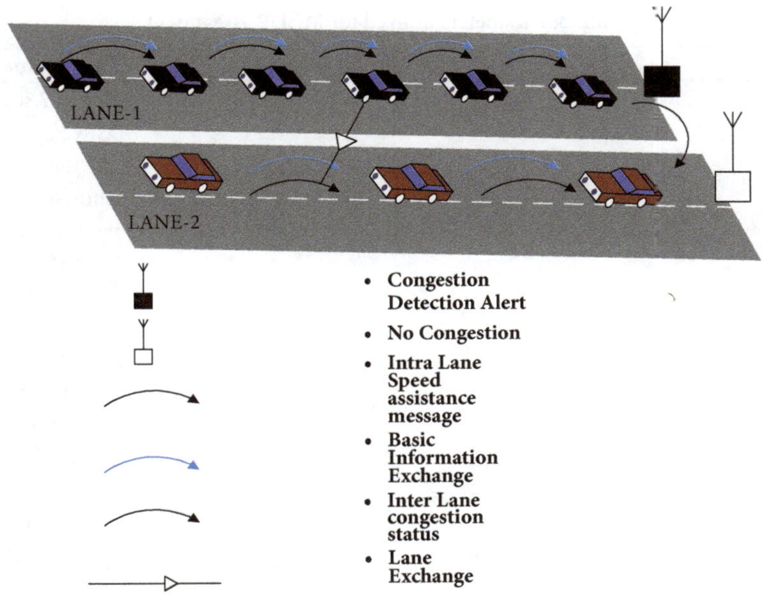

FIGURE 3: The MCDP enabled ITS illustration.

The vehicles collect transportation information from beacon messages and estimate traffic density. Basically, density estimation can be carried out after basic parameters exchange. Each vehicle on specific lane calculates distinct neighbors N_c and estimates intradistance between vehicles (Spacing) on the basis of N_c. Finally, each vehicle within DSRC communication range can calculate the time-gap/Headway T_g as

$$T_g = \frac{1}{\text{Speed}_{\text{highway-limit}}}\left(\frac{\text{DSRC}_{\text{range}}}{N_c} - V_{\text{length}}\right) \quad (10)$$

where $Speed_{highway-limit}$ corresponds to the speed limit of the corresponding highways [32], $DSRC_{range}$ is the communication standard range [37], N_c is the neighbor count, and V_{length} is the length of the vehicle [38]. In the case of the heterogeneous traffic flow, we may consider the following mean length of vehicle:

$$\mu_{\text{V-length}} = \frac{\sum_{v=1}^{N_c} V_{\text{length}_v}}{N_c} \quad (11)$$

And the time headway in (10) can be rewritten as

$$T_g = \frac{1}{\text{Speed}_{\text{highway-limit}}}\left(\frac{\text{DSRC}_{\text{range}}}{N_c} - \mu_{\text{V-length}}\right) \quad (12)$$

If T_g is less than the safety transportation limit τ_0, congestion will be detected. According to travel guide instructions, two vehicles should have at least two seconds time headway. But in certain circumstances, the safety time can be extended due to weather situations [24]. In a word, the transportation congestion detection c can be summarized as follows:

$$c = \begin{cases} 1 & if\ T_g < \tau_0 \\ 0 & otherwise \end{cases} \quad (13)$$

3.3. Driver Assistance. Another notable advantage of the MCDP is its driver assistance capability. The vehicle will analyze the traffic condition based on the calculated distinct neighbors N_c and the time headway T_g. And the appropriate speed can be estimated according to the underlying traffic status.

$$V_{\text{received}} = \frac{1}{\tau_C}\left(\frac{DSRC_{range}}{N_c} - \mu_{V\text{-}length}\right) \quad (14)$$

Here $V_{received}$ is the suggested speed to the driver. The suggested speed can be multicasted intralane on the highway when c=1. The complete pictorial representation of MCDP is given in Figure 3, which shows the entire steps from information dissemination to congestion detection and lane alternation. Additionally, the proposed MCDP scheme can be easily extended to two-way road scenarios by separately counting the vehicles in two opposite directions on the basis of the proposed periodic beacon message exchange. Since the direction information is contained in the transportation control domain for periodic beacon messages, the two-way vehicles can be handled by two independent procedures.

4. Experimental Study

4.1. Analytical Assessment. Different Chinese roads are considered for analytical assessment and we also try to calculate their congestion boundaries in different scenarios. The experiments are performed with a road segment of 1 km having one to three lanes. The traffic is generated by Monte Carlo in Matlab R2015b [39]. To assess the performance, the proposed MCDP is compared with the Green-Shield's car-following model [23, 34]. Green-Shield's car-following model is a well-known density estimation model. Table 3 summarizes all the assumed parameters for analytical assessment. The safety time headway is considered as a performance metric.

TABLE 3: Parameters utilized for MCDP assessment.

Parameter	Value
Safety time/Headway	τ_0 = 2 second
Transmission range	$DSRC_{range}$ = 1000 m
Vehicle's length	5m, 10m
Highway-type	Chinese roads [26]
Number of lanes	City road=1,National highway=1,Express road=2,Expressway=3
Number of Vehicles	2 to 70 randomly generated
Probability of Cars	0.80
Probability of Buses	0.20
Speed limit	40, 80,120,130 km/h [32]
Lane 1 Speed	100 km/h
Lane 2 Speed	120 km/h
Lane 3 Speed	130 km/h
Simulation runs	50
Simulation time	200s
Propagation model	Two ray ground

First, spacing among vehicles is calculated under various vehicle densities and then headways are calculated under highway speed limit. Due to the microscopic feature, every individual vehicle calculates its own headway as an indicator of congestion.

The impact of vehicle density on safety time headway is shown in Figure 4. It can be observed that vehicle density and safety time have an inverse relation. The congestion boundaries of the proposed MCDP over *China city road* and *China national highway* are shown in Figures 4(a) and 4(b), where 37 and 21 vehicles are the beginning of congestion threshold, respectively. On the other hand, Green-Shield model congestion threshold begins after 37 and 21 as shown in Figures 4(a) and 4(b). The exceeding of congestion boundary causes $\tau_0 < 2$ (safety time goes below than transportation limit), which further causes c=1 (congestion on the road). As illustrated in Figures 4(a) and 4(b), the congestion threshold of the proposed MCDP in single lane highway is less than that of the macroscopic Green-Shield's model, which implies that proposed MCDP is more sensitive to congestion detection and provides more safety traveling environment. Due to the macroscopic nature, Green-Shield model considered many irrelevant parameters (exact free flow speed and jam density), which causes communication and computational overhead. Secondly, Green-Shield model relies on expensive road monitoring devices, which can be utilized to cover specific road segment only. The congestion analysis of China express road and China expressway with multilane scenarios is shown in Figures 4(c) and 4(d). The congestion boundary of our proposed MCDP over two-lane China express road is 30, while there are 18 vehicles in case of *Green-Shield model*, as shown in Figure 4(c). In two-lane highway, even in the high-speed limit, the proposed MCDP model is able to accommodate a larger number of vehicles, which is desirable in order to fully utilize both lanes. Hence in the multilane highway, MCDP outperforms the reference *Green-Shield model* in terms of its better utilization of all available lanes. The similar observations can be found in Figure 4(d) for three-lane China expressway, where the congestion threshold of our MCDP is quite more than Green-Shield. In short, by observing the numerical results of MCDP, we can conclude that our proposed MCDP protocol provides a promising approach to detect traffic congestion on the specific highway and to assist the driver.

4.2. Simulation Analysis. In this subsection, the efficiency of MCDP is analyzed under different scenarios. Two well-known simulators such as NS2 (release 3.25) and SUMO (release 0.25.0) are used for assessment. More specifically, NS2 is used for performance evaluation of MCDP while "Sumo 0.25.0" is used for VANETs scenarios generation. Two different types of scenarios (busy congested road and freeway) with varying numbers of connections, transmission time, and simulation time are generated. To precisely simulate such a traffic monitoring system, a federated (Ubuntu 14LTS) framework is required, which combines these two simulators through generic traffic control interfaces.

4.2.1. Traffic Simulation Setup. Here SUMO is used for road traffic generation. Firstly, we designed road traffic scenario with SUMO. Roads of lengths 1 km and 7 km with two lanes are considered as an input for network simulation. The maximum speed was set to 40km/h for busy road and 50km/h for freeway scenarios, respectively. During the simulation, the red lights are turned ON to monitor all the traffic inside the fixed length of the road. The main traffic simulation parameters are summarized in Table 4.

4.2.2. Network Simulation Setup. After the generation of the mobility traces, we set up the network simulation in the NS2 framework, wherein the network simulation parameters are given in Table 5. First, busy road scenario was analyzed through MCDP by varying transmission range and simulation time. Secondly, the freeway scenario was analyzed for the same purpose with the same parameters. And we also

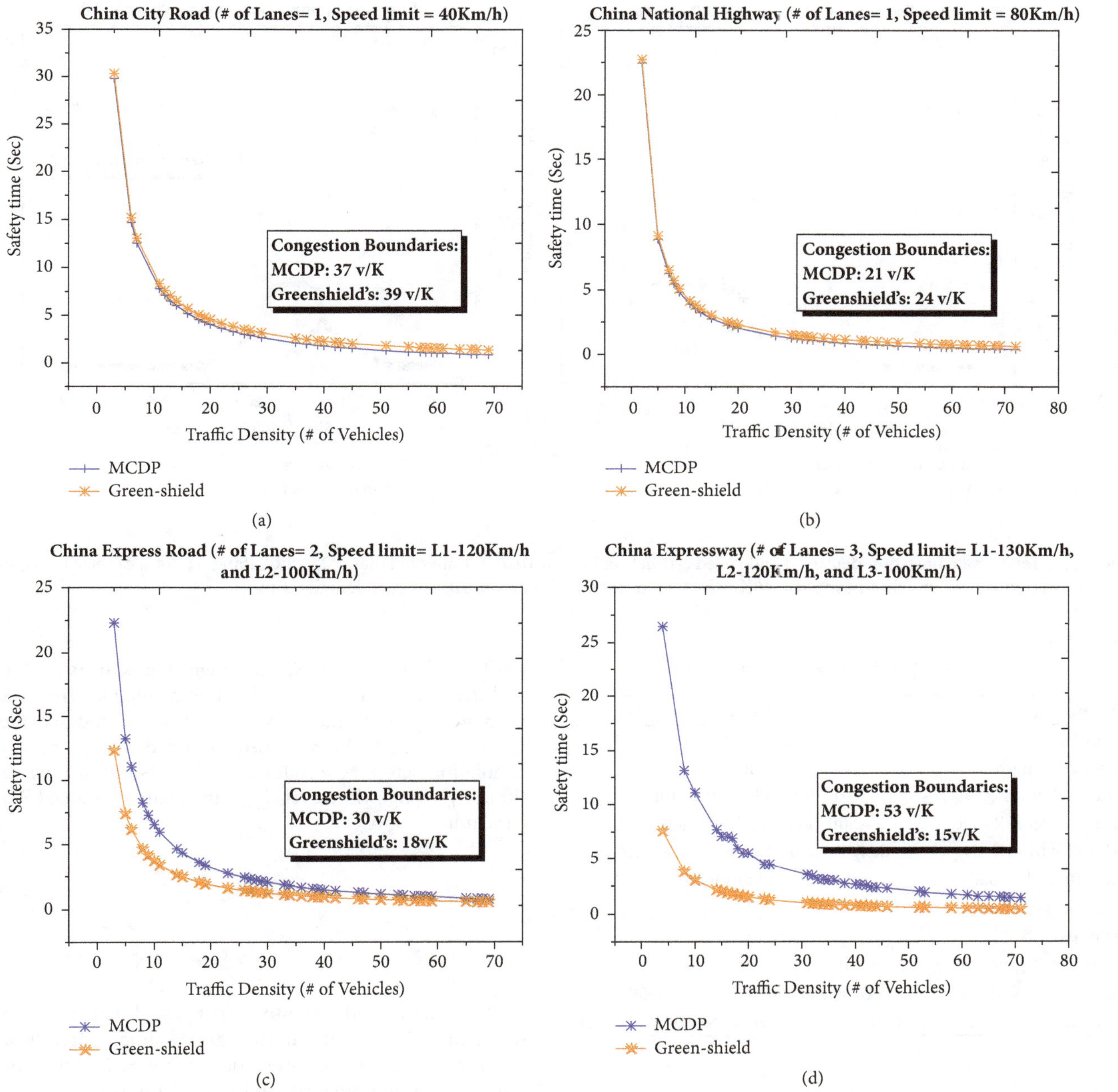

FIGURE 4: Congestion analysis in different single and multilane Chinese highways (DSRC Rang = 1000m, car size = 5m, bus size = 10m, and safety time threshold = 2 sec).

examined the performance of MCDP, DSR, and AOMDV over IEEE802.11P and IEEE802.11ac in terms of throughput, Packet delivery ratio (PDR) and end-to-end delay.

Let us briefly summarize the performance metrics utilized in the simulation analysis.

(i) Congestion Level: Congestion level is measured as a percentage of the additional travel time compared to normal traffic (or free flow situation).

(ii) Estimated Speed: Congestion is a function of the reduction in speed and vice versa. Therefore, the setting of an estimated speed that is directly related to congestion level can be used to assess the traffic congestion.

(iii) Packet Delivery Ratio (PDR): PDR is the ratio of data packets reliably delivered to the destination, *i.e.*, $PDR = P_r/P_t$, where P_r is the total number of received data packets and P_t stands for the total number of transmitted data packets.

(iv) Throughput: Throughput can be utilized to assess the performance of a network by providing the average rate of successful delivery of packets towards the destination.

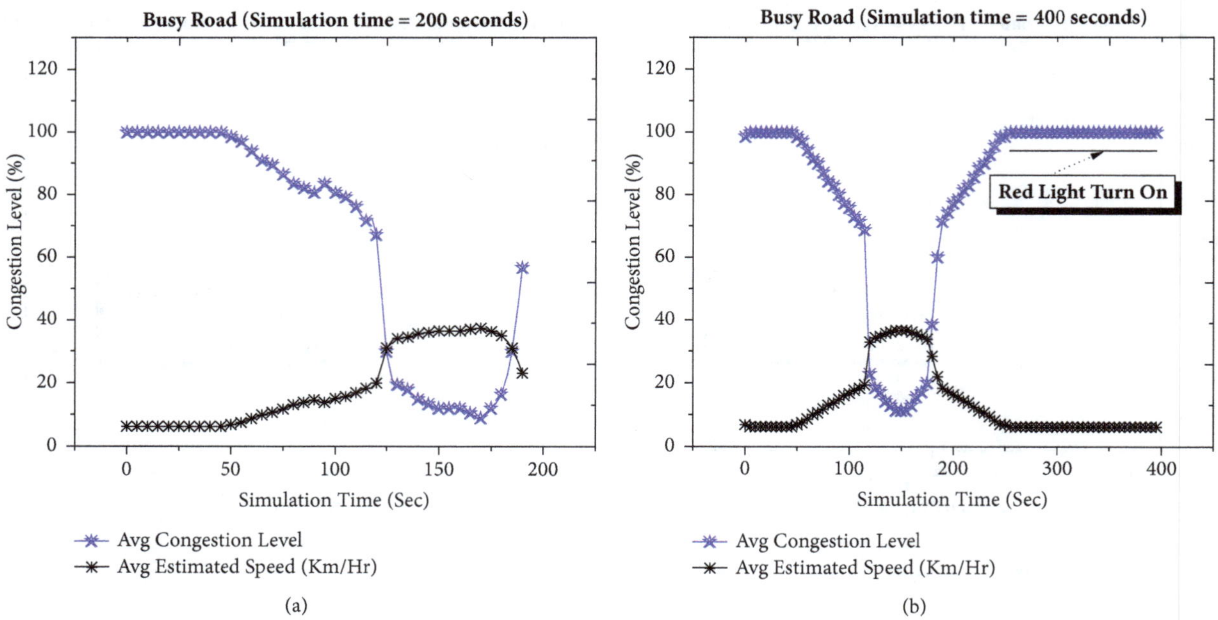

FIGURE 5: Temporal congestion detection and speed estimation in busy road, number of lanes = 2, DSRC Rang = 500m, car size = 5m, safety time threshold = 2 sec, maximum speed = 40Km/H, total numbers of vehicles=70, and road length=1Km.

TABLE 4: Road traffic generation parameters.

Parameter	Value
Simulator	Sumo 0.25.0
Vehicle's length	5m
Simulation time	200, 400, 500, 1000 seconds
Transmission Range	300, 500, 600,1000 meters
Mobility Model	Busy Road and Freeway Scenarios
Length of the roads	1 and 7 kilometers
Numbers of lanes	2
Traffic lights	2
Number of Junctions	2
Number of Nodes	Busy Road =70, Freeway = 50
Vehicle's Max speed	Busy Road =40 Km/h, Freeway = 50 Km/h

(v) End-to-End Delay: End-to-end delay is the average duration that each packet can be received by the last node in the network.

4.2.3. Congestion Detection Analysis. The proposed protocol assumes that each vehicle monitors local traffic congestion detection by analyzing the received beacon messages with detailed information from other vehicles, while each vehicle shares information about its position, direction, speed, and so on. To see the overall performance of the MCDP protocol, the average temporal congestion level and average estimated speed of all vehicles under two different traffic scenarios are shown in Figures 5 and 6, respectively. It can be observed that when the safety time goes below the safety time threshold value τ_0, the congestion level starts to increase and the estimated speed starts to decrease. The results show that when red lights are turned ON, the congestion is at its peak level and the estimated speed tends to be negligible. The average estimated speed to the drivers in varying traffic densities is shown as well in both Figures 5 and 6. In all traffic and simulation setups, the following relationships among safety time t_{safe}, congestion level C_{level}, and estimated speed S_{est} are concluded:

$$\mathrm{C_{level}} \propto \frac{1}{\mathrm{t_{safe}}} \tag{15}$$

$$\mathrm{S_{est}} \propto \mathrm{t_{safe}} \tag{16}$$

In [40], the road gets into the congested state if the vehicle travel time exceeds the normal travel time at free flow. In order to do that, the scheme requires that each section of the road should be under surveillance all the time; meanwhile all the vehicles need to report their traversal time over each section to a centralized entity. In MCDP, the congestion detection can be realized without such an infrastructure. It should be addressed that the speed of a vehicle is directly affected by the level of traffic congestion in its surroundings, and it is reasonable to use it for traffic congestion detection and speed estimation. Additionally, MCDP quantifies the level of congestion locally from the information in the beacon message, which is remarkably important for delay-sensitive applications.

We illustrate the impact of transmission range (TR) for the MCDP in Figure 7. Here three different TRs of 300m, 600m, and 1000m are considered and the congestion detection and speed estimation technique are confined to the vehicle's visibility based on its transmission (DSRC) range. It can be noted that congestion detection level gets decreased

Table 5: Network simulation parameters.

Parameter	Value
Simulator	NS-2.35
Traffic type	UDP
Number of Nodes	50, 70
Mac Protocols	IEEE 802.11p, 802.11ac
Routing Protocols	MCDP, DSR, AOMDV
Number of Connections	5, 10, 15, 20, 25% of the nodes
Packet size	1000 Bytes
CBR rate	64Kb
Parameters	No. of connections, Transmission Time, Simulation Time
CBR Max Packets	2280000
Performance metrics	Delay, PDR, Throughput, Congestion Level, Speed Estimation

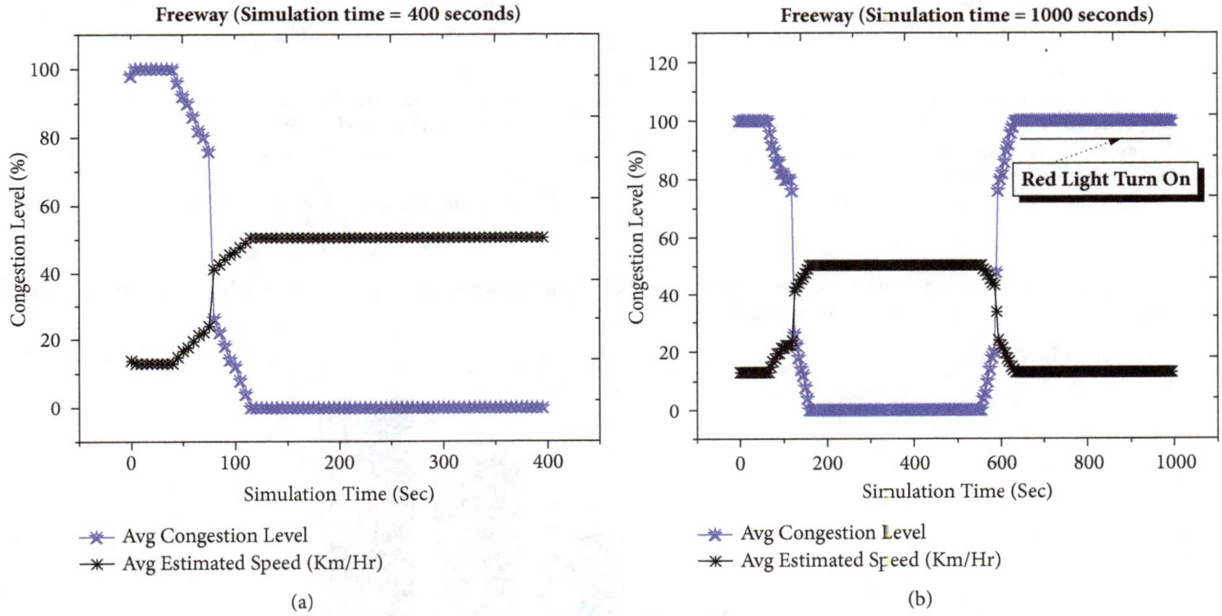

Figure 6: Temporal congestion detection and speed estimation in freeway, number of lanes = 2, DSRC Rang = 500m, car size = 5m, safety time threshold = 2 sec, maximum speed = 50Km/h, total numbers of vehicles = 50, and road length=7Km.

with the increase in transmission range. As illustrated in Figure 7(a), if 1000m transmission range is assumed, even a traffic congestion happens, it may be ignored by the average congestion level detection assessment scheme. This suggests us that a reasonable setup of TR is important to achieve the reasonable congestion detection in the MCDP framework. A very small TR setup may make the system too sensitive to the traffic variation on road, while a too large TR will ignore the real traffic congestion. A reasonable TR setup can realize the proper tradeoff.

4.2.4. Communication Performance Analysis. MCDP is a full-fledge protocol, which is easily implementable and executable to any sort of environment, but the performance of routing protocol also depends on how better the routing takes place in the network. Here we focus on the achieved throughput, end-to-end delay, and PDR performance to assess the routing capability and efficiency of MCDP. The communication performance simulation results of MCDP, DSR, and AOMDV protocols are shown in Figure 8 with a different numbers of UDP connections over IEEE802.11p and IEEE802.11ac, respectively. From Figure 8 it can be easily concluded that, with the increase in a number of connections, the throughput of all routing protocols gets increased. While the achieved throughput of both MCDP and AOMDV protocols is almost the same for all connection numbers. In fact, due to multipath nature, their throughout is better than that of the DSR protocol. The end-to-end delay is illustrated in Figure 9 by considering UDP connections over IEEE802.11p and IEEE802.11ac. It is shown that IEEE802.11p always outperforms IEEE802.11ac in all cases, which complies with the design objective of IEEE802.11p to support the communication link between the vehicles that might exist only for a short amount of time. At the same time, it

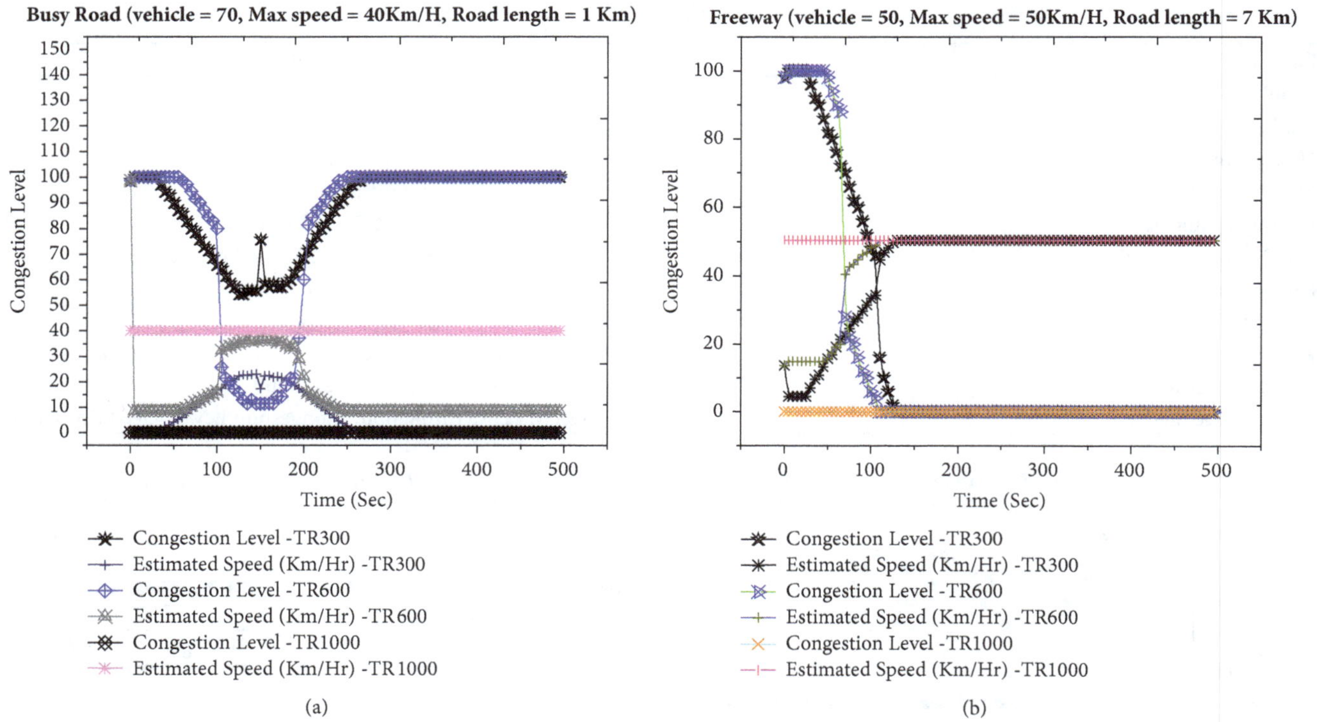

FIGURE 7: Temporal congestion detection and speed estimation using different transmission ranges, number of lanes = 2, car size = 5m, safety time threshold = 2 sec, and simulation time = 500 seconds.

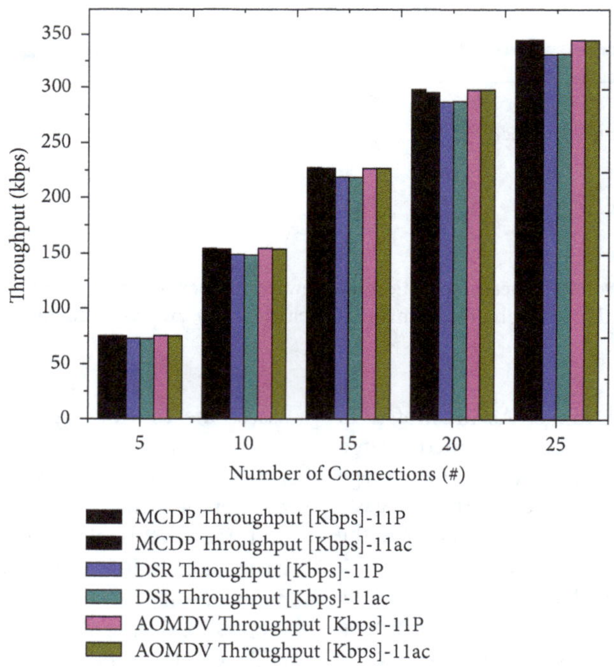

FIGURE 8: Throughput comparison of MCDP, DSR, and AOMDV protocols over 802.11P and 802.11ac with different numbers of connections.

can be observed that the proposed MCDP protocol can achieve a comparable good end-to-end delay and throughput performance while having congestion detection and speed estimation feature.

Finally, the PDR performances of MCDP, DSR, and AOMDV are illustrated in Figure 10 for a different number of connections. It is shown that the proposed MCDP protocol can realize comparable PDR performance as the AOMDV

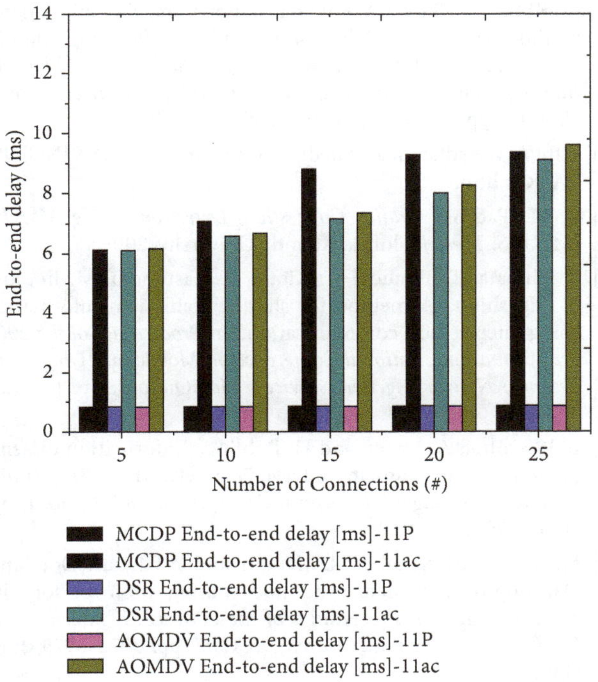

FIGURE 9: End-to-end delay comparison of MCDP, DSR, and AOMDV protocols over 802.11P and 802.11ac with different numbers of connections.

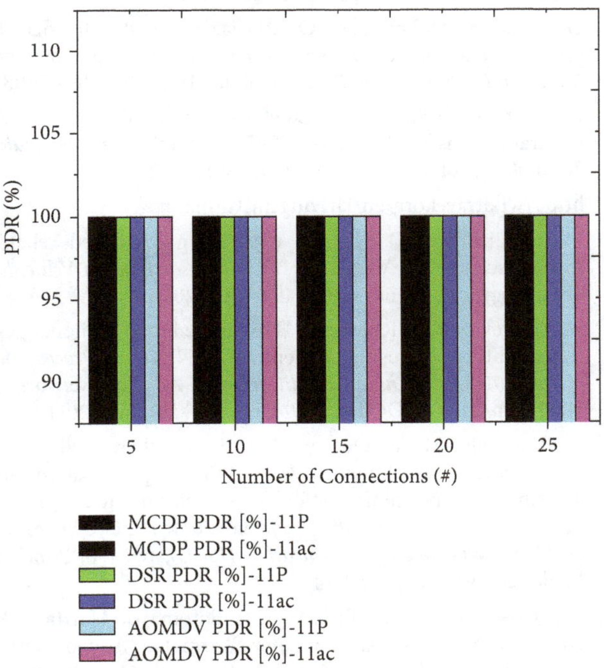

FIGURE 10: PDR comparison of MCDP, DSR, and AOMDV protocols over 802.11P and 802.11ac with different numbers of connections.

and DSR protocols in a different number of connections. Meanwhile, it can also be noted that when there is sufficient connectivity, we can almost guarantee successful packet delivery for the aforementioned three protocols.

5. Conclusions

VANETs technology plays an important role in safety transportation. Microscopic congestion detection protocol (MCDP) is an interesting application of VANET technology to identify road congestion. At the same time, MCDP provides us with a new approach to estimate vehicles density and to assist driver assistance. In a word, MCDP is an inexpensive approach, which integrates basic microscopic vehicle properties with some external road safety parameters to accurately monitor road status. Moreover, MCDP is quite simple protocol and it works both in a single lane and in multilane highway. An appropriate speed suggestion to the driver at every moment is also a potential application of MCDP, which will be left for future investigation.

Conflicts of Interest

The authors declare that they have no conflicts of interest.

Acknowledgments

This work was partly supported by the National Natural Science Foundation of China (NSFC) under Grant no. 61771406.

References

[1] The world's worst traffic jams. http://content.time.com/time/world/article/0,8599,1733872,00.html

[2] J. Fukumoto, N. Sirckane, Y. Ishikawa, T. Wada, K. Ohtsuki, and H. Okada, "Analytic method for real-time traffic problems by using Contents Oriented Communications in VANET," in *Proceedings of the 2007 7th International Conference on ITS Telecommunications*, pp. 1–6, Sophia Antipolis, France, June 2007.

[3] S. Dornbush and A. Joshi, "Streetsmart traffic: discovering and disseminating automobile congestion using vanet's," in *Proceedings of the 65th IEEE Vehicular Technology Conference (VTC '07)*, pp. 11–15, Dublin, Republic of Ireland, April 2007.

[4] S. Inoue, K. Shozaki and Y. Kakuda, "An automobile control method for alleviation of traffic congestions using inter-vehicle ad hoc communication in lattice-like roads," in *Proceedings of the IEEE Globecom Workshops*, pp. 1–6, IEEE, Washington, DC, USA, November 2007.

[5] R. Morla, Vision of Congestion-Free Road Traffic and Cooperating Objects, citeulike: 3806024, 2005.

[6] S. Roy, R. Sen, S. Kulkarni, P. Kulkarni, B. Raman, and L. K. Singh, "WirelessAcrossRoad: RF based road traffic congestion detection," in *Proceedings of the 2011 3rd International Conference on Communication Systems and Networks, COMSNETS 2011*, pp. 1–6, i, January 2011.

[7] D. Huang, S. Shere, and S. Ahn, "Dynamic highway congestion detection and prediction based on shock waves, dynamic highway congestion detection and prediction based on shock waves," in *Proceedings of the ACM VANET '10*, pp. 11–20, 2010.

[8] P. Pongpaibool, P. Tangamchit, and K. Noodwong, "Evaluation of road traffic congestion using fuzzy techniques," in *Proceedings of the IEEE Region 10 Conference, TENCON 2007*, pp. 1–4, November 2007.

[9] E. Hossain, G. Chow, V. C. M. Leung et al., "Vehicular telematics over heterogeneous wireless networks: a survey," *Computer Communications*, vol. 33, no. 7, pp. 775–793, 2010.

[10] M. Seredynski and P. Bouvry, "A survey of vehicular-based cooperative traffic information systems," in *Proceedings of the 14th IEEE International Intelligent Transportation Systems Conference, ITSC 2011*, pp. 163–168, October 2011.

[11] B. Schunemann, J. Wedel, and I. Radusch, "V2X-based traffic congestion recognition and avoidance," *Tamkang Journal of Science and Engineering*, vol. 13, no. 1, pp. 63–709, 2010.

[12] T. Zhong, B. Xu, and O. Wolfson, "Disseminating real-time traffic information in vehicular ad-hoc networks," in *Proceedings of the 2008 IEEE Intelligent Vehicles Symposium, IV*, pp. 1056–1061, June 2008.

[13] G. Korkmaz, F. Özgüner, E. Ekici, and Ü. Özgüner, "Urban multi-hop broadcast protocol for inter-vehicle communication systems," in *Proceedings of the First ACM International Workshop on Vehicular Ad Hoc Networks, Held in Conjunction with MOBICOM 2004*, pp. 76–85, October 2004.

[14] M. Bani Younes and A. Boukerche, "A performance evaluation of an efficient traffic congestion detection protocol (ECODE) for intelligent transportation systems," *Ad Hoc Networks*, vol. 24, pp. 317–336, 2015.

[15] B. S. Kerner, *Introduction to Modern Traffic Flow Theory and Control*, Springer, Berlin, Germany, 2009.

[16] M. Jerbi, S.-M. Senouci, T. Rasheed, and Y. Ghamri-Doudane, "Towards efficient geographic routing in urban vehicular networks," *IEEE Transactions on Vehicular Technology*, vol. 58, no. 9, pp. 5048–5059, 2009.

[17] S. Bilal, S. Madani, and I. Khan, "Enhanced junction selection mechanism for routing protocol in VANETs," *International Arab Journal of Information Technolog*, vol. 8, no. 4, pp. 422–429, 2011.

[18] T. Umer, Z. Ding, B. Honary, and H. Ahmad, "Implementation of microscopic parameters for density estimation of heterogeneous traffic flow for VANET," in *Proceedings of the 2010 7th International Symposium on Communication Systems, Networks and Digital Signal Processing, CSNDSP 2010*, pp. 66–70, July 2010.

[19] J. Ploeg, E. Semsar-Kazerooni, G. Lijster, N. Van De Wouw, and H. Nijmeijer, "Graceful degradation of CACC performance subject to unreliable wireless communication," in *Proceedings of the 2013 16th International IEEE Conference on Intelligent Transportation Systems: Intelligent Transportation Systems for All Modes, ITSC 2013*, pp. 1210–1216, October 2013.

[20] P. Fernandes and U. Nunes, "Platooning with IVC-enabled autonomous vehicles: Strategies to mitigate communication delays, improve safety and traffic flow," *IEEE Transactions on Intelligent Transportation Systems*, vol. 13, no. 1, pp. 91–106, 2012.

[21] J. Zhou and H. Peng, "Range policy of adaptive cruise control vehicles for improved flow stability and string stability," *IEEE Transactions on Intelligent Transportation Systems*, vol. 6, no. 2, pp. 229–237, 2005.

[22] P. Seiler, A. Pant, and K. Hedrick, "Disturbance propagation in vehicle strings," *Institute of Electrical and Electronics Engineers Transactions on Automatic Control*, vol. 49, no. 10, pp. 1835–1841, 2004.

[23] M. Kimura, Y. Taoda, Y. Kakuda, S. Inoue, and T. Dohi, "A novel method based on VANET for alleviating traffic congestion in urban transportations," in *Proceedings of the 2013 IEEE Eleventh International Symposium on Autonomous Decentralized Systems (ISADS)*, pp. 1–7, Mexico City, Mexico, March 2013.

[24] http://www.sdt.com.au/safedrive-directory-FOLLOWINGDISTANCE.htm.

[25] F. M. Padron, *Traffic Congestion Detection Using VANET [M.S.C.S. Thesis]*, Florida Atlantic University, 2009.

[26] N. Shibata, T. Terauchi, T. Kitani, K. Yasumoto, M. Ito, and T. Higashino, "A method for sharing traffic jam information using inter-vehicle communication," in *Proceedings of the 2006 3rd Annual International Conference on Mobile and Ubiquitous Systems: Networking and Services, MobiQuitous*, pp. 1–7, July 2006.

[27] L. Wischhof, A. Ebner, and H. Rohling, "Information dissemination in self-organizing intervehicle networks," *IEEE Transactions on Intelligent Transportation Systems*, vol. 6, no. 1, pp. 90–101, 2005.

[28] M. Jerbi, S.-M. Senouci, R. Meraihi, and Y. Ghamri-Doudane, "An improved vehicular ad hoc routing protocol for city environments," in *Proceedings of the 2007 IEEE International Conference on Communications, ICC'07*, pp. 3972–3979, June 2007.

[29] L.-D. Chou, D. Chunhu Li, and H.-W. Chao, "Mitigate traffic congestion with virtual data sink based information dissemination in Intelligent Transportation System," in *Proceedings of the 3rd International Conference on Ubiquitous and Future Networks, ICUFN 2011*, pp. 37–42, June 2011.

[30] O. Asmaa, K. Mokhtar, and O. Abdelaziz, "Road traffic density estimation using microscopic and macroscopic parameters," *Image and Vision Computing*, vol. 31, no. 11, pp. 887–894, 2013.

[31] F. Knorr, D. Baselt, M. Schreckenberg, and M. Mauve, "Reducing traffic jams via VANETs," *IEEE Transactions on Vehicular Technology*, vol. 61, no. 8, pp. 3490–3498, 2012.

[32] http://wikitravel.org/en/Driving_in_China.

[33] M. H. Eiza and Q. Ni, "An evolving graph-based reliable routing scheme for VANETs," *IEEE Transactions on Vehicular Technology*, vol. 62, no. 4, pp. 1493–1504, 2013.

[34] M. Milojevic and V. Rakocevic, "Distributed road traffic congestion quantification using cooperative VANETs," in *Proceedings of the 2014 13th Annual Mediterranean Ad Hoc Networking Workshop, MED-HOC-NET 2014*, pp. 203–210, June 2014.

[35] G. B. Araujo, M. M. Queiroz, F. d. Duarte-Figueiredo, A. I. Tostes, and A. A. Loureiro, "CARTIM: A proposal toward identification and minimization of vehicular traffic congestion for VANET," in *Proceedings of the 2014 IEEE Symposium on Computers and Communication (ISCC)*, pp. 1–6, Funchal, Madeira, Portugal, June 2014.

[36] F. Terroso-Sáenz, M. Valdés-Vela, C. Sotomayor-Martínez, R. Toledo-Moreo, and A. F. Gómez-Skarmeta, "A cooperative approach to traffic congestion detection with Complex Event processing and VANET," *IEEE Transactions on Intelligent Transportation Systems*, vol. 13, no. 2, pp. 914–929, 2012.

[37] X. Ma, X. Chen, and H. H. Refai, "Performance and reliability of DSRC vehicular safety communication: a formal analysis," *EURASIP Journal on Wireless Communications and Networking*, vol. 1, 2009.

[38] Wikipedia, "Car Dimension" [EB/OL], 2015, http://en.wikipedia.org/wiki/car_dimension.

[39] P. Brémaud, *Markov Chains: Gibbs Fields, Monte Carlo Simulation, and Queues*, vol. 31, Springer Science & Business Media, 2013.

[40] G. Marfia and M. Roccetti, "Vehicular congestion detection and short-term forecasting: a new model with results," *IEEE Transactions on Vehicular Technology*, vol. 60, no. 7, pp. 2936–2948, 2011.

10

Occluded Street Objects Perception Algorithm of Intelligent Vehicles based on 3D Projection Model

Wei Liu, Longsheng Wei, and Yongbo Li

School of Automation, China University of Geosciences, Wuhan, Hubei 430074, China

Correspondence should be addressed to Wei Liu; liuwei@cug.edu.cn

Academic Editor: Yair Wiseman

We present a super perception system of intelligent vehicles for perceiving occluded street objects by collecting images from neighbor front vehicles V2V (vehicle to vehicle) video streams based on 3D projection model. This super power can avoid some serious accidents of driver-assistant systems or automatic driving systems which can only detect visible objects. Our street perception system can "see through" the front vehicles to realize detecting of the occluded street objects only by analyzing the pair images received from front and host (back) vehicles. Upon the 3D projection model based on the pair images, the system uses affine transformation to achieve augmented reality method to increase the visibility perspective of driver system. Experimental results on different datasets are shown to validate our approach. Evaluation method was first introduced into our perception system.

1. Introduction

According to the World Health Organization 2016 Global's report on road traffic injuries, about 1.25 million people die each year as a result of road traffic crashes. Between 20 and 50 million more people suffer nonfatal injuries [1]. The statistics from NHTSA show that 31% of traffic accidents are due to rear-end collisions which happens when front vehicles suddenly slow down or stop because of the chain reaction of the road objects ahead of them [2]. This is mainly because the unawareness of the situation ahead of the front vehicles or incapability of perception of the existence of street objects which are occluded by the front vehicles, which are especially serious in urban street. These can also lead to other kinds of accidents. For example, a car or a pedestrian may burst into the street at the head of other vehicles at any time, while the driver cannot have enough reaction time and will definitely cause serious traffic accident. Another example is when cars execute a passing maneuver through trucks or buses. This will cause one of the most severe road crashes when a vehicle shifts into an opposing traffic lane and crashes head on with an oncoming vehicle.

The WHO report also mentioned that improving the safety features of vehicles will be an efficient way to prevent the road traffic injuries [1]. By analyzing the driving aid technologies of advanced vehicles, it is obvious that an enhanced forward road perception system is an efficient way to decrease the traffic incidents mentioned above. For mostly forward street perception system can only perceive things that vehicles can "see" but failed when street objects are partly or fully occluded. As human beings tend to believe in what they see with their own eyes, the drivers can have enough reaction time if they can see through the front vehicles and can realize the situation happened in front of the front vehicles. In this way, accidents will be decreased sharply in urban street.

The idea of V2V in our paper is similar to the truck platoons [3–5]. The difference is that the truck platoon link trucks or cars in train-like line which can save fuel, fit more cars on road, and potentially improve safety. However, in our systems, vehicles can drive on both the same lanes and different lanes and connect to each other to improve the safety based on sharing the video data. Using V2V communication based on DSRC, a video stream of vehicles' forward-looking cameras can be transmitted between vehicles with low delay [6]. Our perception system of the host vehicle (vehicle A in Figure 1) can be augmented by combining the video information from the front vehicles (vehicle B in Figure 1).

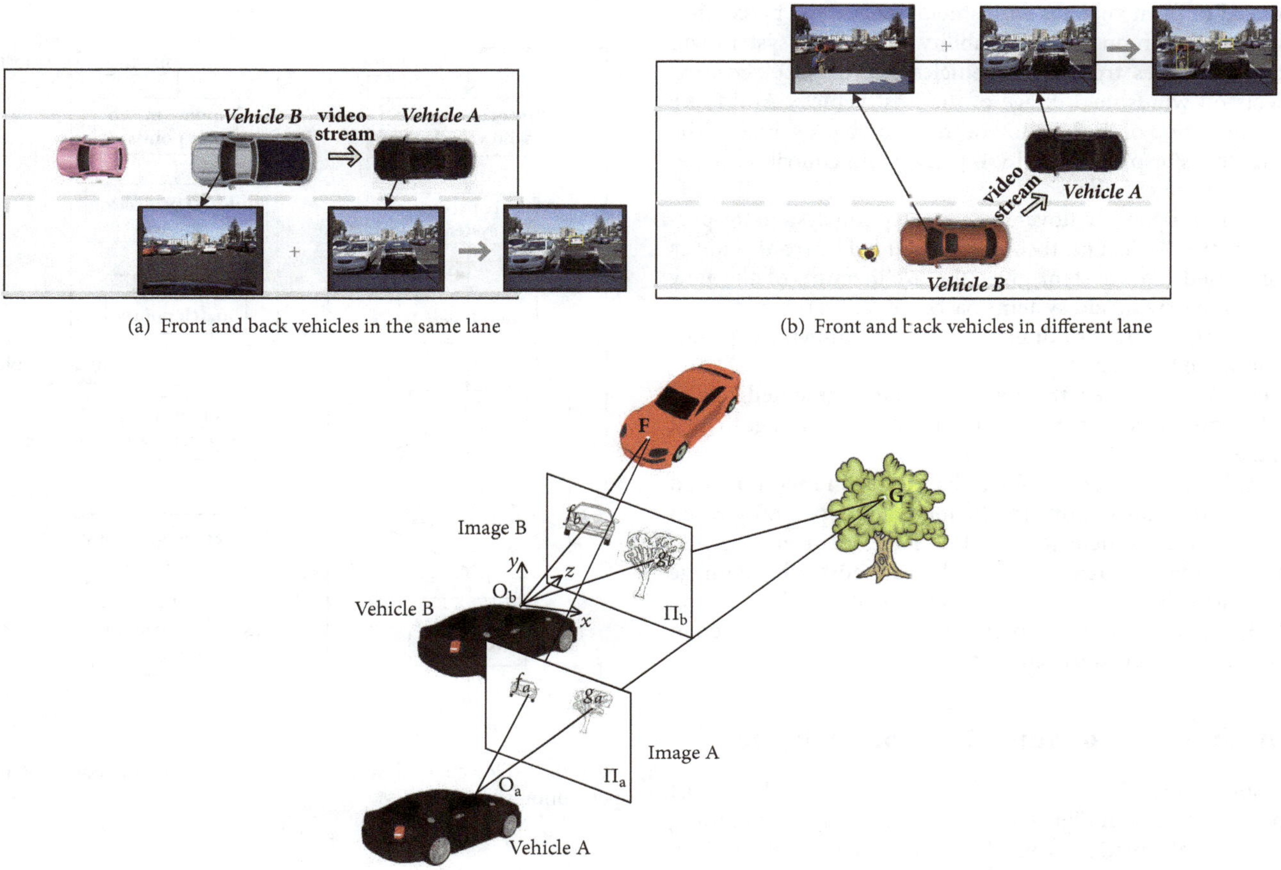

(a) Front and back vehicles in the same lane

(b) Front and back vehicles in different lane

(c) 3D model of perception system

Figure 1: Street super perception system.

The same object will project in different location, scale, and shape in image A (vehicle A) and image B (vehicle B) as described in Figure 1(c). F represents the 3-space point of car, and corresponding image points f_a and f_b are the image of F. We only know car's image points f_b in image B, and image points f_a of the same car in image A need to be calculated. G is the 3-space point of another object (such as tree) and the corresponding image points in image A and image B are g_a and g_b. This object can be seen by both vehicles A and B. Our purpose is trying to find the 3D projection model between two images based on matching points in image A and image B (such as g_a and g_b). Then, points f_a can be computed based on 3D projection model and point f_b.

Few works have been done on front street perception based on V2V video stream. Most of the forward perception systems cannot be deployed as see-through windows [7–11]. These perception systems can detect their distance to other vehicles and can only perceive visible street objects via sensors, such as laser, radar, and camera. As to the collaborative method, [12, 13] employed location information of vehicles which is periodically exchanged to prevent potential danger in advance. Similarly, [14, 15] adopted the route data coming from a conventional navigation system. But drivers are not sensitive to digital information. Human beings tend to believe in what they see with their own eyes. The works [6, 16] propose vehicles transparent method based on V2V video streams in order to deal with passing maneuvers. Their method need accurate distance information gathered from radar sensor to realize the object projection between two images. The work [17] proposes a vehicle blind spot elimination system based on videos captured from other vehicles. But the linear projection fusion causes superior deformation when front and back vehicles are in different lanes.

Great progress has been made on modern object detector based on convolution neural network (CNN) recently, such as Fast-RCNN [18], R-FCN [19], Multibox [20], SSD [18], and YOLO [19], which are good enough to detect most of the objects on road but still failed on perception invisible things (including the fully occluded objects or objects more than 80% occluded).

Our street perception system can "see through" the front vehicles to realize detecting of the "invisible" objects (such as vehicles, pedestrians, and motorcycles) by analyzing the images received from front and back vehicles. We name it super perception system due to its super ability to see things through, shown in Figure 1. Only when street objects are detected in front vehicle's windshield images, the perception system will turn on its "super ability." Then, the system will combine images from front vehicles and project occluded objects on windshield image of the host vehicle. And there is no limitation of the location of front and back vehicles. The main idea is inspired from [6, 17]. The main contributions of this paper are as follows.

(1) In our paper, a novel super perception system for perceiving street object partly or fully occluded by front vehicles is proposed. Our system gives the architecture of the new generation driving aid systems based on V2V stream.

(2) The system also drafts simple communication protocol between two vehicles.

(3) We improved the results of the augmented reality method on projecting objects from front vehicle image to host vehicle image.

(4) We also proposed object-based fusion method based on affine transformation. The fusion will happen only when street objects are detected on the images of front vehicles. And only object's areas were fused to the host vehicle image to augment the perception of the intelligent vehicles.

(5) A performance evaluation method is first proposed in this paper to show the accuracy of the projection.

2. Architecture of Super Perception System

The super perception system, shown in Figure 1, of the host vehicle (vehicle A in Figure 1) can be enhanced by combining the video information from the front vehicles (vehicle B in Figure 1). Vehicle B periodically sends beacon signal to surrounding vehicles based on DSRC equipment after detecting street objects based on CNN. If street objects were detected, B sends beacon signal to A and began to receive the video stream from B. The perception system on A began the fusion process based on 3D projection model. The 3D projection model parameters were computed depending on the synchronize images from A and B. The flowchart in Figure 2 describes the processing procedure of the perception system.

The system is constituted of four function blocks boxed in blue dashed line. Function blocks includes communication protocol block, object detection block, 3D projection model block, and object-based fusion block. In the following sections, these blocks will be described in detail.

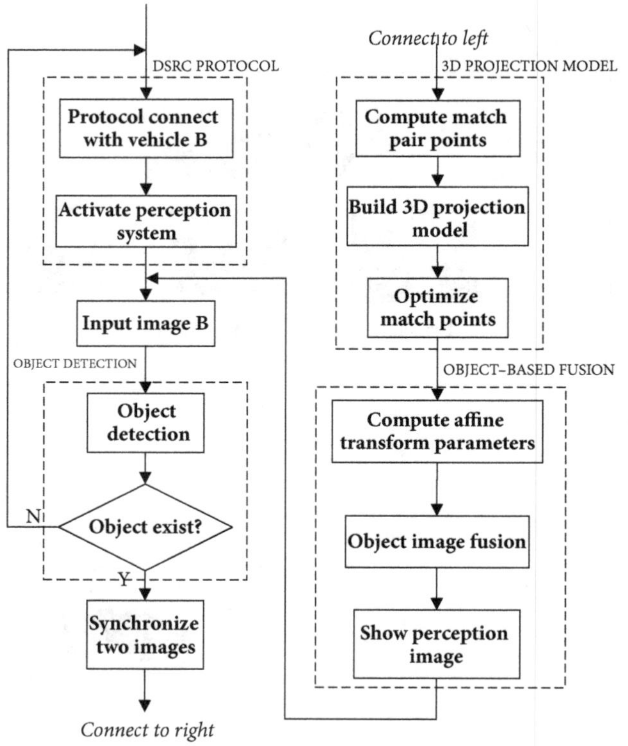

FIGURE 2: Flowchart describing the processing procedure of the perception system.

3. DSRC-Based V2V Communication

A lot of advanced driver-assistance system applications are available based on V2V communication standards DSRC [21]. V2V communication plays a decisive role in several of these cooperative approaches. Figure 3 shows a flowchart describing the communication process (in black) of each vehicle and the communication protocol (in blue) between two vehicles.

Every vehicle periodically sends beacon signals to nearby vehicles after street objects are detected in its forward images. In Figure 3 B detected cars or pedestrian in its image and then sent beacon signals to vehicle A. A receives beacon signal from B and the perception system is activated. The cooperative protocol between two vehicles is initiated, with vehicle A request for video stream and camera intrinsic parameter from vehicle B. Then, B sends those data to A. The delay caused by transmission will be lower than 100ms for per frame image. And the effect under 100ms can be ignored in perception systems, which is proved in experiments in [3]. If no objects were detected, vehicle B will stop sending videos and will send stop signal to A, which will terminate the communication between A and B.

4. SSD Based Street Objects Detection

Before the perception system is activated, the image of vehicle B needs to be analyzed first. Detection must be applied before pursuing projection, as the projection only happens when there exist some occluded street objects (which mean there are some street objects in front of B). Here we adopt an end-to-end single deep neural network to realize detection algorithm, named SSD [22]. The architecture of SSD network is shown in Figure 4. VGG16 form the early network layers (truncated before any classification layer). The layers added to the truncated VGG16 are 5 convolutional feature layers which progressively decrease in size and can produce feature maps in different scale. A small set of default boxes slide on several feature maps. For each default box, the shape offset and the confidence score were calculated for all street object categories (here we have 3 kinds of street objects; they are vehicle, pedestrian, and bicycle). The objective loss function

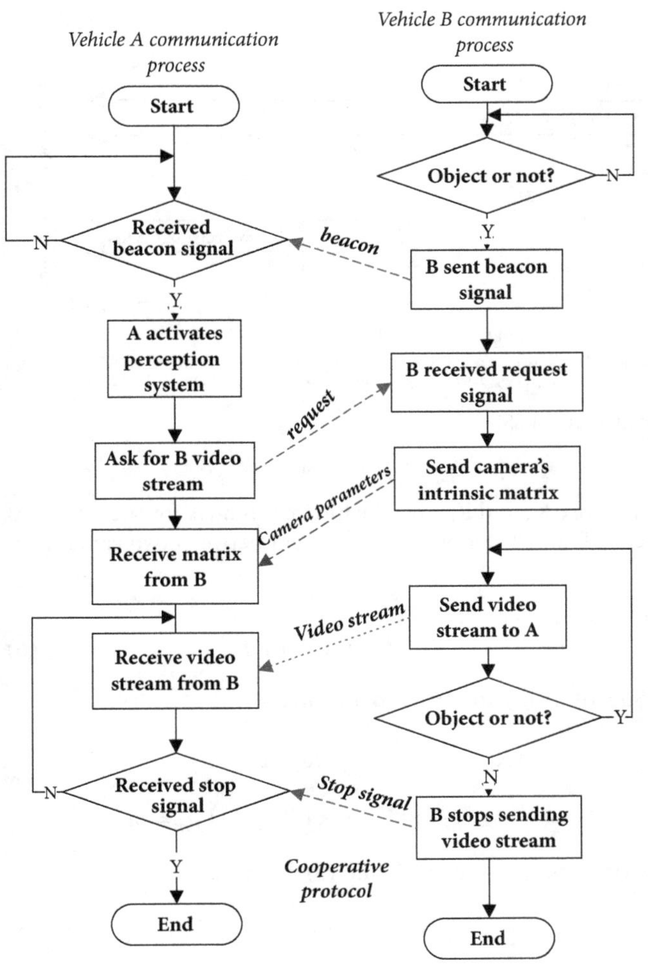

FIGURE 3: Flowchart showing the communication protocol between two vehicles and processing flow of each vehicle.

consists of the confidence loss (conf) and the localization loss (loc):

$$L(x,c,l,g) = \frac{1}{N}\left(L_{conf}(x,c) + \alpha L_{loc}(x,l,g)\right). \quad (1)$$

N is the number of matched default boxes. l means the location of predicted box and g presents the ground-truth box. $x = \{1, 0\}$ indicate the matching of default box and the ground-truth box. Then, a set of default boxes with each feature map cell are associated at the top of the network. Finally, model supplies nonmaximum suppression algorithm to find the most confidence level boxes depending on the score and region. This algorithm can achieve 90.27% mAP on our testing street dataset at 49 FPS on a Nvidia Titan GTX 1080i GPU.

SSD achieves good performance both on accuracy and on speed which can meet our request. Though SSD performs poor on small object detection, small object means far distance from vehicle's perception system which will not cause accident in this situation.

5. 3D Front-and-Back Projection Model

If street objects are detected in image of vehicle B, two images from both front and back vehicles are used to construct the 3D projection model, shown in Figure 5. Different from the projection model proposed by Yair [23–25], our 3D projection model is based on the epipolar geometry which only depends on the cameras' internal parameters and their relative pose, but independent of scene structure. As the locations of the two cameras are different, the same object will have different size, location, and deformation after projection into two cameras. Founding the transformation function between two images is the key to projection. Feature pairs in both images achieve the realization of 3D projection model. The processing can be divided into 3 parts: (1) feature pair selection; (2) camera epipolar geometry estimation; (3) object-based fusion. More details are discussed below.

5.1. Feature Pair Selection. In order to provide a representative description of the object, we can extract characteristic feature points on the object in an image. We make use of these descriptions to find the correspondence points of the same object in two images. To perform trustful points matching, it is very important that the description of the points should be invariant to the changes in scale, noise, illumination, and deformation. Feature Pairs Selection includes feature detection and feature matching.

(1) Feature Detection. Here, we adopt Lowe's SIFT method [26] as feature selection and description method. SIFT method uses a 128-element-long feature vector descriptor to characterize the gradient pattern in a properly oriented neighborhood surrounding a SIFT feature. These features are (semi-)invariant to incidental environmental changes in lighting, viewpoint, and scale. Here, we used a public-domain SIFT implementation (http://www.vlfeat.org/).

(2) Feature Matching. Emphatically, matching SIFT features in front and back images is trying to search the similarity of the those descriptors, as lacking of the relative pose of two cameras. Brute-force algorithm is adopted here to match feature pairs in the front and back images. The Euclidean distance was computed between feature vectors as the matching score. The selected matching pair needs to meet this equation (2) [26].

$$ratio = \frac{\max\left(dis\left(f_a, f_b\right)\right)}{\max \sec\left(dis\left(f_a, f_b\right)\right)} > 0.8. \quad (2)$$

$\max(dis(f_a, f_b))$ means the best matching pair and $\max \sec(dis(f_a, f_b))$ is the second best one. f_a, f_b represent points from feature map of vehicle images A and B. Figure 7(a) depicts the matching results between front and back images. The results reveal that there exists error pair matching only based on similarity. In the following section, geometrical constraints will be introduced in filter error matches.

5.2. Camera Epipolar Geometry Estimation. No matter what relative position two vehicles are at, if two images acquired by

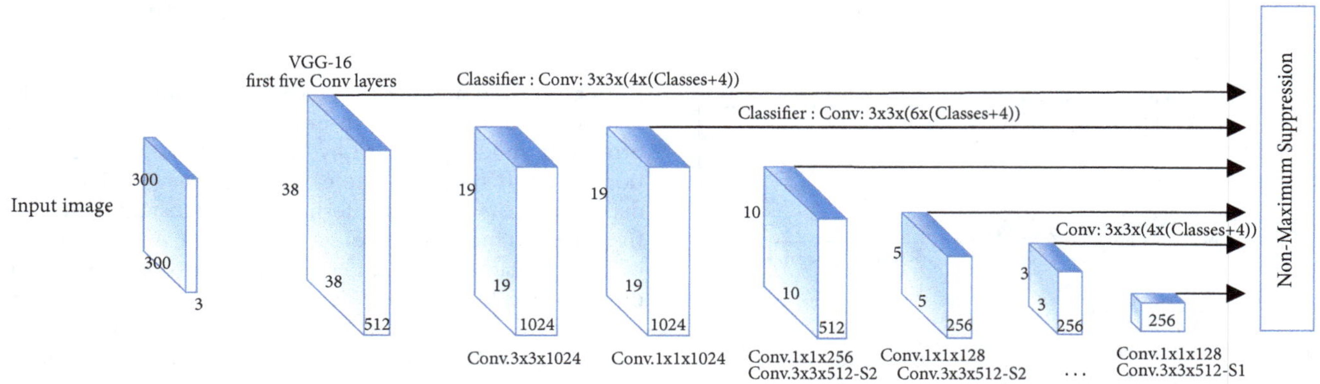

FIGURE 4: Network architecture of SSD.

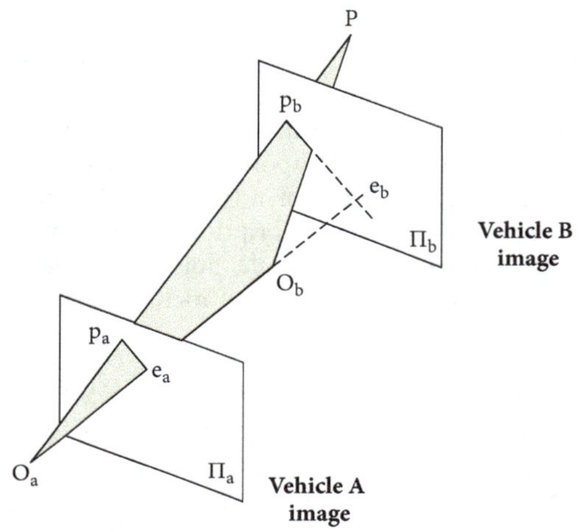

FIGURE 5: Front and back 3D projection model.

two vehicles have noncoincident centers, then the fundamental matrix F is the unique 3∗3 rank 2 homogeneous matrix which satisfies [27]

$$p_a F p_b = 0 \tag{3}$$

where (p_a, p_b) is any pair of the corresponding points in two images from vehicles A and B. This enables F to be computed from image correspondences alone. Here, we used the five-point algorithm [28]. The Normalized Eight-Point Algorithm [29] can also be used to improve the performance while lowering the efficiency.

Based on the fundamental matrix F, we can calculate the parameters rotation R and direction of translation T, which shows the relative pose between two cameras. As shown in Figure 5, a point 3-space P is imaged as p_a in the view of vehicle A and p_b in the view of vehicle B. O_a and O_b denote two cameras' optical centers, and $\prod_a$ and $\prod_b$ are the corresponding image A and image B. Geometrically, points O_a, O_b, p_a, p_b, and P all lie on the same epipolar plane, which gives the expression

$$O_b p_b \cdot (O_b O_a \times O_a p_a) = 0. \tag{4}$$

Suppose K_a and K_b give the intrinsic matrix for the two cameras. R and T denote the movement between two cameras.

$$T = O_b O_a \tag{5}$$

$$Op = RK^{-1}p + T. \tag{6}$$

Substitute (5) and (6) into formula (4):

$$\begin{gathered} \left(K_b^{-1} p_b\right)^T \left(T \times \left(RK_a^{-1} p_a + T\right)\right) = 0 \\ p_b^T \left(K_b^{-T} T \times \left(RK_a^{-1}\right)\right) p_a = 0. \end{gathered} \tag{7}$$

Comparing formulas (7) and (3), we have

$$\left(K_b^{-T} T \times \left(RK_a^{-1}\right)\right) = F. \tag{8}$$

As F is calculated, we can compute R and T with formulas (5) and (8).

5.3. Feature Pair Optimization. Generally, five pairs of points or eight pairs of points are enough to compute fundamental matrix F. In fact, we often match more features than that. Hence, we can iteratively use these features to gain an optimistic result. Here, RANSAC algorithm is employed to improve the robustness in camera motion inference. Randomly selected n small subsets "seed" (n pairs of matching points), fundamental matrix F is calculated in n times. The value of $|p_a F p_b|$ calls the residual error, which is, ideally, supposed to be zero. An F will be computed by those outlier-free seeds and will produce small residual errors in $|p_a F p_b|$ for mostly inlier matching pairs. We preserve those seeds that produce the minimum median $|p_a F p_b|$ residual errors.

After filtering the error pair points, five-point algorithm is then performed to compute the precise value of F, R, and T by using all remaining pairs in a least-square way. Also, epipolar e_a and e_b can be solved by calculating the standard Singular Value Decomposition (SVD) of F [27].

5.4. Object-Based Data Fusion

(1) Estimation of Transformation Parameters. In order to realize fusing objects on image B to image A, we need to

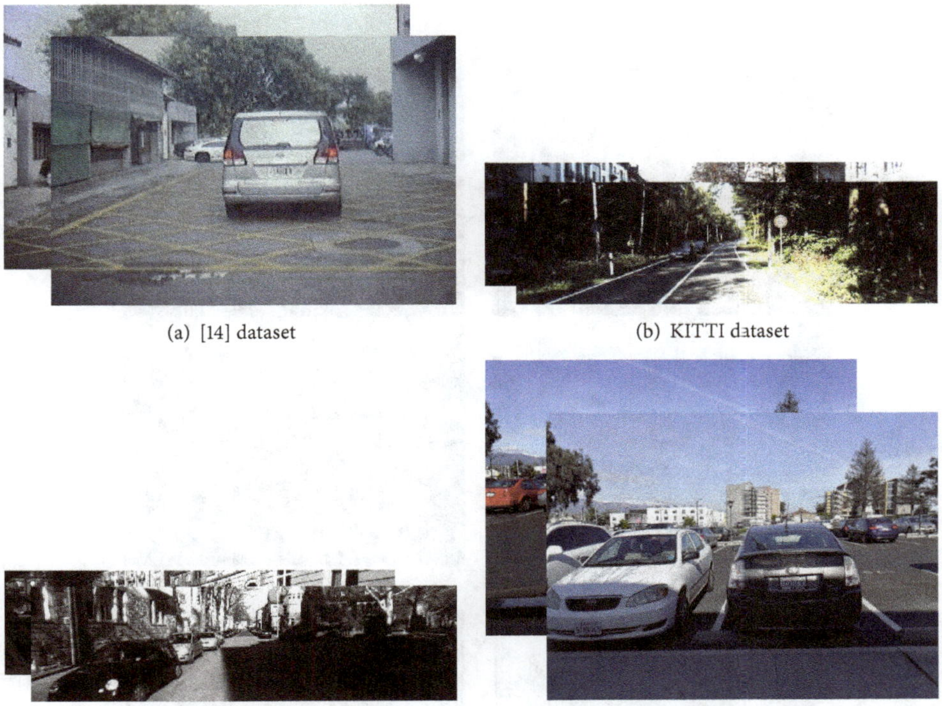

(a) [14] dataset (b) KITTI dataset

(c) Karlsruhe (d) Our dataset

FIGURE 6: Image dataset used in this paper.

figure out some information related to detected objects. The information includes size, shape, and location of the fusion region. Hence, the mapping parameters between two images need to be estimated. The work [17] used polar coordinates to approximately simulate the mapping relation between two images in a global, linear form to get visually appealing color fusion result. This method can achieve good performance in vehicles in the same lane (shown in Figure 1(a)) but failed in vehicles in different lane (shown in Figure 1(b)). As an affine transformation is a nonsingular linear transformation followed by a translation, we regard the mapping relationship between two images as an affine transformation. It has the matrix representation

$$P_B = HP_A. \tag{9}$$

P_B and P_A, respectively, represent matching pair point matrix in two images. H is the parameter matrix of affine transformation. The homogeneous formula is as follows:

$$\begin{pmatrix} x_b \\ y_b \\ 1 \end{pmatrix} = \begin{pmatrix} a_{11} & a_{12} & t1 \\ a_{21} & a_{22} & t2 \\ 0 & 0 & 1 \end{pmatrix} \begin{pmatrix} x_a \\ y_a \\ 1 \end{pmatrix}. \tag{10}$$

a_{11}, a_{12}, a_{21}, a_{22}, t_1, and t_2 are six parameters in H matrix. Five pairs of point are enough to calculate the parameters of affine transformation. However, we have more feature pairs than that. Hence, RANSAC is also used to optimize the parameter of the transformation.

Affine transformation can offer better performance than linear method in [17], but the projection of the objects from image B to image A is still not precisely correct. Because we use matching points g_a and g_b of object G to estimate the parameters of affine transformation of points f_a and f_b of object F, this rough method can only be applied to realize visualization, which helps drivers to "see" objects occluded and perceive the approximate situation ahead of the front vehicles. In order to get precise results, more information should be adopted, such as deep information of every pixel. Or we can use other transformation to replace affine transformation.

(2) Images Fusion. The fusing region, where the fusing process is applied to, is a circle area in the image of vehicle A. The center and radius of the circle depend on the detected object region location and size. Epipolar e_a and e_b can be used to eliminate those objects that are not occluded by vehicle B. The blending method is similar to [14]. The blending weight is adjusted to use more color from the front image B close to fusion center and more color from the back image A away from center toward the edge of the circle. The transparency parameter controls the mixture of two images.

6. Experiment Results

Our proposed system runs on a server with a Nvidia Titan GTX 1080i GPU.

6.1. Datasets. Experiments were performed on dataset from [17], Karlsruhe dataset, KITTI dataset, and our own dataset. Four Datasets were shown in Figure 6.

(a) Matching pairs based on similarity

(b) Optimized matching pairs based on geometry constraint

FIGURE 7: Matching pairs features between front and back vehicles.

(a) (b) (c)

FIGURE 8: Affine transformation results based on front and back vehicle images.

In [17], two minvans were equipped with forward-looking cameras and video capturing system which capture on-board video and analyze the store videos. Two vehicles were driven in a front-and-back style following configuration on many different roads. Two vans were driven in a front-and-back style in the same lane. The data were captured on many different streets.

Our perception system was applied in two situations: front and back vehicles in the same lane and vehicles in the difference lanes. The [17] dataset includes images in which vehicles are in the same lane. However, our dataset contains both situations. Our database images were acquired from videos captured by mobile phones camera. We equipped three cars with forward-looking cameras. Most of our data now were captured in parking lot area.

Karlsruhe dataset [30] contains high-quality stereosequences recorded from a moving vehicle in Karlsruhe. The sequences, which are captured by Pointgrey Flea2 firewire cameras, are saved as rectified images in ∗.png format. Ground-truth odometry from an OXTS RT 3000 GPS/IMU system is provided in a separate text file. Here, we use two frames ($\triangle$t) in the video to simulate the front and back vehicle images. And we used Karlsruhe dataset and KITTI dataset [31] to evaluate the accuracy of the projection results. However, it is hard to evaluate the results by using dataset in [17] and our dataset because the objects in back images are occluded by the front vehicles.

6.2. Feature Matching and Optimization Results. Feature matching and optimized matching results are shown in Figure 7. Results in column (a) are the matching results after pursuing Brute-force algorithm and results in column (b) show the matching results after being optimized by adopting geometrical constraint of projection model. The experiment results reveal the optimized results, in which error matches were deleted.

6.3. Affine Transformation Results. In our system, the objects in two images are supposed to meet the affine transformation. The affine transformation results are shown in Figure 8 and the quantitative evaluation is performed on KITTI and

FIGURE 9: IoU of two methods compared with ground-truth.

(a) Fusion based on polar linear method [14]　　(b) Fusion based on affine transformation

FIGURE 10: The results of two projection methods.

Karlsruhe dataset. The images in these two datasets are captured by one vehicle; here we use two frames with interval $\triangle t$ in the data frames to simulate the front and back vehicle images. $\triangle t$ is a random value within 3~10.

The (a) images represent the front vehicle images and the (b) pictures are supposed to be the back vehicle images. The (c) pictures show the results that transform the front images to the images of the back vehicles based on affine transforming.

The (b) images were taken as the ground-truth images and the affine results (c) are compared to the ground-truth images. The evaluation results are represented in Figure 9 and Table 1 shows the improving results of our method compared to [17] especially if the locations of the front and back vehicles do not follow the linear relationship. Method in [17] supposes that the front vehicle and back vehicle meet the linear model, but, in fact, most of the situations do not meet this hypothesis.

IoU (Intersection over Union) is an evaluation metric used to measure the accuracy of an object detector.

$$IoU = \frac{ObjectBox \cap GroundTruth}{ObjectBox \cup GroundTruth}. \tag{11}$$

We use IoU here to evaluate affine results which is key to the accuracy of perception. In Figure 9, red box is the ground-truth bounding box, yellow box is the result of [14] method, and green box is the result of our method.

Figure 10 reveals the projection results based on different fusion methods. Result of (a) adopts the method used in [17] which supposes the projection between front and back vehicle images satisfied the polar linear relationship, whereas result in (b) adopts the affine transform to hypothesis of the relationship between two images. Obvious improvement can be seen from the compared results. However, the results of affine transformation still were not accurate according to the facts, because the pair features choose from images that belong to different objects (most of them belong to background) in different depth. And these pair features were used to calculate the affine parameters of one object, which results in the deviation results. So we need more complicated model to imitate the relationship between the objects.

TABLE 1: Results on the KITTI and Karlsruhe dataset for different IoU thresholds.

Methods	0.6	0.7	0.8	0.9
[14]	80.43%	79.16%	65.22%	56.12%
Our method	82.52%	80.32%	67.01%	62.82%

6.4. Object Fusion Results. Affine transformation is supposed to satisfy the relationship between front and back vehicles, so the augmented fusion can be done based on the above calculated affine parameters. Figure 11 reveals the final fusion results. If the front vehicle (vehicle B) detected the object on street, it will send its image B to back vehicle (vehicle A) and the detected objects are fused in image A. The yellow rectangular shows the detected objects which are occluded by front vehicle B. The fusion process is to blend the pixels color in image A with the corresponding pixels of objects' area in image B. The fusion region is a circle where the center is the center of the rectangular and we set the transparency parameter to be 1. The blending weight is adjusted to use more color from the front image close to the center and more color from the back image away from the center.

FIGURE 11: The object-based fusion results.

7. Conclusion

In this paper, we introduce a super perception system which can "see through" vehicle and detect the fully occluded street objects. Our perception system is a good example of advanced driver-assistance system (ADAS) that can collect information from sensors in neighbor vehicles. Our future research will focus on the algorithm on how to improve the accuracy of the projection model, which means correct location and correct size of the occluded objects after projecting from front image to the back image. We also will invent a performance evaluation method to evaluate the projection results of the system.

Conflicts of Interest

The authors declare that they have no conflicts of interest.

Acknowledgments

This work was supported in part by the National Nature Science Foundation of China under Grants 61503349 and 61603357. The authors would like to thank Professor Shi Jinlong for helpful discussions. They also thank Professor Yuanfang Wang for supplying dataset of [17] and helpful comments.

References

[1] "Global Status Report on Road Safety," http://www.who.int/mediacentre/factsheets/fs358/en/, 2016.

[2] "Traffic Safety Facts," https://crashstats.nhtsa.dot.gov/#/, 2015.

[3] K. Liang, J. Mårtensson, and K. H. Johansson, "When is it fuel efficient for a heavy duty vehicle to catch up with a platoon?" *IFAC Proceedings Volumes*, vol. 46, no. 21, pp. 738–743, 2013.

[4] K.-Y. Liang, J. Martensson, and K. H. Johansson, "Fuel-saving potentials of platooning evaluated through sparse heavy-duty vehicle position data," in *Proceedings of the 25th IEEE Intelligent Vehicles Symposium, IV 2014*, pp. 1061–1068, Dearborn, MI, USA, June 2014.

[5] K.-Y. Liang, J. Mårtensson, and K. H. Johansson, "Heavy-duty vehicle platoon formation for fuel efficiency," *IEEE Transactions on Intelligent Transportation Systems*, vol. 17, no. 4, pp. 1051–1061, 2016.

[6] P. Gomes, C. Olaverri-Monreal, and M. Ferreira, "Making vehicles transparent through V2V video streaming," *IEEE Transactions on Intelligent Transportation Systems*, vol. 13, no. 2, pp. 930–938, 2012.

[7] G. Burnett, "A road-based evaluation of a head-up display for presenting navigation information," in *Proceedings of the HCI International Conference*, vol. 3, pp. 180–184.

[8] S. Ammoun and F. Nashashibi, "Design and efficiency measurement of cooperative driver assistance system based on wireless communication devices," *Transportation Research Part C: Emerging Technologies*, vol. 18, no. 3, pp. 408–428, 2010.

[9] M. A. Akhlaq, *Smart-Dashboard Augmenting Safe & Smooth Driving*, Blekinge Institute of Technology, Karlskrona, Sweden, 2010.

[10] A. Amditis, M. Bimpas, G. Thomaidis et al., "A situation-adaptive lane-keeping support system: overview of the SAFELANE approach," *IEEE Transactions on Intelligent Transportation Systems*, vol. 11, no. 3, pp. 617–629, 2010.

[11] C. Lin, C. Lin, D. Huang, and Y. Chen, "Design a support vector machine-based intelligent system for vehicle driving safety warning," in *Proceedings of the 2008 11th International IEEE Conference on Intelligent Transportation Systems (ITSC)*, pp. 938–943, Beijing, China, October 2008.

[12] Q. Xu, T. Mak, J. Ko, and R. Sengupta, "Vehicle-to-vehicle safety messaging in DSRC," in *Proceedings of the the first ACM workshop*, p. 19, Philadelphia, PA, USA, October 2004.

[13] T. ElBatt, S. K. Goel, G. Holland, H. Krishnan, and J. Parikh, "Cooperative collision warning using dedicated short range wireless communications," in *Proceedings of the 3rd ACM International Workshop on Vehicular Ad Hoc Networks (VANET '06)*, pp. 1–9, September 2006.

[14] M. Ferreira, R. Fernandes, and H. Conceicao, "Self-orgnized traffic control," in *Proceedings of the 7th ACM International Workshop on VehiculAr Inter-NETworking*, pp. 85–90, 2010.

[15] M. Ferreira and P. M. D'Orey, "On the impact of virtual traffic lights on carbon emissions mitigation," *IEEE Transactions on Intelligent Transportation Systems*, vol. 13, no. 1, pp. 284–295, 2012.

[16] "Transparent Trucks Show the Road Ahead," https://www.thecarconnection.com/news/1040575_transparent-trucks-show-the-road-ahead, 2009.

[17] C.-T. Lin, Y.-C. Lin, L.-T. Chen, and Y.-F. Wang, "Front vehicle blind spot translucentization based on augmented reality," in *Proceedings of the 2013 IEEE 78th Vehicular Technology Conference, VTC Fall 2013*, September 2013.

[18] S. Ren, K. He, R. Girshick, and J. Sun, "Faster R-CNN: towards real-time object detection with region proposal networks," in *Advances in Neural Information Processing Systems*, pp. 91–99, 2015.

[19] J. Dai, Y. Li, K. He, and J. Sun, "R-fcn: object detection via region-based fully convolutional networks," https://arxiv.org/abs/1605.06409, 2016.

[20] C. Szegedy, S. Reed, D. Erhan, and D. Anguelov, "Scalable, high-quality object detection," *Computer Science > Computer Vision and Pattern Recognition*, pp. 1412–1441, 2014.

[21] R. Penazzi, P. Capozio, M. Duncan, A. Scuderi, M. Siti, and E. Merli, "Cooperative safety: a combination of multiple technologies," in *Proceedings of the Design, Automation and Test in Europe, DATE 2008*, pp. 959–961, March 2008.

[22] W. Liu, D. Anguelov, D. Erhan et al., "SSD: single shot multibox detector," in *European Conference on Computer Vision*, pp. 21–37, Springer, 2016.

[23] Y. Wiseman and I. Grinberg, "Circumspectly crash of autonomous vehicles," in *Proceedings of the 2016 IEEE International Conference on Electro Information Technology (EIT)*, pp. 0387–0392, Grand Forks, ND, USA, May 2016.

[24] Y. Wiseman and I. Grinberg, "When an inescapable accident of autonomous vehicles is looming," *International Journal of Control and Automation*, vol. 9, no. 6, pp. 297–308, 2016.

[25] Y. Wiseman and I. Grinberg, "Autonomous vehicles should not collide carelessly," *Advanced Science and Technology Letters*, vol. 133, pp. 223–228, 2016.

[26] D. Lowe, "Distinctive image feature from scale-invariant key points," *International Journal Computer Vision*, vol. 60, pp. 91–100, 1997.

[27] R. Hartley and A. Zisserman, *Multiple view geometry in computer vision*, Cambridge University Press, 2nd edition, 2003.

[28] D. Nistér, "An efficient solution to the five-point relative pose problem," *IEEE Transactions on Pattern Analysis and Machine Intelligence*, vol. 26, no. 6, pp. 756–770, 2004.

[29] R. I. Hartley, "In defense of the eight-point algorithm," *IEEE Transactions on Pattern Analysis and Machine Intelligence*, vol. 19, no. 6, pp. 580–593, 1997.

[30] A. Geiger, J. Ziegler, and C. Stiller, "StereoScan: dense 3D reconstruction in real-time," in *Proceedings of the IEEE Intelligent Vehicles Symposium (IV '11)*, pp. 963–968, Baden-Baden, Germany, June 2011.

[31] A. Geiger, P. Lenz, C. Stiller, and R. Urtasun, "Vision meets robotics: the KITTI dataset," *International Journal of Robotics Research*, vol. 32, no. 11, pp. 1231–1237, 2013.

Potentialities of Autonomous Vehicles for Online Monitoring of Motorway Traffic Volume

Hyun-ho Chang[1] and Byoung-jo Yoon[2]

[1]*Graduate School of Environmental Studies, Seoul National University, Seoul, Republic of Korea*
[2]*Department of Urban Engineering, Incheon National University, Incheon, Republic of Korea*

Correspondence should be addressed to Byoung-jo Yoon; bjyoon63@inu.ac.kr

Academic Editor: Md. A. S. Kamal

The fact that real-time autonomous vehicle (AV) traffic volume can be collected without a field detector by virtue of advanced global positioning system (GPS) and wireless communication technologies can render a promising solution to online monitoring of traffic volume in the upcoming AV era. To demonstrate this opportunity, this paper proposes a new method to monitor real-time motorway traffic volumes for road locations where no detector is installed using AV traffic volume. The modeling concept is based on the obvious fact that AV traffic volume is a direct portion of total traffic volume. The capabilities of the method are demonstrated through an experimental study using real-world GPS-enabled smartphone vehicle navigation data. The results show that online motorway traffic volume can be effectively monitored throughout the day with 5.69% average error at the 14.91% penetration rate of AVs during the daytime. Therefore, it is expected that AVs can at least be used as complementary means for the role of vehicle detectors in the near future due to the fact that the detection range of AVs is not spatially constrained.

1. Introduction

Real-time traffic volume is essential for traffic control and management in intelligent transportation systems (ITS). Since the introduction of ITS, various vehicle detectors that are based on in-roadway and over-roadway sensor technologies have been utilized to monitor traffic flow variables (i.e., volume, speed, and density). Typically, the spatially consecutive and dense deployment of vehicle detectors is utilized for the instant and accurate monitoring of the variables due to the fact that the spatial coverage of current detectors ranging from conventional inductive loop to radar sensing is constrained to fixed point or fixed short length. This surveillance strategy requires extensive budgets and resources in order to guarantee the reliability of monitored information. In addition, vehicle detectors are operated without the change of their locations in many cases after they are installed in the field.

To address these obstacles from the perspective of ITS infrastructure management and the constrained spatial coverage of vehicle detectors, several investigations to produce the three traffic variables for unmeasured points or road sections using advanced data have been introduced in our literature review. The existing studies have been focused on travel speed or density, showing remarkable and distinguished estimation accuracy according to the data used. Despite the importance of traffic volume, however, any method for dynamic traffic volume still has not been reported even under the condition of advanced data. As such, dynamic traffic volume estimation for unmeasured locations is a new research topic for solving the spatially constrained and fixed coverage of vehicle detectors and then for reducing the budgets and resources for surveillance infrastructure in modern and near future ITS.

Fortunately, it is expected that AVs play a key role as a new moving probe source by virtue of an advanced global positioning system (GPS) and communication technologies for their driving. To mine this promising opportunity, the aim of this study is to initiatively demonstrate the potential of autonomous vehicles (AVs) for producing dynamic traffic volumes for an unmeasured location by using a novel method. The proposed method in this paper is to develop a new concept that contracts and then expands AV traffic volumes into overall traffic volumes. In addition, based on the analysis results, some findings and research directions for the online monitoring of traffic volume in the present and near future era are discussed.

2. Literature Review

Despite advanced vehicle detection technologies in modern ITS, it seems that the measurements are insufficient due to constrained space coverage (i.e., fixed point or short-length detection area) and the high costs of installation and maintenance. To address these hindrances, academic investigations to dynamically estimate three traffic flow variables (i.e., volume, speed, and density) have been conducted using advanced (probe) data. Note that the literature review of this paper is focused on academic research in which real-world advanced data was employed to produce the three variables.

For the dynamic measurement of travel speed or time, four types of probes have been employed: automatic vehicle identification (AVI) [1], vehicle-based GPS mobile sensor [2, 3], cellular phone (CP) [4–6], and GPS-enabled CP [7, 8].

Probe data that is collected from AVI or vehicle-based GPS mobile sensor systems was employed as a direct portion of vehicle travels in [1–3]. AVI probe data from the San Antonio TransGuide system was employed for estimating average link travel time in [1]. The study demonstrated that a low sampling rate, less than 1%, can represent average link travel time effectively. Travel speed was estimated using GPS-based taxi probe data in [2, 3]. The two studies indicated that imperfect probe data can be used for monitoring traffic status in practice. The results of the two studies in [1, 2] are similar to [9, 10] in terms of a minimum sampling rate of 1~3%.

CP probe data based on tracking CP footprints in a cellular communication network was used to measure travel speed and time between freeway locations in [4–6]. The potentialities of a CP-based system for monitoring traffic status were demonstrated. Despite the reliable measurement accuracy of the CP probe, research indicated in [4] that the noise of the CP probe should be addressed successfully. The noise problem occurs when travel speed is low. Thus, CP-based monitoring systems are suitable for a high-speed roadway such as a motorway [5]. The measurement accuracy was highly improved through an advanced tracking algorithm in [6]. The CP-based system only measures traffic status between two fixed cells (i.e., antennas), as the system totally relies on the geographical configuration of cells. Therefore, the CP-based system is not suitable for measuring traffic status in an urban road network.

Noticeably, two studies demonstrated that the spatiotemporal dotted trajectory of an individual GPS-enabled smartphone can be employed as an effective traffic probe to measure accurate traffic status in [7, 8]. The two studies proved that a 2-3% penetration of GPS probes in the driver population is sufficient to measure accurate traffic speed [7]. The uncertainty of speed measurements caused by the noise of a GPS signal was effectively addressed in [8]. The study showed that the behaviors of speed measured by smartphone trajectory are statistically the same as those of actual speed. The strength of the trajectory probe is suitable for various measurements of traffic status without spatial constraints.

A few studies for estimating traffic density [11] and hourly traffic volume [12, 13] have been reported. A combination method of a kinematic wave model and a probability model for dynamically estimating traffic density was developed based on reconstructing individual vehicle trajectory using heterogeneous data (loop detector, AVI probe, and GPS probe data) in [11]. An innovative methodology for estimating hourly traffic volumes using cellular-phone call count and its probability crossing intercell boundaries was proposed in [12, 13]. The two studies demonstrated that cellular-phone call data can be employed for inferring hourly traffic volumes, where the resulting estimation error was 20%. Despite the initiative efforts, it can be seen that more accuracy should be achieved with more short time length such as a level of five-minute data aggregation for the applications of ITS.

Based on the literature review on the dynamic estimation of the traffic variables using advanced probe data, several studies for traffic speed and density have reached an acceptable level in terms of estimation accuracy. However, more investigations of traffic volume need to be performed at a level of acceptable accuracy. In a strict sense, research for estimating dynamic traffic volumes has not been reported or highly correlated using advanced probe data that is a direct portion of traffic volume. In addition, the dynamic evolution of traffic volume behaves like a chaotic system [14, 15]. That is, a time-series of traffic volume data naturally reveals intensive and wide fluctuations in ergodic and nonperiodic manners. This fact makes it difficult to directly estimate reliable traffic volumes by using either direct low probe data from a GPS-enabled vehicle or indirect probe data from a GPS-enabled mobile phone. More importantly, a new approach should be developed, which is capable of adaptively recognizing the temporal evolution of traffic volumes even in the case when the penetration rate of probe data available to traffic volume is low. In this context, the direct monitoring of dynamic traffic volume using the trajectory probe data of AV is one of new research directions in modern ITS.

3. Methodology

3.1. Approach Concept. The operation of AVs should be supported with advanced GPS and communication technologies. That is, AVs that are on a driving service can be considered a moving GPS probe in a road network. It is also surely expected that detailed point-to-point operation trajectory data can be stocked by virtue of the GPS technology, and then the trajectory data can be transferred to an advanced data centre through the communication technology online. This develops the assumption that AV probe volume is a direct portion of total traffic volume or at least is very highly related to that in some way. If this assumption is reasonable, then traffic volume at a target road location can be produced using a suitable relationship between probe volumes and traffic volumes that are collected from the nearby locations of the target road location. Moreover, the AV trajectory data is not spatially constrained unlike existing vehicle detectors, and thus, the probe volume can be accurately monitored at any road location. This implies in turn that traffic volume at any road location (used by AVs) can be effectively monitored using the probe volume. Therefore, it is expected that AVs render golden opportunities against the traffic surveillance system of modern and future ITS.

To demonstrate the potential of AVs for the real-time monitoring of traffic volume, a new method to produce dynamic traffic volume at any road location using AV probe volume data is proposed in this study. The method is developed on the basic concept that probe volume data provide key information to solve the uncertainty problem in the direct monitoring of traffic volume. This modeling approach is also supported by the fact that probe volume is at least a direct portion of traffic volume. The method consists of two modeling concepts: contraction and expansion of probe volume. In the contraction method, time-series probe volume data are adjusted into suitable data by eliminating its unnecessary random noise. The method also adaptively interpolates zero probe volume values into useful values, because zero probe volume values inevitably occurs when either traffic volume and/or the penetration rate of AV is low. In the expansion method, an adjusted probe volume at a road location where a traffic volume estimation is desired is converted into a traffic volume value by using a relationship between adjusted probe volumes and traffic volumes. As such, the combination of the two methods can solve the problem of direct monitoring traffic volume efficaciously by diminishing the number of uncertainties that inevitably occurs in solving this problem.

3.2. Contraction Method. It is natural that the temporal evolution of AV probe volume reveals more wide relative variations than that of overall traffic volume under the condition that the penetration rate (0.0~1.0) of AV to traffic volume is less than 1.0. This is because the probe volume is a sort of random sample with a given penetration rate. In this case, undesirable estimation results (i.e., over- and under-estimation problems) are unavoidable when a time series of probe volume data that includes the random-sampling variation in itself is directly used for the monitoring of traffic volume without any filtering process. This problem becomes more serious when the sampling variability of collected temporal probe volumes increases under the condition of low AV penetration rates and/or low traffic volumes. Zero probe volumes can also occur frequently, even when traffic volumes are low and AV penetration rates are not low.

To address this problem effectively and to ensure the reliability of traffic volume monitoring, two processes that adjust raw probe volumes into suitable probe volumes are essentially required as follows: unnecessary random variations that intrinsically exist in raw probe volumes should be filtered; and zero probe volume values should also be interpolated with useful values. The two processes are concurrently conducted by a contraction method proposed in this study.

The contraction method is devised based on the assumption that the temporal variation of actual traffic volumes is highly related to that of raw probe volumes. This assumption could be reasonable if the raw probe volume that include random-sampling variation is a part of the traffic volume.

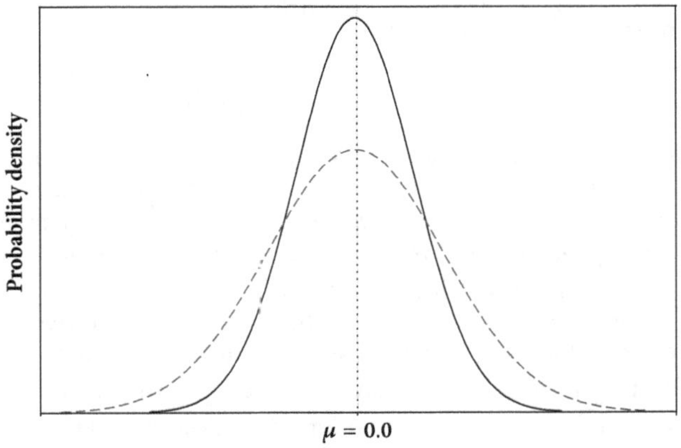

FIGURE 1: Distributions of relative variation.

Thus, the unnecessary variation of raw probe volumes can be removed by using the distribution of relative variation (RV) of raw temporal probe volumes and that of RV of temporal traffic volumes. To measure time-series RV values, time-series values and moving-average values for traffic volume and probe volume are defined as follows. Let $s = \{tg, up, dn\}$ be the target location (tg) and the upstream and downstream of tg, respectively. Let $x = \{q, p\}$ be a set of traffic volume (q, vehicles per length of time interval) and probe volume (p, vehicles per length of time interval). Let $\mathrm{X} = \{\mathrm{Q}, \mathrm{P}\}$ be time-series sets of x. Note that a form of time series (i.e., a series of time intervals) at the present time interval (t) toward the past is defined as $\mathrm{T} = [(t), (t-1), \ldots, (t-d)]$, where d is the embedding size of time series. Let $\mathrm{X}_s = [x_s(t), x_s(t-1), \ldots, x_s(t-d)]$ be a time series of x for s. Let $\mathrm{X}_s^b = [x_s^b(t), x_s^b(t-1), \ldots, x_s^b(t-d)]$ be a time series of moving-average values for X and s, where each element of X_s^b (i.e., $x_s^b(i)$) is calculated as $[\sum_{k=0}^{m} x_s(i-k)]/[m+1]$ with m (≥ 1), ($\forall x, s$, and $i \in \mathrm{T}$). Let $\mathrm{R}_s^X = [r_s^x(t), r_s^x(t-1), \ldots, x_s^x(t-d)]$ be a time series of RV values for X and s. Thus, each element of R_s^X is computed using each element of X_s and that of X_s^b (where if $s = tg$, then $\mathrm{X} \neq \mathrm{Q}$), as follows:

$$r_s^x(i) = \frac{x_s(i) - x_s^b(i)}{x_s^b(i)}, \quad \forall s, x, i, \ i \in \mathrm{T} \tag{1}$$

Figure 1 shows the RV distributions of traffic and probe volumes (i.e., R_s^Q and R_s^P), where the variance of probe volume is greater than that of traffic volume. The RV distribution of probe volume can be adjusted similar to that of traffic volume using the standard deviation. This statistical principle is employed to modify a variation of temporal probe volumes. Let σ_s^X be the standard deviation of R_s^X (where if $s = tg$, then $\mathrm{X} \neq \mathrm{Q}$). Let $\mathrm{P}_s^a = [p_s^a(t), p_s^a(t-1), \ldots, p_s^a(t-d)]$ be a time series of adjusted probe volumes for s. Finally, each element of P_s^a is estimated based on each component of P_s^b and the rate of σ_s^Q to σ_s^P as

$$p_s^a(i) = p_s^b(i) \times \left[1.0 + r_s^P(i) \times \frac{\sigma_s^Q}{\sigma_s^P}\right], \quad \forall s, i, \; s \neq tg, \; i \in \mathrm{T} \tag{2}$$

Thus, each probe volume (i.e., $p_s(i)$) is contracted into $p_s^a(i)$ by removing unnecessary random-sampling variation. Additionally, $p_s^a(i)$ is effectively generated in the case of a low value of $p_s(i)$, and $p_s(i)$ is also interpolated with a useful value even when $p_s(i)$=0.0, if $p_s^b(i) > 0.0$, $\sigma_s^P > 0.0$, and $\sigma_s^Q > 0.0$.

Due to the fact that no traffic volume is collected at the target location, it is impossible to directly compute the adjusting factor (i.e., the rate of σ_s^Q to σ_s^P). Therefore, it is assumed that σ_{tg}^P (i.e., the standard deviation of R_{tg}^P) is more similar to σ_s^P when P_{tg}^b is nearer to P_s^b, $(s \neq tg)$, from the viewpoint of RV distribution of probe volume. This can be supported by the rational reasoning that penetration rates of AV between the three locations are similar or at least not significantly different, and then the similarity of P_s^b values between different locations highly relies on Q_s^b values even though Q_{tg}^b cannot be estimated. Based on this assumption, closeness (c_s) between P_{tg}^b and P_s^b can be a direct solution to combine the two adjusting-factor values of upstream and downstream locations into an adjusting-factor value for tg. Euclidean distance metric that is one of direct and widely applied techniques to determine the degree of similarity in the time-series analysis of discrete dynamical system [15] is used in this study, and c_s between P_{tg}^b and P_s^b is calculated as

$$c_s = \left[\left|p_{tg}^b(t) - p_s^b(t)\right|^2 + , \dots , + \left|p_{tg}^b(t-d) - p_s^b(t-d)\right|^2\right]^{1/2}, \quad s \neq tg \tag{3}$$

The inverse of c_s (i.e., $c_s^{-1} > 0.0$) that is effectively utilized in pattern selection-based traffic volume forecasting [15, 16] is employed to weight the two adjusting-factor values of upstream and downstream locations. According to this consideration, a value of adjusting factor (i.e., the rate of σ_{tg}^Q to σ_{tg}^P) for tg at (t) is estimated using a weighted average function, which combines the two adjusting-factor values by c_s^{-1}. Finally, a contracted probe volume ($p_{tg}^a(t)$) for tg at (t) is computed as

$$p_{tg}^a(t) = p_{tg}^b(t) \times \left[1.0 + r_{tg}^P(t) \times \frac{\sum_s\left(\left(\sigma_s^Q/\sigma_s^P\right) \times c_s^{-1}\right)}{\sum_s c_s^{-1}}\right], \quad s \neq tg \tag{4}$$

As such, $p_{tg}(t)$ is modified into $p_{tg}^a(t)$ by using the two adjusting-factor values (i.e., σ_s^Q/σ_s^P) and the inverse of c_s (i.e., c_s^{-1}). Furthermore, $p_{tg}^a(t)$ is robustly estimated when a value of $p_{tg}(t)$ is very low, and $p_{tg}(t)$ is also interpolated with a useful value even when $p_{tg}(t)$=0.0, if $p_{tg}^b(t) > 0.0$, $r_{tg}^P(t) > 0.0$, and $\sigma_s^P > 0.0$.

3.3. Expansion Method. To expand the contracted probe volume (i.e., $p_{tg}^a(t)$) to a traffic volume (i.e., $\hat{q}_{tg}(t)$) for the target location (tg) at time interval (t), a weighted power curve is employed to determine a relationship between contracted probe volumes (i.e., $p_s^a(i)$, , $s \neq tg, i \in \mathrm{T}$) and traffic volumes (i.e., $p_s(i)$, , $s \neq tg, i \in \mathrm{T}$) in this study. The expansion method is stated according to two parts: a weighting function and the determination of an optimal fitting curve. From the perspective of temporal development of traffic flow, it is self-evident that the temporal evolution of traffic flow nearer to (t) is more related to traffic flow at (t) [14–16]. This is considered with the bisquare weighting function that was introduced to explain nonstationary relationships between spatial elements in [17]. Let $W_s = [w_s(t), w_s(t-1), \dots, w_s(t-d)]$ be a series of weight values (0.0~1.0) for T. The variation of weight value according to the proximity of time is illustrated in Figure 2(a), which can be efficaciously used in the case that the penetration rate of AV varies according to the time periods of day (e.g., peak and off-peak time). The bisquare function for temporal nonstationarity can be expressed as

$$w_i = \left[1 - \left(\frac{i}{d}\right)^2\right]^2, \quad i = [0, 1, \dots, d] \tag{5}$$

In order to find an optimal power curve, traffic volume data and contracted probe volume data for upstream and downstream locations are used as a dependent variable and an independent variable, respectively. The dependent and independent variables are defined as follows. Let $\mathrm{Q} = [\mathrm{Q}_{up}, \mathrm{Q}_{dn}]$ and $\mathrm{P} = [\mathrm{P}_{up}^a, \mathrm{P}_{dn}^a]$ be sets of traffic volumes and adjusted probe volumes for the upstream and downstream locations, respectively. In addition, let $\mathrm{W} = [\mathrm{W}_{up}, \mathrm{W}_{dn}]$ be a set of weight values for the two locations. For the convenience of the description of the expansion method, these definitions are redefined with the number of observations (N=2×(d+1)). Let $\mathrm{Q} = [q_1, q_2, \dots, q_N]$ and $\mathrm{P} = [p_1, p_2, \dots, p_N]$ be dependent and independent variables, and let $\mathrm{W} = [w_1, w_2, \dots, w_N]$ be a set of weight values.

The temporal evolution of traffic volume states reveals intensive variation in ergodic and nonperiodic manners [14, 15]. Hence, it is natural that the temporal variation of probe volumes varies more steeply and widely than that of traffic volumes, even though the probe is a direct part of traffic volume. Thus, if a linear regression model is employed, then unacceptable results (e.g., repetitive overestimations and underestimations, and even negative estimations) can occur by failing the directionality of relationship between P and Q. To prevent these undesirable results, a power curve with versatility in curve fitting (ranging from logarithmic, linear, to positively exponential types) is used to understand the relationship between the two variables as shown in Figure 2(b). The power curve for the members of [P, Q] (i.e., $[p_i, q_i]$, $i \in N$) is defined as

$$q_i = \alpha \cdot p_i^{\beta} + \gamma \tag{6}$$

where α (>0.0) and β (>0.0) are the coefficient and exponent of p_i, respectively; γ ($0.0 \leq \gamma \leq \gamma_{max}$) is a constant term; $\gamma_{max} = \min\{\overline{q}_s\}$; and $\overline{q}_s$ is an average of elements of Q_s, where $s =$

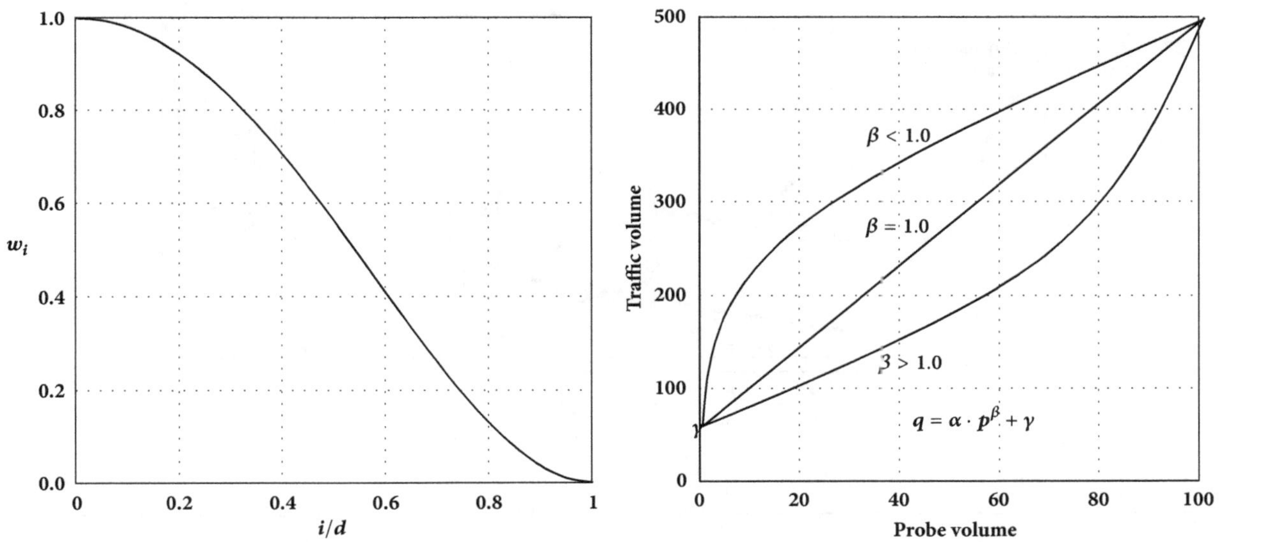

(a) Weight curve by time proximity

(b) Versatility of curve fitting

FIGURE 2: Power curve for expansion of probe volume.

$\{up, dn\}$. To prevent negative estimations, α is greater than 0.0 and γ is greater than or equal to 0.0. β is greater than 0.0 and γ is less than or equal to γ_{max}, since traffic volumes do not decrease according to the increase of probe volumes. For an optimal curve that minimizes total estimation error, a local error for each observation can be expressed as

$$\epsilon_i = q_i - \left(\widehat{\alpha} \cdot p_i^{\widehat{\beta}} + \widehat{\gamma}\right) \tag{7}$$

where $\widehat{\alpha}$, $\widehat{\beta}$, and $\widehat{\gamma}$ are optimal α, β, and γ values, respectively, and ϵ_i is the estimation error for observation i, $i \in N$.

As mentioned, temporal traffic volumes fluctuate widely, and then the quantity of traffic volume varies from low to high levels. Thus, a family of the residual sum of squares that are widely employed as an objective function for determining an optimal fitting can have a 'bias and variation problem' in the case of low traffic volumes [18]. That is, the low traffic volume can be over- or underestimated, due to its low contributions in decision-making of an optimal curve fitting.

To handle this problem effectively, the sum of a weighted absolute relative error is used as an objective function of a minimization problem to determine an optimal power curve. The absolute relative error also provides an unbiased basis [18], and thus, it is widely used as a performance measure in the area of time-series estimation and prediction. Here, a minimization problem for determining an optimal expansion curve is defined as

$$\begin{aligned} \text{Min.} \quad & \frac{\sum_{i=1}^{N} w_i \times \left| q_i - \left(\widehat{\alpha} \cdot p_i^{\widehat{\beta}} + \widehat{\gamma}\right)\right| / q_i}{\sum_{i=1}^{N} w_i} \\ \text{S.T.} \quad & 0.0 < \widehat{\alpha}, \\ & 0.0 < \widehat{\beta}, \\ & 0.0 \le \widehat{\gamma} \le \gamma_{max} \end{aligned} \tag{8}$$

Once, the estimated values of $\widehat{\alpha}$, $\widehat{\beta}$, and $\widehat{\gamma}$ for an optimal curve are identified through solving the minimization problem, a traffic volume for tg at (t) is directly produced as

$$\widehat{q}_{tg}(t) = \widehat{\alpha} \cdot p_{tg}^{a}(t)^{\widehat{\beta}} + \widehat{\gamma} \tag{9}$$

where $\widehat{q}_{tg}(t)$ and $p_{tg}^{a}(t)$ is the estimated traffic volume and the adjusted probe volume for tg at (t), respectively.

4. Results and Potentialities

4.1. Study Design. In order to demonstrate the potential of GPS probe data collected by autonomous vehicles, a case study was conducted using two types of data: GPS probe volume data and traffic volume data. The GPS probe data that was collected by a smartphone vehicle navigation system is most similar to the probe data of autonomous vehicles under the present conditions, due to the fact that the vehicle-probe volume is a direct portion of traffic volume. In this context, it seems at least that the used probe data contains the characteristics of probe data that are collected through autonomous vehicles, even though the features of mixed traffic flow with general vehicles and autonomous vehicles have not been investigated with real-world data so far.

The used motorway data is shown in Figure 3. The test bed is a part of the Seoul External Circulation Motorway 100, one of the main motorways in South Korea. The target road location consists of four lanes, and the upstream road section includes one interchange and one junction and the distance is 11.6 km. The downstream road section contains two interchanges and four junctions and the distance is 25.6 km. It can be seen that the test-bed conditions are unfavorably severe for the experimental condition, whereas the test-bed conditions are desirable to demonstrate the potential of the

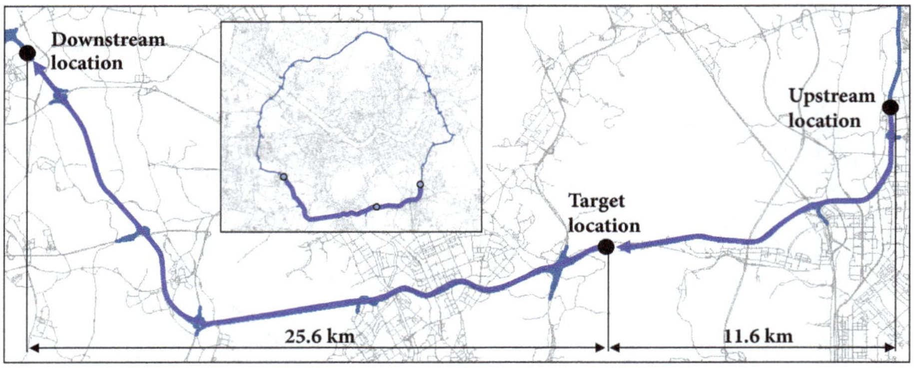

FIGURE 3: Test bed.

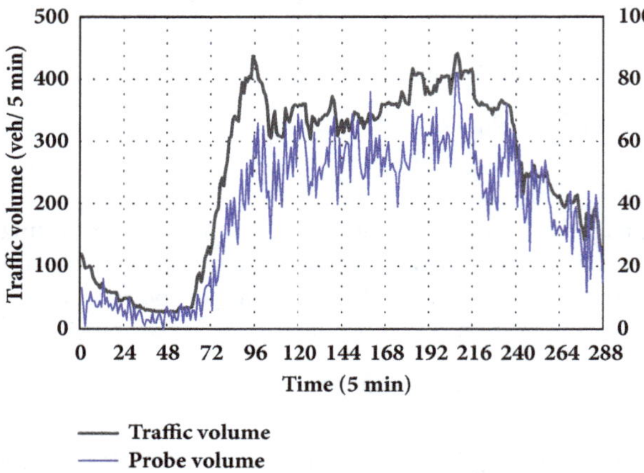

FIGURE 4: Temporal variation of traffic and probe volumes.

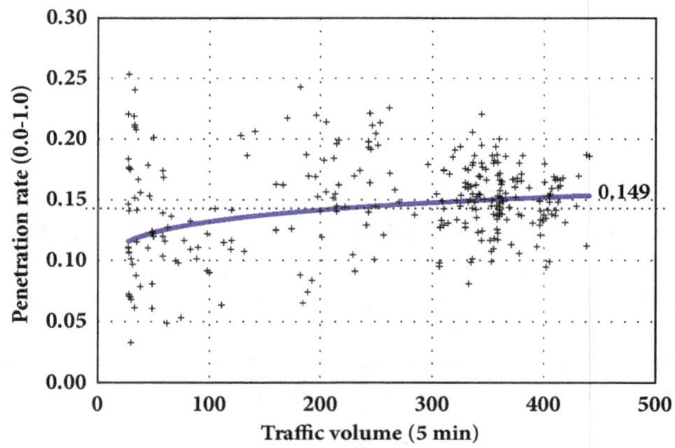

FIGURE 5: Penetration rate according to traffic volume.

proposed method in this paper. One-week individual point-to-point trajectory data was collected on December 24-30, 2016. The individual data was aggregated with a 5-minute interval. In addition, traffic volume data with a 5-minute aggregation was collected by an electronic toll collection system.

The temporal evolution of the two types of data are shown in Figure 4. The traffic volume varies from 27 to 442, and its temporal development reveals intensive variation in terms of relative percentage difference (RPD) (%), $[x(t+1) - x(t)]/x(t) \times 100$. RDP values range widely from -21.10 to 39.27. Regarding the probe volume ranges from 0 to 82, the RDP values vary from -100.0 to 800.0 except for a zero probe volume. As such, it is obvious that the temporal evolution of probe volumes exhibits more intensive and steep variation than that of traffic volumes by means of RDP. Despite these intensive variations, the statistical correlation between the two types of data is up to 0.933. This fact directly indicates that the probe volume is a direct portion of traffic volume and can reflect the features of traffic volume in some way.

Figure 5 shows the penetration rate (PR) of probe volume to traffic volume, where PR = [probe volume / traffic volume]. PR widely varies spanning from 0.06 to 0.25 with an average of 0.149. It seems that the evolution behavior of PR according to traffic volume is near to a mixed state but also has a closed boundary condition. The width of variation becomes narrower when traffic volumes increase, due to the fact that the sampling variability decreases when either the sample size (i.e., probe volume) increases or the variance of population (i.e., traffic volume) decreases. Moreover, a trend curve of PR shows logarithmic growth and increases when traffic volumes increase. This is because drivers have a tendency to use more vehicle navigation systems when their travel distance increases and because traffic congestion usually occurs at daytime.

To measure the performance of the propose method in this paper, the following four performance measures were carefully selected. Absolute percentage error (APE, %) and relative percentage error (RPE, %) provide a useful basis for comparison when traffic volume varies widely [15, 16]. APE and RPE have a weakness in the case of low traffic volume, as the relative temporal variation of low traffic volume is high. In this vein, straight error for lane (SEL, vehicles per lane), which can be useful in practice, is introduced in this study. The hit rate, one of crucial performance measures for real-life applications, was also utilized with RPE and SEL. In addition, the mean of APEs was employed to analyze and identify the optimal parameter values (i.e., d and m values)

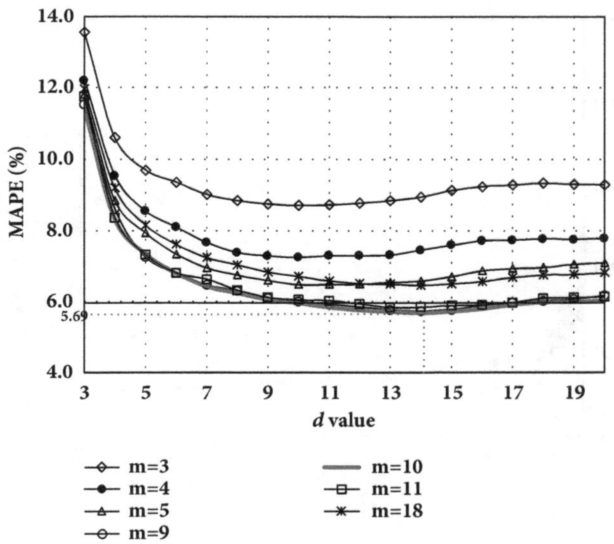

FIGURE 6: Effects of m and d values on estimation error rates.

of the presented method. APE, RPE, and SEL are expressed as

$$\text{APE}\,(\%) = \frac{|\hat{y}_i - y_i|}{y_i} \times 100, \quad y_i > 0.0 \tag{10}$$

$$\text{RPE}\,(\%) = \frac{(\hat{y}_i - y_i)}{y_i} \times 100, \quad y_i > 0.0 \tag{11}$$

$$\text{SEL}\,(\text{veh}) = \frac{(\hat{y}_i - y_i)}{l} \tag{12}$$

where y_i and $\hat{y}_i$ is the observed value and the estimated value of sample i, respectively, and l is the number of lanes.

4.2. Results and Findings. The proposed model was developed based on the combination of the contraction and conversion method (C2C). Hence, the performances of the C2C method are highly dependent on the m and d values in terms of estimation accuracy. The m and d values contribute key roles to contract the temporal variation of probe volumes, and the d value highly influences the determination of optimal curve fitting. The effects of the combination of the two parameter values on estimation accuracy are shown with estimation error rates in Figure 6. As for the m value, the estimation error curve steeply decreases (m=3⟶9) and then stays (m=9⟶11) at the optimal error space and then increases (m=11⟶18) when the d values are greater than 9. This indicates that a locality of temporal evolution of probe or traffic volumes exists in terms of moving average, even though the temporal development of probe and traffic volumes reveals intensive and steep variations. Regarding the d value, the estimation error exponentially decreases to the optimal error space and then gradually increases when the d value increases with little variation. This indirectly implies that a locality of temporal evolution of probe or traffic volumes exists in terms of determination of optimal relationship between probe and traffic volumes, whether the boundary condition is obvious or not. The optimal error space is very stable within a minimal error +0.5%, which indicates that suitable parameter values can be analyzed and determined within the margin of error on a daily or monthly basis in advance.

In addition, the optimal m and d values of 10 and 14, respectively, were selected for more analysis.

Two relationships between α and β values according to γ values are shown in Figures 7(a)-7(b), where the explanatory power of probe volume data is divided into two regimes with an obvious boundary condition. As for γ=0.0, the β value exponentially decreases when the α value increases, showing a high relationship with the R^2 value of 0.93. The cases of $\beta < 1.0$ reach to 83.02%. This means that the relationships between probe and traffic volumes are logarithmic in many cases. This fact also indicates that negative estimations inevitably arise in the case of very low probe volume if a linear relationship is used, which is directly connected to the prediction failure. Regarding the case of γ >0.0, the β value steeply decreases according to the increment of the α value, and the two parameters are more highly connected than the upper relationship by means of R^2. The cases of $\beta > 1.0$ are up to 91.90%. This indicates that the relationships between probe and traffic volumes are upward in many cases. This fact also implies that the underestimation problem unavoidably occurs when probe volume is very low if a single linear relationship is employed. Therefore, it can be seen in our case that if a linear model is used, estimation failure inevitably occurs in the case of low traffic volume except for a few cases of γ >0.0.

Figure 8 demonstrates the time-series variations of raw probe volumes and contracted probe volumes. Extreme variations, which can cause undesirable estimation results, are adjusted within the range of temporal variations of traffic volumes. The standard deviation of the RDP (SDRDP) of raw probe volume is 76.94%, whereas that of filtered probe volume is 7.73%. Similarly, adjustment gain is up to 89.95% [=(76.94-7.73)/76.94×100]. The SDRDP value of adjusted probe volume is also similar to that of traffic volume (7.45%). This suggests that extreme estimations can be effectively prevented through the contraction of temporal variation of probe volumes.

The relationships between probe volumes and traffic volumes for the before and after cases are shown in Figure 9. The contraction method effectively improves the relationship of the two variables in terms of R^2, where the R^2 value increases from 0.84 to 0.95. It can be seen at least that this result is acceptable, even considering that the natural attribute of R^2 increases when the number of observations increases. Specifically, the effect of variation contraction is distinguished in the case when traffic volumes are less than 100. This is because temporal traffic volumes exhibit less variation than that of probe volumes as shown in Figure 4. In the same context, the explanatory power of probe volumes is remarkably improved when traffic volumes are greater than 300.

The analysis results are summarized with three traffic volume regimes in Table 1, showing noticeable performances.

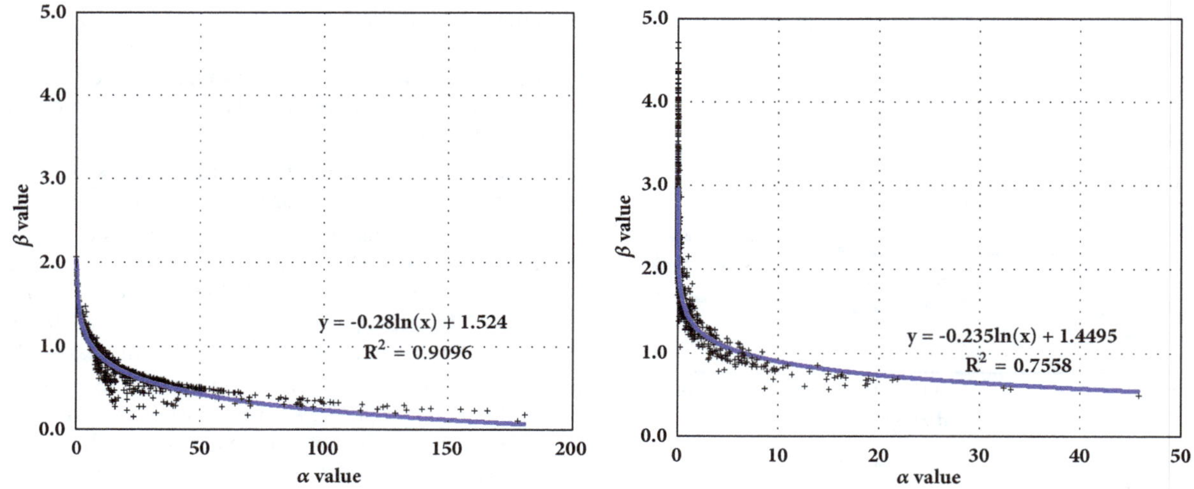

(a) γ =0.0

(b) γ >0.0

FIGURE 7: Relationships between α and β values according to γ values.

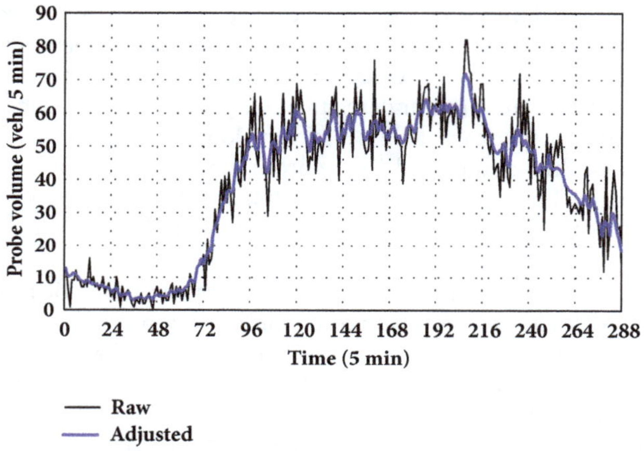

FIGURE 8: Temporal comparison of before and after contraction.

For all regimes, it can be seen that the accuracy performance of the C2C method is at least comparable to those of modern vehicle detectors in terms of the mean of APE (MAPE, %), 5.69%. Note that the accuracy performances of traffic counting for inductive loop, laser scanner, weight-in-motion (WIM) piezoelectric, and WIM quartz detectors in the case of 5-minute data aggregation were reported as 10.6, 24.1, 7.4, and 17.6% by means of MAPE, respectively [19]. The worst performances for APE and RPE measures are shown in the low-volume regime, excluding SEL as shown in Figures 10 and 11. The APEs are greater than 20% for several cases, which is undesirable from the standpoint of forecasting. Note that the tolerable detection error in the case of vehicle detectors should not vary from actual volumes by more than 20.0% [12]. Despite these undesirable performances, forecasting for the low regime can be also acceptable with the maximal SEL of 4.73, which is almost equal to one vehicle per one minute in practice. The hit rate within RPE±10.0% does not reach 90.0% for all regimes, whereas the hit rate within SEL±10 vehicles are up to 98.86%. On the contrary, in the cases of middle and heavy volume regimes, the APE values are less than 10.0% in most cases as shown in Figure 10, where the temporal variation of estimations concurs with that of the observations. The hit rate within RPE±10% is also up to 91.27%. In addition, the hit rate within RPE±20.0% for the middle and heavy regimes reaches 99.61% (Figure 11(a)). Moreover, the worst cases that span to -32.32% or +32.60% occur in the late-night hours, even though they are acceptable in terms of SEL within ±3.0 vehicles (Figure 11(b)). Note that the accuracy performances of the proposed method for the case of low traffic volume are comparable to those of pattern selection-based single-interval forecasting [15, 16] in terms of MAPE. Therefore, traffic volumes estimated from AV probe volumes can also be regarded as a promising option for traffic volume detection.

4.3. Present and Future Potentialities. For more real-world applications in the present and near future, more analysis for the potential of the C2C method was conducted through both the data-aggregation level and the penetration rate of AVs. Figure 12 shows the performances of the hit rate within RPE±10% according to data-aggregation levels. Note that accuracy performances (MAPE) for inductive loop, video image, laser scanner, WIM piezoelectric, and WIM quartz detectors in the case of 15-minute data aggregation were reported as 9.4, 34.1, 19.8, 5.7, and 12.3, respectively [19]. As for the 10-minute aggregation level, the hit rate reached 88.39% with the MAPE of 4.85% for all regimes and was up to 94.65% with the MAPE of 3.77% when traffic volume (vehicles/ 10 min) is greater than 200. In regard to the 15-minute and 30-minute aggregation levels, it can be seen that the performance of the C2C method is obviously comparable to the required detection accuracy for modern vehicle detectors. Accordingly, it is expected that the C2C method for directly monitoring traffic volumes can at least be feasible

TABLE 1: Summary of the results.

Performance Measures		All regimes Cases (volume)	Low regime 481 (<100)	Middle regime 532 (<300)	Heavy regime 1003 (300≤)
APE (%)	Mean	5.69	9.93	6.08	3.45
	Max.	32.60	32.60	21.62	15.74
	Median	3.98	8.41	5.04	2.73
RPE (%)	Mean	0.17	0.47	0.22	-0.01
	Max.	32.60	32.60	21.62	15.74
	Min.	-32.32	-32.32	-19.53	-12.53
	SD	7.91	12.51	7.67	4.42
	HR±10%	83.28	57.80	80.45	97.01
	HR±20%	96.83	87.94	98.87	100.00
SEL (veh)	Mean	-0.01	0.02	0.12	-0.10
	Max.	14.00	4.73	12.87	14.00
	Min.	-12.07	-6.02	-12.07	-11.89
	SD	3.45	1.49	3.74	3.92
	HR±5 veh	85.07	99.79	82.52	79.36
	HR±10 veh	98.86	100.00	98.87	98.31

Note: SD stands for standard deviation.

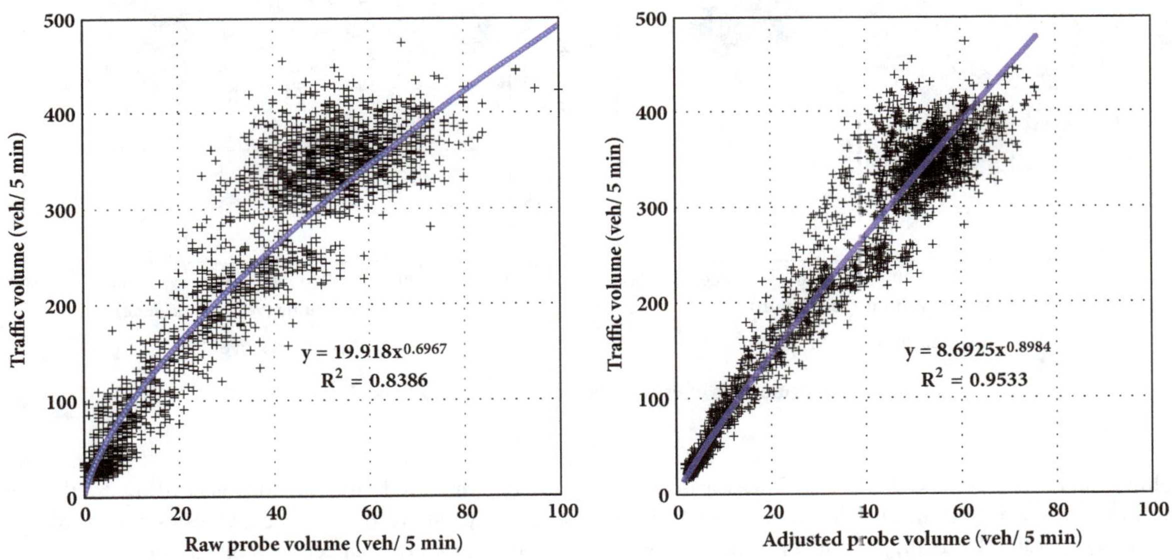

(a) Before adjustment

(b) After adjustment

FIGURE 9: Efficacy of the contraction method.

as complementary means for the role of vehicle detectors if vehicle trajectory data is available with the penetration rate of 0.15%. Moreover, the levels of estimation reliability can be flexibly considered and employed according to the various tactics of traffic operation and control.

In order to demonstrate the potentialities of AV probe volume according to the penetration rate of AVs in the near future, we conducted a random simulation to generate temporal probe volume data. The penetration rate of the probe data used in our case study was employed for the basis of the probability of random selection (rather than a simple random sampling) to consider actual sampling rate. In addition, it is not easy for a random-sampling method to realistically mimic the temporal evolution of probe volumes while considering that of traffic volumes, due to the chaotic behaviors of traffic flow as mentioned before. The probability of selection was computed as

$$p_{sel} = \frac{PR_d}{PR_r} \tag{13}$$

where p_{sel} (0.0~1.0) is the probability of selection, PR_r (0.0-1.0) is the real-world penetration rate of probe volume to

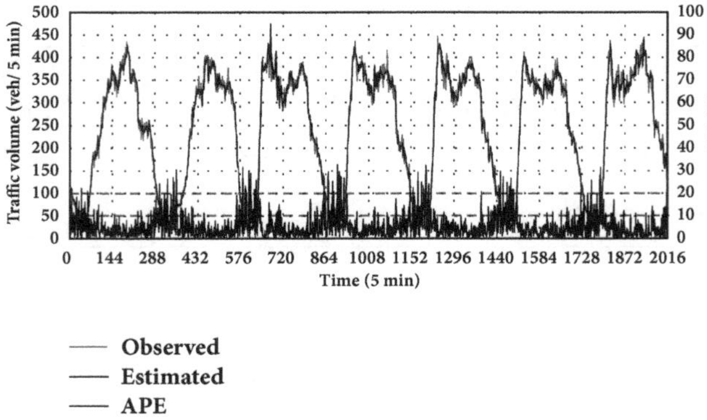

FIGURE 10: Temporal comparison of actual and estimated traffic volumes.

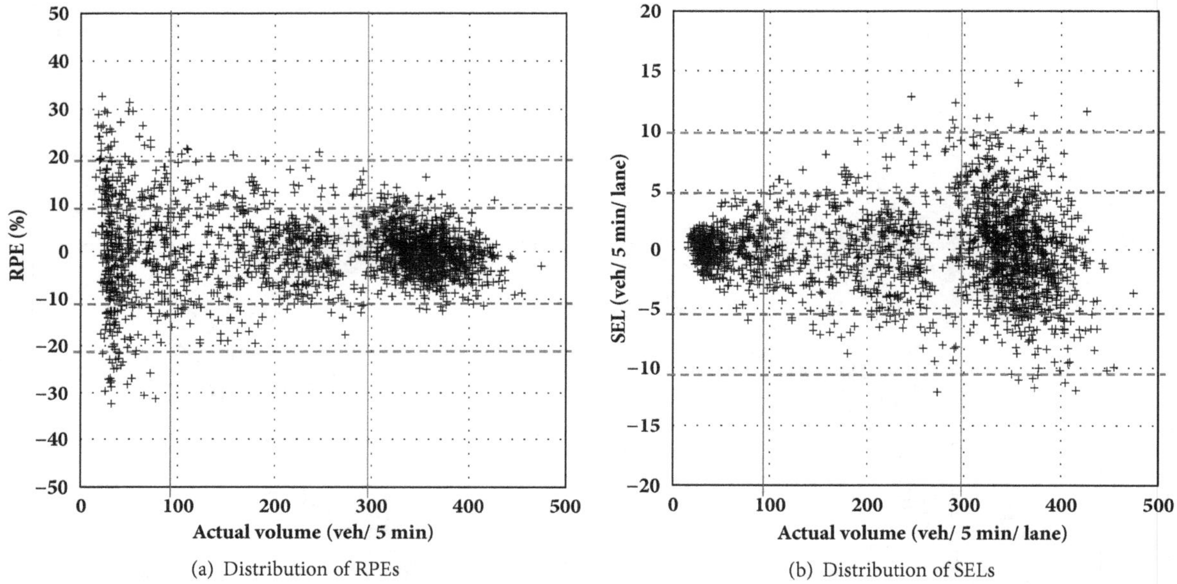

FIGURE 11: Distributions of estimation errors according to traffic volumes.

traffic volume, and RP_d ($0.0 \leq PR_d \leq PR_r$) is a desired penetration rate. Thus, a random sample value for each time interval (t) was generated by using p_{sel} and a probe volume at (t). 14 scenarios of p_d from 0.01 to 0.14 were repeated twenty times for each probe volume data for the day time (06:00-24:00).

Figure 13 shows the performances of the C2C method according to each scenario with the median of 20 MAPE values. The estimation errors exponentially decrease when the penetration rate increases. The span of errors also decreases from 1.80% to 0.16% as the penetration rate increases. Based on the results, it can be seen that the monitoring accuracy of 93.77% can be accomplished within the maximum average error of 6.63% at the penetration rate of 0.05. In addition, it seems that the penetration rate of 0.10 can yield the monitoring accuracy of 94.78%. These analysis results suggest that the direct monitoring of real-time traffic volumes can be realized since the introduction of AVs to real roadways. There are also obvious possibilities that the monitoring accuracy can be improved dramatically according to the results of this study, where the error curve does not converge within a minimum error space. Furthermore, the probe volume of autonomous vehicles can be combined with that of a smartphone car navigation system (or a vehicle-GPS system) to guarantee the monitoring accuracy of real-time traffic volume until the market occupancy of AVs reaches a suitable level.

5. Conclusion Remarks

It is expected that autonomous vehicles can render new solutions to fundamental hindrances and unsolved academic issues in modern ITS. One of the fundamental hindrances is vehicle detectors that are essential for real-time traffic surveillance, which requires extensive budgets and resources in order to guarantee the reliability of monitored information. In addition, their spatial coverage of detection is constrained to fixed point or fixed short length.

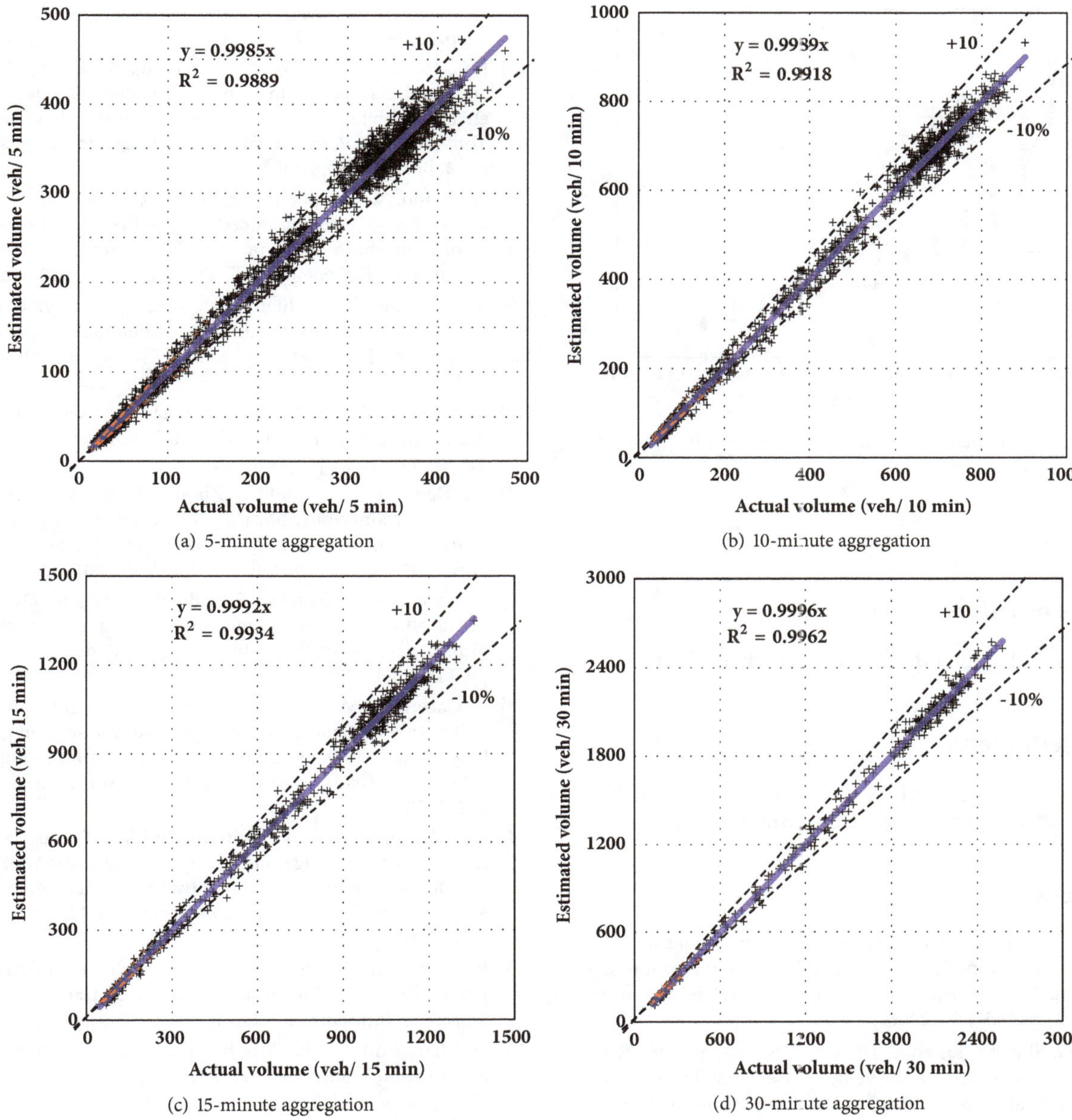

FIGURE 12: Predictability of the proposed model according to the aggregation level.

To realize this opportunity, a new concept for direct real-time monitoring was initiatively introduced in this paper. Using real-world probe volume data collected from a smartphone car navigation system, the potentialities of autonomous vehicles for direct monitoring of traffic volumes for road locations where real-time traffic volumes are desired were demonstrated with a novel and practical approach. The results were noticeable in terms of explanation of temporal variation of real-life traffic volumes. It turned out that the monitoring accuracy of the developed method is at least comparable to the actual detection accuracy of modern vehicle detectors, and it can reliably meet the required detection accuracy of vehicle detectors. Therefore, it can be seen that the direct monitoring of traffic volume is one of promising approaches to solve the current hindrance of traffic volume surveillance. In addition, the developed method is instantly feasible when probe volume data is available at least with the penetration rate of 0.05.

This study contributes a first step in proposing a promising solution to the direct monitoring of real-time traffic volumes. Despite the meaningful results of this research, there are other opportunities in direct real-time monitoring of traffic flow for unobserved road locations with advanced methodologies. We are still conducting investigations to improve the performance of the method and are searching for new potentialities in modern ITS.

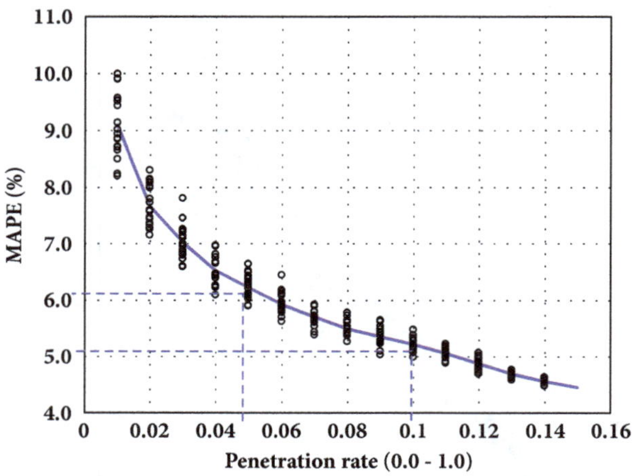

FIGURE 13: Predictability according to penetration rate.

Conflicts of Interest

The authors declare that they have no conflicts of interest.

Acknowledgments

This work was supported by the University of Incheon (International Cooperative) Research Grant in 2014.

References

[1] F. Dion and H. Rakha, "Estimating dynamic roadway travel times using automatic vehicle identification data for low sampling rates," *Transportation Research Part B: Methodological*, vol. 40, no. 9, pp. 745–766, 2006.

[2] X. Li, W. Shu, M. Li, H.-Y. Huang, P.-E. Luo, and M.-Y. Wu, "Performance evaluation of vehicle-based mobile sensor networks for traffic monitoring," *IEEE Transactions on Vehicular Technology*, vol. 58, no. 4, pp. 1647–1653, 2009.

[3] X. Zhan, S. Hasan, S. V. Ukkusuri, and C. Kamga, "Urban link travel time estimation using large-scale taxi data with partial information," *Transportation Research Part C: Emerging Technologies*, vol. 33, pp. 37–49, 2013.

[4] H. Bar-Gera, "Evaluation of a cellular phone-based system for measurements of traffic speeds and travel times: a case study from Israel," *Transportation Research Part C: Emerging Technologies*, vol. 15, no. 6, pp. 380–391, 2007.

[5] K. Sohn and K. Hwang, "Space-based passing time estimation on a freeway using cell phones as traffic probes," *IEEE Transactions on Intelligent Transportation Systems*, vol. 9, no. 3, pp. 559–568, 2008.

[6] A. Janecek, D. Valerio, K. A. Hummel, F. Ricciato, and H. Hlavacs, "The Cellular Network as a Sensor: From Mobile Phone Data to Real-Time Road Traffic Monitoring," *IEEE Transactions on Intelligent Transportation Systems*, vol. 16, no. 5, pp. 2551–2572, 2015.

[7] J. C. Herrera, D. B. Work, R. Herring, X. Ban, Q. Jacobson, and A. M. Bayen, "Evaluation of traffic data obtained via GPS-enabled mobile phones: the Mobile Century field experiment," *Transportation Research, Part C: Emerging Technologies*, vol. 18, no. 4, pp. 568–583, 2010.

[8] G. Guido, V. Gallelli, F. Saccomanno, A. Vitale, D. Rogano, and D. Festa, "Treating uncertainty in the estimation of speed from smartphone traffic probes," *Transportation Research Part C: Emerging Technologies*, vol. 47, no. 1, pp. 100–112, 2014.

[9] M. Chen and S. I. J. Chien, "Determining the number of probe vehicles for freeway travel time estimation by microscopic simulation," *Transportation Research Record*, no. 1719, pp. 61–68, 2000.

[10] M. Chen and S. I. J. Chien, "Dynamic freeway travel time prediction using probe vehicle data," *Transportation Research Record*, no. 1768, pp. 157–161, 2001.

[11] W. Deng, H. Lei, and X. Zhou, "Traffic state estimation and uncertainty quantification based on heterogeneous data sources: A three detector approach," *Transportation Research Part B: Methodological*, vol. 57, pp. 132–157, 2013.

[12] N. Caceres, L. M. Romero, F. G. Benitez, and J. M. Del Castillo, "Traffic flow estimation models using cellular phone data," *IEEE Transactions on Intelligent Transportation Systems*, vol. 13, no. 3, pp. 1430–1441, 2012.

[13] N. Caceres, L. M. Romero, and F. G. Benitez, "Inferring origin-destination trip matrices from aggregate volumes on groups of links: A case study using volumes inferred from mobile phone data," *Journal of Advanced Transportation*, vol. 47, no. 7, pp. 650–666, 2013.

[14] E. I. Vlahogianni, M. G. Karlaftis, and J. C. Golias, "Statistical methods for detecting nonlinearity and non-stationarity in univariate short-term time-series of traffic volume," *Transportation Research Part C: Emerging Technologies*, vol. 14, no. 5, pp. 351–367, 2006.

[15] B. Yoon and H. Chang, "Potentialities of data-driven nonparametric regression in urban signalized traffic forecasting," *Journal of Transportation Engineering*, vol. 10, 2014.

[16] H. Chang and B. Yoon, "High-speed data-driven methodology for real-time traffic flow predictions: practical applications of ITS," *Journal of Advanced Transportation*, vol. 2018, Article ID 5728042, 11 pages, 2018.

[17] F. Fotheringham, C. Brunsdon, and M. Charlton, *Geographically weighted regression: the analysis of spatially varying relationship*, Wiley, Chichester, England, 2003.

[18] D. Mohamad, K. Sinha, T. Kuczek, and C. Scholer, "Annual average traffic prediction model for county roads," *Transportation Research Record*, no. 1617, pp. 69–77, 1998.

[19] P. Bellucci and E. Cipriani, "Data accuracy on automatic traffic counting: The SMART project results," *European Transport Research Review*, vol. 2, no. 4, pp. 175–187, 2010.

12

Optimal Control of Overtaking Maneuver for Intelligent Vehicles

Balázs Németh,[1] **Péter Gáspár**,[1] **and Tamás Hegedűs**[2]

[1] *Systems and Control Laboratory, Institute for Computer Science and Control, Hungarian Academy of Sciences, Kende Utca 13-17, H-1111 Budapest, Hungary*
[2] *Department of Control for Transportation and Vehicle Systems, Budapest University of Technology and Economics, Budapest, Hungary*

Correspondence should be addressed to Balázs Németh; balazs.nemeth@sztaki.mta.hu

Academic Editor: Aboelmaged Noureldin

In the paper a hierarchical overtaking strategy, which is a driver assistance function or rather an autonomous function in electric/autonomous vehicles, is proposed. The solution uses speed and acceleration signals from the surrounding vehicles. These signals are processed with clustering methods in order to achieve probability density functions and predict their expected motion. The strategy includes several additional layers, such as decision making concerning the maneuver, the computation of the required trajectory, and the tracking control of the vehicle. Trajectory generation is formed as an optimization task, which is able to include the prediction model of the surrounding vehicles in the constraints. A robust Linear Parameter Varying (LPV) control design method is proposed to guarantee the tracking of the computed reference. The proposed strategy is able to guarantee the safe motion of the vehicles and handle the interactions with the other traffic participants.

1. Introduction and Motivation

Overtaking and lane changing maneuvers are critical on roads due to various types of human errors. The necessary distance required by the maneuver must be estimated accurately and the vehicle should return to the lane as fast as possible and the maneuver must be safe regarding the other participants in the traffic.

The appropriate handling of overtaking maneuvers is also difficult for several reasons:

(i) The motion of the vehicle ahead must be monitored and predicted

(ii) The motions of vehicles in the environment, especially vehicles coming from the opposite lane and traveling in the return lane, must be monitored and predicted

(iii) The overtaking maneuver must be designed

(iv) Before the maneuver the decision concerning the necessary overtaking must be made

(v) Several factors, such as lateral/longitudinal acceleration and jerks in relation to comfort, must be considered

(vi) steering, braking, and driving must be coordinated

Several different control approaches in the field of overtaking maneuvers of vehicles have been developed. An optimal control design of the overtaking trajectory using polynomial equations to minimize the lateral jerk was proposed by [1]. The model predictive control (MPC) method for path planning together with collision avoidance with the application of overtaking was found in [2]. In [3] it has been shown that the nonconvex optimal control problem of the vehicle positioning can be transformed to a convex quadratic program, which can be solved efficiently. A stochastic model predictive control, in which the velocities of the surrounding vehicles are considered, is designed [4]. A mixed integer programming method, which uses a low number of variables, was presented by [5]. Another solution to reduce the numerical difficulties was to use linguistic variables. For example, a two-level fuzzy logic control was proposed by [6] and the application of the Q-learning method was demonstrated by [7]. The fuzzy logic with the visual system, differential GPS, and inertial measurements was integrated by [8]. Real-time measurement from the overtaken vehicle to generate the vehicle trajectory was used in [9]. A nonlinear adaptive controller design using third-order reference trajectory for overtaking scenarios was proposed by [10].

The prediction of vehicle motions is strongly linked to the overtaking in the field of autonomous vehicles; see, e.g., [11]. With the precise estimation of the environment with vehicles, pedestrians, and cyclists the safe operations of the autonomous systems can be significantly enhanced. Several methods have also been designed to predict the motion of human-driven vehicles; see, e.g., [12]. Probabilistic approaches based on the dynamic Bayesian network and Markov chain models were found in [13, 14]. A method which used the past similarities in vehicle motion between the current vehicle and predefined number of test vehicles was proposed by [15]. The factor of driver aggression and the motion of vehicles in unorganized traffic were considered in the motion planning of an overtaking maneuver in [16].

In this paper an overtaking strategy of our own vehicle is developed. In the following our vehicle will be referred to as ego vehicle. The overtaking strategy is formed in a hierarchical structure with different layers. The core of the autonomous strategy is the motion prediction of the preceding vehicle and surrounding vehicles (e.g., in the opposite lane). Speed and acceleration signals are used to generate probability density functions, which are built in a constrained optimization structure of the trajectory generation. The result of the calculation is a clothoid trajectory, which enhances smooth driving and traveling comfort. Moreover, the paper proposes a strategy, with which further vehicle motions can be considered, e.g., following and overtaking the preceding vehicles. The tracking of the designed trajectory is based on the robust LPV vehicle control, which has the important role in guaranteeing safety in all scenarios, such as overtaking, lane changing, and following the preceding vehicle.

1.1. Architecture of the Robust Overtaking Control Strategy. The overtaking control has been composed of a hierarchical architecture with three different layers, as found in Figure 1. It contains the layer of the surrounding vehicle motion estimation, the trajectory optimization with an overtaking decision logic, and the vehicle control, which generates the steering angle. The advantage of the hierarchical design is the variation in the design method of the layers. The different tasks of the layers require different approaches; e.g., the estimation is based on clustering and probability-based methods, and the trajectory design is based on a constrained optimization, while the vehicle control should guarantee robustness. Therefore, it is sufficient to define the interfaces and the layers can be designed independently. In the architecture the interfaces of the layers are the following:

(a) The inputs of the surrounding vehicle estimation block are the longitudinal velocity and the acceleration signals of the ego and the surrounding vehicles, while the outputs are the lateral constraints in the vehicle motion Y^{min}, Y^{max}

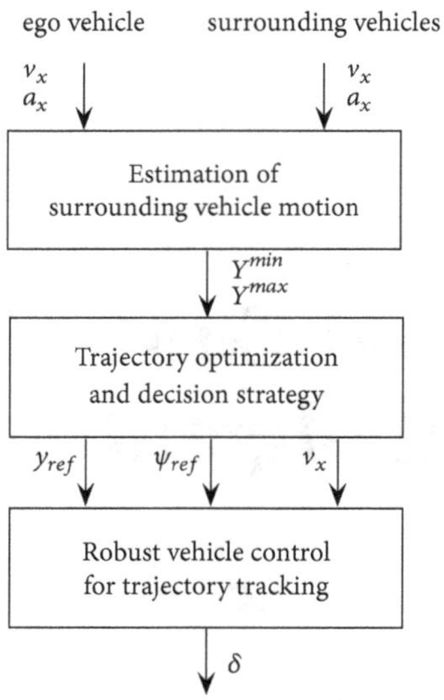

FIGURE 1: Architecture of the control system.

(b) Based on the constraints the actual reference signals y_{ref}, ψ_{ref} are computed together with the desired longitudinal velocity v_x

(c) Finally, the robust vehicle control computes the actual steering angle δ

In the rest of the paper the layers of the automated overtaking strategy are presented. The estimation of the surrounding, especially the preceding vehicle motion, which includes the processing of vehicle data and the prediction of its position, is presented in Section 2. The trajectory design is formed in a constrained optimization problem, which is based on a clothoid path of the vehicle; see Section 3. It also shows the decision logic of the overtaking and lane changing maneuvers. The robust vehicle control based on the generated reference signals is presented in Section 4. The operation of the control system is illustrated through the CarSim simulation environment in Section 5. Finally, Section 6 summarizes the contributions of the paper.

2. Estimation of the Motion of Surrounding Vehicles

Safe overtaking and lane changing maneuvers require information about the motion of the preceding vehicle. In this paper its motion is estimated using the current longitudinal velocity v_x and acceleration a_x signals. It is assumed that the motion of the preceding vehicle in the future will be similar to what it has been so far. Moreover, it is also assumed that the required signals v_x and a_x are available for the ego vehicles.

The proposed estimation has two main steps. First, the preceding vehicle signals are processed through a clustering procedure and the probability density function is generated. Second, the future position of the vehicle is predicted based on the calculated density functions.

2.1. Processing of Preceding Vehicle Signals. The preceding vehicle is considered to be driven by a human, whose velocity

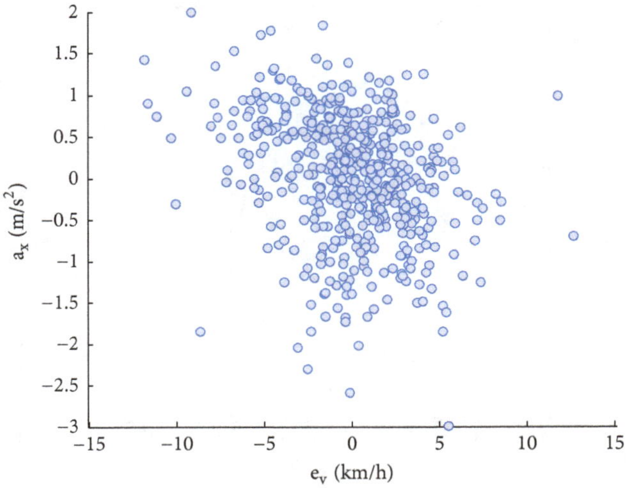

FIGURE 2: Data on preceding vehicle.

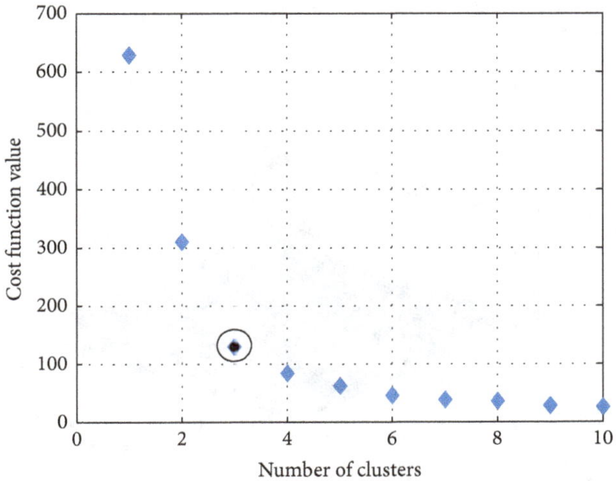

FIGURE 3: Selection of cluster number.

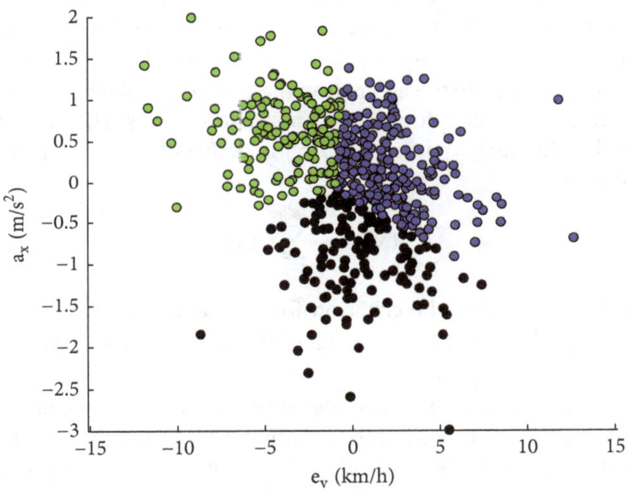

FIGURE 4: Clustered data on preceding vehicle.

selection is determined by a reference velocity $v_{x,ref}$. The value of $v_{x,ref}$ can depend on several factors, e.g., the speed limitation, the quality of the road surface, the curvature and the number of lanes on the road, and the traffic situation. In the method the velocity of the human-driven vehicle is compared to the reference velocity, and the difference is computed as $e_v = v_x - v_{x,ref}$. If several measurements are available from a longer time period, the required signals of the preceding vehicle can be estimated. As an example the relationship between the speed differences e_v and the longitudinal acceleration a_x is illustrated in Figure 2. In this simplified illustrative scenario $v_{x,ref}$ is considered as the speed limit on the course of the vehicle. It is shown that most of the data are around $e_v = 0$, $a_x = 0$, which means that the vehicle is generally cruising close to $v_{x,ref}$. However, the signals have a variance; e.g., if the velocity changes the vehicle is accelerated to reach the speed limit again, which leads to $e_v < 0$ and $a_x > 0$. The purpose of the analysis is to derive the probabilities of the different driving motions of the preceding vehicle.

The generation of probability density functions requires the clustering of data. In this vehicle dynamic examination clustering means that the collected data are ordered in groups, depending on their locations in the plane e_v, a_x. Numerous algorithms for this problem have been developed in recent decades; see, e.g., partitioning methods [17], hierarchical algorithms [18], and density based methods [19]. In this paper a fast and efficient improved k-means method is used as presented and analyzed in [20].

The purpose of the algorithm is to minimize the Euclidean distance between the objects and the centre of the selected cluster, such as

$$J_{cl} = \sum_{i=1}^{K} \sum_{j=1}^{N} \left\| x_j - c_i \right\| \quad (1)$$

where K represents the number of clusters, x_j contains $[e_v, a_x]$ object data, N is the number of the objects, and $c_i \in C_i$ is the centre of the cluster C_i. The candidate clusters must be nonempty and nonoverlapping.

In the estimation process the number of the clusters significantly influences the results. Several methods have been developed for the determination of the cluster number; see, e.g., [21]. Below the elbow method is applied. In this case it is necessary to find the cluster number with which the value of the cost function J_c drops significantly. In our case the elbow method proposes $K = 3$ clusters; see Figure 3. Result of the clustering is shown in Figure 4, where the clusters are marked with different colours.

Results of the clustering show that the e_v, a_x data have variance inside the clusters. Thus, in what follows probability density functions are fitted to the data of each cluster. The function requires a two-dimensional Gaussian form:

$$f_i(x) = \frac{1}{(2\pi)^{n/2} \left|\sum_i\right|^{0.5}} e^{-0.5(x-\mu_i)^T \sum_i (x-\mu_i)}, \quad (2)$$

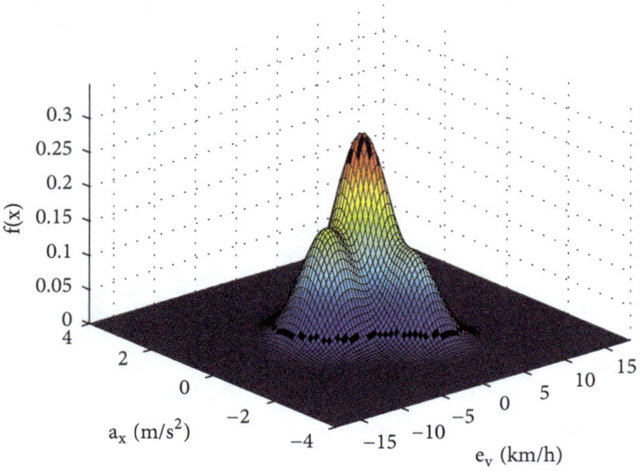

FIGURE 5: Probability density function of preceding vehicle.

where i is related to the cluster number, μ_i represents the mean value within a cluster, $\sum_i$ is the covariance matrix, and $x = [e_v, a_x]^T$ contains the value of the elements in the current cluster. In the estimation the μ_i and $\sum_i$ values must be computed for all f_i functions of all the K clusters. The overall probability density function $f(x)$ is derived from each $f_i(x)$ such as

$$f(x) = \frac{1}{K}\sum_{i=1}^{K} f_i(x), \tag{3}$$

which guarantees the criterion for probability density functions $\int_{e_v}\int_{a_x} f(x)dx = 1$. Result of the method for our case is illustrated in Figure 5.

Based on the proposed algorithm the preceding vehicle data are transformed into a probability density function, which depends on both e_v and a_x. In the following the function $f(x)$ is used to predict the future position of the vehicle considering the probability of driver actuation.

2.2. Prediction of Vehicle Position. The future position of the vehicle is predicted based on the longitudinal kinematic equations

$$v_x(t_{i+1}) = v_x(t_i) + \int_{t_i}^{t_{i+1}} a_x(t)\,dt, \tag{4a}$$

$$s(t_{i+1}) = s(t_i) + \int_{t_i}^{t_{i+1}} v_x(t)\,dt, \tag{4b}$$

where v_x is the longitudinal velocity and s is the displacement. In the estimation the relations (4a) and (4b) are transformed into a discrete form with the time step $T = t_{i+1} - t_i$. Thus, in the following the discrete longitudinal velocity $v_{x,i}$ and displacement s_i are used, where i represents the step index. The purpose of the prediction is to predict $s_1 \dots s_n$ values based on the probability density function $f(x)$. For each prediction step i, slices of the function $f(x)$ are generated as $f(x)|_{e_v(i)}$; see, e.g., Figure 6.

The computations of $v_{x,i}$ and s_i require the value of $a_{x,i}$, which is generated from the gridding of $f(x)|_{e_v(i)}$. The aim

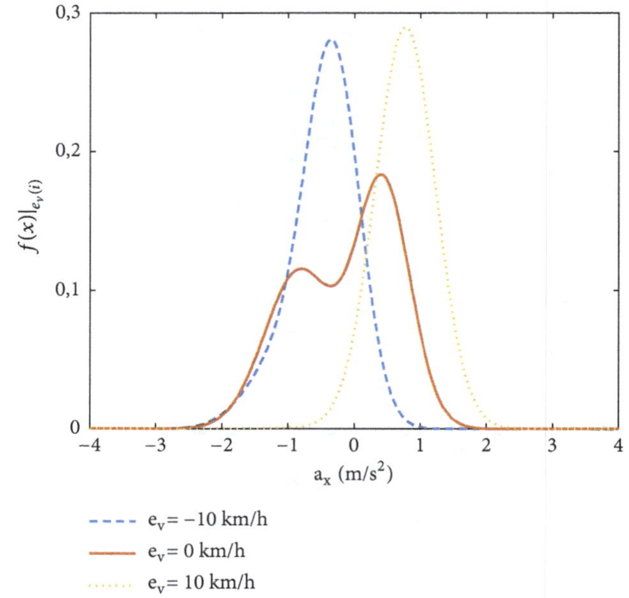

FIGURE 6: Slices of probability density function at different velocities.

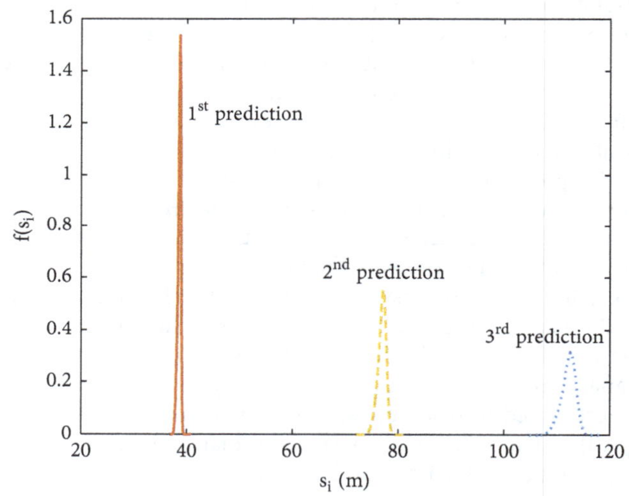

FIGURE 7: Illustration of the s_i probability density function in the predictions.

of the computation is to determine the probability density function of the vehicle position $f(s_i)$, depending on s_i for each time step. The computation is based on the relations of (4a) and (4b), in which the probability density function of $f(x)|_{e_v(i)}$ is used. The computation of v_1 and s_1 for $i = 1$ requires the deterministic initial velocity v_0 and position s_0, which can be measured.

Due to $f(x)|_{e_v(i)}$ the speed and positions v_1, s_1 are also represented by probability density functions $f(s_1), f(v_1)$. Therefore, in the computation of v_2, s_2 for $i = 2$ the previous velocity and position of $i = 1$ are considered with their probability density functions. As an illustration, Figure 7 shows an example for $i = 1 \dots 3$ predictions. Here the initial velocity $v_{x,0} = 140\ km/h$, the initial position $s_0 = 0$, and the sampling time 1 *sec* are selected. The illustration shows

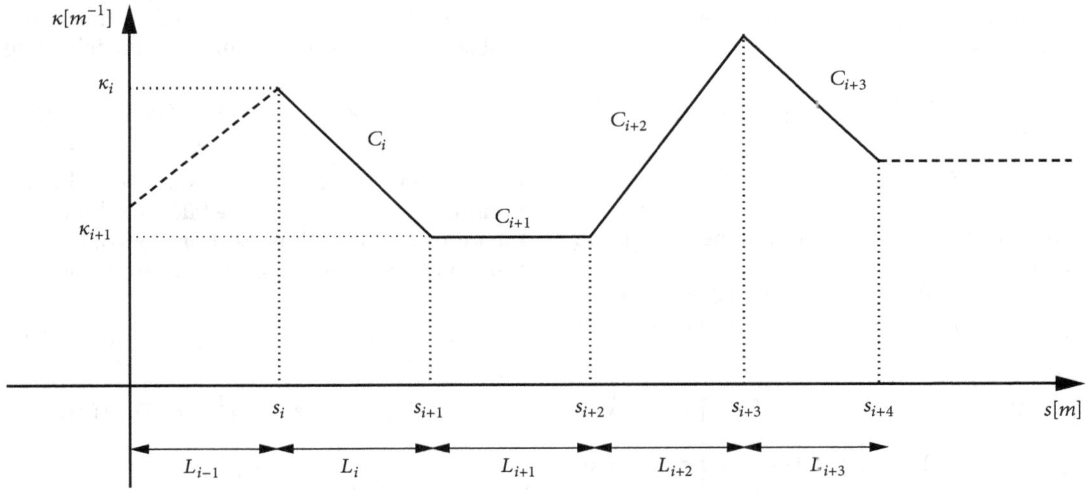

FIGURE 8: Piecewise linear formulation of the curvature.

that for $i = 1$ the probability density function of s_1 has high values in a small range, which means that the position of the vehicle can be estimated with high probability. However, e.g., for $i = 3$ the probability density function of s_3 has high values in a broad range, which means that the forthcoming position of the vehicle has higher uncertainty.

In the prediction of the vehicle position the probability density function has fundamental importance. The probability $P(s_{i,-} < s_i < s_{i,+})$ that the vehicle in the prediction step i is between the positions $s_{i,-} \dots s_{i,+}$ is computed as follows:

$$P\left(s_{i,-} < s_i < s_{i,+}\right) = \int_{s_{i,-}}^{s_{i,+}} f\left(s_i\right) ds_i \tag{5}$$

A safe maneuver to avoid a collision requires a guarantee that the probability of the vehicle in the critical position range $s_{i,-} < s_i < s_{i,+}$ is smaller than a predefined p_{max} value. If $P(s_{i,-} < s_i < s_{i,+}) > p_{max}$, the position $s_{i,-} < s_i < s_{i,+}$ must not be reached by the ego vehicle. Consequently, based on the safe maneuver the vehicle must perform the overtaking or decelerate to the speed of the preceding vehicle. Thus, the trajectory of the ego vehicle must be designed in such a way that both the estimation of the preceding vehicle motion and the maximum probability level p_{max} are taken into consideration. Note that if p_{max} is reduced, the safety of the maneuver increases, but the avoidable regions also increase. It can lead to a conservative maneuver, which requires too long traveling in the opposite lane to overtake the preceding vehicle.

Finally, it is necessary to mention that the proposed estimation method is also used for the estimation of the motions of the vehicles in the opposite lane. If the data e_v, a_x of the closest vehicle in the opposite lane are available, the prediction of the motion can be performed.

3. Formulation of Trajectory Design

In the overtaking and lane changing strategy the result of the estimation is used for the computation of the vehicle trajectory. In the design two criteria are considered. First, the results of the estimation must be incorporated in the trajectory design to guarantee safe cruising. Second, the generated trajectory must guarantee a comfortable maneuver. It means that the motion of the vehicle must be smooth, which is achieved by applying a clothoid trajectory. The advantages of the clothoid trajectory were presented in [22, 23]. In the following an optimal trajectory design method which guarantees both requirements is proposed.

3.1. Modelling of Trajectory Tracking. The lateral motion of the vehicle is formulated based on the kinematic model of the vehicle, such as

$$\frac{dy(t)}{dt} = v_x \sin \psi(t), \tag{6a}$$

$$\frac{d\psi(t)}{dt} = \frac{v_x}{D} \tan \delta(t), \tag{6b}$$

where y is the lateral displacement, v_x is the longitudinal velocity in the vehicle coordinate system, ψ is the yaw angle, δ is the front steering angle of the front wheels, and D is the distance between the front and the rear axle. Translate the motion equation to space domain by making $v_x = ds(t)/dt$ and assume that v_x is a continuous function in the rearrangement of (6a) and (6b):

$$\frac{dy(s)}{ds} = \sin \psi(s) \tag{7a}$$

$$\frac{d\psi(s)}{ds} = \frac{\delta(s)}{D}. \tag{7b}$$

The term $\delta(s)/D$ in (7a) and (7b) represents the curvature $\kappa(s)$ of the path; see also [23].

To guarantee a smooth trajectory for the vehicle, the curvature $\kappa(s)$ is constrained so as to generate a clothoid form; see Figure 8. It leads to the following formulation [23]:

$$\kappa_{i+1} = \kappa_i + c_i L_i, \tag{8}$$

where $L_i = T v_{x,i}$ is the distance between two section points and c_i is the ratio of the clothoid section.

The motion equations (7a) and (7b) are transformed into a discrete form such as

$$y_{i+1} = y_i + L_i \sin\psi \tag{9a}$$

$$\psi_{i+1} = \psi_i + \kappa_{i+1} L_i = \psi_i + \kappa_i L_i + \frac{L_i^2}{2} c_i \tag{9b}$$

For small yaw angle differences from the path the assumption $\sin\psi \approx \psi$ is acceptable.

Using (8), (9a), and (9b) the system is transformed into a state-space representation $x_{i+1} = Ax_i + Bu_i$ in the following way:

$$\begin{bmatrix} y_{i+1} \\ \psi_{i+1} \\ \kappa_{i+1} \end{bmatrix} = \begin{bmatrix} 1 & L_i & 0 \\ 0 & 1 & L_i \\ 0 & 0 & 1 \end{bmatrix} \begin{bmatrix} y_i \\ \psi_i \\ \kappa_i \end{bmatrix} + \begin{bmatrix} 0 \\ \dfrac{L_i^2}{2} \\ L_i \end{bmatrix} c_i \tag{10}$$

where the system depends on the values of c_i. The purpose of the trajectory computation is to determine these values.

3.2. Optimization of Trajectory Tracking. In the trajectory design the lateral position of the vehicle y_{i+1} must be controlled. It is necessary to guarantee that y_{i+1} tends to the reference trajectory $y_{i+1,ref}$ as accurately as possible in order to avoid the preceding vehicle and keep the required lanes during the overtaking maneuver. This task leads to a constrained tracking optimization problem, which is formed in the following way.

The trajectory design of the overtaking maneuver is formed in a finite horizon length L_n ahead of the vehicle. The lateral position error $y_{i+1} - y_{i+1,ref}$ in the horizon is described in the following way:

$$\begin{aligned} Y(\mathscr{C}) &= \begin{bmatrix} y_{i+1} \\ y_{i+2} \\ \vdots \\ y_{i+n} \end{bmatrix} - \begin{bmatrix} y_{i+1,ref} \\ y_{i+2,ref} \\ \vdots \\ y_{i+n,ref} \end{bmatrix} \\ &= \begin{bmatrix} CA \\ CA^2 \\ \vdots \\ CA^n \end{bmatrix} x_i - \begin{bmatrix} y_{i+1,ref} \\ y_{i+2,ref} \\ \vdots \\ y_{i+n,ref} \end{bmatrix} \\ &+ \begin{bmatrix} CB & 0 & \cdots & 0 \\ CAB & CB & \cdots & 0 \\ \vdots & \vdots & \ddots & \vdots \\ CA^{n-1}B & CA^{n-2}B & \cdots & CB \end{bmatrix} \begin{bmatrix} c_i \\ c_{i+1} \\ \vdots \\ c_{i+n-1} \end{bmatrix} \\ &= \mathscr{A} - \mathscr{R} + \mathscr{B}\mathscr{C} \end{aligned} \tag{11}$$

where $\mathscr{A}$ contains the current states of the system, $\mathscr{R}$ is built by the reference values, $\mathscr{B}$ is built by the state matrices, and $\mathscr{C}$ is built by the ratios of clothoid sections. In the tracking problem it is necessary to minimize the following function:

$$J(\mathscr{C}) = \frac{1}{2} Y^T(\mathscr{C}) Q Y(\mathscr{C}) + \mathscr{C}^T R \mathscr{C}, \tag{12}$$

where Q and R are weighting matrices. The role of Q is to minimize the tracking error, while the role of R is to minimize the ratios of the clothoid sections. Substituting (11) into the function (12), the function is transformed as

$$\begin{aligned} J(\mathscr{C}) &= \frac{1}{2}(\mathscr{A} - \mathscr{R} + \mathscr{B}\mathscr{C})^T Q (\mathscr{A} - \mathscr{R} + \mathscr{B}\mathscr{C}) \\ &+ \mathscr{C}^T R \mathscr{C} = \frac{1}{2}\left(\mathscr{A}^T - \mathscr{R}^T + \mathscr{C}^T\mathscr{B}^T\right) Q (\mathscr{A} - \mathscr{R} \\ &+ \mathscr{B}\mathscr{C}) + \mathscr{C}^T R \mathscr{C} = \frac{1}{2}\left(\mathscr{A}^T Q \mathscr{A} - \mathscr{A}^T Q R \right. \\ &+ \mathscr{A}^T Q \mathscr{B}\mathscr{C} - \mathscr{R}^T Q \mathscr{A} + \mathscr{R}^T Q \mathscr{R} - \mathscr{R}^T Q \mathscr{B}\mathscr{C} \\ &\left. + \mathscr{C}^T\mathscr{B}^T Q \mathscr{A} - \mathscr{C}^T\mathscr{B}^T Q \mathscr{R} + \mathscr{C}^T\mathscr{B}^T Q \mathscr{B}\mathscr{C}\right) \\ &+ \mathscr{C}^T R \mathscr{C} = \frac{1}{2}\mathscr{C}^T\left(\mathscr{B}^T Q \mathscr{B} + 2R\right)\mathscr{C} + \frac{1}{2}\left(\mathscr{A}^T Q \mathscr{B} \right. \\ &\left. - \mathscr{R}^T Q \mathscr{B}\right)\mathscr{C} + \mathscr{C}^T \frac{1}{2}\left(\mathscr{B}^T Q \mathscr{A} - \mathscr{B}^T Q \mathscr{R}\right) + \varepsilon = \frac{1}{2} \\ &\cdot \mathscr{C}^T \phi \mathscr{C} + \beta^T \mathscr{C} + \mathscr{C}^T \gamma + \varepsilon \end{aligned} \tag{13}$$

where

$$\phi = \mathscr{B}^T Q \mathscr{B} + 2R, \tag{14a}$$

$$\beta^T = \frac{1}{2}\left(\mathscr{A}^T Q \mathscr{B} - \mathscr{R}^T Q \mathscr{B}\right), \tag{14b}$$

$$\gamma = \frac{1}{2}\left(\mathscr{B}^T Q \mathscr{A} - \mathscr{B}^T Q \mathscr{R}\right), \tag{14c}$$

$$\varepsilon = \frac{1}{2}\left(\mathscr{A}^T Q \mathscr{A} - \mathscr{A}^T Q R - \mathscr{R}^T Q \mathscr{A} + \mathscr{R}^T Q \mathscr{R}\right). \tag{14d}$$

In (13) ε contains all the constant components. Since ε is independent of the effect of $\mathscr{C}$ on $J(\mathscr{C})$, it can be canceled from the further optimization problem.

Through the minimization of the cost function $J(\mathscr{C})$ the tracking of the reference trajectory $y_{i+1,ref}$ can be guaranteed. In order to increase the safety of the maneuver both collision avoidance and lane keeping are built into the design. Therefore, constraints are incorporated into the trajectory optimization problem. The constraints are related to both the minimum and the maximum values of y_{i+1}. Practically, y_{i+1}^{min} and y_{i+1}^{max} are determined by the lanes, which guarantees lane keeping. However, in the overtaking maneuver the constraints must be modified with the current position of the preceding vehicle. The constraints of the minimum values of the positions are formed as $y_{i+j} \geq y_{i+j}^{min}$, $j \in \{1, n\}$. The inequalities are transformed into a matrix representation:

$$\mathscr{A} + \mathscr{B}\mathscr{C} \geq Y^{min}, \tag{15}$$

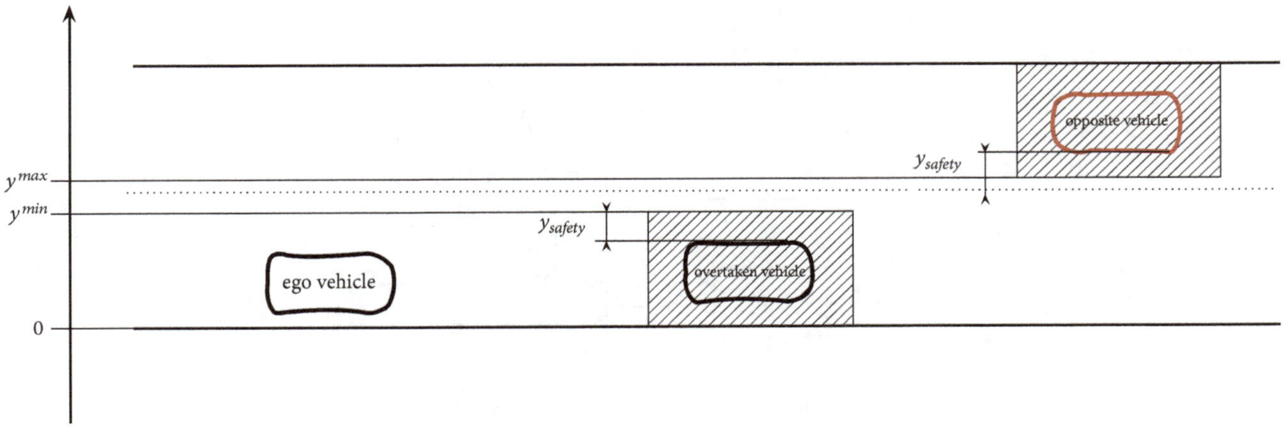

FIGURE 9: Determination of Y^{min} and Y^{max}.

where $Y^{min} = \left[y_{i+1}^{min}\ y_{i+2}^{min}\ \cdots\ y_{i+n}^{min}\right]^T$. Thus, the constraint on the clothoid ratios $\mathscr{C}$ is formed as

$$\mathscr{A} - Y^{min} \geq -\mathscr{B}\mathscr{C}. \tag{16}$$

Similarly, the constraints of the maximum values of y_{i+1} are derived as

$$Y^{max} - \mathscr{A} \geq \mathscr{B}\mathscr{C}, \tag{17}$$

where $Y^{max} = \left[y_{i+1}^{max}\ y_{i+2}^{max}\ \cdots\ y_{i+n}^{max}\right]^T$. The minimum Y^{min} and maximum Y^{max} values are determined by the edges of the lanes and the surrounding vehicles, especially the preceding vehicle and the vehicle in the opposite lane. If a vehicle in the opposite lane performs a lane change maneuver, it can limit the lateral motion of the ego vehicle. This information is also incorporated in Y^{max}, with which the safe cruising of the ego vehicle is guaranteed against collisions with the vehicles from the opposite lane. The selections of Y^{min} and Y^{max} are illustrated in Figure 9, where y_{safety} describes a safety zone of the vehicles.

Finally, from (13), (16), and (17) the constrained trajectory optimization problem is formed as

$$\min_{u_i \ldots u_{i+n-1}} \frac{1}{2}\mathscr{C}^T\phi\mathscr{C} + \beta^T\mathscr{C} + \mathscr{C}^T\gamma \tag{18}$$

such that the following constraints are also guaranteed:

$$\mathscr{A} - Y^{min} \geq -\mathscr{B}\mathscr{C} \tag{19a}$$

$$Y^{max} - \mathscr{A} \geq \mathscr{B}\mathscr{C} \tag{19b}$$

$$\mathscr{C} \in \mathbf{C} \tag{19c}$$

where $\mathbf{C}$ contains the achievable clothoid ratios. The optimization problem can be solved using standard quadratic programming methods, e.g., [24, 25]. The solution of (18) leads to a series of clothoid ratios on the horizon L_n. The current ratio c_i can be computed online during the cruising of the vehicle.

3.3. Decision Strategy of Overtaking and Lane Changing. The overtaking and lane changing strategy is based on the constrained trajectory optimization method (18). Although the constraints and the achievable clothoid ratios are considered, the calculated tracking must be modified for safety reasons.

Before the maneuver the ego vehicle reaches a preceding vehicle which is traveling at a slower speed. The vehicle must follow the preceding vehicle. If the preceding vehicle accelerates or decelerates the ego vehicle must strictly track the velocity within the speed limit. Meanwhile, it calculates a safe trajectory for the overtaking and lane changing. The optimization method (18) results in a clothoid curve, which guarantees the cruising of the ego vehicle between the minimum and the maximum limits.

However, if there is a follower vehicle in the inner lane which is traveling at a higher velocity, a conflict with the ego vehicle may occur during the overtaking maneuver. Moreover, if there is a vehicle in the opposite lane, a conflict with the ego vehicle in the maneuver may also occur. These maneuvers are considered unfeasible. Infeasibility means that it is impossible to find an appropriate trajectory which guarantees both the minimization (18) and the constraints (19a), (19b), and (19c). In this case the ego vehicle must follow the preceding vehicle and modify the longitudinal dynamics.

When the traffic situation changes and the optimization becomes feasible, the overtaking maneuver can be performed. The maneuver is realized through the actuation of steering and driving/braking systems; see, e.g., [26].

Note that the proposed method can be used not only for overtaking, but also for simple lane changing maneuvers, e.g., changing the route in a highway intersection. In this scenario the reference signal of the road $y_{i+1,ref}$ must be modified according to the parameters of the requested lane. Moreover, in a lane changing maneuver the constraints Y^{min} and Y^{max} must be modified for the new lane. Furthermore, the proposed optimization (18) and decision problem can be formed as a MPC task, in which the steering intervention can also be incorporated. However, the entire control system must guarantee the robustness of the system, which can lead to high complexity in the computation of the vehicle control. In the paper a robust LPV-based control design method is proposed

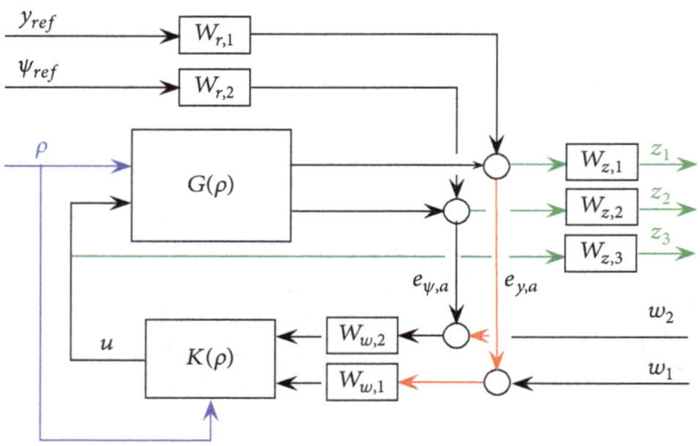

FIGURE 10: Closed-loop interconnection structure.

to guarantee the tracking of the generated trajectory and the robustness of the controlled system.

4. Robust LPV Control for Autonomous Overtaking

The goal of the robust LPV control is to guarantee the tracking of the trajectory, which has been generated by optimization (18) together with the decision algorithm. The results of the optimization are $u_i \dots u_{i+n-1}$, which are equal to the ratios of the clothoid sections $c_i \dots c_{i+n-1}$. In the case of the tracking control the vehicle must follow the reference path $y_{ref} = y_{i+1} = y_1$ and the reference heading angle $\psi_{ref} = \psi_{i+1} = \psi_1$. The reference signals are computed from (10) such as

$$y_{ref} = y_0 + L_0 \psi_0 \tag{20a}$$

$$\psi_{ref} = \psi_0 + L_0 \kappa_0 + \frac{L_0^2}{2} c_0 \tag{20b}$$

where y_0, ψ_0 are related to $y(t), \psi(t)$ in discrete time, while L_0 is the length of the forthcoming segment and κ_0 is the result in the previous trajectory computation step. The parameter c_0 is the first element among the results of the optimization u_0.

After the computation of the reference signals the tracking performances together with the control performance are defined as

$$z_1 = y_{ref} - y(t) \quad |z_1| \longrightarrow min, \tag{21a}$$

$$z_2 = \psi_{ref} - \psi(t) \quad |z_2| \longrightarrow min, \tag{21b}$$

$$z_3 = \delta(t) \quad |z_3| \longrightarrow min. \tag{21c}$$

The performance vector is as follows: $z = \begin{bmatrix} z_1 & z_2 & z_3 \end{bmatrix}^T$.

The model of the vehicle is described by the dynamical bicycle model; see [27]:

$$m\ddot{y} = C_f\left(\delta - \frac{\dot{\psi} l_f + \dot{y}}{v_x}\right) + C_r\left(\frac{\dot{\psi} l_r - \dot{y}}{v_x}\right) \tag{22a}$$

$$J\ddot{\psi} = C_f l_f\left(\delta - \frac{\dot{\psi} l_f + \dot{y}}{v_x}\right) - C_r l_r\left(\frac{\dot{\psi} l_r - \dot{y}}{v_x}\right) \tag{22b}$$

where J is the yaw inertia of the vehicle, m is the vehicle mass, C_f, C_r are the cornering stiffness coefficients, and l_f, l_r are geometric parameters. The signal $\dot{y}$ is the lateral velocity and $\dot{\psi}$ is the yaw rate. The longitudinal velocity v_x may vary along the route of the vehicle; therefore it is chosen as a scheduling variable. Equations (22a) and (22b) can be transformed into a state-space representation, where the state vector is $x = \begin{bmatrix} \dot{\psi} & \dot{y} & \psi & y \end{bmatrix}^T$, the control input is the steering angle $u = \delta$, and the measured output vector is $y_m = \begin{bmatrix} \psi & y \end{bmatrix}^T$. The state equation and the performances as well as the measurements are written as

$$\dot{x} = A(\rho) x + Bu \tag{23a}$$

$$z = C_1 x, \tag{23b}$$

$$y_m = C_2 x + Du, \tag{23c}$$

where $\rho = v_x$ is selected as a scheduling variable. Its value is modified through the decision strategy.

The control design is based on a weighting strategy, which is formulated through a closed-loop interconnection structure; see Figure 10. The interconnection structure contains several weighting functions, whose roles are to guarantee the trade-off between the performances and to scale the signals. The weights $W_{w,1}, W_{w,2}$ are related to the sensor characteristics on the lateral error and yaw error measurement. These weights are related to the sensor dynamics. $W_{z,1}, W_{z,2}$, and $W_{z,3}$ are the weights for the performances, which represent the minimization of z_1, z_2, and z_3. The weights are formed as $W_{z,i} = A_i/(T_i s + 1)$ where A_i, T_i are design parameters. These parameters influence the balance between the performances z_1, z_2, and z_3. Moreover, the role of the weight $W_{r,i}$ is to consider the dynamics of the reference signals y_{ref}, ψ_{ref}. The blocks also represent the dynamics of their variations.

The design of the control is based on robust LPV methods. The advantage of these methods is that the controller meets stability and performance demands by using affine parameterized Lyapunov functions in the entire operational interval, since the controller is able to adapt to the current operational conditions; see [28, 29]. The solution of an LPV problem is governed by the set of infinite dimensional LMIs being satisfied for all $\rho \in \mathcal{F}_{\mathcal{P}}$; thus it is a convex problem. In

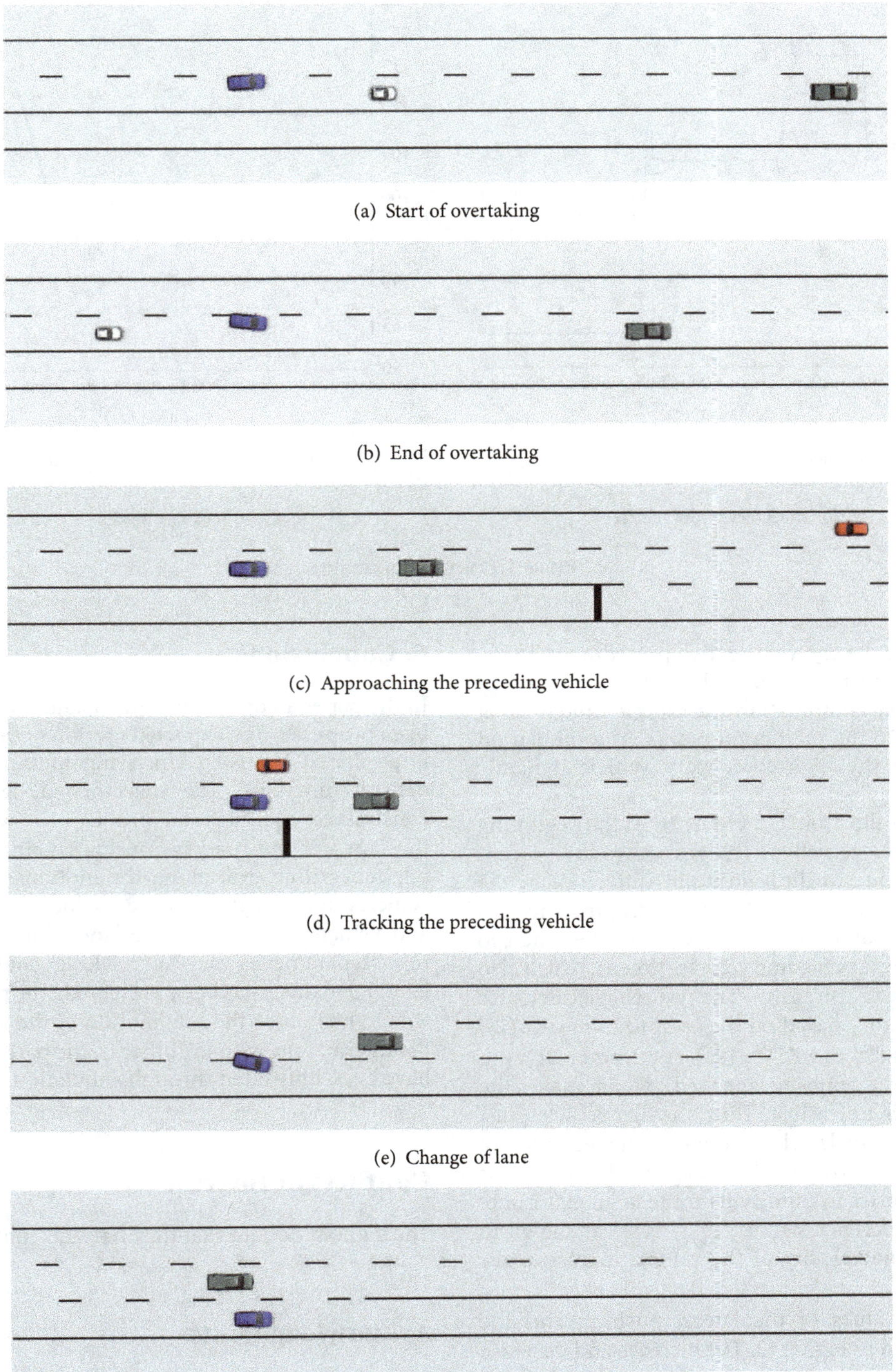

(a) Start of overtaking

(b) End of overtaking

(c) Approaching the preceding vehicle

(d) Tracking the preceding vehicle

(e) Change of lane

(f) Cruising with its own reference velocity

FIGURE 11: Simulation scenario.

practice, this problem is set up by gridding the parameter space and solving the set of LMIs that hold on the subset of $\mathscr{F}_{\mathscr{P}}$. The induced $\mathscr{L}_2$ norm of parameter-dependent stable LPV systems with zero initial conditions is defined as

$$\inf_{K} \sup_{\varrho \in \mathscr{F}_{\mathscr{P}}} \sup_{\|w\|_2 \neq 0, w \in \mathscr{L}_2} \frac{\|z\|_2}{\|w\|_2}, \tag{24}$$

where w is the disturbance and z is the performance. The result of the optimization is the steering angle δ.

5. Simulation Results

To illustrate the efficiency of the proposed method a complex simulation scenario is presented. The traffic scenario contains

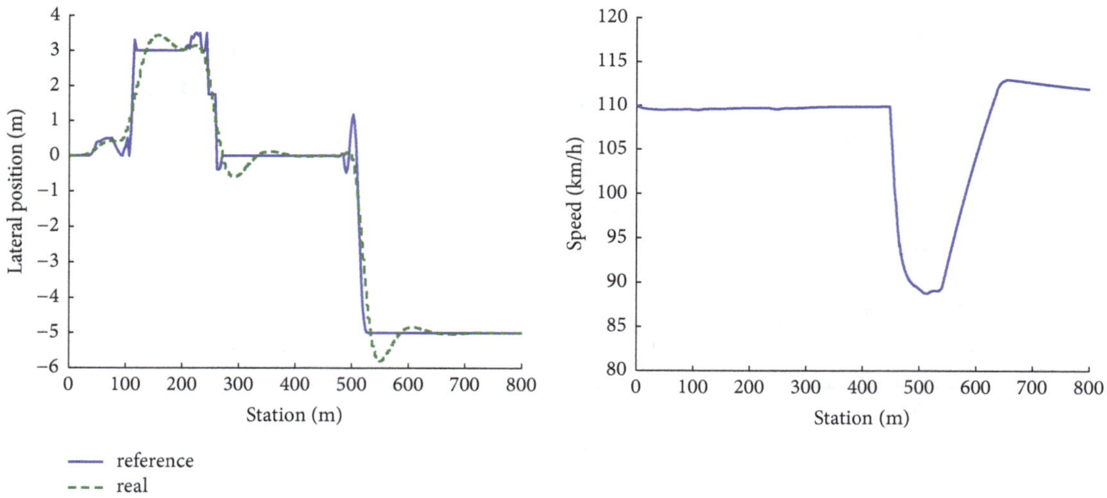

(a) Trajectory of the vehicle

(b) Velocity profile

Figure 12: Simulation results.

four vehicles, such as the ego vehicle, two preceding vehicles, and another vehicle in the opposite lane. In the setting of the optimal trajectory computation the sample time is set at $T = 0.1s$ with $n = 20$ the prediction points. The simulation is performed using the high-complexity vehicle dynamic software CarSim.

The simulation is illustrated in Figure 11. At the beginning of the simulation the ego vehicle (blue) reaches the preceding vehicle (white) due to the significant difference in the velocities; see Figure 11(a). Since the overtaking maneuver is feasible, it is carried out as Figure 11(b) shows. Then the ego vehicle reaches the next preceding vehicle (green), which also has lower velocity; see Figure 11(c). The ego vehicle calculates the overtaking trajectory based on the optimization task (18) and the constraints Y^{min} and Y^{max} (19a), (19b), and (19c). But there is a vehicle in the opposite lane (red), which makes the overtaking maneuver unfeasible. Thus, the ego vehicle must follow the preceding vehicle, which results in the reduction of its velocity; see Figure 11(d). Finally, the ego vehicle performs a lane changing in order to modify its route at an exit ramp. In this case a new reference signal $y_{i+1,ref}$ is set as shown in Figure 11(e). Then the velocity of the vehicle increases; see Figure 11(f).

The numerical values of the lateral position and the velocity are illustrated in Figure 12. The increase and decrease of the lateral positions at 100 m and 200 m, respectively, illustrate the overtaking of the first slower vehicle (white) and then the tracking of the second preceding vehicle (green) due to the third vehicle (red) coming from the opposite lane; see Figure 12(a). In the proposed scenario the safe distance between the vehicles is $20m$, which has been kept by the ego vehicle. The initial velocity is 110 km/h, which must be reduced to keep the velocity of the green preceding vehicle ($90km/h$); see Figure 12(b). At 500 m the lane change is carried out, which results in the changes of the lateral position and the increases in the velocity compared to the original value.

6. Conclusions

In the paper an overtaking and a lane changing strategy have been proposed. The expected motion of the preceding vehicle is predicted by using clustering methods and probability density functions. The trajectory design is formed as a constrained optimization problem, in which the traffic in the lanes of the road is considered. In the decision making concerning overtaking the motions of the surrounding vehicles such as the follower vehicle with higher velocity and vehicles in the opposite lane is also incorporated. The robust parameter varying tracking control design of the lateral dynamics has been presented. The proposed strategy is able to guarantee the safe motion of the vehicles and handle the interactions with the other traffic participants. The results have been illustrated through simulation examples.

Conflicts of Interest

The authors declare that they have no conflicts of interest.

Acknowledgments

The research reported in this paper was supported by the Higher Education Excellence Program of the Ministry of Human Capacities within the frame of Artificial Intelligence research area of Budapest University of Technology and Economics (BME FIKPMI/FM). The research was supported by the Hungarian Government and cofinanced by the European Social Fund through the project "Talent Management in Autonomous Vehicle Control Technologies" (EFOP-3.6.3-VEKOP-16-2017-00001). The work of Balázs Németh was partially supported by the János Bolyai Research Scholarship of the Hungarian Academy of Sciences and the ÚNKP-18-4 New National Excellence Program of the Ministry of Human Capacities.

References

[1] T. Shamir, "How should an autonomous vehicle overtake a slower moving vehicle: design and analysis of an optimal trajectory," *Institute of Electrical and Electronics Engineers Transactions on Automatic Control*, vol. 49, no. 4, pp. 607–610, 2004.

[2] K. Berntorp, "Path planning and integrated collision avoidance for autonomous vehicles," in *Proceedings of the 2017 American Control Conference, ACC 2017*, pp. 4023–4028, USA, May 2017.

[3] N. Murgovski and J. Sjoberg, "Predictive cruise control with autonomous overtaking," in *Proceedings of the 2015 54th IEEE Conference on Decision and Control (CDC)*, pp. 644–649, Osaka, December 2015.

[4] N. A. Nguyen, D. Moser, P. Schrangl, L. del Re, and S. Jones, "Autonomous overtaking using stochastic model predictive control," in *Proceedings of the 2017 11th Asian Control Conference (ASCC)*, pp. 1005–1010, Gold Coast, QLD, December 2017.

[5] F. Molinari, N. N. Anh, and L. Del Re, "Efficient mixed integer programming for autonomous overtaking," in *Proceedings of the 2017 American Control Conference, ACC 2017*, pp. 2303–2308, USA, May 2017.

[6] J. E. Naranjo, C. González, R. García, and T. De Pedro, "Lane-change fuzzy control in autonomous vehicles for the overtaking maneuver," *IEEE Transactions on Intelligent Transportation Systems*, vol. 9, no. 3, pp. 438–450, 2008.

[7] D. C. K. Ngai and N. H. C. Yung, "A multiple-goal reinforcement learning method for complex vehicle overtaking maneuvers," *IEEE Transactions on Intelligent Transportation Systems*, vol. 12, no. 2, pp. 509–522, 2011.

[8] V. Milanés, D. F. Llorca, J. Villagrá et al., "Intelligent automatic overtaking system using vision for vehicle detection," *Expert Systems with Applications*, vol. 39, no. 3, pp. 3362–3373, 2012.

[9] G. Usman and F. Kunwar, "Autonomous vehicle overtaking - An online solution," in *Proceedings of the 2009 IEEE International Conference on Automation and Logistics, ICAL 2009*, pp. 596–601, China, August 2009.

[10] P. Petrov and F. Nashashibi, "Modeling and nonlinear adaptive control for autonomous vehicle overtaking," *IEEE Transactions on Intelligent Transportation Systems*, vol. 15, no. 4, pp. 1643–1656, 2014.

[11] A. Carvalho, S. Lefévre, G. Schildbach, J. Kong, and F. Borrelli, "Automated driving: the role of forecasts and uncertainty—a control perspective," *European Journal of Control*, vol. 24, pp. 14–32, 2015.

[12] S. Lefèvre, D. Vasquez, and C. Laugier, "A survey on motion prediction and risk assessment for intelligent vehicles," *ROBOMECH Journal*, vol. 1, no. 1, 2014.

[13] T. Gindele, S. Brechtel, and R. Dillmann, "A probabilistic model for estimating driver behaviors and vehicle trajectories in traffic environments," in *Proceedings of the 13th International IEEE Conference on Intelligent Transportation Systems (ITSC '10)*, pp. 1625–1631, IEEE, Funchal, Portugal, September 2010.

[14] M. Althoff and A. Mergel, "Comparison of Markov chain abstraction and Monte Carlo simulation for the safety assessment of autonomous cars," *IEEE Transactions on Intelligent Transportation Systems*, vol. 12, no. 4, pp. 1237–1247, 2011.

[15] K. Okamoto, K. Berntorp, and S. Di Cairano, "Similarity-based vehicle-motion prediction," in *Proceedings of the 2017 American Control Conference, ACC 2017*, pp. 303–308, USA, May 2017.

[16] R. Kala and K. Warwick, "Motion planning of autonomous vehicles in a non-autonomous vehicle environment without speed lanes," *Engineering Applications of Artificial Intelligence*, vol. 26, no. 5-6, pp. 1588–1601, 2013.

[17] L. Kaufman and P. Rousseeuw, *Finding Groups in Data: An Introduction to Cluster Analysis*, John Wiley & Sons, New York, NY, USA, 1990.

[18] S. Guha, R. Rastogi, and K. Shim, "CURE: An efficient clustering algorithm for large databases," in *Proceedings of the ACM SIGMOD International Conference on Management of Data*, pp. 73–84, 1998.

[19] M. Ester, H.-P. Kriegel, J. Sander, and X. Xu, "A density-based algorithm for discovering clusters in large spatial databases with noise," in *Proceedings of the in 2nd Intenational Conference Knowledge Discovery and Data Mining*, pp. 226–231, Portland, 1996.

[20] H. S. Park, J. Lee, and C. Jun, "A k-meanslike algorithm for k-medoids clustering and its performance," in *Proceedings of the ICCIE*, 2006.

[21] T. M. Kodinariya and P. R. Makwana, "Review on determining number of cluster in K-means clustering," *International Journal of Advance Research in Computer Science and Management Studies*, vol. 1, pp. 90–95, 2013.

[22] D. K. Wilde, "Computing clothoid segments for trajectory generation," in *Proceedings of the 2009 IEEE/RSJ International Conference on Intelligent Robots and Systems, IROS 2009*, pp. 2440–2445, October 2009.

[23] P. F. Lima, M. Trincavelli, J. Martensson, and B. Wahlberg, "Clothoid-based model predictive control for autonomous driving," in *proceedings of the European Control Conference*, pp. 2983–2990, 2015.

[24] P. E. Gill, W. Murray, and M. H. Wright, *Practical Optimization*, Academic Press, 1981.

[25] C. Schmid and L. T. Biegler, "Quadratic programming methods for reduced hessian SQP," *Computers & Chemical Engineering*, vol. 18, no. 9, pp. 817–832, 1994.

[26] P. Gáspár and B. Németh, "Integrated control design for driver assistance systems based on LPV methods," *International Journal of Control*, vol. 89, no. 12, pp. 2420–2433, 2016.

[27] R. Rajamani, *Vehicle Dynamics and Control*, Springer, New York, NY, USA, 2012.

[28] F. Wu, X. H. Yang, A. Packard, and G. Becker, "Induced L_2-norm control for LPV systems with bounded parameter variation rates," *International Journal of Robust and Nonlinear Control*, vol. 6, no. 9-10, pp. 983–998, 1996.

[29] J. Yu and A. Sideris, "H_∞ control with parametric Lyapunov functions," *Systems & Control Letters*, vol. 30, no. 2-3, pp. 57–69, 1997.

13

A Separation Strategy for Connected and Automated Vehicles: Utilizing Traffic Light Information for Reducing Idling at Red Lights and Improving Fuel Economy

Lin-heng Li, Jing Gan, and Wen-quan Li

School of Transportation, Southeast University, China

Correspondence should be addressed to Lin-heng Li; leelinheng@seu.edu.cn

Academic Editor: Md. A. S. Kamal

Vehicle platoon composed of a group of connected and automated vehicles (CAVs), a coordinated movement strategy, has been widely proposed to address a range of traffic problems. The motion of vehicle in the platoon passing signalized intersections can significantly affect their total trip time and fuel consumption. With the development of advanced communication technology such as V2V and V2I, CAVs can automatically obtain and use the upcoming traffic light timing information to find optimal velocity profiles that can avoid idling at red lights. This paper proposes an optimal velocity control and separation strategy for the platoon to minimize the trip time and reduce fuel consumption as much as possible. Simulation results show that with the introduction of the velocity control and separation strategy, the total trip time and fuel consumption decrease by 19.2% and 18.1%, respectively. Thus the effectiveness of the proposed strategy is demonstrated.

1. Introduction

In the connected and automated vehicles (CAVs) system, vehicles are capable of sharing information and sensing local environment with each other via the advanced communication technologies (e.g., V2V and V2I). The vehicles' information (e.g., location and velocity) and the road transportation infrastructure information (e.g., the traffic light timing, including the phase cycle length, the green phase length, and the start of the first green phase) will be obtained by every vehicle. After receiving such information, the internal decision-making mechanism will make corresponding driving decisions and then achieve the level of automatic driving. Under this circumstance, all CAVs will be platooned through communication and automated control technologies [1]. With CAV platooning, consecutive vehicles are similar to two concatenated carriages of a train and thus shall have much less time headway compared with a pair of conventional human driven vehicles.

All these potential benefits are linked to the expectation that CAVs can significantly improve traffic capacity, efficiency, and safety [2–5]. Studies by Lioris et al. have shown that traffic capacity at signalized intersections could be doubled when vehicles on the road are connected to an intelligent network without changing the signal control [6, 7]. Chang and Edara also examined whether the road traffic efficiency could be further improved under CAVs environment [8].

World Oil Outlook 2016, issued by Organization of Petroleum Exporting Countries (OPEC), predicted that most of the oil consumed today and in the future will come from the road transportation sector. By 2040, the road transportation sector will represent 44% of global oil demand [9]. All the benefits provided by vehicle platooning are also linked to reduce fuel consumption. Lammert et al. conducted a comprehensive investigation on the effect of platooning on fuel consumption of class 8 vehicles, and they found saving of up to 6% for the leading vehicle and 10% for the following vehicle [10]. Alam et al. proposed a particular test to study the fuel reduction that heavy duty vehicle platooning enables and the analysis with respect to the influence of a commercial adaptive cruise control on the fuel consumption [11]. Tsugawa et al. presented an automated truck platoon that has been developed under a national ITS project named Energy ITS;

the results in their study show that the fuel can be saved by about 14 % [12].

Even if vehicle platooning has a certain advantage in traffic capacity improvement and fuel-saving, a huge wastage of traffic capacity and fuel will occur due to the stoppage at signalized intersection during its red phase. Idling at red lights will decrease traffic capacity and increase fuel consumption from many aspects:

(1) **Stop-and-go motion.** X. Zhang et al. proved that fuel consumption and exhaust gas emission can remarkably decrease when the acceleration and deceleration of vehicles are pretty gentle; idling at red light or traveling in a mode of stop-and-go will consume more fuel and emit more greenhouse gases comparing with the vehicles in free flow [13]. Research by Wan N et al. emphasized that, due to vehicles' stop-and-go motion, they need to consume more energy than that during cruising [14].

(2) **Intersection delays.** Intersection delays may include queue delay and control delay. Ch. Ravi Sekhar et al. estimated delay and fuel loss during idling at signalized intersections. The simulation results showed that heavy delays and a huge wastage of fuel at intersections are caused due to stoppage of vehicles during the red phase of the signals, because many vehicles need to stop as a consequence of their arrival either during the red interval or during the green interval when the queue of vehicles that had formed during the previous red interval has not yet fully dissipated [15].

(3) **Congestion.** On the other hand, heavy delays at intersections may result in traffic congestion especially in heavy volume arterial corridor, which will cause a large amount of financial loss, including more traffic capacity and fuel loss. INRIX, a joint traffic data company in London, quantitatively analyzed the impact of traffic congestion on Britain, France, Germany, and the United States. In 2013, the four countries lost 200 billion dollars caused by traffic congestion, accounting for 0.8% of the total GDP of the four countries.

Therefore, a number of benefits can be obtained in limiting the idling time at red light. These benefits include increasing traffic efficiency, saving in fuel use, reduction in exhaust emissions, and even vehicle life extension. In recent years, the exponential increase in the number of vehicles in urban city has resulted in congestion and more fuel consumption at intersections. Traffic efficiency improvement and fuel economy have been paid more attention than ever. Thus, from the perspective of traffic efficiency improvement, travel comfort, and traffic energy conservation, it is of great importance to keep the traffic flow smooth and reduce red light idling.

Besides the fuel wastage at intersection due to the operation of signals, fuel consumption can also be affected by other factors along the entire trip, such as cruising speed, the intervehicle distance, and traffic conditions [16]. With the advanced control system installed on CAVs, all the vehicles can realize autonomous velocity control along the entire trip. Motivated by the problems mentioned above, this paper investigates one optimal platoon velocity control and separation strategy defined to find the optimal velocity profiles on signalized arterials that can avoid idling at red lights and improve fuel economy along the entire trip.

The rest of this paper is organized as follows. Section 2 introduces the conceptualization of vehicle platoon and analyzes the optimization goals in this study. Section 3 proposes the optimal platoon separation and velocity control strategy used in this paper to minimize the trip time and reduce the fuel consumption. The simulation results are presented and discussed in Section 4, and conclusion is given in Section 5.

2. Problem Statement

In this section, the conceptualization of the vehicle platoons in a short length is firstly introduced. Then the objective function of reducing the platoons idling at red lights is formulated. Finally, the formula for the minimum fuel consumption is introduced.

2.1. Conceptualization of Vehicle Platoons. Maiti et al. provided a detailed concept of vehicle platoon [1]. A vehicle platoon generally consists of one leader vehicle and a number of follower vehicles. The leader vehicle takes all decisions on behalf of the whole platoon and controls all the platoon members accordingly; all vehicles in the same platoon share the same velocity. Each platoon has its own ID. And the size of platoon is a dynamic property, which implies the current number of vehicles in platoon; the maximum size of a platoon means the maximum number of vehicles grouping together in a platoon. Each vehicle in a platoon also has its vehicle ID. As mentioned above, the role of a vehicle can be divided into leader vehicle and follower vehicle; the role may get updated by platoon operations (e.g., separation or merging), which means follower vehicle may change into leader vehicle when doing the separation operation, and leader vehicle will also become follower vehicle when the merging operation happens.

A platoon of CAVs is actually a network of dynamical systems, S. E. Li et al. presented a four-component framework to model, analyze, and synthesize a platoon of CAVs from the perspective of multiagent consensus control [17]. When a platoon is driving on the road, the leader vehicle can communicate with the transportation infrastructure relying on the V2I technology. The leader vehicle transmits the platoon's position, size, destination, and other traffic parameters to the transportation infrastructure, to avoid idling at red lights; the leader vehicle can obtain feedback velocity information from the control center who can give feedback on optimal velocity to the leader vehicle according to the green light duration and traffic flow information. When the leader vehicle receives the velocity information, it will transfer this velocity information to other follower vehicles immediately through V2V technology. Therefore vehicles in the same platoon can share the same velocity and then achieve unified operation; the platoon driving schematic diagram can be shown in Figure 1. As for the quality of information flow exchange among vehicles, some scholars addressed the internal stability and scalability issue for platoon under different information topologies; e.g., Y. Zheng et al. have studied the influence of information flow topology on the

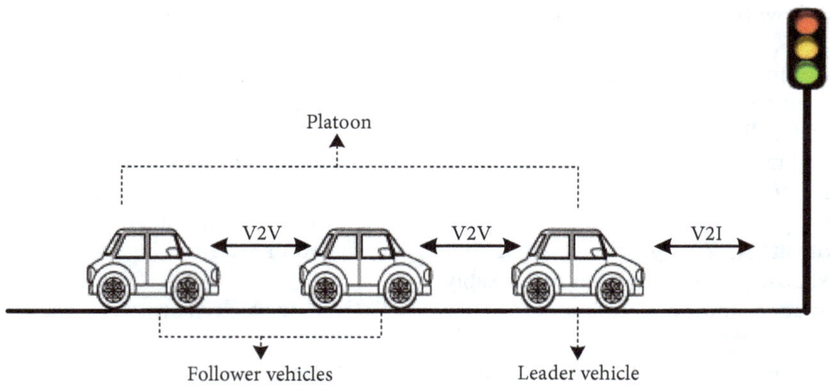

FIGURE 1: Platoon driving schematic diagram.

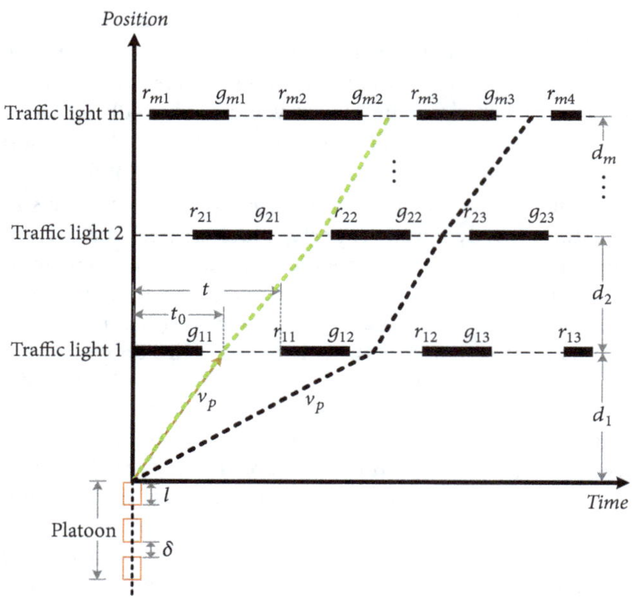

FIGURE 2: Schematic of the trajectory and velocity of leader vehicle.

internal stability and scalability of homogeneous vehicular platoon [18].

2.2. Idling at Red Light. The process of vehicle stopping at the red light and leaving when the light turns green is actually the stop-and-go motion, as mentioned in Section 1; in order to enhance vehicles' safety, energy efficiency, and mobility, the idling at the red light should be reduced as much as possible.

The schematic of the candidate trajectory and velocity of the leader vehicle at each intersection is shown in Figure 2; the green and the blue dotted lines denote the candidate trajectories of the leader vehicle in a platoon at different velocity, with the parameters defined as follows:

(1) The traffic light information of ith intersection is represented by $\{r_{ij}, g_{ij}\}; i = 1, 2, \cdots m, j = 1, 2 \cdots \infty$, where r_{ij} is the start of jth red phase and g_{ij} is the start of jth green phase at ith signalized intersection.

(2) d_i is the distance between consecutive intersections, that is, the distance between the i-1st traffic light and the ith traffic light (called ith segment); in particular, for a platoon, d_1 is the distance to the first upcoming traffic light and the d_2 is the distance between the first upcoming traffic light and the second traffic light, which can be estimated from the vehicle's GPS and the traffic light's location information.

(3) Let l denote average length of a vehicle and δ represent Gap distance between a pair of consecutive vehicles in a platoon

Our goal is to find a permit velocity of platoon which aids in minimizing idling at the red light. The problem of idling at the red light can be transformed into the platoon's waiting time at the traffic light. Based on the signal timing information and platoon information, total waiting time of all vehicles at all traffic lights can be calculated as

$$Z = \sum_{k=1}^{n} \sum_{i=1}^{m} \left\{ t_r^i - \text{mod} \left[\frac{d_i + \theta v_p^i(k) + (k-1)(l+\delta)}{v_p^i(k)\, t_{cycle}^i} \right] \right\} \quad (1)$$

where

Z is the sum of the platoon's waiting time at all the traffic lights;

k is the vehicles' ID number in a platoon, $k = 1, 2, \cdots n$;

i is the traffic lights' number during the trip, $i = 1, 2, \cdots m$;

t_r^i is the red light duration in the ith traffic light, $t_r^i = g_{ij} - r_{ij}$;

θ is the compensation time, for the case of $r_{i1} \neq 0$;

$v_p^i(k)$ is the passing velocity of kth vehicle in a platoon at ith signalized intersection;

t_{cycle}^i is the cycle of traffic signals at the ith traffic light; $t_{cycle}^i = g_{i,j} - g_{i,j-1}$ or $r_{i,j} - r_{i,j-1}$.

Vehicles in the same platoon share the same velocity; *i.e.*, $v_p^i(k) = v_p^i$, where v_p^i denotes the passing velocity of a platoon at ith signalized intersection whose unit is m/s. Specifically, Z only takes a result greater than 0; that is, when the operation result of Z is less than 0, the final result of Z is equal to 0.

2.3. Fuel Consumption. Before studying the influence of idling at red light and velocity on fuel consumption, we introduce the fuel consumption model. Many efforts have been made to understand the relationship between traffic activities and fuel consumption rate; many researchers modeled fuel consumption as a function of vehicle load and average speed [19–22]. There are a number of microscopic fuel consumption models; we use the model proposed by Kamal et al. [23]. This is because of its simplicity; calculating the fuel consumption only uses the instantaneous velocity of platoon. Based on this model, optimal velocity could be found to minimize fuel consumption. Kamal et al. sampled sufficient data from a passenger size vehicle and fitted it into a third order polynomial curve that approximates the relation between fuel consumption rate and velocity. In this model, the fuel consumption rate $\dot{m}$ is estimated as

$$\dot{m} = \alpha_0 + \alpha_1 v_p^i + \alpha_2 \left(v_p^i\right)^2 + \alpha_3 \left(v_p^i\right)^3 \quad (2)$$

where α_0, α_1,α_2, and α_3 are corresponding coefficients, whose values are presented in Table 3.

Hence, during the whole trip, for each vehicle in the same platoon, the fuel consumption can be calculated as

$$J = \left[\alpha_0 + \alpha_1 v_p^i + \alpha_2 \left(v_p^i\right)^2 + \alpha_3 \left(v_p^i\right)^3\right] t_i(k) \quad (3)$$

where $t_i(k)$ represents the trip time of kth vehicle of platoon in ith segment.

Assuming that all the vehicles run at constant speed in each segment ignoring the acceleration or deceleration process, then, for all vehicles in a platoon passing all the intersections without idling, the total fuel consumption can be calculated as

$$J = \sum_{k=1}^{n} \sum_{i=1}^{m} \left[\frac{\alpha_0}{v_p^i} + \alpha_1 + \alpha_2 v_p^i + \alpha_3 \left(v_p^i\right)^2\right] \cdot \left[d_i + (k-1)(l+\delta)\right]. \quad (4)$$

2.4. Optimization Problem and Complexity Analysis. The objective of this paper is to minimize two performance indexes $Z = f(v_p^i(k))$ and $J = f(v_p^i(k))$, *i.e.*, the waiting time at red lights and fuel consumption, by determining velocity profiles $v_p^i(k) = \{v_1, v_2, \ldots v_m\}$ for each vehicle at each segment subject to certain constraints, which will be introduced in detail in Section 3. Obviously, the total waiting time at all the traffic light can be reduced to 0 as long as all the vehicles pass with a suitable velocity to ensure no idling at red light.

Y. Zheng et al. analyzed the complexity of a known green light optimal velocity (GLOV) problem, finding, e.g., optimal velocity that can avoid idling at red lights and minimize the trip time [24]. The complexity analysis shows that GLOV with binary velocity choices belongs to NP-complete, which means it cannot be numerically solved in polynomial time unless P=NP. Intuitively, it is much more difficult to solve this problem with more velocity choices; the number of possible solutions will increase exponentially.

In order to ensure effective solution to our proposed problem, approximation algorithm is proposed in the following optimization strategy. We consider finding optimal velocity profiles for each platoon to save fuel consumption as much as possible while ensuring the improvement of traffic efficiency, that is, to let the maximum number of vehicles pass without idling at red light as much as possible even if the current velocity does not guarantee the lowest fuel consumption.

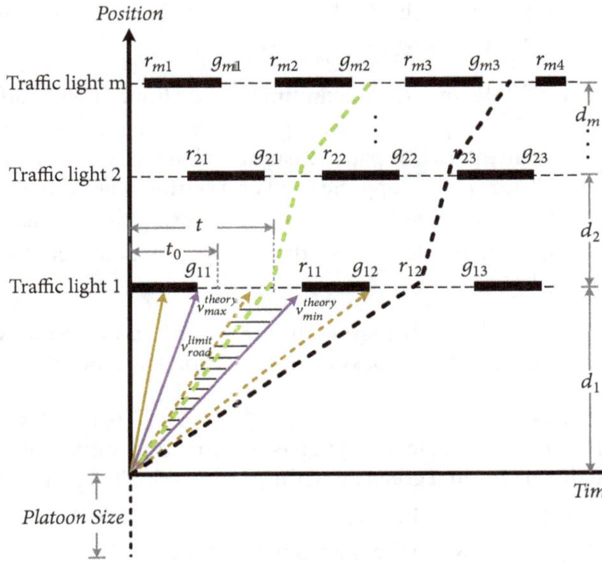

FIGURE 3: Allowable velocity bound for platoon.

3. Optimization Strategy

This section focuses on the introduction and analysis of optimization strategy to solve the problem defined in Section 2.

3.1. Allowable Bounds Analysis for Optimal Velocity Selection. To ensure that all the vehicles in the platoon could pass all the traffic lights without idling at red lights, it must be firstly met that the leader vehicle in a platoon can pass through the intersection during its green phase. The allowable velocity bound for a platoon is shown in Figure 3, with the parameters defined as follows:

(1) $v_{max}^{theory}(i, j)$ and $v_{min}^{theory}(i, j)$ represent the maximum and minimum theory velocity of the leader vehicle of a platoon to pass the ith traffic light without idling during its jth green phase, respectively. Assuming that the vehicles run at constant speed in each segment ignoring the acceleration or deceleration process, then the maximum and minimum theory

velocity for the leader vehicle can be calculated as $v_{max}^{theory}(i,j) = d_i/g_{ij}$, $v_{min}^{theory}(i,j) = d_i/r_{ij}$.

(2) v_{road}^{limit} denotes the maximum velocity limit of the road, which is specified by the government agency on each segment. In this paper, assuming that it is the same for all segments along the road and setting it as 20m/s in following simulation, it is worth noting that its value may be greater or less than v_{max}^{theory}, or even equal to v_{max}^{theory}.

(3) Let $v_{max}^{permit}(i,j)$ present the maximum passing velocity for a platoon to pass the intersection without idling.

For example, if $[v_{min}^{theory}(1,1), v_{max}^{theory}(1,1)] \cap [0, v_{road}^{limit}] \neq \emptyset$ and the leader vehicle wants to pass the first upcoming traffic light during the first green light, the passing velocity v_p should belong to the set intersection$[v_{min}^{theory}(1,1), v_{max}^{theory}(1,1)] \cap [0, v_{road}^{limit}]$, as shown in the shaded part in Figure 2. In this case,

$$v_{\max}^{permit}(i,j) = \begin{cases} \min\left\{v_{\max}^{theory}(i,j), v_{road}^{\text{limit}}\right\}, & \text{if } v_{\max}^{theory}(i,j) \neq v_{road}^{\text{limit}} \\ v_{\max}^{theory}(i,j) = v_{road}^{\text{limit}}, & otherwise. \end{cases} \quad (5)$$

However, if $v_{road}^{limit} < v_{min}^{theory}(1,1)$ which means$[v_{min}^{theory}(1,1), v_{max}^{theory}(1,1)] \cap [0, v_{road}^{limit}] = \emptyset$, the leader vehicle has to decrease current velocity and pass the first upcoming traffic light in its second green phase only if $[v_{min}^{theory}(1,2), v_{max}^{theory}(1,2)] \cap [0, v_{road}^{limit}] \neq \emptyset$, as shown by the blue dotted line in Figure 3; the passing velocity should also follow the restriction of $v_p \in [v_{min}^{theory}(1,2), v_{max}^{theory}(1,2)] \cap [0, v_{road}^{limit}]$.

To sum up, the leader vehicle will find the possibility of passing during jth green phase at ith signalized intersection until the set intersection $[v_{min}^{theory}(i,j), v_{max}^{theory}(i,j)] \cap [0, v_{road}^{limit}]$ is not empty. That is, to ensure that all vehicles can pass all the traffic lights without idling, our optimal velocity solution v_p^i for a platoon should satisfy

$$v_p^i \in \left[\frac{d_i}{r_{ij}}, \frac{d_i}{g_{ij}}\right] \cap \left[0, v_{raod}^{limit}\right] \neq \varnothing, \quad j = 1, 2, \cdots \infty. \quad (6)$$

3.2. Velocity Control and Separation Strategy of Platoon. The detailed process of separation strategy of platoon will be introduced in this section.

According to the concept of vehicle platooning, all vehicles in the same platoon share the same velocity, which means the follower vehicles' trajectory will be parallel to the trajectory of the leader one. If we draw a line parallel to the v_p^i velocity line of the leader vehicle, the intercept on the position axis is the maximum platoon size that can pass the ith intersection during one of its green phases, as shown in Figure 4. It is easy to find that the number of vehicles of a platoon that can pass without idling will increase with the increase of passing velocity.

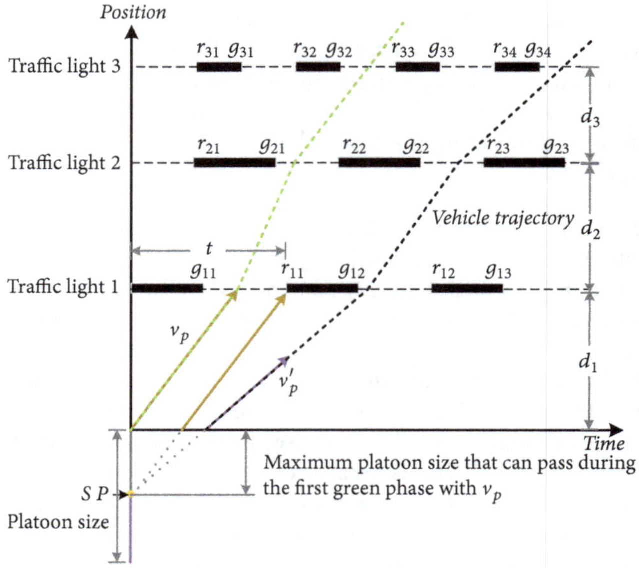

FIGURE 4: Schematic of the separation strategy.

Let $L(v_p)$ denote the maximum platoon size at velocity v_p and $L(platoon)$ represent the real platoon size.

$$\begin{aligned} L(v_p) &= v_p r_{ij} - d_i \\ L(platoon) &= (n-1)(l+\delta) \end{aligned} \quad (7)$$

If $L(v_p)$ is less than $L(platoon)$, some follower vehicles will stop at the red light if they keep the current velocity. To avoid idling at red light for the rest vehicles, they need to decelerate to pass during the next green phase. In other words, the rest vehicles should separate from the original platoon and recombine to a new platoon, that is, our proposed velocity and separation strategy. A new replanning velocity v'_p will be given to this new platoon to ensure that the maximum number of vehicles in this new platoon can pass the traffic light in the next green light.

The maximum number of vehicles that can pass during the first green phase at the first upcoming intersection can be calculated as

$$n_p = \left\lceil \frac{v_p t - d_1}{l+\delta} \right\rceil. \quad (8)$$

As introduced in Section 2.1, each platoon has its own ID, and the vehicles in each platoon also have their own IDs, respectively (e.g., the leader vehicle, 2nd vehicle,. . ., n_pth vehicle, (n_p+1)th vehicle,. . .,mth vehicle). When the platoon needs to do the separation strategy, the leader vehicle will send the separation command to the (n_p+1)th vehicle, who will become the new leader vehicle of a new platoon and obtain a replanning velocity v'_p. Therefore, there exist a separation point (SP) between n_pth vehicle and(n_p+1)th vehicle, as shown in Figure 4.

Vehicles before the separation point (SP) belong to the original platoon, which will pass the first intersection during

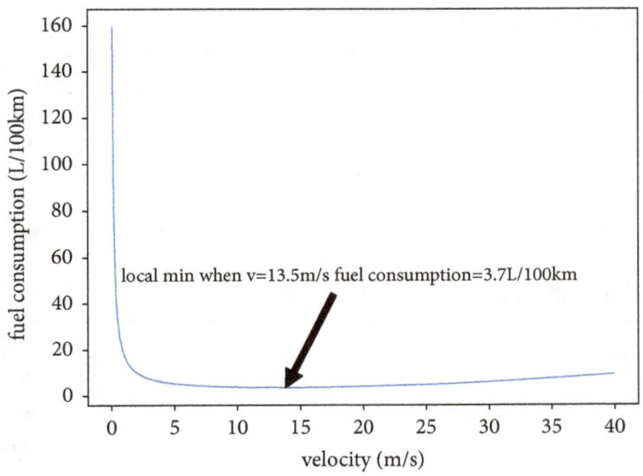

FIGURE 5: The fuel consumption at different velocity.

its first green phase. The new platoon consists of those vehicles behind SP that need to decelerate to v_p' to avoid idling at red light. The velocity space for the replanning velocity v_p' is similar to the passing velocity v_p except $j = 2, 3, \cdots \infty$.

$$v_p' \in \left[\frac{v_p t}{r_{ij}}, \frac{v_p t}{g_{ij}}\right] \cap \left[0, v_{road}^{limit}\right] \neq \varnothing, \quad j = 2, 3, \cdots \infty \tag{9}$$

Let $v_{optimal}^{fuel}$ denote the optimal velocity in case of lowest fuel consumption. According to the fuel consumption formula (4) with the parameters' value in Table 3, we can see that when the velocity is equal to 13.5m/s, the fuel consumption is minimal, as shown in Figure 5. That is $v_{optimal}^{fuel} = 13.5m/s$.

On the one hand, our first goal in this paper is to improve traffic efficiency by avoiding idling at red light; at each section for a platoon, we can choose the maximum passing velocity v_{max}^{permit} for a platoon, which can ensure that maximum number of vehicles can pass without idling. On the other hand, a goal to minimize fuel consumption as much as possible is also taken into consideration in this paper as mentioned in Section 2. Obviously, the optimal fuel velocity does not necessarily satisfy the need for letting maximum number of vehicles pass at each green phase. In other words, the maximum passing velocity v_{max}^{permit} is not necessarily the most fuel-efficient. Hence, a velocity control strategy is also considered based on the separation strategy. The velocity control and separation strategy to obtain the optimal platoon velocity profiles for all vehicles in the platoon is shown in Figure 6.

Take the first intersection, for example; one optimal solution to the problem defined in Section 2 can be constructed by the following steps:

Step 1. Check if $v_{optimal}^{fuel}$ belongs to the interval$[v_{min}^{theory}(1,1), v_{max}^{permit}(1,1)]$, where $v_{max}^{permit}(1,1)$ satisfies (5); if it belongs to it, turn to Step 2; otherwise, turn to Step 3.

Step 2. Compare the maximum platoon size $L(v_{optimal}^{fuel})$ at velocity $v_{optimal}^{fuel}$ and the actual platoon size $L(platoon)$, which can be calculated according to (7). If $L(v_{optimal}^{fuel}) \geq L(platoon)$, which means that all vehicles in the original platoon can pass the first intersection during its first green phase with velocity $v_{optimal}^{fuel}$, then choose the passing velocity $v_p = v_{optimal}^{fuel}$. Otherwise, turn to Step 3.

Step 3. Compare the platoon size $L(v_{max}^{permit})$ with velocity v_{max}^{permit} and the actual platoon size $L(platoon)$. If $L(v_{max}^{permit}) \geq L(platoon)$, which means that the platoon can pass the intersection with a velocity below v_{max}^{permit}, taking into account fuel economy, the platoon pass with velocity v_p that minimizes the fuel consumption:

$$v_p = \begin{cases} v_{min}^{all} & \text{if } J\left(v_{min}^{all}\right) \leq J\left(v_{max}^{permit}\right) \\ v_{max}^{permit} & \text{if } J\left(v_{min}^{all}\right) > J\left(v_{max}^{permit}\right) \end{cases} \tag{10}$$

where v_{min}^{all} means the minimum velocity that can ensure all vehicles pass without idling when $L(v_{max}^{permit}) \geq L(platoon)$.

$$v_{min}^{all} = \frac{(n-1)(l+\delta) + d_1}{r_{11}} \tag{11}$$

If $L(v_{max}^{permit}) < L(platoon)$, which means that all vehicles cannot pass the intersection during its first green phase, even with the max velocity, then turn to Step 4.

Step 4. The original platoon is separated into two new platoons: platoon 1.1 and platoon 1.2. To ensure that as many vehicles as possible can pass the intersection within a green phase, the new platoon 1.1 will pass the intersection with velocity v_{max}^{permit}. And for the new platoon 1.2, return to Step 1 again to find an optimal passing velocity v_p' to pass the intersection during its next green phase.

For the remaining intersections, the process to find the optimal velocity solution is almost similar to the above steps, except that at Step 1, we check if $v_{optimal}^{fuel}$ belongs to $[v_{min}^{theory}(i,1), v_{max}^{permit}(i,1)]$ for ith intersection.

4. Simulation Case Studies

The route is assumed to have 4 intersections, and the parameters of the traffic light location and timing information are shown in Table 1. The platoon and road information is shown in Table 2. The parameter values of fuel economy model (5) are shown in Table 3. The simulations are run in MATLAB on an Intel® Core™i7 processor with 3.40 GHz processing speed per core, 8 GB of RAM.

4.1. Simulation 1: Conventional Strategy without Separation Strategy. Firstly, we study the case of conventional strategy; the separation strategy is not activated.

In the simulation, we consider a platoon of 20 vehicles. Figure 7 shows the simulation results of the velocity profile of

TABLE 1: Traffic light location and timing information.

	d_i (m)	t_r^i (s)	t_{cycle}^i (s)	r_{i1} (s)
$i = 1$	500	30	50	10
$i = 2$	500	39	59	9
$i = 3$	500	30	50	45
$i = 4$	500	40	65	22

TABLE 2: Platoon and road information.

Parameter	Value	Unit
l	5	m
δ	1	m
$n_{vehicle}$	20	veh
v_{road}^{limit}	20	m/s

TABLE 3: Parameters value of fuel consumption.

Coefficient	Value	Unit
α_0	0.1569	mL/s
α_1	2.450×10^{-2}	mL/m
α_2	-7.415×10^{-4}	mLs/m^2
α_3	5.975×10^{-5}	mLs^2/m^3

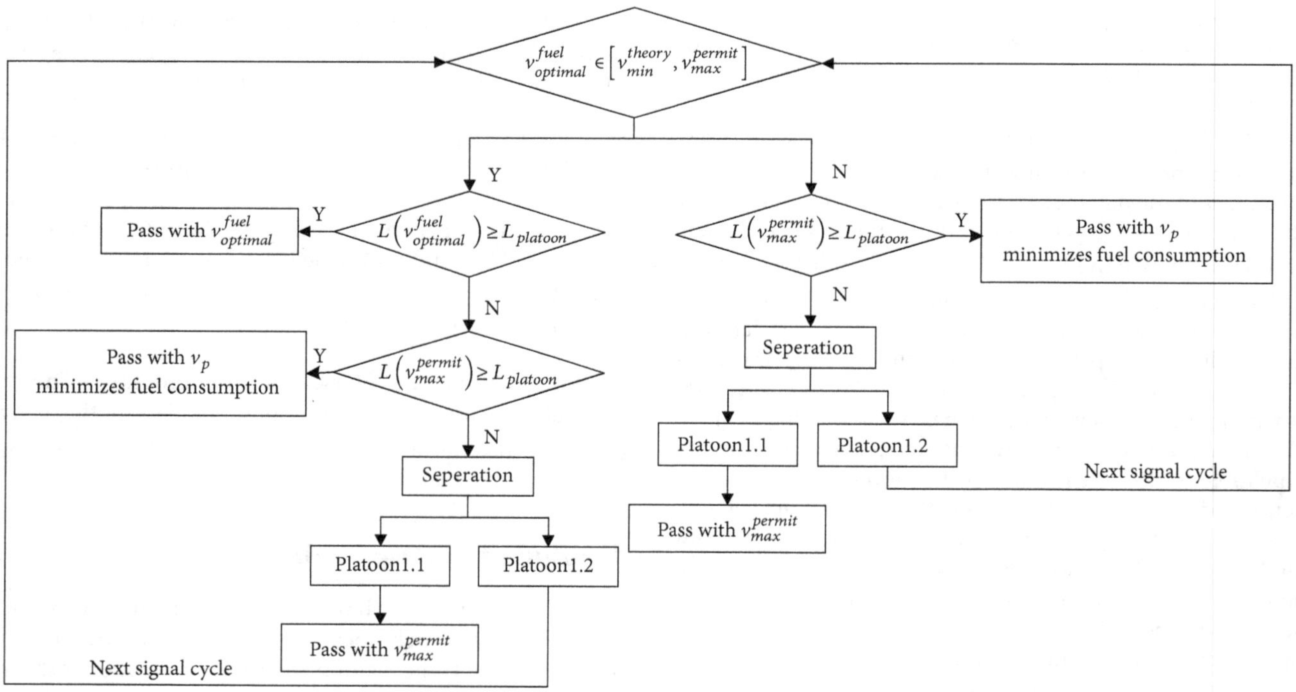

FIGURE 6: The process of our proposed velocity control and separation strategy.

each vehicle in the platoon under conventional circumstance. It is not difficult to see that most of vehicles in this platoon will be idle at all intersections, except that a small number of vehicles do not have to wait for the red light only at the third intersection. This is because of the lack of velocity control and separation strategy.

4.2. Simulation 2: Separation Strategy with Minimum Fuel Consumption. To solve the deficiency of conventional strategy, we propose a velocity control and separation strategy that takes into account the traffic efficiency and fuel-saving simultaneously as introduced in Section 3.2. The velocity profile of each vehicle in the platoon under this strategy is

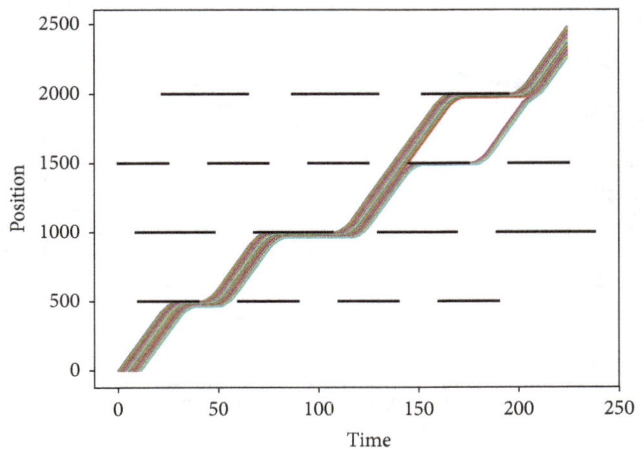

FIGURE 7: Trajectory of vehicles of platoon without separation strategy.

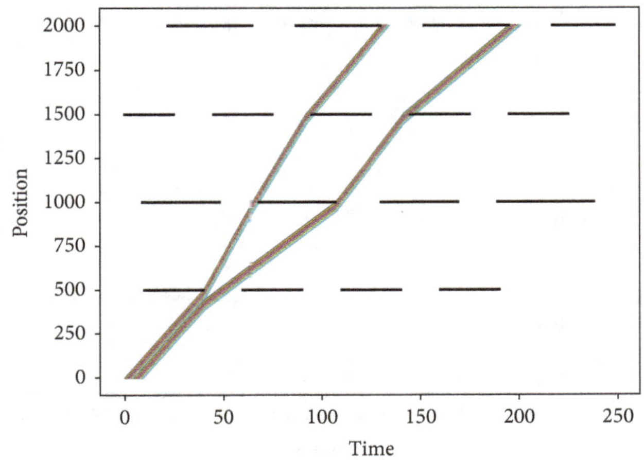

FIGURE 9: Vehicles' trajectory of platoon with separation strategy (*maximum passing velocity*).

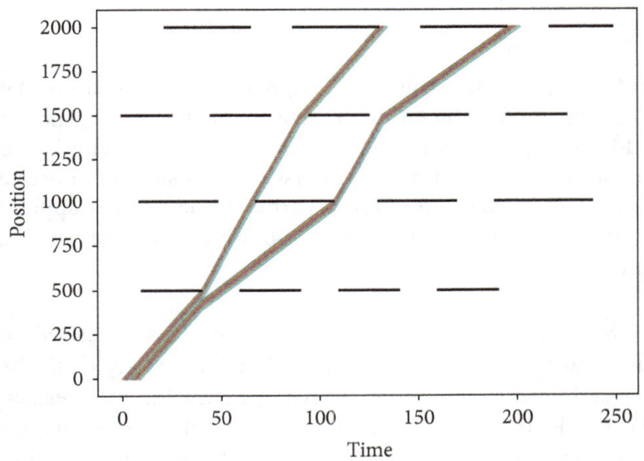

FIGURE 8: Trajectory of vehicles of platoon with separation strategy (minimum fuel consumption).

shown in Figure 8. One can see that the separation of original platoon occurs at the first intersection, dividing it into two new platoons. The separation point is located at the 11th vehicle, and the velocity profile of 1st vehicle and 11th vehicle is

$$v_p(1) = \{12.5m/s, 20m/s, 18.6m/s, 13.1m/s\}$$
$$v_p(11) = \{12.5m/s, 8.3m/s, 14.5m/s, 9.3m/s\}. \quad (12)$$

4.3. Simulation 3: Separation Strategy without Considering Fuel-Saving. In order to better understand how our proposed velocity control and separation strategy achieves fuel-saving under the premise of guaranteeing that maximum number of vehicles can pass the intersection without idling, we implement another simulation in which all the vehicles choose the maximum velocity to pass intersections without considering fuel-saving. This simulation is almost similar to simulation 2 except that all vehicles pass all the intersections with maximum passing velocity v_{max}^{permit}. Figure 9 shows velocity profile of each vehicle in the platoon under this circumstance. The velocity profile of 1st vehicle and 11th vehicle with this strategy is as follows:

$$v_p(1) = \{12.5m/s, 20m/s, 20m/s, 12.5m/s\}$$
$$v_p(11) = \{12.5m/s, 8.3m/s, 20m/s, 7.9m/s\}. \quad (13)$$

Comparing with our proposed separation strategy that considers fuel-saving, we find that the main difference between these two different situations is reflected at the third intersection. Under the premise of ensuring that all vehicles can pass without idling, our proposed separation strategy chooses the velocity that can minimize the fuel consumption, but this strategy chooses the maximum velocity to pass the intersection.

4.4. Evaluation of Proposed Velocity Control and Separation Strategy. Figure 10 shows that the fuel consumption and total travel time decrease with the introduction of our proposed velocity control and separation strategy.

One can see that the fuel consumption of each vehicle decreases dramatically with our proposed separation strategy. For the platoon with 20 vehicles, the total fuel consumption decreases by 18.1% comparing with the conventional strategy. By minimizing the fuel costs, we also implicitly increase some of the societal benefits of our proposed platoon separation strategy. Minimizing fuel consumption is equivalent to minimizing emissions [25]. Also, when we minimize the fuel costs, longer platoons are preferred as the total savings will be higher with more following vehicles in the system.

With our proposed separation strategy, the total travel time decreases by 19.2% compared with the conventional strategy. The first ten cars are particularly noticeable thanks to velocity control and separation strategy. In other words, traffic efficiency has improved. And longer platoons are associated with more efficient road utilization since the vehicles within a platoon drive closer together. The reduced space utilization as a result of platooning might help improve the traffic throughput.

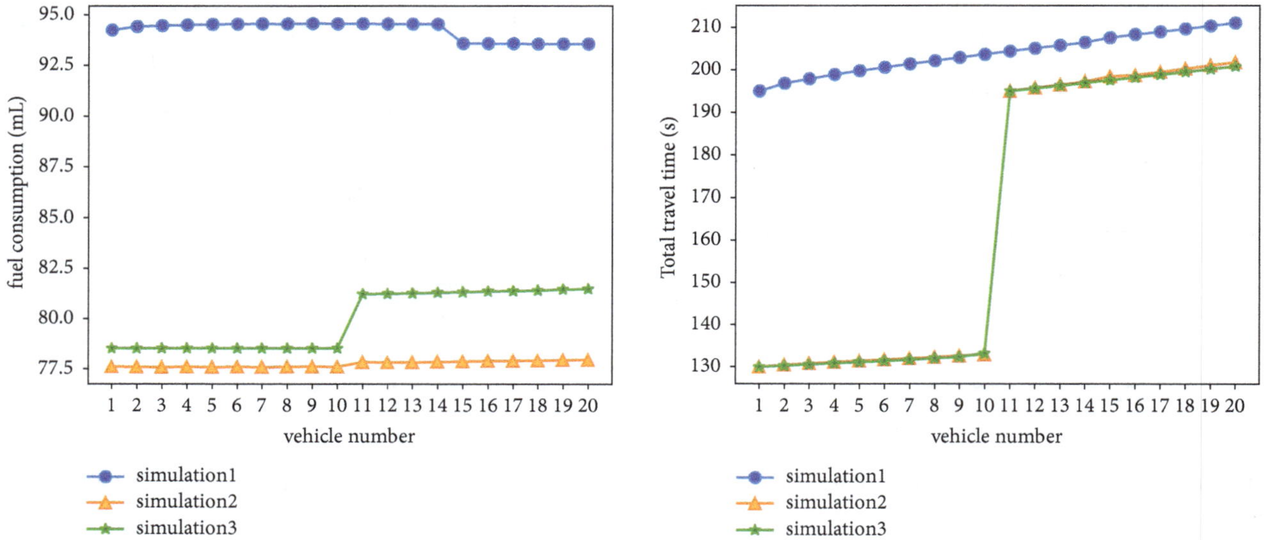

FIGURE 10: The total fuel consumption and the total travel time of each vehicle.

Intuitively, passing with the maximum velocity means that the travel time is the minimum. It is worth noting that this conclusion is only valid for a single intersection. Interestingly, for multiple intersections, the total travel time is not necessarily the smallest even if the maximum velocity is selected at each intersection, which is determined by the difference of signal phase between two consecutive signalized intersections. If the green phase difference of two adjacent intersections is very gentle, vehicles cannot pass these two intersections continuously during the same green phase due to the maximum velocity limit. In other words, even if the maximum speed is selected at the previous intersection, the platoon can only pass the consecutive intersection until its next green phase by reducing more velocity. This will probably lead to a decrease in the overall average speed, which in turn increases the total travel time. As we can see in Figure 10, the travel time of 10th–20th vehicle in simulation 2 is slightly less than that in simulation 3, although maximum velocity is selected at each intersection in simulation 3. This is because the green phase difference between 3rd and 4th intersection is very gentle. Therefore, passing with the maximum velocity may not necessarily improve traffic efficiency; on the contrary, choosing the optimal fuel consumption may even improve traffic efficiency in some cases.

5. Conclusion

A velocity control and separation strategy aimed at avoiding idling at red light and reducing fuel consumption as much as possible was proposed in this paper. The simulation results suggested that our proposed strategy effectively improves the performance of the platoon. The total travel time and the fuel consumption were reduced by 19.2% and 18.1%, respectively. The ultimate objectives of platooning are to enhance highway safety, improve traffic utility, and reduce fuel consumption. The main novelty and contribution of this work is providing an optimal platoon velocity control method and a separation strategy at signalized intersection that considers both traffic utility improvement and fuel economy. Additionally, it is worth noting that the using scenarios of this strategy involve multiple intersections instead of only one single signalized intersection, and this strategy can be applied to full autonomous or semiautonomous vehicles in the future.

Several extensions to the present study are desired in the future. Some assumptions made in this study could be violated, and we caution against generalizing the results. We would like to mention that the simulation results are based on the assumption that all vehicles run at a constant velocity ignoring the acceleration or deceleration process. Actually, the fuel consumption and the state of the platoon system may change with the acceleration or deceleration process. Additionally, this paper only considers the separation strategy of a single static platoon and ignores the dynamics of platooning process between multiple platoons. The dynamics increase the complexity of the decision-making process.

Therefore, each problem discussed above presents an important and very challenging research topic. In the future work, the impact of acceleration and deceleration process on fuel consumption and travel time needs to be investigated to examine the validation of the simulation results. Nevertheless, this paper provides an explicit strategy to better improve the traffic efficiency and fuel-saving in a vehicle platoon.

Conflicts of Interest

The authors declare that they have no conflicts of interest.

Acknowledgments

This work was supported by the National Natural Science Foundation of China (Grant no. 61573098).

References

[1] S. Maiti, S. Winter, and L. Kulik, "A conceptualization of vehicle platoons and platoon operations," *Transportation Research Part C: Emerging Technologies*, vol. 80, pp. 1–19, 2017.

[2] A. Ghiasi, O. Hussain, Z. Qian, and X. P. Li, "A mixed traffic capacity analysis and lane management model for connected automated vehicles: A Maekov chain method," *Transportation Research Part B*, vol. 106, pp. 266–292, 2017.

[3] D. Chen, S. Ahn, M. Chitturi, and D. A. Noyce, "Towards vehicle automation: Roadway capacity formulation for traffic mixed with regular and automated vehicles," *Transportation Research Part B: Methodological*, vol. 100, pp. 196–221, 2017.

[4] L. Ye and T. Yamamoto, "Modeling connected and autonomous vehicles in heterogeneous traffic flow," *Physica A: Statistical Mechanics and its Applications*, vol. 490, pp. 269–277, 2018.

[5] A. Olia, S. Razavi, B. Abdulhai, and H. Abdelgawad, "Traffic capacity implications of automated vehicles mixed with regular vehicles," *Journal of Intelligent Transportation Systems: Technology, Planning, and Operations*, pp. 1–19, 2017.

[6] J. Lioris, R. Pedarsani, F. Y. Tascikaraoglu, and P. Varaiya, "Doubling throughput in urban roads by platooning," *IFAC-PapersOnLine*, vol. 49, no. 3, pp. 49–54, 2016.

[7] J. Lioris, R. Pedarsani, F. Y. Tascikaraoglu, and P. Varaiya, "Platoons of connected vehicles can double throughput in urban roads," *Transportation Research Part C: Emerging Technologies*, vol. 77, pp. 292–305, 2017.

[8] Y. Chang and P. Edara, "Evaluation of a reservation-based intersection control algorithm for hurricane evacuation with autonomous vehicles," *International Journal of Disaster Risk Reduction*, 2017.

[9] Organization of Petroleum Exporting Countries, "World Oil Outlook," 2016.

[10] M. P. Lammert, A. Duran, J. Diez, K. Burton, and A. Nicholson, "Effect of Platooning on Fuel Consumption of Class 8 Vehicles Over a Range of Speeds, Following Distances, and Mass," *SAE Commercial Vehicle Engineering Congress (COMVEC)*, vol. 7, no. 2, pp. 626–639, 2014.

[11] A. Al Alam, A. Gattami, and K. H. Johansson, "An experimental study on the fuel reduction potential of heavy duty vehicle platooning," in *Proceedings of the International IEEE Conference on Intelligent Transportation Systems*, pp. 306–311, 2010.

[12] S. Tsugawa, S. Kato, and K. Aoki, "An automated truck platoon for energy saving," in *Proceedings of the IEEE/RSJ International Conference on Intelligent Robots and Systems*, vol. 32, pp. 4109–4114, 2011.

[13] X. Zhang, B. Jia, and R. Jiang, "Impact of safety assistance driving systems on oscillation magnitude, fuel consumption and emission in a car platoon," *Physica A: Statistical Mechanics and its Applications*, vol. 505, pp. 995–1007, 2018.

[14] N. Wan, A. Vahidi, and A. Luckow, "Optimal speed advisory for connected vehicles in arterial roads and the impact on mixed traffic," *Transportation Research Part C: Emerging Technologies*, vol. 69, pp. 548–563, 2016.

[15] C. R. Sekhar, P. Raj, P. Parida, and S. Gangopadhyay, "Estimation of delay and fuel loss during idling of vehicles at signalized intersection in Ahmedabad," *Procedia - Social and Behavioral Sciences*, vol. 104, pp. 1178–1187, 2013.

[16] K. Shi, B. Di, K. Zhang, C. Feng, and L. Svirchev, "Detrended cross-correlation analysis of urban traffic congestion and NO2 concentrations in Chengdu," *Transportation Research Part D: Transport and Environment*, vol. 61, pp. 165–173, 2018.

[17] S. E. Li, Y. Zheng, K. Li et al., "Dynamical Modeling and Distributed Control of Connected and Automated Vehicles: Challenges and Opportunities," *IEEE Intelligent Transportation Systems Magazine*, vol. 9, no. 3, pp. 46–58, 2017.

[18] Y. Zheng, S. E. Li, J. Wang, D. Cao, and K. Li, "Stability and scalability of homogeneous vehicular platoon: Study on the influence of information flow topologies," *IEEE Transactions on Intelligent Transportation Systems*, vol. 17, no. 1, pp. 14–26, 2016.

[19] O. Jabali, T. Van Woensel, and A. G. de Kok, "Analysis of travel times and CO_2 emissions in time-dependent vehicle routing," *Production Engineering Research and Development*, vol. 21, no. 6, pp. 1060–1074, 2012.

[20] T. Bektaş and G. Laporte, "The pollution-routing problem," *Transportation Research Part B: Methodological*, vol. 45, no. 8, pp. 1232–1250, 2011.

[21] Y. Suzuki, "A new truck-routing approach for reducing fuel consumption and pollutants emission," *Transportation Research Part D: Transport and Environment*, vol. 16, no. 1, pp. 73–77, 2011.

[22] Y. Y. Kuo, "Using simulated annealing to minimize fuel consumption fort the time-dependent vehicle routing problem," *Computers & Industrial Engineering*, vol. 59, pp. 157–165, 2010.

[23] M. A. S. Kamal, M. Mukai, J. Murata, and T. Kawabe, "Ecological driving based on preceding vehicle prediction using MPC," *IFAC Proceedings Volumes*, vol. 44, no. 1, pp. 3843–3848, 2011.

[24] Y. Zheng, S. Eben, B. Xu, K. Q. Li, and J. Wang, "Complexity analysis of green light optimal velocity problem: an NP-complete result for binary speed choices," in *Proceedings of the 14th Intelligent Transportation System Asia Pacific Forum*, 2015.

[25] G. Scora and M. Barth, "Comprehensive Modal Emissions Model (CMEM), Version 3.01 User guide," Riverside Centre for Environmental Research and Technology, University of California, 2006.

14

Hybrid Optimization-Based Approach for Multiple Intelligent Vehicles Requests Allocation

Ahmed Hussein, Pablo Marín-Plaza, Fernando García, and José María Armingol

Intelligent Systems Lab (LSI) Research Group, Universidad Carlos III de Madrid (UC3M), Calle Butarque 15, Leganés, 28911 Madrid, Spain

Correspondence should be addressed to Ahmed Hussein; ahussein@ing.uc3m.es

Academic Editor: Aboelmaged Noureldin

Self-driving cars are attracting significant attention during the last few years, which makes the technology advances jump fast and reach a point of having a number of automated vehicles on the roads. Therefore, the necessity of cooperative driving for these automated vehicles is exponentially increasing. One of the main issues in the cooperative driving world is the Multirobot Task Allocation (MRTA) problem. This paper addresses the MRTA problem, specifically for the problem of vehicles and requests allocation. The objective is to introduce a hybrid optimization-based approach to solve the problem of multiple intelligent vehicles requests allocation as an instance of MRTA problem, to find not only a feasible solution, but also an optimized one as per the objective function. Several test scenarios were implemented in order to evaluate the efficiency of the proposed approach. These scenarios are based on well-known benchmarks; thus a comparative study is conducted between the obtained results and the suboptimal results. The analysis of the experimental results shows that the proposed approach was successful in handling various scenarios, especially with the increasing number of vehicles and requests, which displays the proposed approach efficiency and performance.

1. Introduction

Self-driving cars could revolutionize how people get around; with the introduction of automation into roads, it will contribute to solving the issues related to the traffic accidents, congestion, and energy consumption. The driverless vehicle technologies are advancing on a great scale, specifically the multiple sensors fusion techniques, deep learning, and computational intelligence. Together, they enable these vehicles to understand the nearby surroundings and take appropriate actions to navigate on their own from one point to another. The technology does not stop at vehicle being fully automated; however multiple of them cooperate together to achieve smoother driving operation [1–3].

Researchers in the Intelligent Transportation Systems (ITS) field are investing more time on the self-driving cars research and development [4–6]. However, during the last decade, Multirobot Systems (MRS) fell under the research attention of the ITS community. This increased interest comes from the significant advantages and higher potential provided by MRS over single robot systems. MRS can be simply understood to be a group of robots cooperating together for accomplishing a certain task or mission [7, 8]. With reference to the literature review survey, the coordination and cooperation among multiagent systems can be modeled as Multirobot Task Allocation (MRTA) problem. MRTA is a NP-hard problem, which concerns the use of the available resources in an efficient manner. Accordingly, the decision of which robot will do which task strongly affects the performance of the system [9, 10].

Accordingly, this paper introduces a hybrid optimization-based approach to solve the MRTA problem. This approach can coordinate among several intelligent vehicles and aid in the cooperation for better overall performance. The algorithm is implemented on automated golf carts and tested in off-road environments in real-life scenarios. Moreover, in order to prove the proposed solution efficiency, several experiments are carried out over well-known benchmarks. The benchmarks optimal solution is obtained after hundreds of hours of computations. The proposed approach was able to obtain a near-optimal solution in a matter of seconds. This proves the proposed approach high performance.

The remainder of the paper is organized as follows: Section 2 presents the previous work in the field of multiple automated vehicles and the allocation problem, followed by Section 3, which introduces the problem formulation, solution constraints, objective function, and the proposed solving hybrid approach. In Section 4, the used platform is described, along with the experimental setup in the designed architecture, selected scenarios, and the evaluation metrics. Results and discussion are illustrated in Section 5. Finally, the conclusion and future work are summarized in Section 6.

2. Related Work

Autonomous cars are becoming more frequent in nowadays scientific and industrial context, since Google launched their driverless car project in 2011 [4]. From then onwards, many other approaches proposed different architectures and solutions, all of them moving towards the development of the autonomous vehicle. For instance, Mercedes with the Bertha project proved the viability of autonomous vehicles in German roads based on advance sensing capabilities [11]. The V-Charge project researches in the direction of allowing automated valet parking for self-driving cars [12]. Moreover, several vehicle manufacturers have proposed different solutions in the field of autonomous vehicles which are close to markets, such as BMW and Audi [13], Mercedes-Benz [14], and Volvo [15]. Furthermore, several other proposals offer the possibility of including autonomous vehicles in public transportation systems [5, 6].

The self-driving technology is advancing rapidly, but in order to safely deploy vehicles on public roads, cooperation with other road users is mandatory. This cooperation would allow the safe interaction with other vehicles, whether with a human driver or driverless. Although some of the proposed solutions already handle the presence of other vehicles in the road [16–18], by handling static and dynamic obstacles and adapting the trajectory accordingly, the cooperative driving is not yet achieved. Cooperative driving demands the information exchange and control strategies deployed in all the vehicles involved following a two-way communication. In this sense, during recent years many works have addressed this topic, trying to provide solutions based on different configuration and solutions. The Grand Cooperative Driving Challenge 2016 was created with the purpose of boosting cooperative automated vehicles in the form of a cooperative based competition [19]. On the one hand, authors in [20] presented an auction-based cooperative control for autonomous vehicles; on the other hand, authors in [21] proposed an approach to enhance common motion planning algorithms; this proposal allows cooperation with human-driven vehicles. Additionally, a novel concept is presented, based on a centralized strategy, using maneuver templates, which are formalized collaborative maneuvers, to select cooperative driving strategies [22]. All this work proves the importance of collaboration and communication among vehicles to allow a safe and efficient deployment of autonomous vehicles really in driving scenarios.

Since cooperation is essential among multiple vehicles on the road, coordination becomes the first issue to solve. In the MRS, the question of which robot is going to execute which task is answered through task allocation problem, which is commonly known as MRTA problem. By reviewing the literature, it was found that different optimization approaches have been used in order to solve the general task allocation problem and were also used in order to solve the MRTA problem. In [23], a mixed integer linear programming optimization approach was used in order to allocate heterogeneous robots for maximizing the coverage area of the regions of interest. In [24], a simulated annealing approach was used to solve the allocation of MRS through formulating the MRTA problem as multi-Traveling Salesmen Problem (mTSP). Then in [25], a market-based approach was proposed to solve the MRTA problem for heterogeneous robots and task formulated as mTSP. Moreover, the task allocation problem was also solved using hybrid optimization approaches such as the tabu search with random search method in [26] and tabu search with noising method in [27].

3. Methodology

In this section, the MRTA problem modeling and formulation are introduced, followed by the selected objective cost function. Afterward, the proposed algorithm is described, highlighting the main contribution. MRTA problem [7] is formulated to allocate multiple robots, in this case, vehicles, to numerous tasks, in this case, requests. The procedures are summarized as follows:

(1) Given a set of n vehicles, $V = \{V_1, V_2, \ldots, V_n\}$.

(2) Given a set of m requests, $R = \{R_1, R_2, \ldots, R_m\}$.

(3) Allocation of the requests to the vehicles occurs, $A : R \rightarrow V$.

(4) Output set S is the best allocation of the requests to the vehicles:

$$S = \{(V_1 \;\; R_1) \;\; (V_2 \;\; R_2) \;\; \cdots \;\; (V_k \;\; R_k)\} \quad \text{for } 1 \leq k \leq m. \tag{1}$$

(5) Allocation S minimizes or maximizes a certain objective function in order to get the best performance of the system.

3.1. Problem Formulation. A variant of mTSP is used to model MRTA problem. In the standard mTSP formulation, n nodes are defined with the edges distances and m salesmen are known. The salesmen are required to cover all the available nodes and return back to their starting node, such that each salesman makes a round trip. The mTSP can be formally defined on a graph $G = (V, E)$ where V is the set of n nodes and E is the set of edges. Let $c = (c_{ij})$ be the distance matrix associated with E. Assuming the more general case which is an asymmetric mTSP, thus $c_{ij} \neq c_{ji}$ $\forall(i, j) \in E$. The mTSP can be formulated as follows [9]:

$$x_{ij} = \begin{cases} 1 & \text{if edge } (i, j) \text{ is used in the tour} \\ 0 & \text{otherwise} \end{cases} \quad (2)$$

$$\text{minimize} \quad \sum_{i=1}^{n}\sum_{j=1}^{n} c_{ij} \times x_{ij} \quad (3)$$

$$\sum_{j=2}^{n} x_{1j} = m \quad (4)$$

$$\sum_{j=2}^{n} x_{j1} = m \quad (5)$$

$$\sum_{i=1}^{n} x_{ij} = 1, \quad j = 2, \ldots, n \quad (6)$$

$$\sum_{j=1}^{n} x_{ij} = 1, \quad i = 2, \ldots, n \quad (7)$$

$$x_{ij} \in \{0, 1\}, \quad \forall (i, j) \in E \quad (8)$$

$$\sum_{i \in S}\sum_{j \in S} x_{ij} \leq |\text{SubTour}| - 1, \quad \forall \text{SubTour} \subseteq V \setminus \{1\}, \ \text{SubTour} \neq \phi, \quad (9)$$

where (3) represents the objective function which is the summation of the total distance traveled, (4) and (5) ensure that exactly m salesmen departed their starting node and returned back. Equations (6), (7), and (8) are the usual assignment constraints. Finally, (9) is the subtour elimination constraint.

The proposed formulation is extended and adapted to the problem, where vehicles represent the salesmen and requests represent the cities. Therefore, vehicles capabilities and requests requirements are considered and included in the mTSP implementation. The added features of the vehicles are capacity, velocity, energy, efficiency, and sensors, and the ones for the requests are timestamp, priority, and passengers.

Where the vehicle capacity is the maximum number of passengers that it can hold, the velocity is a representation of the maximum speed it can reach, the energy is a representation of the battery level, the efficiency is a representation of the aging factor, and finally the sensors are a set of on-board devices to consider the vehicles as heterogeneous robots. On the other hand, the request timestamp is the date and time of the request creation, the priority is a representation or the request urgency based on the request type, and passengers are a representation of the number of users for the request.

3.2. Solution Construction. The solution is constructed as a set that includes a list of all vehicles in the system, followed by their assigned requests. The order of the list defines the quality of the solution according to the objective function. For example, for a problem with three vehicles and five requests, one of the possible candidate solutions is represented as follows:

$$\text{Candidate Solution} = [V1 \ \ R1 \ \ R2 \ \ V2 \ \ R3 \ \ R4 \ \ V3 \ \ R5]. \quad (10)$$

Any combination of this list presents another candidate solution, and since the mTSP is a permutation problem, therefore the order of this list affects the quality of the solution as per the objective function. The order implies that each vehicle is going to execute all requests succeeding it. In candidate solution (10), requests 1 and 2 are executed by vehicle 1, requests 3 and 4 are executed by vehicle 2, and finally request 5 is executed by vehicle 3.

3.3. Objective Function. Although the MRTA problem is formulated as an instance of the mTSP, the same objective function of the mTSP previously explained in (3) cannot be straightforwardly used as the objective function for the MRTA problem. Therefore, some variations had to be introduced to the objective function of the mTSP in order to be effectively used for the MRTA problem [28].

There are three main variations of the MRTA problem objective function compared with the mTSP objective function. First, it is a multiobjective function instead of a single objective function, second, the variable to be minimized is the time rather than the distance, and, third, the time of the maximum subtour is minimized rather than the total time, thus dealing with it as a MinMax problem. Then, for k subtours and r requests in each subtour, the total traveling time is calculated as follows:

$$f(\mathbf{x}) = \underset{j \in \{1,2,\ldots,k\}}{\arg\max} \frac{\sum_{i=1}^{r-1} \text{distance}\left(\text{subtour}_{j_i}, \text{subtour}_{j_{i+1}}\right)}{\text{vehicle velocity}_j} + \underset{j \in \{1,2,\ldots,k\}}{\arg\max} \frac{\sum_{i=1}^{r} \text{execution time}\left(\text{subtour}_{j_i}\right)}{\text{vehicle efficiency}_j}. \quad (11)$$

3.4. Solution Constraints. Although any arrangements of the vehicles and requests solution set are considered as a candidate solution for the mTSP problem, this does not guarantee that this solution is feasible for the MRTA problem. Therefore, few constraints are applied to the obtained solution, to ensure its validity and feasibility, which means checking whether the vehicle capabilities and request requirements are matching.

The first constraint is related to the capacity; for example, a vehicle with a maximum capacity of 4 passengers cannot handle a request of 6 passengers in one go; thus the request is decomposed into several requests and reallocated accordingly. The next constraint is the energy; the vehicle battery level is always taken into consideration before the final allocation of the request, since if the vehicle does not have sufficient energy for a specific request, it is reallocated to another vehicle. Another constraint is the priority level of the request, which implies that some requests, maintenance request, for instance, must be executed first. Last but not least, there is a constraint related to the mounted on-board sensors, which check if the request requires the presence of a specific sensor in the vehicle.

These applied constraints strongly affect the search space of the problem through decreasing the number of candidate solutions that can be accepted as feasible solutions. One may think that this decrease of the number of feasible solutions among the candidate solutions may make it easier for the applied algorithm to find the best solution than the case without the constraints. However, these applied constraints, in fact, make it more difficult and more time consuming to find the best solution. This is mainly because the algorithm will be visiting a large number of solutions that are candidate solutions of the mTSP but are not feasible to solve the MRTA problem.

3.5. Proposed Algorithm. The proposed solution is designed as metaheuristic optimization algorithm. It is a hybrid approach, which is based on both trajectory-based and population-based techniques. The trajectory-based one is the family of optimization techniques that use a single solution throughout the algorithm, in order to find the near-optimal solution. While the population-based one is the family of optimization techniques that iteratively transforms a set of solutions, in order to generate a new population of solutions with the aim of finding the near-optimal solution [29].

On the one hand, SA was selected as a trajectory-based approach, where the neighboring operator is randomly chosen at each iteration for diversity. The mutation operators are swapping, deletion and insertion, inversion, and scrambling. On the other hand, genetic algorithm (GA) was selected as the population-based approach, where the selected mutation operators are the same; however, additional crossover operators are selected, which are partially mapped crossover and order crossover.

The swapping operator chooses two random elements of the solution list and swaps them with each other. Deletion and insertion operator chooses a random element and deletes it from its current position and randomly inserts it in a new position. The inversion operator chooses two random positions and inverts the order of elements between these two positions. The scrambling operator picks two random positions and scrambles the elements between these two positions. Figure 1 illustrates an example of all proposed operators, where positions 1 and 6 are selected as the two random elements. At each iteration, one of the four mutators is randomly chosen in order to generate a neighboring solution of the current solution. The four methods vary in their diversification and intensification effect on the generated neighboring solution.

The crossover operator is the mimicking of the biological recombination between chromosomes, when some portion of the genetic material is swapped between chromosomes producing a new offspring chromosome. On the one hand, in the partially mapped crossover, two points are selected at random in both parents solutions and the offspring is created by exchanging the in-between chromosomes. On the other hand, in the order crossover a portion of one parent is mapped to a portion of the other parent; then from

Candidate solution

V1	R1	R2	V2	R3	R4	V3	R5

Swapping

V1	V3	R2	V2	R3	R4	R1	R5

Deletion & insertion

V1	R2	V2	R3	R4	V3	R1	R5

Inversion

V1	V3	R4	R3	V2	R2	R1	R5

Scrambling

V1	R3	R4	V3	R2	R1	V2	R5

FIGURE 1: Example for the mutation operators.

Parent solution 1

V1	R1	R2	V2	R3	R4	V3	R5

Parent solution 2

V2	R2	R5	V3	R1	R3	R4	V1

Partially mapped crossover

V3	R5	R2	V2	R3	R4	R1	V1

Order crossover

V3	R1	R2	V2	R3	R4	V1	R5

FIGURE 2: Example for the crossover operators.

the replaced portion onwards, the rest is filled up by the remaining genes, where already present genes are omitted and the order one is preserved. Figure 2 illustrates an example of proposed crossover operators, where positions 2 and 5 are selected as the two random elements. At each iteration, one of the two crossover mutators is randomly chosen in order to generate a neighboring solution of the current solution. Additionally, the random choice of the used operator gives the algorithm both the explorative and exploitative features that are useful in escaping local minimum and finding a better solution through searching in the neighbors of elite solutions, respectively.

The algorithm is used to solve the formulated model of the MRTA problem. Inputs are the list of the requests with their requirements, the list of vehicles with their capabilities, and the matrix of the distances between the points of interests of the requests locations. At each iteration of the algorithm, a solution is constructed to be evaluated; initially, it is random. The initial solution set is always filled with random elements of the requests list and the vehicles list until both lists are empty; the only constraint is that the start of the solution must be an element of the vehicles list. After the construction of the initial candidate solution or any other neighboring solution through the algorithm iterations, the feasibility of this solution must be checked, according to the solutions constraints that are previously explained in Section 3.4. As the algorithm progresses, neighboring solutions of the current solution must be generated in order to explore the search space of the problem. Algorithm 1 presents the proposed algorithm used to solve the MRTA problem in this paper.

Here the parameters can be defined as follows:

(i) *parentsList* is the list with parents solutions.

(ii) *childrenList* is the list with children solutions.

(iii) *nextGenerationList* is the list with next generation solutions.

(iv) *iterationsNumber* is the number of iterations.

(v) *elitismPercent* is the elitism percent.

(vi) *populationSize* is the population size.

(vii) *initialTemperature* is the initial temperature.

(viii) *finalTemperature* is the final temperature.

(ix) *currentTemperature* is the current temperature.

(x) *iterationsPerTemperature* is the number of iterations per temperature decrement.

(xi) α is the geometric coefficient.

(xii) *transitionProbability* is the transition probability.

(xiii) *currentAllocation* is the current allocation solution.

(xiv) *currentCost* is the current solution cost.

(xv) *neighborAllocation* is the neighbor allocation solution.

(xvi) *neighborCost* is the neighbor solution cost.

(xvii) *bestAllocation* is the best allocation solution.

(xviii) *bestCost* is the best solution cost.

4. Experimental Work

In this section, the real-world platform is described, followed by the proposed ROS-based architecture for the multiple vehicle cooperation. Moreover, it presents the selected scenarios along with the chosen evaluation metrics.

4.1. Platform Description. In this work, two automated vehicles were selected as the platforms for the real-world experiments. Figure 3 shows the vehicles; they are part of the Intelligent Campus Automobile (iCab) project. They are electric golf carts, which are modified mechanically and electronically for the purpose of fully automated vehicles.

Each vehicle is equipped with multiple on-board sensors, such as GPS and compass module, stereo-camera, laser-rangefinder, 3D LiDAR, optical encoders, and ultrasonic sensors. All sensors are connected to the on-board embedded computers, which are connected to a 4G router with a constant Internet connection [30].

The on-board embedded computers use Robot Operating System (ROS) architecture to carry out numerous technologies. The architecture is divided into three layers. (1) Low-level commands are executed in the reactive layer, such as communicating with sensors and actuators. (2) Communication between the layers and decomposing the complex task

FIGURE 3: iCab 1 and iCab 2 automated vehicles platforms.

to simple ones are executed in the sequencer layer. (3) Finally, high-level commands are executed in the deliberative layer, which is mainly responsible for environment perception, navigation, planning, localization, and mapping among others [31].

4.2. Communication and Cooperation Architecture. Communication and cooperation among a team of unmanned vehicles is a crucial task. Since in case of long duration missions, or vehicle failure, it is unlikely that a single vehicle is enough to complete the operation [10], therefore, for the communication among vehicles, the *multimaster_fkie* package is used [32]. It combines the required nodes to establish and manage a multimaster network over the implemented ROS-based architectures. In order to use the aforementioned package, vehicles must operate under a common network. Thus, for secure and stable paradigm, a Virtual Private Network (VPN) is used to overlay the commonly available one, which is implemented following the ITS Cooperative ITS (C-ITS) ETSI TR 101 607 V1.1.1 (2013-05) protocol. Detailed description for using the adopted communication schemes is explained in [33].

For the cooperation, *mrta* node is implemented as the one responsible for handling the task allocation problem. The inputs for the node are vehicle status topic, vehicle pose topic, and list of requests topic. The allocation output is published to the *task_executor* node of each vehicle. The idea of this node is to decompose the requests allocation into executable tasks, according to the vehicles capabilities. The objective is to execute the user transport requests by multiple automated vehicles, using coordination and cooperation mechanisms. Passengers use their own smartphones to create a request by selecting the number of the passengers and the pick-up and drop-off locations. The transport requests are then stored on a webserver, which is accessible by every vehicle in the system [34].

To avoid the centralized paradigm, leader token approach is proposed. Thus, at each iteration, the vehicles run a token selection algorithm, which determines the leader token holder for this loop. The algorithm decision is based on the vehicle status and current computational load. Only the leader vehicle communicates with the webserver for the updates of the requests lists and then publishes it to all other vehicles in the system. Accordingly, this mechanism ensures

```
Input: Requests list requests, Vehicles list vehicles, Distances matrix distances
Output: Best allocation bestAllocation
for i ← 1 to populationSize do
    parentsList ← generateValidSolution(requests, vehicles, distances)
end
currentAllocation ← minimumOf(parentsList)
for i ← 1 to iterationsNumber do
    if i ≤ 25% of iterationsNumber then
        elitismPercent = 20%
    else if i > 25% of iterationsNumberAND i ≤ 50% of iterationsNumber then
        elitismPercent = 30%
    else if i > 50% of iterationsNumberAND i ≤ 75% of iterationsNumber then
        elitismPercent = 40%
    else
        elitismPercent = 50%
    end
    childrenList ← crossover(least 20% of parentsList)
    childrenList ← mutation(top 80% of parentsList)
    nextGenerationList ← minimum elitismPercent of parentsList
    nextGenerationList ← minimum elitismPercent of childrenList
    for j ← 1 to (100% − elitismPercent × 2) of iterationsNumber do
        currentAllocation ← generateValidSolution(requests, vehicles, distances)
        currentCost ← getAllocationCost(currentAllocation)
        bestCost ← currentCost
        while currentTemperature < finalTemperature do
            for i ← 1 to iterationsPerTemperature do
                neighborAllocation ← generateNeighborSolution(currentAllocation)
                neighborCost ← getAllocationCost(neighborAllocation)
                if neighborCost < currentCost then
                    currentAllocation ← neighborAllocation
                    currentCost ← neighborCost
                    if neighborCost < bestCost then
                        bestAllocation ← neighborAllocation
                        bestCost ← neighborCost
                    end
                else
                    Generate: random number randomNumber ∈ ]0, 1[
                    transitionProbability = exp(−(neighborCost − currentCost) / currentTemperature)
                    if transitionProbability > randomNumber then
                        currentAllocation ← neighborAllocation
                        currentCost ← neighborCost
                    end
                end
            end
            currentTemperature = currentTemperature * α
        end
        nextGenerationList ← currentAllocation
    end
    if minimumOf(parentsList) < currentAllocation then
        currentAllocation ← minimumOf(parentsList)
    end
    parentsList ← nextGenerationList
end
bestAllocation ← currentAllocation
```

ALGORITHM 1: Proposed hybrid optimization-based algorithm.

the consistency and synchronization of the requests through a distributed paradigm.

4.3. Selected Scenarios. In order to test the proposed approach, several scenarios are selected in both simulation and real-world. The simulation scenarios are selected from well-known benchmarks of mTSP; this is in order to have the optimal cost available for comparison. Each scenario consists of a different number of cities, which are distributed over the environment in various locations.

Figure 4 shows the four selected scenarios, where the top left one is Christofides/Eilon with 51 cities, top right one is Berlin (Groetschel) with 52 locations, bottom left one is Christofides/Eilon with 76 cities, and bottom right one is Rattled Grid (Pulleyblank) with 99 cities [35, 36]. Each scenario has only one depot, represented by the red marker in the graphs.

On the other hand, the real-world scenario was designed in a way to evaluate the functionality of the proposed solution and the architecture in the platforms. The scenario involved three users using the application to create transportation requests. The three users had different starting points and different destinations; moreover the request time was close to others to evaluate how the vehicles are going to respond. Figure 5 shows the environment map with the vehicles and passengers locations, the two vehicles are represented with the golf-cart clip-art, and the passengers representation is marked in three different colors, with a circular shape of the same color for the desired destination. The experiment was video recorded and its results discussion is in the next section.

4.4. Evaluation Metrics. The proposed hybrid optimization-based solution is used to solve each scenario of the MRTA problem and the results are recorded for evaluation. In order to evaluate the quality, two evaluation metrics are introduced, which are the allocation cost and computational time.

The first evaluation metric is the allocation cost of the best allocation found. The allocation cost is calculated based on the objective function. Thus two allocations costs are computed, one is the MinMax cost, which represents the length of the longest subtour in the allocation, and the other is the total overall cost of all subtours.

The second metric is the computational time required by the algorithm to reach the best solution. The timer starts after the databases of the vehicles and requests are read and stops when the algorithm stops; then the elapsed time is reported. Since the computational time calculation is based on the machine, the computer used for all experiments has the specifications in Table 1.

5. Results

In this section, a comparative study is conducted between the proposed approach results and the reference optimal results presented in [35, 36]. The optimal results are obtained for the four well-known benchmarks, which are described in the scenarios subsection. The authors in [35] adjusted the benchmark settings, to have all vehicles located at the depot, where the vehicles are required to visit all locations in the scenario. With the condition that each location is visited exactly once, vehicles must return to the depot afterward.

Table 1: System specifications.

Processor	4 cores, 4 threads @3.8 GHz
Memory	16 GB DDR4-2133
Storage	HDD 1 TB 7200 RPM @3 Gb/s

Table 2 summarizes all the results from both the optimal solutions costs, compared to the obtained solutions costs, in addition to the computational time for the obtained solutions costs. The optimal costs are obtained using IBM CPLEX; it took 96 hrs for the eil-51 dataset, 120 hrs for the berlin-52 dataset, 168 hrs for the eil-76 dataset, and 216 hrs for the rat-99 dataset. On the other hand, the reported values of the obtained costs are the mean of the 25 experiments of each scenario.

These results show that the proposed approach was able to converge to the near-optimal values in much less computational time, which can actually be considered running in real-time. For the MinMax costs, Table 3 shows the deviation errors to the optimal costs, presenting the fact that all errors are less than 10%. However, the more visible contribution is that the error is actually decreasing with the increasing number of vehicles. This proves that the proposed approach is more capable of handling multiple vehicles and obtaining more accurate solutions in all tested benchmarks.

Since there are several possibilities for the solutions permutations, when the MinMax solution is optimized, this does not guarantee the optimization for the total cost as well. However, the proposed approach objective function was designed to take this into consideration. Table 4 shows the deviation errors from the optimal total costs. The same behavior of the MinMax costs is quite obvious, where the more the vehicles introduced to the system, the more capable the proposed approach to find near-optimal allocations. Moreover, in the case of berlin-52 dataset, the proposed approach obtained allocations better than the suboptimal ones in the case of 5 and 7 vehicles.

The conducted comparative study highlighted the high performance of the proposed approach in different scenarios and its scalability in handling multiple vehicles. Therefore, the proposed approach was able to not only obtain near-optimal allocations in much less computational time, but also outperform the CPLEX approach in one of the scenarios, under the same set of constraints and conditions.

Additionally, the real-world experiment was successful in terms of allocation of the requests and the vehicles performance to pick up and drop off the passengers, along with the automated navigation from the starting points to the destinations. The two iCab platforms were connected via the V2V communication and once the first request was created on the webserver, the token-holder vehicle collected the requests and the available vehicles to start the proposed task allocation algorithm. The first request was allocated to iCab-1 and while the vehicle is navigating to the first destination point, the second request came in, at which the allocation algorithm output was to allocate it to iCab-1, since this

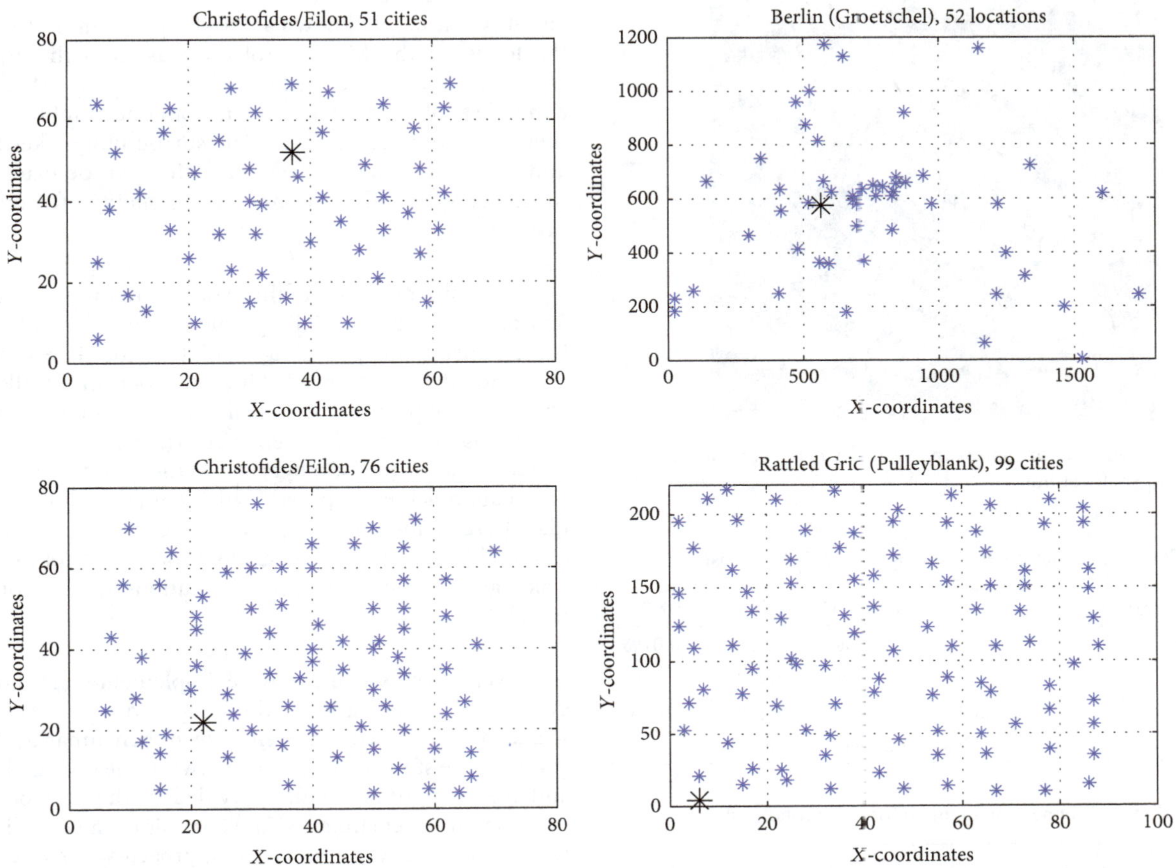

FIGURE 4: mTSP selected benchmark scenarios. Red marker is the depot.

TABLE 2: Benchmarks comparative results.

Benchmark	Vehicles	Optimal MinMax	Optimal total	Obtained MinMax	Obtained total	Comp. time (sec)
eil-51	2	222.73	444.33	236.49	470.97	124.86
	3	159.57	477.15	167.11	498.64	137.06
	5	123.96	615.19	129.36	629.64	158.86
	7	112.07	762.83	116.87	764.24	182.67
berlin-52	2	4110.21	8217.94	4315.83	8630.19	131.94
	3	3244.37	9591.15	3387.35	10115.18	140.97
	5	2441.39	12084.90	2509.84	**11969.61**	163.10
	7	2440.92	16768.79	2491.83	**15628.95**	194.21
eil-76	2	280.85	561.48	307.68	613.75	281.89
	3	197.34	587.65	210.97	632.07	292.83
	5	150.30	748.43	158.24	772.34	333.45
	7	139.62	964.69	143.74	970.42	386.62
rat-99	2	728.71	1456.95	801.68	1603.31	442.96
	3	587.17	1751.95	636.41	1903.04	495.36
	5	469.25	2336.22	487.71	2399.87	555.85
	7	443.91	3074.30	462.59	3135.78	645.43

FIGURE 5: Environment map with the real-world scenario vehicles and passengers locations.

TABLE 3: Deviation errors to optimal MinMax costs.

Benchmark/vehicles	eil-51	berlin-52	eil-76	rat-99
2 vehicles	6.18%	5.00%	9.55%	10.01%
3 vehicles	4.73%	4.41%	6.90%	8.39
5 vehicles	4.36%	2.80%	5.29%	3.93%
7 vehicles	4.28%	2.09%	2.95%	4.21%

TABLE 4: Deviation errors to optimal total costs.

Benchmark/vehicles	eil-51	berlin-52	eil-76	rat-99
2 vehicles	6.00%	5.02%	9.31%	10.05%
3 vehicles	4.50%	5.46%	7.56%	8.62
5 vehicles	2.35%	**−0.95%**	3.19%	2.72%
7 vehicles	0.18%	**−6.80%**	0.59%	2.00%

optimizes the overall solution. Therefore, iCab-1 continues the navigation to the first destination point and then picks up the second request to continue afterward to the second destination point. The last request was communicated while the vehicle is navigating and this time the allocation algorithm assigned it to iCab-2, which was free and able to pick up the passenger. All computations are executed on the on-board vehicles computers in real-time, and since the problem was addressing 2 vehicles and 3 requests in an environment with 12 points of interests, the computational time was approx. 1860 milliseconds for each allocation solution.

6. Conclusion and Future Work

In recent years, the MRS has received a significant consideration from various researchers in the ITS field. This is due to the fact that MRS can improve the concept of cooperative driving of automated vehicles. Moreover, since the number of automated vehicles on the public roads is increasing, the necessity of MRS is also increasing. One of the main challenges for the MRS is the task allocation, to identify which task should be executed by which vehicle, and it is commonly known as MRTA problem. In this work, the focus on the MRTA problem was from the aspect of transportation requests, where users make requests through a mobile application to be driven from one point to another; then an automated vehicle in the surrounding takes the job and so on. The idea of task allocation can be extended to much more than that to improve the quality of cooperative driving.

The main intentions of this work are to propose a generic approach for solving MRTA problem, which not only should be responsible for providing a solution that is feasible, but also should be optimized. Since the optimized allocation enables the appropriate use of all available resources and thus increasing the overall system performance and decreasing the costs, accordingly, a hybrid metaheuristic optimization-based approach is proposed, which combines both the simulated annealing and the genetic algorithm approaches, to ensure the exploitation and exploration aspects of the search space and obtain near-optimal solutions in less computational time.

Several test scenarios were implemented in order to evaluate the efficiency of the proposed approach. These scenarios are based on well-known benchmarks; thus a comparative study is conducted between the obtained results and the suboptimal results provided in the previous work over the same benchmarks. The analysis of the experimental results shows that the proposed approach was successful in handling various scenarios, especially with the increasing number of vehicles and requests, which displays the proposed approach efficiency and scalability. Additionally, the proposed approach was tested on automated vehicles platforms in a real-world scenario, at which the vehicles performed the assigned tasks flawlessly, which proves the functionality of the approach outside the simulation environment.

While the proposed approach was successful in solving the MRTA problem and optimizing the solutions of the benchmarks in significantly less time, there are some aspects that should be considered for future improvement. The cooperation architecture is designed to handle heterogeneous tasks, which means that tasks have specific requirements that constrain specific vehicles to complete it. Moreover, heterogeneous vehicles, not only in terms of capabilities, but also in terms of type, for instance, carry out experiments with both ground and aerial automated vehicles. Last but not least, it was assumed that the distance matrix includes the shortest free paths between the points of interests for the tasks; however, since the vehicles have already on-board trajectory planning algorithms, the approach should consider incorporating this work in the distance calculation.

Conflicts of Interest

The authors declare that there are no conflicts of interest regarding the publication of this paper. The iCab platforms, which were utilized in the real-world experiments, are property of Intelligent Systems Lab (LSI) Research Group at Universidad Carlos III de Madrid (UC3M), which is the affiliation of all authors.

Acknowledgments

This research is supported by Madrid Community Project SEGVAUTO-TRIES (S2013-MIT-2713) and by the Spanish Government CICYT Projects (TRA2015-63708-R and TRA2016-78886-C3-1-R).

References

[1] S. Kato, S. Tsugawa, K. Tokuda, T. Matsui, and H. Fujii, "Vehicle Control Algorithms for Cooperative Driving with Automated Vehicles and Intervehicle Communications," *IEEE Transactions on Intelligent Transportation Systems*, vol. 3, no. 3, pp. 155–160, 2002.

[2] J. Jin, J. Gubbi, S. Marusic, and M. Palaniswami, "An information framework for creating a smart city through internet of things," *IEEE Internet of Things Journal*, vol. 1, no. 2, pp. 112–121, 2014.

[3] D. L. Bock, D. Kettles, and J. Harrison, "Automated, autonomous and connected vehicle technology assessment," *Electric Vehicle Transportation Center (EVTC '16)*, 2016.

[4] S. L. Poczter and L. M. Jankovic, "The google car: driving toward a better future," *Journal of Business Case Studies*, vol. 10, no. 1, pp. 1–7, 2014.

[5] A. Y. S. Lam, Y.-W. Leung, and X. Chu, "Autonomous vehicle public transportation system," in *Proceedings of the 3rd International Conference on Connected Vehicles and Expo, ICCVE 2014*, pp. 571–576, Austria, November 2014.

[6] P. Khaligh and U. Weidmann, "A conceptual framework for the interactions of autonomous public transport systems and urban planning guideline," in *Proceedings of the Swiss Transport Research Conference*, pp. 1–18, 2016.

[7] B. P. Gerkey and M. J. Matarić, "A formal analysis and taxonomy of task allocation in multi-robot systems," *International Journal of Robotics Research*, vol. 23, no. 9, pp. 939–954, 2004.

[8] L. E. Parker, *Multiple Mobile Robot Systems*, Springer, 2008.

[9] M. Badreldin, A. Hussein, and A. Khamis, "A Comparative Study between Optimization and Market-Based Approaches to Multi-Robot Task Allocation," *Journal of Advances in Artificial Intelligence*, vol. 2013, pp. 1–11, 2013.

[10] A. Khamis, A. Hussein, and A. Elmogy, "Multi-robot task allocation: A review of the state-of-the-art," *Cooperative Robots and Sensor Networks*, vol. 604, Springer International Publishing, pp. 31–51, 2015.

[11] J. Ziegler, P. Bender, M. Schreiber et al., "Making bertha drive-an autonomous journey on a historic route," *IEEE Intelligent Transportation Systems Magazine*, vol. 6, no. 2, pp. 8–20, 2014.

[12] P. Furgale, U. Schwesinger, M. Rufli et al., "Toward automated driving in cities using close-to-market sensors: An overview of the V-Charge Project," in *Proceedings of the 2013 IEEE Intelligent Vehicles Symposium, IEEE IV 2013*, pp. 809–816, Australia, June 2013.

[13] DailyNews, "Bmw, audi push self-driving cars closer to reality," in *Daily News - Autos*, 2013.

[14] A. Bragman, *Mercedes-benz tech brings cars closer to self driving*, USA Today, 2016.

[15] L. Laursen, "Volvo to test self-driving cars in traffic," *IEEE Spectrum - Tech Talk*, 2013.

[16] D. González, J. Pérez, V. Milanés, and F. Nashashibi, "A Review of Motion Planning Techniques for Automated Vehicles," *IEEE Transactions on Intelligent Transportation Systems*, vol. 17, no. 4, pp. 1135–1145, 2016.

[17] P. Bender, O. S. Tas, J. Ziegler, and C. Stiller, "The combinatorial aspect of motion planning: Maneuver variants in structured environments," in *Proceedings of the IEEE Intelligent Vehicles Symposium, IV 2015*, pp. 1386–1392, Republic of Korea, July 2015.

[18] X. Qian, F. Altché, P. Bender, C. Stiller, and A. De La Fortelle, "Optimal trajectory planning for autonomous driving integrating logical constraints: An MIQP perspective," in *Proceedings of the 19th IEEE International Conference on Intelligent Transportation Systems, ITSC 2016*, pp. 205–210, Brazil, November 2016.

[19] C. Englund, L. Chen, J. Ploeg et al., "The Grand Cooperative Driving Challenge 2016: Boosting the introduction of cooperative automated vehicles," *IEEE Wireless Communications Magazine*, vol. 23, no. 4, pp. 146–152, 2016.

[20] H. Rewald and O. Stursberg, "Cooperation of autonomous vehicles using a hierarchy of auction-based and model-predictive control," in *Proceedings of the 2016 IEEE Intelligent Vehicles Symposium, IV 2016*, pp. 1078–1084, Sweden, June 2016.

[21] M. Naumann and C. Stiller, "Towards cooperative motion planning for automated vehicles in mixed traffic," in *Proceedings of the IEEE International Conference on Intelligent Robots and Systems (IROS)*, pp. 1–6, 2017.

[22] S. Manzinger, M. Leibold, and M. Althoff, "Driving strategy selection for cooperative vehicles using maneuver templates," in *Proceedings of the 28th IEEE Intelligent Vehicles Symposium, IV 2017*, pp. 647–654, USA, June 2017.

[23] N. Atay and B. Bayazit, "Mixed-integer linear programming solution to multi-robot task allocation problem," Tech. Rep., Washington University, vol. 54, St. Louis, USA, 2006.

[24] A. R. Mosteo, *Multi-robot Task Allocation for Service Robotics: From Unlimited to Limited Communication Range*, Universidad de Zaragoza, 2010.

[25] A. Hussein and A. Khamis, "Market-based approach to Multi-robot Task Allocation," in *Proceedings of the 2013 International Conference on Individual and Collective Behaviors in Robotics, ICBR 2013*, pp. 69–74, Tunisia, December 2013.

[26] W. Kmiecik, M. Wojcikowski, L. Koszalka, and A. Kasprzak, "Task allocation in mesh connected processors with local search meta-heuristic algorithms," *Lecture Notes in Computer Science*, vol. 5991, Springer Berlin Heidelberg, no. 2, pp. 215–224, 2010.

[27] W. Chen and C. Lin, "A hybrid heuristic to solve a task allocation problem," *Computers & Operations Research*, vol. 27, no. 3, pp. 287–303, 2000.

[28] A. Maji and M. K. Jha, "Multi-objective highway alignment optimization using a genetic algorithm," *Journal of Advanced Transportation*, vol. 43, no. 4, pp. 481–504, 2009.

[29] A. Hussein, P. Marin-Plaza, F. Garcia, and J. M. Armingol, "Optimization-based approach for cooperation and coordination of multi-autonomous vehicles," in *Proceedings of the International Conference on Computer Aided Systems Theory (Eurocast)*, pp. 286-287, 2017.

[30] A. Hussein, P. Marin-Plaza, D. Martin, A. De La Escalera, and J. M. Armingol, "Autonomous off-road navigation using stereo-vision and laser-rangefinder fusion for outdoor obstacles detection," in *Proceedings of the 2016 IEEE Intelligent Vehicles Symposium, IV 2016*, pp. 104–109, Sweden, June 2016.

[31] P. Marín-Plaza, J. Beltrán, A. Hussein et al., "Stereo vision-based local occupancy grid map for autonomous navigation in ros," in *Proceedings of the Joint Conference on Computer Vision, Imaging and Computer Graphics Theory and Applications (VISIGRAPP)*, vol. 3, pp. 703–708, Rome, Italy, Feburary 2016.

[32] S. H. Juan and F. H. Cotarelo, *Multi-master ros systems , Institut de Robotics and Industrial Informatics [master, thesis]*, 2015.

[33] A. Hussein, P. Marin-Plaza, F. Garcia, and J. M. Armingol, "Autonomous cooperative driving using v2x communications in off-road environment," in *Proceedings of the IEEE International Conference on Intelligent Transportation Systems (ITSC '17)*, pp. 1–6, 2017.

[34] A. Kokuti, A. Hussein, P. Marin-Plaza, A. de la Escalera, and F. Garcia, "V2X communications architecture for off-road autonomous vehicles," in *Proceedings of the 2017 IEEE International Conference on Vehicular Electronics and Safety (ICVES)*, pp. 69–74, Vienna, Austria, June 2017.

[35] R. Necula, M. Breaban, and M. Raschip, "Tackling the Bi-criteria facet of multiple traveling salesman problem with ant colony systems," in *Proceedings of the 27th IEEE International Conference on Tools with Artificial Intelligence, ICTAI 2015*, pp. 873–880, Italy, November 2015.

[36] R. Necula, M. Breaban, and M. Raschip, "Performance evaluation of ant colony systems for the single-depot multiple traveling salesman problem," in *Proceedings of the 10th International Conference on Hybrid Artificial Intelligent Systems, HAIS 2015*, pp. 257–268, Spain, June 2015.

15

Blind Corner Propagation Model for IEEE 802.11p Communication in Network Simulators

Sanchai Jaktheerangkoon, Kulit Na Nakorn, and Kultida Rojviboonchai

Chulalongkorn University Big Data Analytics and IoT Center (CUBIC), Department of Computer Engineering, Faculty of Engineering, Chulalongkorn University, Bangkok, Thailand

Correspondence should be addressed to Kultida Rojviboonchai; kultida.r@chula.ac.th

Academic Editor: Mihai Dimian

Vehicular Ad Hoc Network (VANET) has been developed to enhance quality of road transportation. The development of safety applications could reduce number of road accidents. IEEE 802.11p is a promising standard for intervehicular communication, which would enable the connected-vehicle applications. However, in the well-known network simulators such as NS3 and Omnet, there is no propagation model that can simulate the IEEE 802.11p communication at blind corner realistically. Thus, in this paper, we conducted the real-world experiments of IEEE 802.11p in order to construct the model to describe the characteristics of the IEEE 802.11p communication at the blind corners. According to the experimental results, we observe that the minimum distance between the vehicle and the corner can effectively be represented as the key parameter in the model. Moreover, we have a variable parameter for adjusting the impact of the obstruction which could be different at each type of blind corners. The simulation results using our proposed model are compared with those using the existing obstacle model. The results showed that our proposed model is much more closely aligned with the real experimental results.

1. Introduction

World Health Organization (WHO) reported that 1.2 million people from all over the world die and 20 to 50 million people suffer from injuries because of road accidents each year [1]. According to [2], a lot of accidents occurred at the intersections. Blind corner is one type of the intersections where the accidents occur easily. This is because one vehicle from one side of the corner cannot see the other vehicles from the other side of the corner. We consider the blind corners as the corners with obstacles and they rarely have space for sidewalk. The blind corners can be generally found in many locations, for example, in the cities of Asian countries, in small alley, in local way, and inside the organization area. These locations are surrounded by buildings that obstruct the driver's line of sight as shown in Figure 1. Not only can the buildings cause the blind corner, but also walls, trees, and construction sites can also cause the blind corner. Furthermore, the traffic lights are rarely found in such locations. That is why the accidents could occur easily at the blind corners.

Even though the line of sight of the driver at the blind corners is blocked by the obstacles, wireless communication can partially pass through the obstacles. As a result, the vehicles can sense other vehicles around. The wireless communication network among vehicles is introduced as Vehicular Ad Hoc Network (VANET). VANET has been developed to enhance the quality of road transportation and Intelligent Transportation System (ITS). VANET consists of 2 types of communications: vehicle-to-vehicle (V2V) and vehicle-to-infrastructure (V2I). One of the major ITS applications is the safety application, in which some warning signals can be sent to other vehicles in case there are vehicles or pedestrians nearby. By receiving the signals, the intelligent vehicle can decide if it can move on or it needs to brake. This could reduce the number of accidents.

The IEEE developed the 802.11p Wireless Access for Vehicular Environment (WAVE) standard as a support for VANET applications [3]. The IEEE 802.11p has the communication range up to 1,000 m if the vehicle speed is less than 200 km/h. As the blind corner is one of the critical locations where the accidents can occur easily, the VANET

FIGURE 1: Sample building that causes blind corner.

communication could help notify the driver. Nevertheless, the obstacles at the blind corner not only obstruct the sight of the driver but also obstruct the wave propagation signal. Consequently, the communication could fail easily. In other words, the performance of IEEE 802.11p can be degraded when the communication occurs at the blind corner. This is one of the vulnerabilities of IEEE 802.11p communications for safety applications [4].

The effect of the obstruction leads to the decreased communication range of IEEE 802.11p and the performance degradation of the protocols and applications that rely on VANET. To consider this issue, researchers and developers evaluate their work using both real-world experiment and simulation. The real-world experiment is the evaluation method that uses real equipment running in real scenarios. Although this method provides the testing performance accurately, it is time-consuming, expensive, and difficult to reproduce the test cases and it is difficult to scale to large scenarios. Thus, the simulation is an alternative method for performance evaluation.

To enable realistic simulation, the propagation models are applied to the nodes in the simulation. Recently, there have been researches about the propagation models for each kind of obstructions such as vehicle obstruction [5, 6] and building obstruction [7–12]. These models have been evaluated by comparison to the results from the real experiments. As the models are applied, the simulation results can become more realistic in each specific scenario. However, to the best of our knowledge, the existing models are not suitable for applying in blind corner scenario.

In our previous works, we conducted the real-world experiment to study the performance of IEEE 802.11p in blind corner scenario [12]. The experiment was conducted using Denso Wireless Safety Unit (WSU), which is IEEE 802.11p communication module. Each vehicle was equipped with the WSU module. The results showed that two vehicles at different sides of the corner can communicate with each other when the minimum distance between the vehicle and the corner is less than 60 m. The results emphasized that the performance at the blind corner should be taken into a serious consideration when developing any safety applications.

In this paper, our contribution is twofold. The first contribution is that we extend our previous work to cover more blind corners in order to investigate and generalize the characteristics of the blind corners regardless of the specific corners. The second contribution is that we propose a novel blind corner model, which is implemented as an extension of the well-known models in the network simulators. This model can represent the characteristics of IEEE 802.11p communication at blind corners realistically.

There are two methods to construct a new model which are ray tracing method and flat propagation method [10]. The ray tracing method has a complex computation, consumes a lot of resources and time, and is difficult to parameterize the parameters. On the other hand, the flat propagation method normally calculates some values and uses some probabilistic functions to represent signal propagation loss. Our design concern is that the model has to be simple in order to consume less time when it is applied in the simulation. Therefore, we use the flat propagation method. Our model can be implemented as an extension of the two-ray ground model and the Nakagami model, the well-known propagation models in the network simulators, so it is easily applied to any network simulators. The network simulator that we use in this work is NS-3 [13], which is open source software for network simulation and is very popular as a network platform in research and education. For performance evaluation, we conduct extensive real-world experiments and compare the results from the experiments to the results from our proposed model.

The following of this paper is organized as follows: the background knowledge and related works about propagation models are described in Section 2. Then, our field experiment settings and results are shown in Section 3. In Section 4, our model is elaborated. The discussion and comparison with other related works are shown in Section 5. Finally, Section 6 concludes our work.

2. Background Knowledge and Related Works

The blind corner is a critical scenario for the safety applications, so the performance of the safety application should be considered. Although we can install the infrastructure such as Roadside Unit (RSU) at the corner to improve the communication performance, it consumes a lot of cost for deployment. Moreover, it is not cost-effective to deploy the RSUs at all corners in the city because there are too many blind corners in most of the countries, especially the countries in Asia. Therefore, the scenario where the communication occurs at the blind corners without any additional infrastructures should be taken into consideration. Such scenario can be considered as beneficial to the safety applications because it can help reduce the accidents at any corners.

One method that can help simulate the communication effectively is to add the propagation loss models or signal attenuation models. There are two methods to construct propagation models, which are ray tracing method [14, 15] and flat propagation method [8–11]. The ray tracing method traces the signal from source to destination using the characteristics of signals like reflection, diffraction, and interference. Then, the loss of the signal is calculated. Although this

Table 1: Comparison of the models.

Issue	Obstacle model [8, 9]	CORNER model [10, 11]	Blind corner model (our model)
(i) Signal attenuation	√	√	√
(ii) Simulator used	NS-3, Omnet	QualNet	NS-3
(iii) Communication module used in the real experiments	IEEE 802.11p	IEEE 802.11b/g	IEEE 802.11p
(iv) Characteristics of the corner used in the experiment	Not blind corner	Not blind corner	Blind corner (the corner with no space for side walk or the side walk less than 1 m)
(v) Parameters used in path loss calculation approach	Number of walls penetrated	Wave characteristics parameter	Minimum distance

method can give an accurate signal strength at the destination as an output of the model, it is not widely used because the obstacle topologies and the signal attenuation characteristics for each obstacle have to be collected and set in advance. Moreover, it consumes a lot of resource and simulation time.

Researchers mostly use the flat propagation method instead of the ray tracing method. The flat propagation method utilizes values that can be obtained from the topology easily as inputs for the formula. The formula will give a result as attenuated signal strength. The most common value that is used as an input is distance. This method consumes less resource and simulation time. The examples of the models using this method are two-ray ground model and Nakagami model. These two models are embedded into most of the network simulators. However, many researchers still introduce new models based on the flat propagation method for more realistic simulation in some specific scenarios.

The obstacle model [8, 9] uses number of walls where the signal has to penetrate as a main variable for formula calculation. As a result, the obstruction amount of the signal depends on number of walls the signal is passing through. This model focuses mainly on building obstacles, while the CORNER model [10, 11] uses the wave characteristics such as reflection and diffraction as a main variable for formula calculation. The CORNER model classifies the scenarios into 3 categories: line of sight (LOS), non-line of sight with 1 corner (NLOS1), and non-line of sight with 2 corners (NLOS2).

Our work focuses on blind corners, which are the corners that have no space for sidewalk or the corner with the sidewalk less than 1 meter. At the corner with no space for sidewalk, the signal from the transmission node will penetrate to the wall, leading to more signal obstruction. Our model is different from the obstacle model and the CORNER model in that we use the minimum distance between the vehicles to the corner as the main parameter to calculate in our formula. We explain the rationale behind our design in Section 4.

Our model, the obstacle model, and the CORNER model are compared as shown in Table 1. All of the models perform signal attenuation when the transmitted signal travels through the obstacles. In order to calculate the path loss, the obstacle model uses number of walls penetrated, the CORNER model uses wave characteristics such as reflection and diffraction effects, and our model uses the minimum distance as the main parameter. The obstacle model and our model are implemented in the NS-3 simulator, whereas the CORNER model is implemented in the QualNet simulator. The obstacle model is also embedded in Veins framework based on Omnet simulator [16]. For the communication module used in the real experiments, IEEE 802.11p is used in our model and the obstacle model, whereas IEEE 802.11b/g is used for the CORNER model.

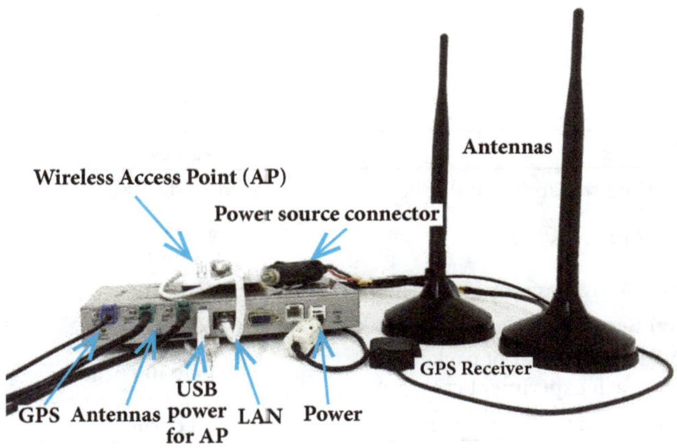

Figure 2: Denso WSU experiment set.

3. Field Experiment

We set up field experiment to study the characteristics of real IEEE 802.11p communication devices in blind corner scenario [12]. Our previous work reveals the performance of IEEE 802.11p where the communication range among 2 vehicles in blind corner scenario is less than 6% of full specification of IEEE 802.11p, which is 1,000 m. In order to investigate the performance in more details, we extend the experiment to cover more samples of blind corners and propose the model for using in the simulators.

3.1. Field Experiment Settings. We set up the experiment scenario using 2 vehicles on different side of the blind corner. Each vehicle is equipped with Denso Wireless Safety Unit (WSU), which is connected to 2 external antennas (see Figure 2). The antennas are placed at 1.2 m from ground. IEEE

(a) Bird's eye view (b) Perspective view

FIGURE 3: The experiment scenario: (a) bird view and (b) perspective view.

TABLE 2: The experiment settings.

Settings	Values
(i) Data transmission device	Denso WSU 5001-T
(ii) Antennas	2 external antennas at 1.2 m from ground
(iii) Transmission power	20 dBm
(iv) Beacon interval	10 Hz
(v) Total number of packets sent each experiment case	Approximately 100 beacons
(vi) Number of experiments	4 blind corners

FIGURE 4: Electrical Engineering Lab.

FIGURE 5: Mechanical Engineering Lab.

802.11p is used as a communication module in Denso WSU. Because GPS does not provide accurate position information, we have measured and recorded the location manually with a standard measuring wheel. As a result, we can obtain an accurate position of the vehicle which leads to more accurate result.

The network traffic generated in the experiment is 10 Hz beacon, which is the minimum transmission frequency required for safety applications [17]. The attached antennas transmit the signal with 20 dBm power. In each experiment case, one vehicle is fixed at distance d_1 on one side of the blind corner. The other vehicle is moving between d_2 − 1 m and d_2 + 1 m on the other side of the blind corner. Then, we calculate the average value of the results for that point. We send approximately 100 beacons for each case and calculate the packet delivery ratio and average RSSI. The experiment scenario is depicted in Figure 3 and the settings are summarized in Table 2.

We did experiments at 4 blind corners: Electrical Engineering Lab (see Figure 4), Mechanical Engineering Lab (see Figure 5), Civil Engineering Lab (see Figure 6), and Centennial Building (see Figure 7) in Faculty of Engineering, Chulalongkorn University. These 4 blind corners represent different types of blind corners. We divide the blind corners into 2 types, which are the corner with large obstruction (Electrical Engineering Lab and Mechanical Engineering Lab) and the corner with small obstruction (Civil Engineering Lab and Centennial Building). The corner with large obstructions is the corner with concrete building, which has a lot of large machines made of metal inside. The corner with small obstruction is the corner with concrete building which has a wide free space inside and mostly contains tables and chairs. The locations of all blind corners in our experiment are shown in bird's eye view in Figure 8. Because the range of each corner is different, the maximum distance between the vehicles and the corner is differently set for each blind corner. The range for each experiment is shown in Table 3. We conducted extensive experiments and found out that the results have the same trend regardless of the day of the experiment and distance step. In order to conduct the experiment with less time, we increase the distance step in our later experiments.

TABLE 3: Experiment distance for each blind corner.

Location	Vehicle 1 distance (d_1)	Vehicle 2 distance (d_2)
Electrical Engineering Lab	0, 5, 10, 15, 20, 30, 40 m	1 – 55 m (step every 2 m)
Mechanical Engineering Lab	0, 5, 10, 15, 20, 30, 40 m	1 – 30 m (step every 3 m) 30 – 58 m (step every 4 m)
Civil Engineering Lab	0, 5, 10, 15, 20, 30, 40 m	1 – 15 m (step every 2 m) 15 – 39 m (step every 3 m) 39 – 55 (step every 4 m)
Centennial Building	0, 5, 10, 20, 30 m	1 – 30 m (step every 3 m) 30 – 42 m (step every 4 m)

FIGURE 6: Civil Engineering Lab.

FIGURE 7: Centennial Building.

FIGURE 8: The bird's eye view of the blind corners in our experiment. (1) Electrical Engineering lab. (2) Mechanical Engineering Lab. (3) Civil Engineering Lab. (4) Centennial Building.

3.2. Field Experimental Results. Packet delivery ratio (PDR) and average Received Signal Strength Indicator (RSSI) are our evaluation metrics. The PDR is calculated from the ratio between number of packets received at the destination node and number of packets sent by the source node. The average RSSI is calculated by averaging RSSI of all the received packets at the destination node. The results for each blind corner are shown in 3-dimensional graph. For the PDR, the 3 dimensions are the distance between the first vehicle and the corner, the distance between the second vehicle and the corner, and PDR. For the average RSSI, the 3 dimensions are the distance between the first vehicle and the corner, the distance between the second vehicle and the corner, and the average RSSI.

Figure 9 shows the real experimental results for all blind corners. Figures 9(a) and 9(b) show PDR and RSSI results of the first blind corner (Electrical Engineering Lab). Figures 9(c) and 9(d) show PDR and RSSI results of the second blind corner (Mechanical Engineering Lab). Figures 9(e) and 9(f) show PDR and RSSI results of the third blind corner (Civil Engineering Lab). Figures 9(g) and 9(h) show PDR and RSSI results of the fourth blind corner (Centennial Building). As can be seen from the results, both PDR and RSSI are an inverse variation to the distance from the corner.

According to the results from 4 blind corners, we can divide the blind corners into 2 types which are the corner with large obstruction (the first and the second blind corners) and the corner with small obstruction (the third and the fourth blind corners). The first and the second blind corners are concrete buildings with a lot of large machines inside, leading to a lot of signal attenuations, while the third and the fourth blind corners are concrete buildings with a wide free space inside, leading to less signal attenuation.

Moreover, it can be noticed that all the results shown in the graphs are quite symmetric. Therefore, PDR and the average RSSI are associated with the minimum distance between the vehicles and the corner. This is because the closer the vehicle to the corner is, the smaller effect of blind corner the communication experiences. More discussion can be found in our previous work [12].

From the experimental results, we also observe the latency of the transmission between 2 nodes. The latency is around 89-95 ms for all distances at all blind corners. This latency value can be considered as a parameter in the simulation setup.

4. Blind Corner Propagation Loss Model

4.1. Study of NS-3 Propagation Model. Network simulator is a tool that is very popular in education and research field. The network simulator allows researchers to simulate various kinds of networks and various kinds of scenarios

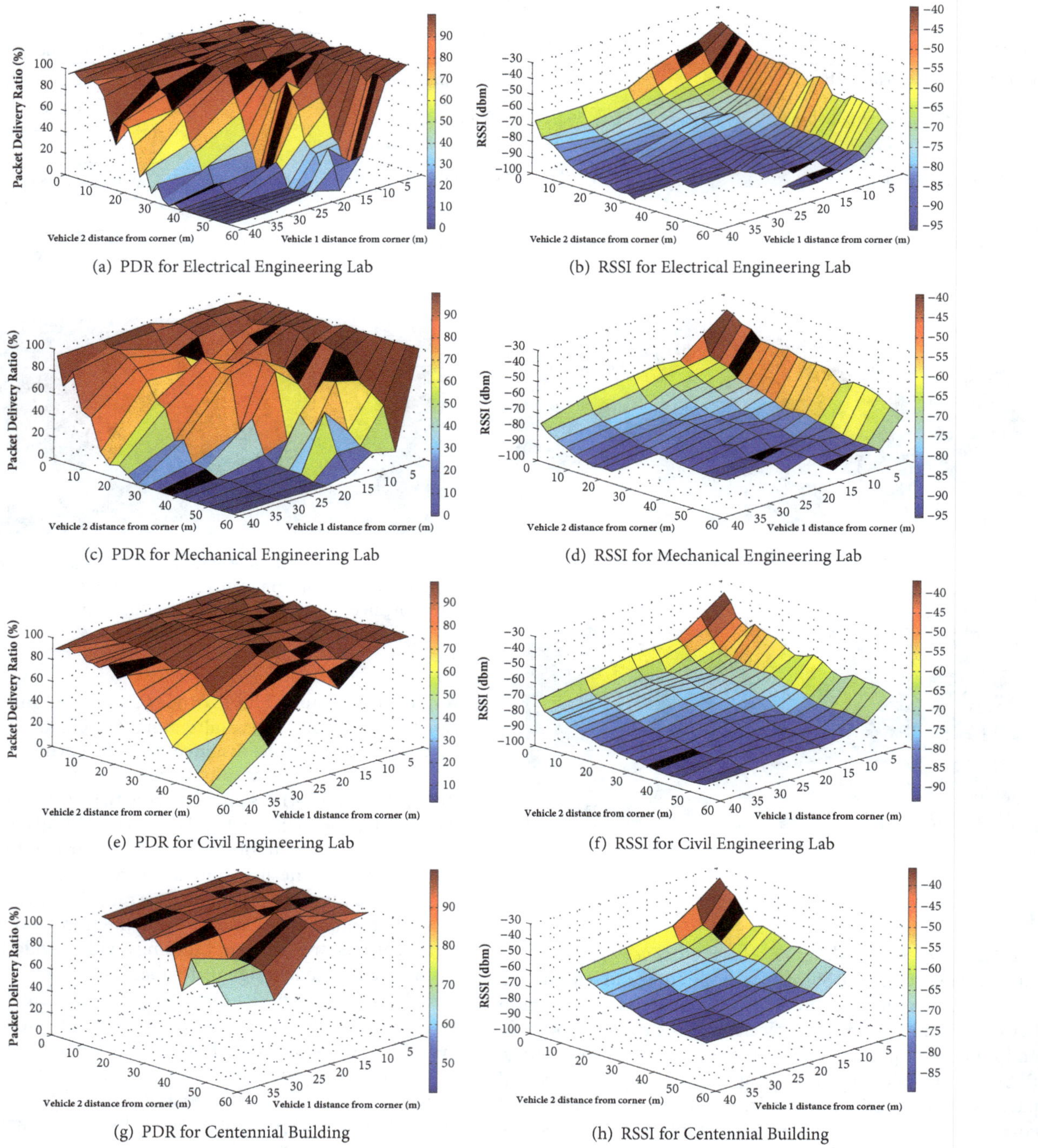

(a) PDR for Electrical Engineering Lab
(b) RSSI for Electrical Engineering Lab
(c) PDR for Mechanical Engineering Lab
(d) RSSI for Mechanical Engineering Lab
(e) PDR for Civil Engineering Lab
(f) RSSI for Civil Engineering Lab
(g) PDR for Centennial Building
(h) RSSI for Centennial Building

FIGURE 9: Results from real experiments. (a) PDR for Electrical Engineering Lab. (b) RSSI for Electrical Engineering Lab. (c) PDR for Mechanical Engineering Lab. (d) RSSI for Mechanical Engineering Lab. (e) PDR for Civil Engineering Lab. (f) RSSI for Civil Engineering Lab. (g) PDR for Centennial Building. (h) RSSI for Centennial Building.

which are convenient for simulating a large-scale network. One of the most popular network simulators is NS-3, which is open-sourced software. Moreover, NS-3 also supports simulation over vehicular network that uses IEEE 802.11p as wireless interface. Using the simulator, we have to consider which propagation model is suitable to simulate packet loss.

Normally, for vehicular network, the well-known propagation model setting is to use two-ray ground model coupled with Nakagami model. These two models result from the Euclidean distance between transmission node and receive node. By using NS-3, the results of PDR and the average RSSI when applying these models are shown in Figure 10. We observe that the graph characteristics of the simulation results

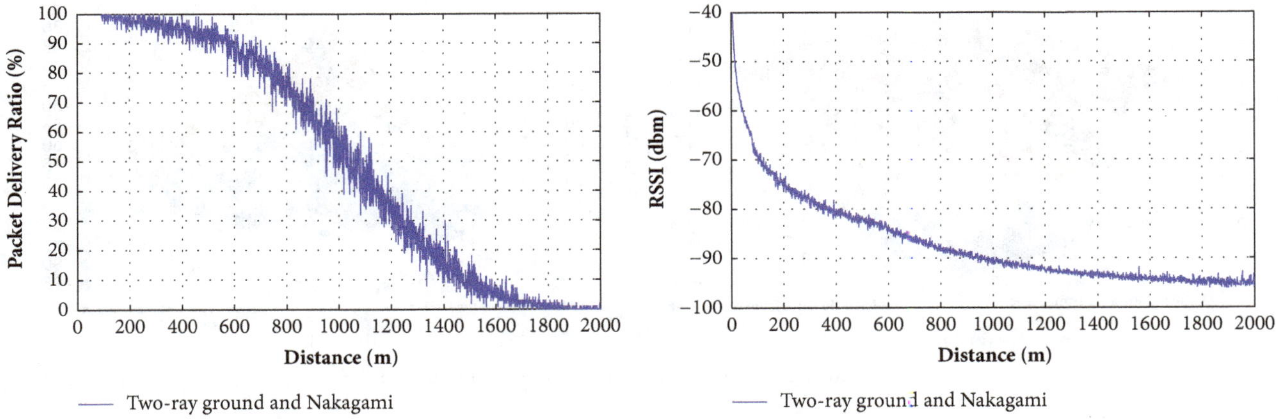

(a) PDR (b) RSSI

FIGURE 10: Simulation results when applying two-ray ground model coupled with Nakagami model. (a) PDR. (b) RSSI.

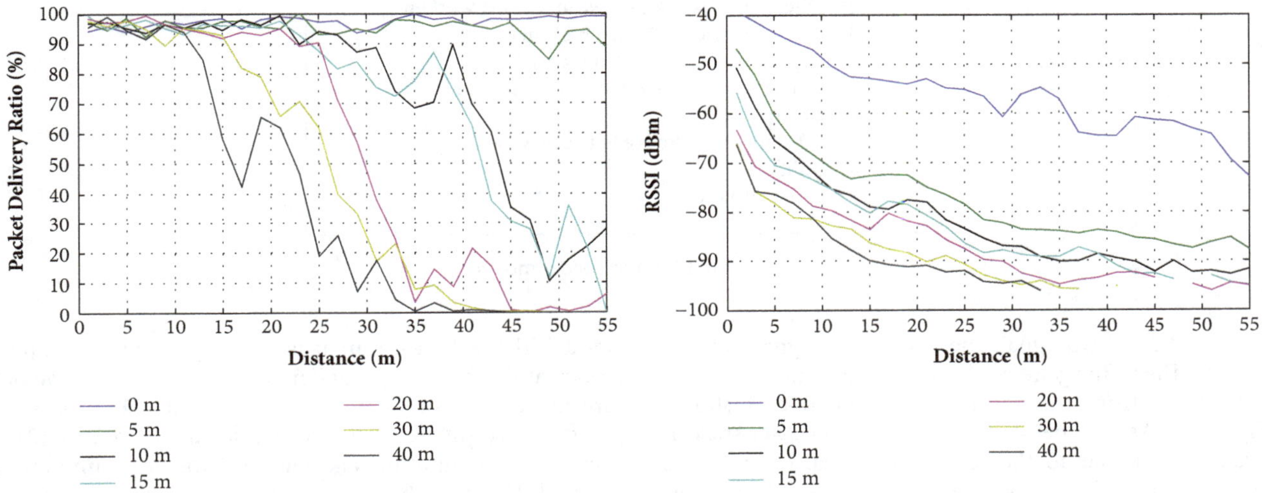

(a) PDR (b) RSSI

FIGURE 11: Results from real experiments for Electrical Engineering Lab. (a) PDR. (b) RSSI.

are similar to the results from real blind corner experiments. For more clarification, we simplify the result from real experiments in Figures 9(a) and 9(b) to 2-dimensional graph as shown in Figures 11(a) and 11(b). Each graph shows the result for different distances of vehicle 1. As can be seen, all the graphs have the same trend, similarly to S-shaped curve. Each point in the real experiment result in Figure 11 can be mapped to the result in Figure 10 by increasing the distance. On the other words, the signal characteristics when travelling through the blind corner behave the same as the signal when the vehicles are in line of sight at longer distance. As a result, we modify the distance calculation when the signals travel through blind corner. Since distance is the most important factor for calculation in the propagation models, modifying distance calculation is like modifying the characteristics of the models applied. This can represent characteristics of the communication at blind corners.

Referring to the variables depicted in Figure 3, we modify the distance calculation and use the summation of distances between the vehicles and the corner $(d_1 + d_2)$ instead of using the Euclidean distance (d). This is because in the case of the blind corner this summation represents the distance the signal really travels. Moreover, as we discussed that the minimum distance is associated with the result, we add the minimum distance factor to the distance calculation. This factor represents that the closer the vehicle to the corner is, the smaller effect of blind corner the communication experiences. This leads to the higher PDR. As a result, the estimated distance is formulated as shown in the following equation:

$$\text{Estimated Distance} = (d_1 + d_2) \times \min(d_1, d_2) \qquad (1)$$

In order to investigate the PDR and RSSI results when applying our estimated distance equation, we set up the simulation scenario the same as the real experiment scenario which is shown in Table 2. There are 2 vehicles on different side of blind corners. The 2 vehicles are transmitting and

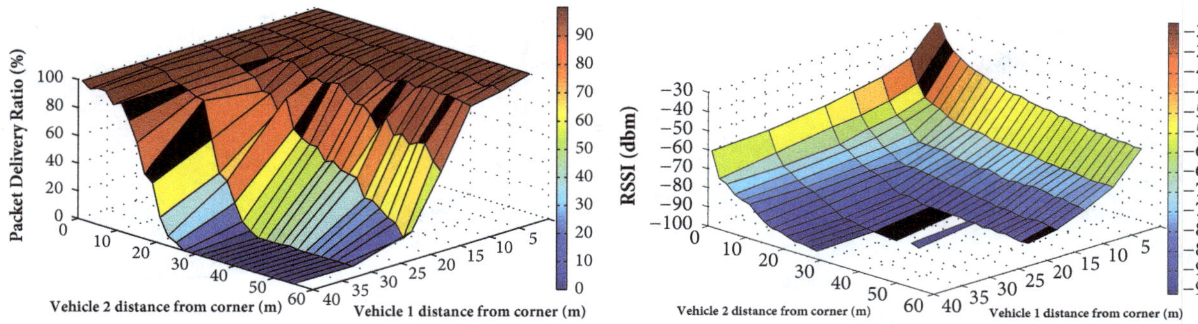

(a) PDR (b) RSSI

FIGURE 12: Results for preliminary simulation. (a) PDR. (b) RSSI.

Input: Location of vehicle V_1 and V_2, Obstacles list, α
Output: Distance
(1) Distance = UNKNOWN
(2) if there are obstacles between V_1 and V_2 then
(3) d_1 = Distance of V_1 from corner
(4) d_2 = Distance of V_2 from corner
(5) Distance = $(d_1 + d_2) \times \min(d_1, d_2) \times \alpha$
(6) else do
(7) Distance = Distance between V_1 and V_2
(8) end if
(9) return Distance

ALGORITHM 1: Blind corner model.

receiving IEEE 802.11p signal. The transmitted signal strength is 20 dBm. The traffic generated is 10 Hz beacon.

The result from the preliminary simulation is shown in Figure 12. As can be seen, the graph characteristics in Figure 12 are similar to the results from real experiments shown in Figure 9. The packet delivery ratio and the average RSSI have inverse variation to the distance. So we consider this method to be used in our model.

4.2. Proposed Blind Corner Model. From the experimental results in Figure 9, each type of blind corners does not obstruct the transmitted signal equally. The PDR ratio and the average RSSI are not the same for all blind corners. According to this reason, it can be seen that only the minimum distance is not enough for the model. Therefore, we add a parameter α in order to adjust the degree of the obstruction. Equation (2) formulates the modified version of the estimated distance from (1).

$$\text{Estimated Distance} = (d_1 + d_2) \times \min(d_1, d_2) \times \alpha \quad (2)$$

The parameter α is used to adjust the degree of the obstruction, which represents low obstruction to high obstruction. α must be greater than or equal to 0.4. If α is less than 0.4, it cannot be used because it makes the estimated distance lower than the real Euclidean distance. This will lead to nonrealistic simulation result.

The distance calculation can be divided into 2 cases: (1) line of sight and (2) non-line of sight. If the vehicles are in line of sight, it is not necessary to use our distance calculation model. The real Euclidean distance can be used as an input in Nakagami or two-ray ground models. However, if the vehicles are in non-line of sight, our distance calculation model is useful. The estimated distance calculated according to (2) can be used as an input in Nakagami or two-ray ground models instead. The algorithm that describes our distance calculation is shown in Algorithm 1.

5. Results and Discussion

In this section, we show the results from the simulation using our model and suggest the appropriate range of the parameter α. We simulate the scenarios using the simulation setting as shown in Table 2, so that all the settings will be the same as the real experiments. Then, we vary the value of α. We compare the results from the simulation and those from the real experiments for each blind corner. We calculate the root-mean-square error (RMSE) between the results from the simulation and those from the real experiments using (3). Equation (3) shows RMSE calculation, where i is the distance between the first vehicle and the corner, j is the distance between the second vehicle and the corner, $x_{i,j}$ is the PDR of the distance pair i, j from the real experiment, $s_{i,j}$ is the PDR from the simulation, and n is number of distance pairs. The RMSE for each α and each blind corner are shown in Figure 13.

$$\text{RMSE} = \sqrt{\frac{\sum_{\text{Each pair of distance } i,j} \left(x_{i,j} - s_{i,j}\right)^2}{n}} \quad (3)$$

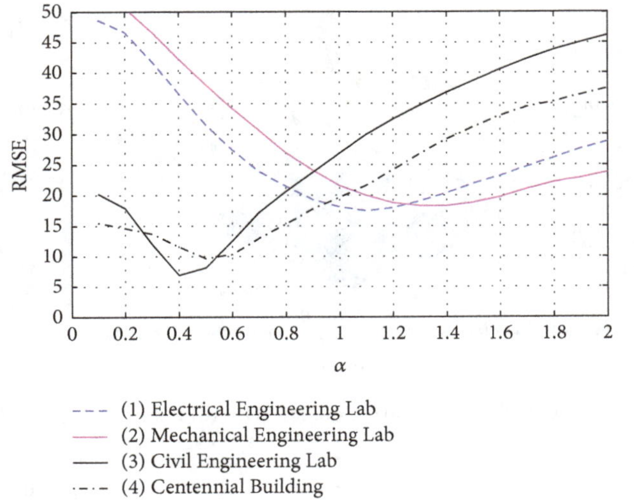

FIGURE 13: RMSE between simulation result and real experiment result.

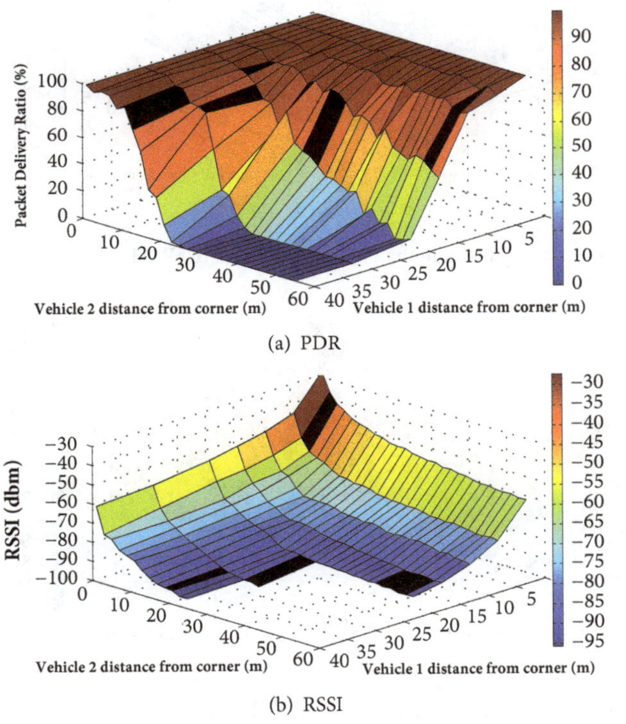

(a) PDR

(b) RSSI

FIGURE 14: Results from the simulation with $\alpha = 1.1$ for blind corner with large obstruction. (a) PDR. (b) RSSI.

According to Figure 13, we suggest the weight factor α for each type of blind corners as follows.

For the first and the second blind corners of the building with large obstruction, we use α between 1.1 and 1.3. The simulation results using the recommended values 1.1 and 1.3 are shown in Figures 14 and 15, respectively. As can be seen, Figure 14 provides similar results to Figures 9(a) and 9(b). Figure 15 provides similar results to Figures 9(c) and 9(d).

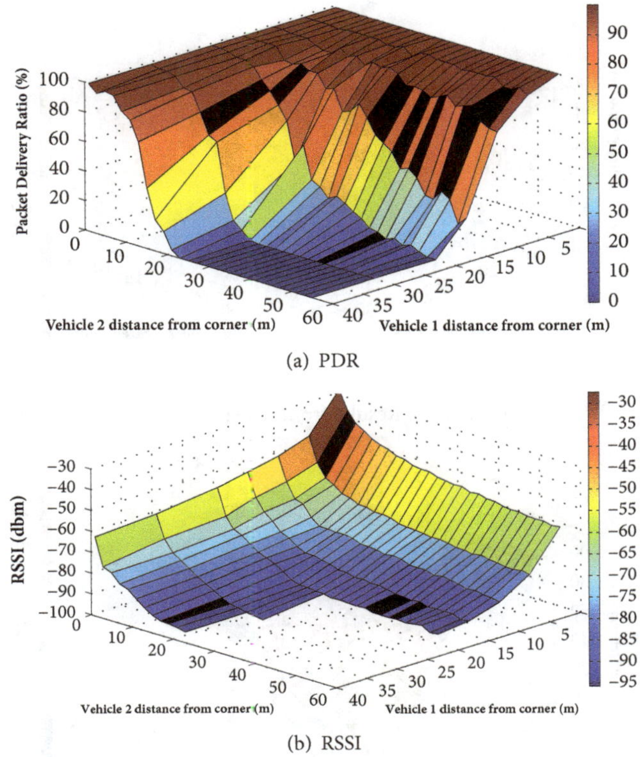

(a) PDR

(b) RSSI

FIGURE 15: Results from the simulation with $\alpha = 1.3$ for blind corner with large obstruction. (a) PDR. (b) RSSI.

For the third and the fourth blind corners of the building with small obstruction, we use α between 0.4 and 0.5. The simulation results using the recommended values 0.4 and 0.5 are shown in Figures 16 and 17, respectively. As can be seen, Figure 16 provides similar results to Figures 9(e) and 9(f). Figure 17 provides similar results to Figures 9(g) and 9(h).

The results from the simulation are close to the results from real experiments. However, the results from the simulation have a higher PDR and average RSSI, and the graph trends are smoother than the graph for real experiments. This is because in real experiments there might be some other factors that we cannot detect or some factors that are difficult to produce in simulations such as environmental interferences.

We also compare our work with the obstacle model [8, 9] which is implemented in NS-3. The obstacle model is also embedded in Veins framework based on Omnet simulator, which is another popular network simulator. Therefore, we use the obstacle model as our baseline. The simulation scenario is set the same as the real experiment scenario, which is shown in Table 2. The results using the obstacle model are shown in Figure 18. As can be seen, the graph characteristics of PDR reduce rapidly at a specific range. The result does not reflect communication in real world, where PDR gradually reduces. Compared to our results shown in Figures 14–17, it can be seen that our model can provide much more similar results to the real world and better represents the real characteristics of IEEE 802.11p at blind corners.

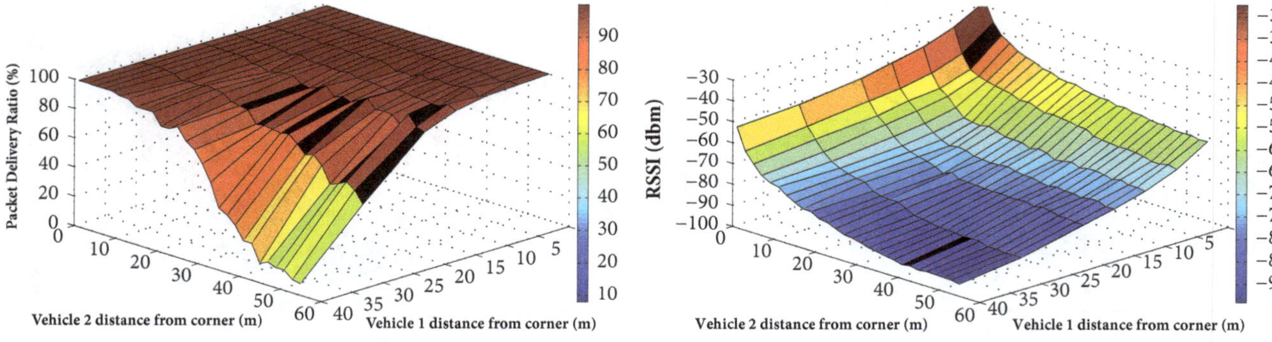

(a) PDR (b) RSSI

FIGURE 16: Results from the simulation with $\alpha = 0.4$ for blind corner with small obstruction. (a) PDR. (b) RSSI.

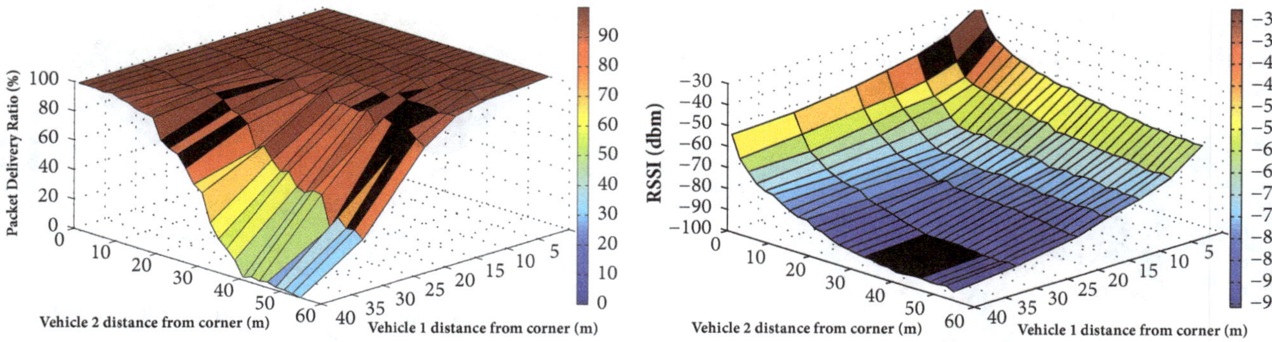

(a) PDR (b) RSSI

FIGURE 17: Results from the simulation with $\alpha = 0.5$ for blind corner with small obstruction. (a) PDR. (b) RSSI.

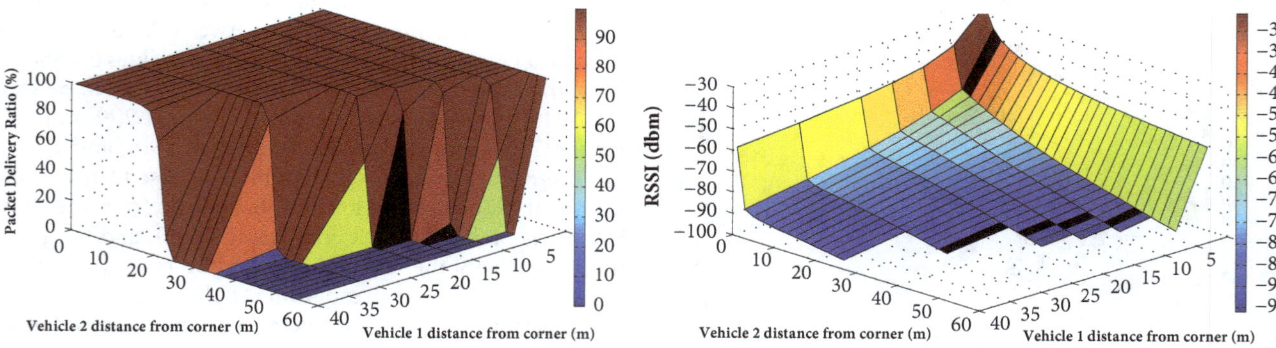

(a) PDR (b) RSSI

FIGURE 18: Results from the simulation when simulating with obstacle model. (a) PDR. (b) RSSI.

6. Conclusion

In this paper, we propose the blind corner model that represents the characteristics of IEEE 802.11p communication at blind corners. We modify the distance calculation when the signals travel through blind corner. Our model utilizes the minimum distance between the vehicles to the corner, which is the most important variable for propagation model calculation. Moreover, our model has a parameter to adjust the degree of the obstruction. We also conduct extensive real experiments with IEEE 802.11p communication devices and do comparison to the simulation result. According to the experimental results and the simulation results, we suggest that the parameter should be set between 1.1 and 1.3 for the buildings with large obstruction and between 0.4 and 0.5 for the buildings with small obstruction. The comparison result shows that our proposed model can represent the characteristics of IEEE 802.11p communication at blind corners better than the obstacle model. Therefore, our research is useful in research field. The protocols and applications can be tested realistically in blind corner scenarios by using our model.

Conflicts of Interest

The authors declare that there are no conflicts of interest regarding the publication of this paper.

Acknowledgments

This research was supported by Chula Computer Engineering Graduate Scholarship for CP Alumni and the Wireless Network and Future Internet Research Unit, Ratchadaphiseksomphot Endowment Fund, Chulalongkorn University. The authors would like to acknowledge ITS Thailand and DENSO for the support of the wireless safety units and National Broadcasting and Telecommunications Commission (NBTC) for partial support and thank Mr. Pawissakan Chirupphapa, Mr. Chanthawat Rattanapongphan, Mr. Teerapat Vongsuteera, Mr. Adsadawut Chanakitkarnchok, and members of ISEL Lab for their help in the experiments.

References

[1] World Health Organization, *Global status report on road safety 2017*, World Health Organization, 2017.

[2] New South Wales Government, *Road traffic casualty crashes in New South Wales 2016*, New South Wales Government, 2016.

[3] *IEEE Standard for Information technology– Local and metropolitan area networks– Specific requirements*, Specifications Amendment 6: Wireless Access in Vehicular Environments, Part 11 edition, 2010, Wireless LAN Medium Access Control (MAC) and Physical Layer (PHY), pp. 1 - 51.

[4] A.-M. Cailean, B. Cagneau, L. Chassagne, V. Popa, and M. Dimian, "A survey on the usage of DSRC and VLC in communication-based vehicle safety applications," in *Proceedings of the 2014 21st IEEE Symposium on Communications and Vehicular Technology in the BeNeLux, IEEE SCVT 2014*, pp. 69–74, Netherlands.

[5] R. Meireles, M. Boban, P. Steenkiste, O. Tonguz, and J. Barros, "Experimental study on the impact of vehicular obstructions in VANETs," in *Proceedings of the IEEE Vehicular Networking Conference (VNC '10)*, pp. 338–345, December 2010.

[6] R. He, A. F. Molisch, F. Tufvesson, Z. Zhong, B. Ai, and T. Zhang, "Vehicle-to-vehicle propagation models with large vehicle obstructions," *IEEE Transactions on Intelligent Transportation Systems*, vol. 15, no. 5, pp. 2237–2248, 2014.

[7] H. Tchouankem, T. Zinchenko, and H. Schumacher, "Impact of buildings on vehicle-to-vehicle communication at urban intersections," in *Proceedings of the 2015 12th Annual IEEE Consumer Communications and Networking Conference, CCNC 2015*, pp. 206–212, USA, January 2015.

[8] S. E. Carpenter and M. L. Sichitiu, "An obstacle model implementation for evaluating radio shadowing with ns-3," in *Proceedings of the the 2015 Workshop*, pp. 17–24, Barcelona, Spain, May 2015.

[9] C. Sommer, D. Eckhoff, R. German, and F. Dressler, "A computationally inexpensive empirical model of IEEE 802.11p radio shadowing in urban environments," in *Proceedings of the 8th International Conference on Wireless On-Demand Network Systems and Services (WONS '11)*, pp. 84–90, January 2011.

[10] E. Giordano, R. Frank, G. Pau, and M. Gerla, "CORNER: A radio propagation model for VANETs in Urban scenarios," *Proceedings of the IEEE*, vol. 99, no. 7, pp. 1280–1294, 2011.

[11] Q. Sun, S. Y. Tan, and K. C. Teh, "Analytical formulae for path loss prediction in urban street grid microcellular environments," *IEEE Transactions on Vehicular Technology*, vol. 54, no. 4, pp. 1251–1258, 2005.

[12] S. Jaktheerangkoon, K. N. Nakorn, and K. Rojviboonchai, "Performance study of IEEE 802.11p in blind corner scenario," in *Proceedings of the 2016 International Symposium on Intelligent Signal Processing and Communication Systems, ISPACS 2016*, Thailand, October 2016.

[13] "Network Simulator 3," https://www.nsnam.org/.

[14] S. A. Hosseini Tabatabaei, M. Fleury, N. N. Qadri, and M. Ghanbari, "Improving propagation modeling in urban environments for vehicular ad hoc networks," *IEEE Transactions on Intelligent Transportation Systems*, vol. 12, no. 3, pp. 705–716, 2011.

[15] J. Maurer, T. Fügen, and W. Wiesbeck, "Physical layer simulations of IEEE802.11a for vehicle-to-vehicle communications," in *Proceedings of the 62nd Vehicular Technology*, pp. 1849–1853, 2005.

[16] "Vehicles in Network Simulation," http://veins.car2x.org/.

[17] Z. Hameed Mir and F. Filali, "LTE and IEEE 802.11p for vehicular networking: a performance evaluation," *Eurasip Journal on Wireless Communications and Networking*, vol. 1, pp. 1–15, 2014.

16

A Robust License Plate Detection and Character Recognition Algorithm based on a Combined Feature Extraction Model and BPNN

Fei Xie,[1,2] **Ming Zhang**,[3] **Jing Zhao**,[2,4] **Jiquan Yang**,[1,2]
Yijian Liu,[1,2] **and Xinyue Yuan**[1,2]

[1]*School of Electrical and Automation Engineering, Nanjing Normal University, Nanjing 210042, China*
[2]*Jiangsu Key Laboratory of 3D Printing Equipment and Manufacturing, Nanjing 210042, China*
[3]*Department of Electronic Engineering, City University of Hong Kong, Hong Kong*
[4]*School of Automation, Nanjing University of Posts and Telecommunications, Nanjing 210023, China*

Correspondence should be addressed to Ming Zhang; mzhang367-c@my.cityu.edu.hk

Academic Editor: Yair Wiseman

The rapid development of the license plate recognition technology has made great progress for its widespread uses in intelligent transportation system (ITS). This paper has proposed a novel license plate detection and character recognition algorithm based on a combined feature extraction model and BPNN (Backpropagation Neural Network) which is adaptable in weak illumination and complicated backgrounds. Firstly, a preprocessing is first used to strengthen the contrast ratio of original car image. Secondly, the candidate regions of license plate are checked to verify the true plate, and the license plate image is located accurately by the integral projection method. Finally, a new feature extraction model is designed using three sets of features combination, training the feature vectors through BPNN to complete accurate recognition of the license plate characters. The experimental results with different license plate demonstrate effectiveness and efficiency of the proposed algorithm under various complex backgrounds. Compared with three traditional methods, the recognition accuracy of proposed algorithm has increased to 97.7% and the consuming time has decreased to 46.1ms.

1. Introduction

Vehicle plate licenses recognition (VPLR) plays a significant role in the field of intelligent transportation system. It has been widely used in traffic management, vehicle monitoring, suspect vehicle tracking, and many other fields. For example, in some cities in China, a new VPLR technology which enables drivers to pay parking fee using electronic wallet in a short time without leaving cars has received widely favourable reception. The parking fee from the drivers can be automatically collected by OCR (Optical Character Recognition) systems which can recognize license plates. However, errors sometimes occur when a vehicle is not identified or when a vehicle is wrongly identified as another vehicle. Therefore, RFID (Radio Frequency Identification) device and Bluetooth equipment can be combined to yield a better recognition performance [1–3]. Many new technologies like that are emerging and rising due to people's changing demands; it is necessary to improve license plate location and recognition algorithms to increase vehicle management efficiency. In general, license plate recognition process consists of three steps: license plate localization, characters segmentation, and characters classification and recognition. Since the license plate localization is the first and essential step of the recognition process, the result has a direct impact on the accuracy of character segmentation and character recognition. However, the license plate can be easily affected by external factors such as lightning conditions, weather, and backgrounds; besides, most VPLR systems do not fully consider the complexity of background and illumination conditions in the practical application, so locating and detecting the license plate from original images accurately and efficiently are still vital steps and the main difficulties for successful license plate recognition [4–6].

At present, the representative license plate classification and recognition task is performed by machine learning methods, such as a Support Vector Machine (SVM) classifier, and the neural network methods including Backpropagation Neural Network (BPNN) [7–11]. In particularly, multilayer neural networks and backpropagation training is adequate for vehicle plate licenses reading and recognition [12–14]. In addition, since the high dimensional original images contain redundant information, it is better to extract useful image characteristics rather than using each pixel value of the images as feature vectors [15–17]. Therefore, the feature extraction method plays a significant role in these license plate recognition algorithms, which need to extract different sets of features as the input layer of the network.

In the past several years, different methods have been exploited to extract features to describe various characters by many researchers. In [18], Afeefa et al. use the Hierarchical Temporal Memory (HTM) spatial pooler to recognize the characters more accurately and efficiently. They present three algorithms to extract features but each of them has its limitations in recognizing some of the characters. In [19], it is found that the Position of Peaks algorithm can recognize English numbers successfully. However, it cannot distinguish between Arabic numbers 0, 1, and 6. In [20], a Pixel Density method is presented, which relies on processing pixels along vertical and horizontal lines taken across the character to achieve recognition. In [21], the proposed algorithm develops the line containing a feature or a set of combined features after the process of quantization, which can identify a specific character in the different training datasets. In [22], a hybrid license plate extraction method based on the edge statistics and morphology can detect the region of license plate quickly and accurately. In [23], a vertical traverse density (VTD) vector and horizontal traverse density (HTD) vector are proposed to describe each character object. This method is convenient to achieve recognition and it is less time-consuming; however, due to the similar structures in different characters, the proposed algorithm has difficulty distinguishing between T and L, Z and E, and some other groups. In summary, a robust character recognition algorithm based on an efficient feature extraction method further needs to be proposed to improve license plate recognition.

In order to overcome the problem that the features extracted from the characters cannot describe the detail clearly, which leads to error recognition, and that the number of features extracted is so large that its implementation costs much time, this paper proposes a robust license plate detection and character recognition algorithm based on a novel combined feature extraction model and BPNN. The combined feature extraction method uses VTD, HTD features, and edge distance features as the learning and training samples of the VPLR classifier. Compared with the traditional license plate detection and character recognition method, it has the following features and advantages:

(1) A robust license plate detection and character recognition algorithm based on a novel combined feature extraction model and BPNN has been proposed in order to improve recognition efficient of VPLR algorithm.

(2) The new combined feature extraction model consists of three sets of features: two sets of traverse density features (VTD, HTD) and a set of edge distance features, which contains more information for the neural network training.

(3) The proposed algorithm can effectively determine and recognize variable license plates and has a good compatibility to regional difference under both weaker illumination and complex backgrounds.

FIGURE 1: Original image of a vehicle.

Aside from this introductory section, the remainder of this paper is composed of 4 more sections. In Section 2 the license detection and character segmentation algorithm using preprocessing, accurately license plate positioning, and histogram of transformation numbers projection is proposed. Then, Section 3 describes the details of a new combined features extraction model with BPNN. Additionally, field test and experimental results analyses are given in Section 4. Finally, Section 5 concludes the paper with a summary of the results achieved in this paper and a discussion of future work.

2. License Plate Detection and Character Segmentation Algorithm

2.1. Preprocessing of Original Car Image. Considering actual conditions, there is much interference in original car images such as the size of the image, lighting, and imaging quality, which influence the recognition performance seriously. In order to locate the license plate quickly and accurately, preprocessing of original images needs to be carried out. Here, the original images are captured at a high resolution (1250×750), which ensures that both the small license plate and small characters on them can be processed and recognized using the proposed algorithms. The original image of a vehicle is shown in Figure 1.

In the first step, grey image conversion is used to transform color image to grey image. Then, considering that the grey level of license plate area is not obviously distinct from the other areas in the image, a grey level stretch processing is applied to enhance contrast between two parts. After that, a set of candidate regions are extracted after edge detection, image erosion operation, and morphological

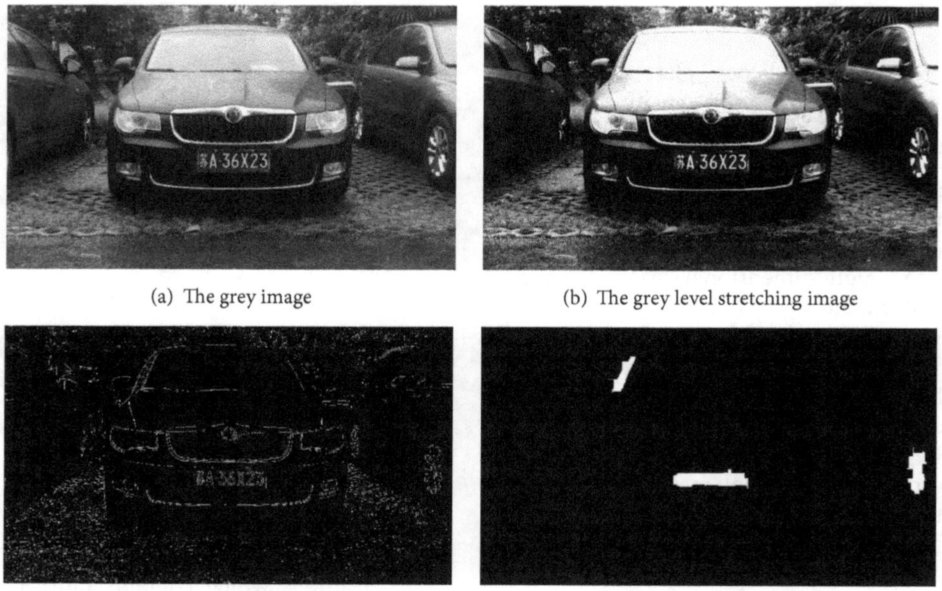

(a) The grey image

(b) The grey level stretching image

(c) The edge detection image

(d) The morphological operation image

FIGURE 2: The result of each step in the original image preprocessing.

closed operation [24]. The whole steps of the original image preprocessing are described as follows:

(1) First of all, original image is converted to grey image which decreases the computational cost.
(2) Then, a grey level stretch processing is applied to enhance contrast between the license plate area and the other parts of image.
(3) Edge is detected using Roberts operator to highlight the difference between the license plate frontier and the background.
(4) Image erosion operation and morphological closed operation are performed, excluding the small parts which are certainly not parts of the license plate region.

The corresponding result of each step is shown in Figure 2.

2.2. License Plate Coarse Extraction of the Candidate Regions. In the previous paragraph, it is shown that the candidate regions (marked with white color) of the license plate are clearly distinguished from the background (marked with black color). Then in the next step, the length and width information of the real license plate will be recorded to check the candidate regions one by one to select the true license plate area. As Chinese license plate outline size is 140mm∗440mm, the ratio is about 440/140≈3.15 [25], where setting the selected requirement is the aspect ratio ranging from 2 to 4. Using this method, the approximate poisoning of the license plate is achieved. The entire steps of license plate coarse extraction of the candidate regions are shown as follows:

(1) Three candidate regions are marked with different colors.
(2) According to prior experience and information mentioned above, the most likely area of the license plate region is verified which is marked with white color while the other candidate regions are all removed from the image.
(3) According to the location of the license plate region, the corresponding position in the grey-scale image is located.

Figure 3 illustrates the result of each step in the processing.

2.3. Accurate Positioning Based on Horizontal and Vertical Integral Projection Method. In order to segment and recognize the character accurately, it is necessary to process the approximately positioning of the license plate to get the accurate positioning of the license plate. In this paper, the horizontal integral projection and vertical integral projection are adopted [26]. A horizontal first order difference is carried on coarse location image f to achieve the horizontal accurate positioning image r. The horizontal integral projection can be described by the following equation:

$$r(i,j) = |f(i,j) - f(i,j-1)|, \quad i = 1,2,3,\dots,m, \; j = 2,3,4,\dots,n \tag{1}$$

where, $r(i,j)$ is the pixel value of the image r, $f(i,j)$ is the pixel value of the image f, and m and n are the height and the width of the image f, respectively. Then the projection value of the row i named $T_1(i)$ can be obtained by accumulating

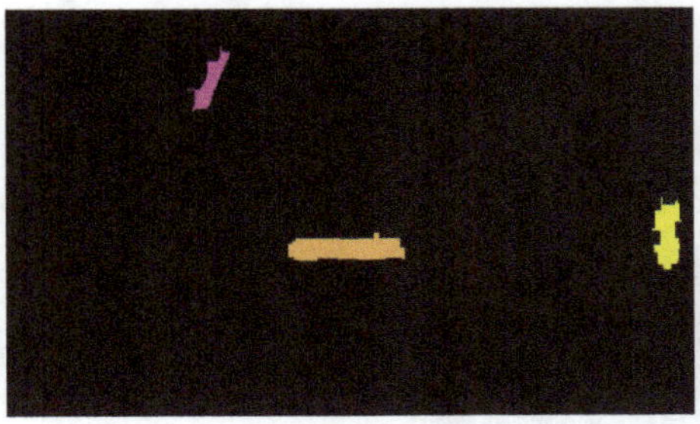

(a) Marked image

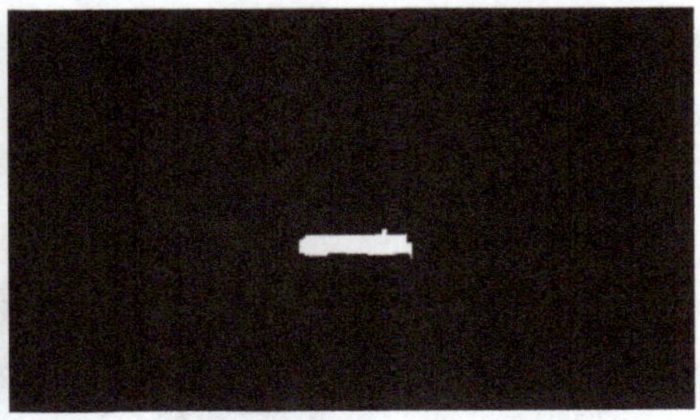

(b) Selected license plate region

(c) The license plate region detection in the original grey image

FIGURE 3: The result of each step in the coarse extraction of the license plate region.

the pixel value of image r per row, which can be depicted as follows.

$$T_1(i) = \sum_{j=2}^{n} r(i, j) \tag{2}$$

Considering that the edge of the image is not smooth enough, the average value is used as $T(i)$ to locate the boundary efficiently. The equation can be expressed as follows.

$$T_1(i) = \frac{(T_1(i-1) + T_1(i) + T_1(i+1))}{3} \tag{3}$$

The horizontal integral projection of the coarse location image is shown in Figure 4.

After this processing, the statistics of each row are saved in $sum[i]$ array, where i is the value of the corresponding row. Analysing horizontal projection value, character regions are generally corresponding to the intermediate dense and high regions which accounted for position, and the upper and lower edge of license plate are corresponding to the left and the right regions which accounted for position. In this way, the license plate can be located accurately in vertical direction. Here, the algorithm is described as follows:

(a) according to experimental results, setting the 0.6 of the maximum value of $sum[i]$ as threshold;

(b) from left to right, researching the array $sum[i]$ to find the first element value which is bigger than threshold, taking down the index of the value as m;

(c) from right to left, researching the array $sum[i]$ to find the first element value which is bigger than threshold, taking down the index of the value as n;

(d) extracting the image between the row m and row n as the accurate poison in vertical direction.

The difference between the coarse image and processing image is shown in Figure 5; it is shown that the two screws have been removed, while the license plate frames in horizontal direction still exist.

Similarly, the vertical integral projection is applied to locate the license plate accurately in horizontal direction. The equation can be depicted as follows:

$$g(i, j) = |r(i, j) - r(i-1, j)|, \quad i = 2, 3, \ldots, m, \; j = 1, 2, 3, 4, \ldots, n \tag{4}$$

where $r(i, j)$ is the pixel value of the image r which has been processed above, $g(i, j)$ is the pixel value of the image g, and m and n are the height and the width of the image g, respectively. Then the projection value of the column i named $T_2(i)$ can be obtained by accumulating the pixel value of image r per column, which can be depicted as follows.

$$T_2(j) = \sum_{i=2}^{n} g(i, j) \tag{5}$$

Considering that the edge of the image is not smooth enough, the average value is used as $T(i)$ to search for the boundary efficiently. The equation can be expressed as follows.

$$T_2(j) = \frac{(T_2(j-1) + T_2(j) + T_2(j+1))}{3} \tag{6}$$

The vertical integral projection of the processed image is shown in Figure 6.

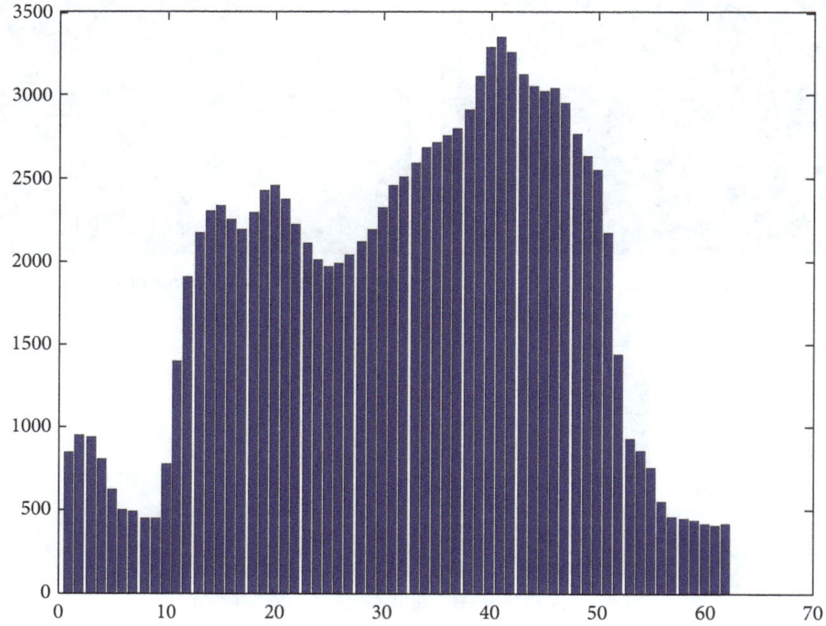

FIGURE 4: Histogram of horizontal integral projection.

(a) Coarse image

(b) Accurate poisoning image in vertical direction

FIGURE 5: Difference in coarse image and processed image.

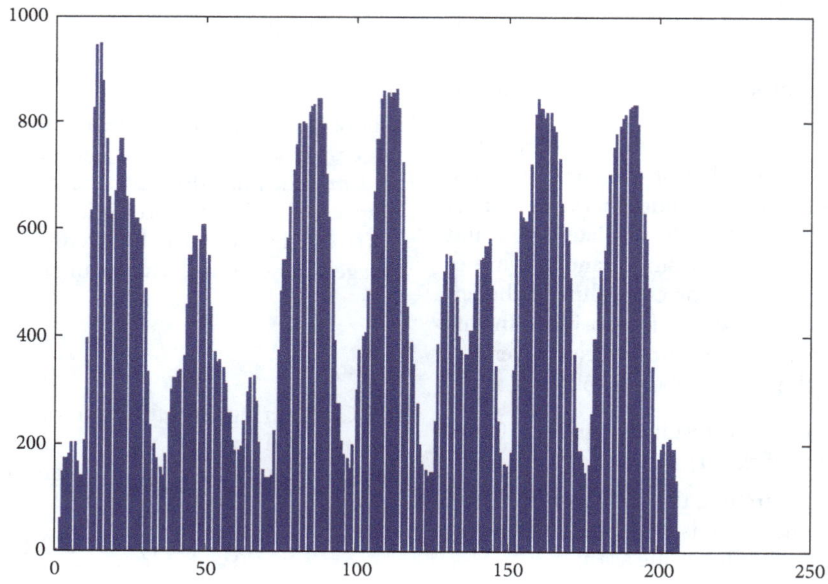

FIGURE 6: Histogram of vertical integral projection.

FIGURE 7: Accurately poisoning image of the license plate.

Here, the statistics of each column are saved in $sum[j]$ array, where j is the value of the corresponding column. From Figure 6, we can conclude that characters regions are generally corresponding to the peaks and the valleys while some small parts in the left and right are corresponding to license plate frames. Then we can use the algorithm to locate the license plate accurately in horizontal direction which can be described as follows:

(a) according to experimental results, setting the 0.7 of the average value of $sum[j]$ as threshold;

(b) from left to right, researching the array $sum[j]$ to find the first element value which is bigger than threshold, taking down the index of the value as m;

(c) from right to left, researching the array $sum[j]$ to find the first element value which is bigger than threshold, taking down the index of the value as n;

(d) extracting the image between the column m and column n as the accurate position in horizontal direction.

The final accurate poisoning image is shown in Figure 7; it can be seen that the image contains exactly the seven characters without any other area.

2.4. Character Segmentation Method Using the Number of Alternating White and Black. After extracting the accurate positioning image of the license plate, the character segmentation is researched at this stage. The license plate image consists of seven characters, a dot, and the space between them. To get the images only containing each character, it is natural to seek for the starting and ending points of each character, so an algorithm based on calculating the numbers of white and black transformations per column is presented. The step of the algorithm is described as follows:

(a) Grey-scale image is converted to binary one by Otsu method [27].

(b) An array $T[j]$ is created to store the transformation value of each column, where $j = 1, 2, \ldots, n$ is the index of each column and n is the width of the license plate image. The initial values are 0.

(c) Each pixel value is searched per column from up to bottom; if the pixel value changes from 1 to 0 or 0 to 1, it is defined as a transformation, and then one is added to the corresponding value of the column $T[j]$; otherwise, the original value stays.

(d) Draw the histogram of $T[j]$ until the last column is completely processed.

As shown in Figure 8, eight simply connected regions stand for seven characters and a dot, respectively. The blank space between each region stands for the interval between two adjacent characters on the license plate. Therefore, the first character starts with point 1 (the left edge) and ends with point A marked in Figure 8. Thus, the first character can be separated ranging from column 1 to column A from the image. The second character starts with point B and ends with point C. Thus, the second character can be separated ranging from the column B to the column C from the image. Noting that the point D represents the dot which needs to be removed, the remaining image can start with the ending point of the dot which is about 2/3 length distance of B to C from the point C. In this way, the remaining five characters can be segmented one after another. The final segmentation result is shown in Figure 9, which has also been introduced in [28].

3. Character Recognition Based on Traverse Density Features with Edge Distance Features

3.1. Features Extraction Algorithm Based on VTD and HTD Methods. The traditional traverse dense features consist of horizontal traverse density (HTD) and vertical traverse density (VTD), which has been explained in [28]. Here, in order to describe procedure of the HTD features extraction clearly, a character object "N" is used.

The character N is selected after the binary operation. Scanning the column and row of the image pixels one by one and recording the number of changes in black and white pixels, the rule is shown as follows:

$$\textit{feature}\ \text{value} = \begin{cases} 1, & \text{while the first pixel of the row or column is white} \\ \textit{feture value} + 1, & \text{while } r(\mathrm{i},\mathrm{j}) = 0\&\&r(\mathrm{i},\mathrm{j}+1) = 1 \text{ in row} \parallel r(\mathrm{i},\mathrm{j}) = 0\&\&r(\mathrm{i}+1,\mathrm{j}) = 1 \text{ in column} \end{cases} \quad (7)$$

where the initial feature value is 0 and $r(\mathrm{i},\mathrm{j})$ is the pixel value of the image. In the case of the column scan of the character "N", from left to right, the first column is alternating between black and white once, so the corresponding value is marked with one small black dot in the box. It can be clearly seen from Figure 10(a) that the most number of alternating black and white pixels is once [28]. These characteristic values are accurately remarked in the box. From left to right, a series of feature values are obtained, which can be recorded as vertical traverse density (VTD). In the same way, the character object "N" will be scanned from top to bottom, the side parts of the image have altered form black to white twice, and the intermediate part of that is three times. A set of feature values will be recorded as the horizontal traverse density (HTD). The

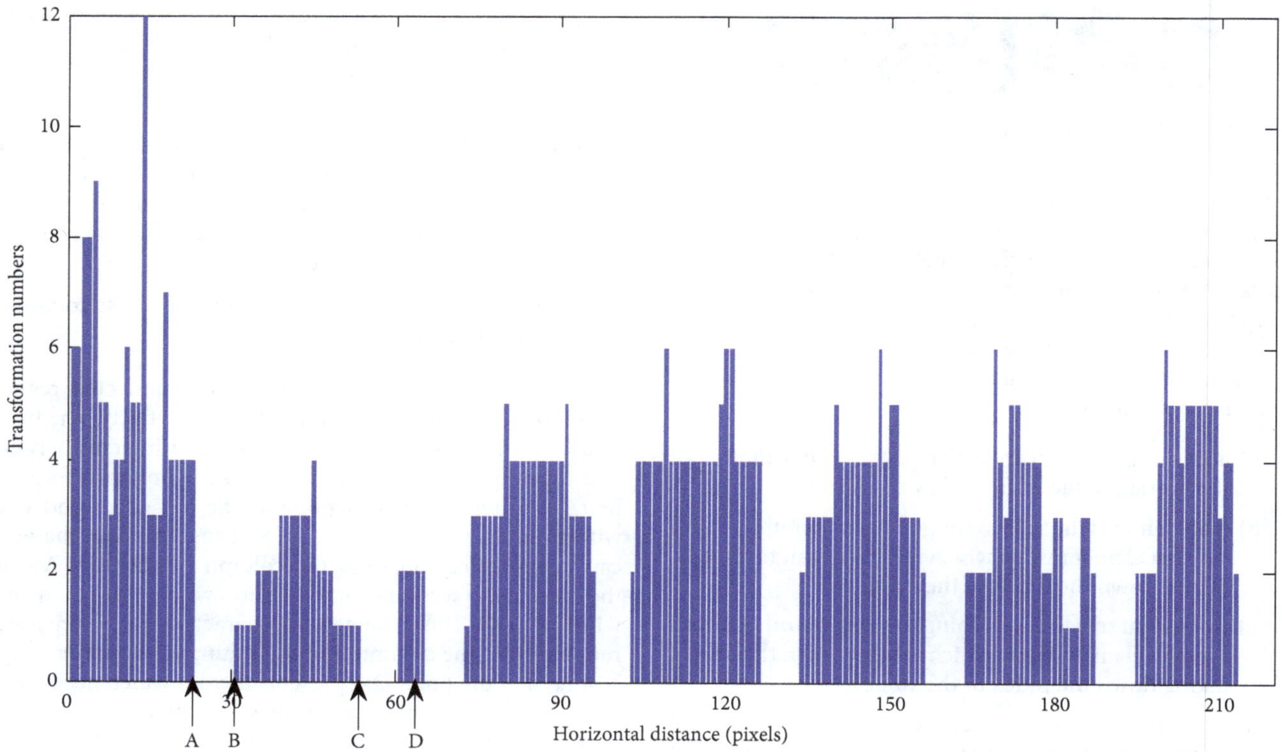

FIGURE 8: Histogram of transformation numbers projection.

FIGURE 9: Character segmentation results.

VTD and HTD are features used for character recognition, and they illustrate the image features of the target characters accurately.

As is shown in Figure 10, the HTD and VTD feature vectors of character "N" are "2222222222223333333333333-2222222222" and "11111111112211111111", respectively. In this method, the feature vectors of all the letters and numbers can be obtained. Since the VTD and HTD feature extraction is based on scanning each column and row of the image, the dimensions of the VTD and HTD feature are the width and height of the character to be measured. It is known to us that the larger the size of target image is, the more useful information it contains. As a result, the recognition rate will be higher in theory while it costs more time in manipulation, so it is important to make a balance between the recognition rate and the computation time of the algorithm. According to the experimental results, the final optimal license plate size is 35×20, i.e., the dimensions of the two feature vectors are 35 and 20, respectively. To improve the recognition rate and the efficiency of the system, all the character images to be measured and the template character images are normalized to size 35×20.

3.2. Improved Features Extraction Algorithm Using Edge Distance Features. The algorithm mentioned above uses VTD and HTD features as two sets of characteristic vectors. However, when handling some words which have similar traverse density structure like the character "E" and the character "Z", the character "T" and the character "L", it does have the same feature vectors which may cause misleading results in recognition.

As can be seen from Figure 11, it is clear that both the character "E" and the character "Z" have the same VTD and HTD feature vectors; hence characters "E" and "Z" are identical in feature vectors which are indistinguishable [28]. In order to address this problem, another set of features is put forward to compensate for the deficiency in traditional feature extraction algorithm.

Here, the proposed feature extraction method focuses on the edge distance features. The step of extracting the edge distance features can be described as follows:

(a) From the up to the bottom, search each pixel of row from the left to the right of the character image.

(b) Find the first pixel of the row i whose value is 1 (white), recording the index of the pixel, setting it as $d[i]$, where $i = 1, 2, \ldots, n$ and n is the height of the character image.

(c) The elements in the array $d[i]$ consist of the feature vector.

The algorithm can be seen in Figure 12. In this way, the edge distance feature values of the character "Z" are changing

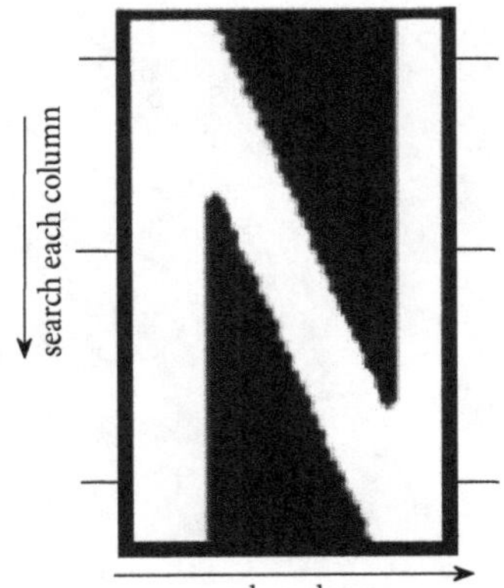

(a) The image of character "N"

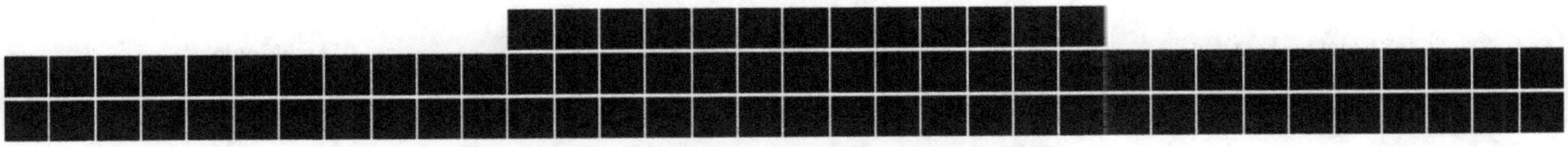

HTD (N) =222222222222333333333333322222222222

(b) The HTD features of character "N"

VTD (N) =1111111111221111111

(c) The VTD features of character "N"

FIGURE 10: The VTD and HTD features of character "N".

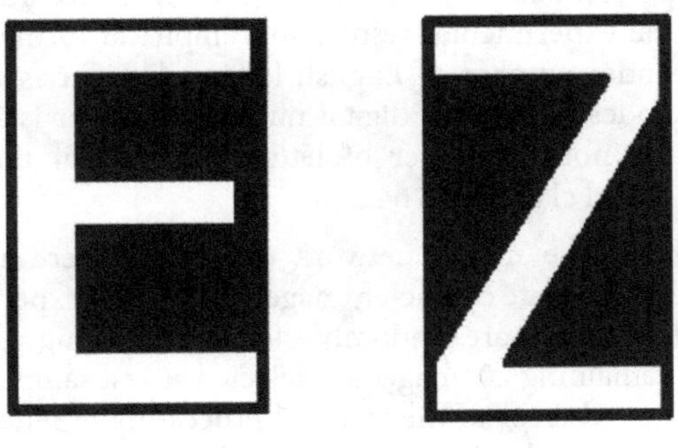

FIGURE 11: The character "E" and the character "Z".

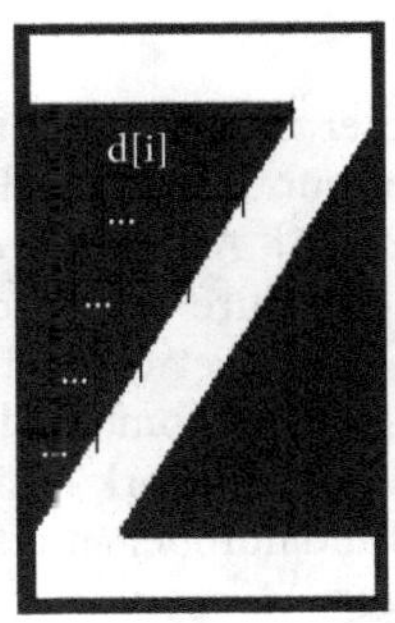

FIGURE 12: The edge distance feature values extraction of the character "Z".

from 1 to m, where m is the width of the image, while the most edge distance feature values of the character "E" are near to 1, so the difference of the feature vector between two characters is obvious when adding this set of features [28].

Thus, three different sets of features have been defined in the improved features extraction algorithm: the vertical traverse density, the horizontal traverse density, and the distance from the left edge to the first white pixel.

3.3. License Plate Characters Classification and Recognition Using Backpropagation Neural Network. In order to train a good recognition performance model, the license plate characters of the training samples must be collected. Our goal is to recognize the Chinese characters, the numbers 0-9, and the English letters A-Z of the license plate. The samples of the license plate character images are stored by different classification in the following folder according to the corresponding correct character information. For instance, the collection of license plate training samples of Chinese character "苏", English letter "E", and the number "6" is, respectively, shown in Figures 13(a), 13(b), and 13(c). They stand for one representation of the Chinese characters, the English letters, and the Arabic numbers, respectively.

The proposed character recognition method based on traverse density features with edge distance features has good real-time performance and the algorithm is convenient to implement. If it is combined with neural network, the recognition accuracy, especially the similar characters recognition

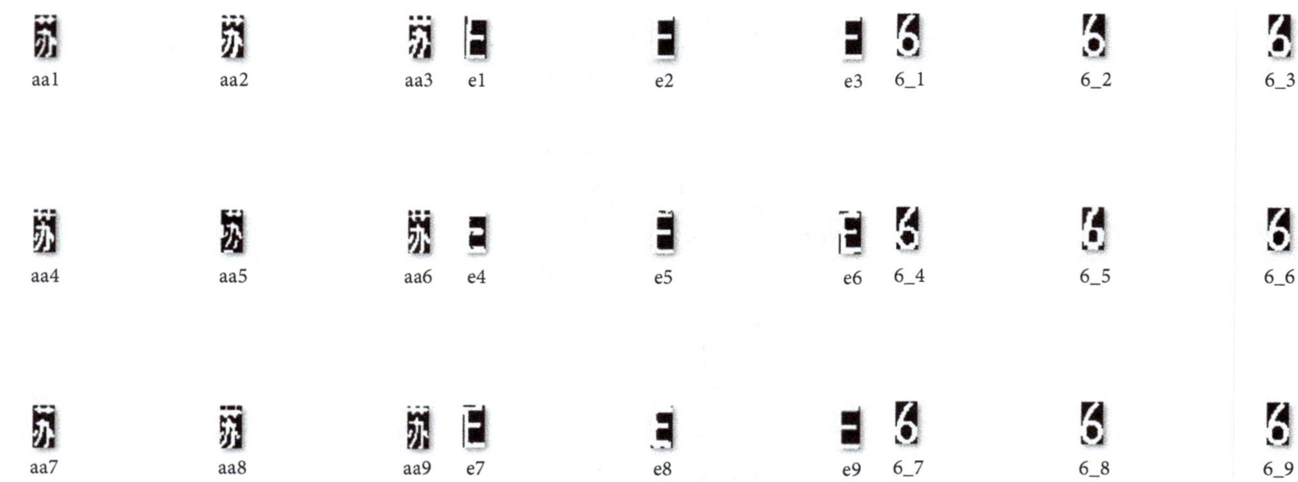

(a) Chinese character (b) English letter (c) Arabic number

FIGURE 13: The license plate training samples.

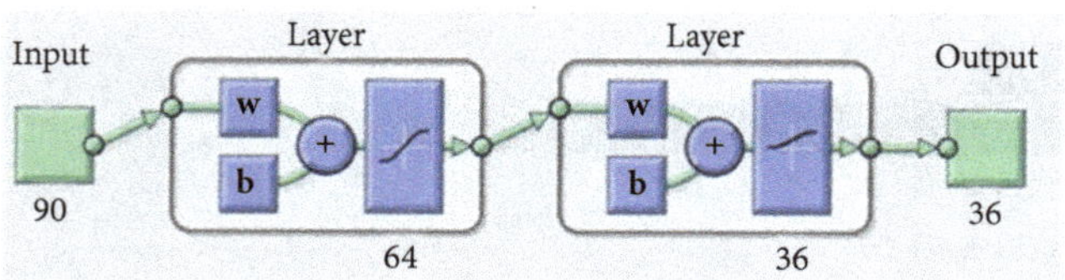

FIGURE 14: Structure of the letters and numbers mixed classifier based on BPNN.

accuracy, will be further improved by utilizing the specific learning process of the neural network. Here, BPNN is chosen as a training model [29–31]. As is researched in the previous paragraph, three sets of feature vectors can describe most of the characteristics of the target characters, and these three sets of feature vectors can be combined as a vector used as input vector for the neural network designing and building. Structure of the letters and numbers mixed classifier based on BPNN is shown in Figure 14.

(1) The number of neuron nodes in the input layer: The neuron nodes in the input layer are determined by the dimension of the features extracted from the images. The size of license plate character images used in this article is 35×20. That is to say, the dimension of the HTD features of the character is 35, the dimension of the VTD features of the character is 20, and the dimension of the edge distance features is 35. Connecting three sets of features into one feature vector, a 90-dimensional vector can be obtained as the input vector of the neural network. Therefore, the input layer has 90 nodes.

(2) The number of neuron nodes in the output layer: The neuron nodes in the output layer are decided by the total numbers of Chinese characters, English letters, and digital numbers waiting to be classified. There are 26 English letters, 10 digital numbers, and 36 alphanumeric characters in the system, so the output layers have 26 nodes, 10 nodes, and 36 nodes corresponding to the English letters classifiers, digital numbers classifiers, and alphanumeric characters classifiers, respectively.

(3) The number of neuron nodes in the hidden layer: The nodes numbers of hidden layer are determined by many factors, having a great impact on the performance of the training model. According to the experimental results and empirical formula, the nodes number of English letters classifier is 48, the nodes number of digital number classifier is 27, and the nodes number of letters and digital numbers mixed classifier is 64.

To train the neural network effectively, there are 100 vehicle license plate character images used in the experiment, of which 80 images are randomly selected as training samples, and the remaining 20 images are selected as test samples. The mean squared error of the training processing of letters and numbers mixed classifier is plotted in Figure 15.

4. Experiment and Analysis

In order to test and verify performance of the proposed algorithm, field tests are provided in this section. To demonstrate the proposed license plate detection algorithm, several experiments using samples with different backgrounds have been conducted. Figures 16(a) and 16(b) show the real license plate detection, segmentation, and recognition results of two small family vehicles. As can be seen from Figure 16, the proposed license plate detection and recognition algorithm can effectively determine license plate location and complete recognition function under complex backgrounds.

Besides, to better present the test results, a vehicle license plate recognition system has been designed. The entire processing results of the original image under more complex and weaker illumination backgrounds are shown in Figure 17. As

Table 1: Identification performance of each character.

Object	Parameter	Value
Letters	Recognition Rate (%)	96.1
	Recognition Times (ms)	45.6
Numbers	Recognition Rate (%)	97.8
	Recognition Times (ms)	48.1
Characters (Letters & Numbers)	Recognition Rate (%)	97.7
	Recognition Times (ms)	50.3

Table 2: Comparison of the proposed method and conventional methods on three factors.

Method	Features dimension	Precision (%)	Computational time (ms)
Proposed method	90	97.7	46.1
Line processing [12]	55	94.5	69.8
Peak position [13]	90	95.6	65.3
Pixel density [14]	55	95.1	57.2

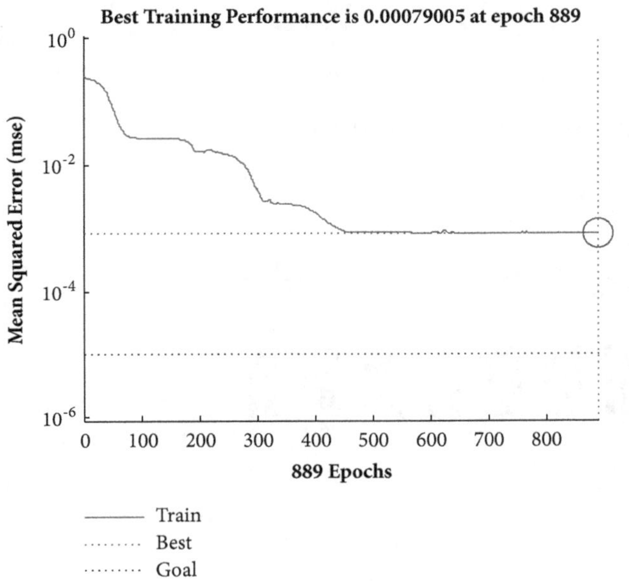

Figure 15: Error curve of the neural network training performance.

illustrated in Figure 17, the proposed license plate detection and recognition algorithm can detect and recognize the license plate under the weak ambient illumination since image under the grey level stretch processing is robust to the illumination information.

Moreover, 100 license plates are collected in different backgrounds and different illumination condition which are used to test license plate detection and characters recognition. As a result, 95 license plates are completely correctly identified. The correct license plate recognition rate is 95 percent. There are total 700 valid characters which have been measured and 684 of them are correctly identified, which shows that the character recognition rate is 97.7 percent. The average recognition time of 100 license plates samples for single license plate is 707ms. Besides, the recognition rate of letters and numbers and the consuming time of them for single character are summarized and shown in Table 1, whose analysis is also given in [28].

Then, to evaluate the proposed features extraction algorithm, 100 license plates, including 50 percent samples collected by the authors and another 50 percent samples collected from some vehicle libraries, are used to compare the accuracy rate and run time of the several introduced methods. Compared with the conventional methods, the proposed algorithm occupies a dominant position in the recognition rate and the computation time. Experimental results are summarized in Table 2. The proposed algorithm shows a precision of 97.7% and consuming time of 46.1ms, outperforming the three conventional methods.

Finally, in order to further verify the adaptivity and compatibility of the proposed algorithm, tests are also implemented using Hong Kong vehicle license plate. Hong Kong license plate is made up of English letters and Arabic numbers without Chinese characters. Due to the difference in size, module, and font between Chinese mainland license plate and Hong Kong SAR license plate, it is still challenging to the presented plate detection and feature extraction method. The license plate detection, segmentation, and recognition results of several Hong Kong vehicle license plates are shown in Figure 18. As can be seen from Figure 18, the algorithm can accurately detect and recognize the Hong Kong license plate, which indicates that the proposed algorithm has a good compatibility to regional difference in license plate.

5. Conclusion

This paper presents a robust license plate detection and character recognition algorithm based on a combined feature extraction model and BPNN. The experimental results demonstrate that the proposed algorithm obtains a good recognition performance and compatibility in identifying different license plates under weaker illumination and complex backgrounds. Compared with the three traditional methods, the proposed method provides a better recognition accuracy and time-consuming performance with 97.7% and 46.1ms. In

(a) A small brown family car

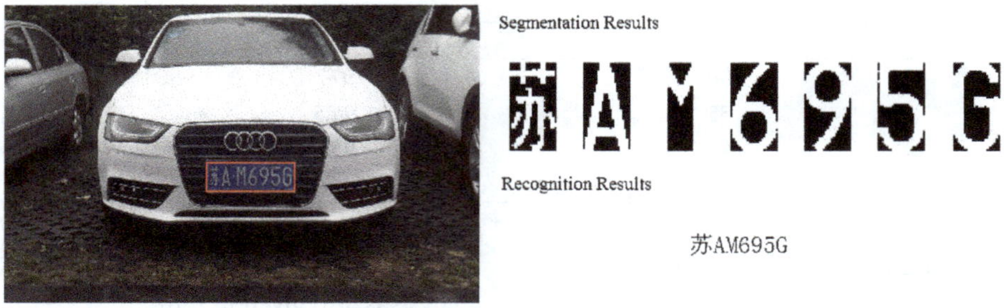

(b) A small white family car

Figure 16: The license plate detection, segmentation, and recognition results.

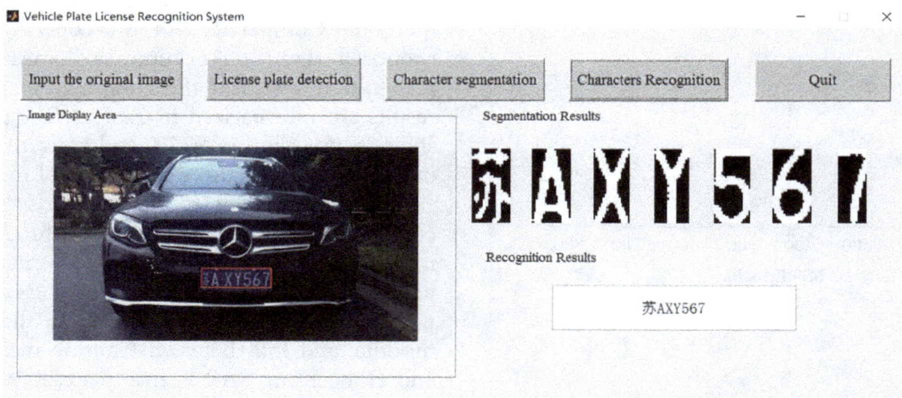

(a) A SUV black family car

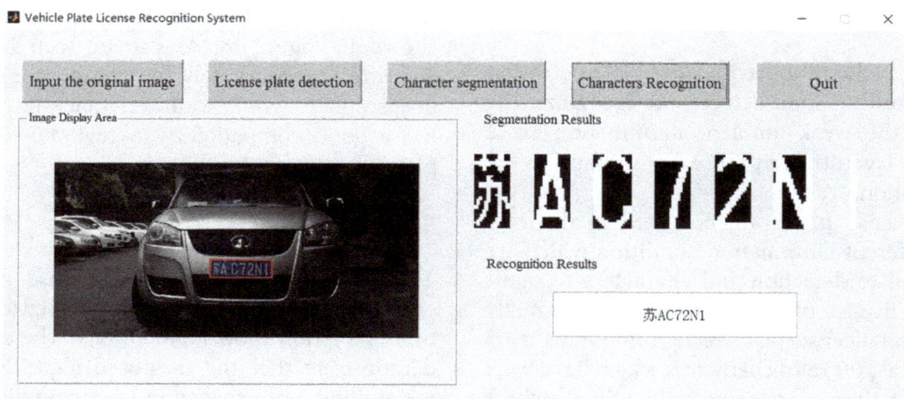

(b) A SUV silver family car

Figure 17: Recognition results of the license plate in weak illumination environment.

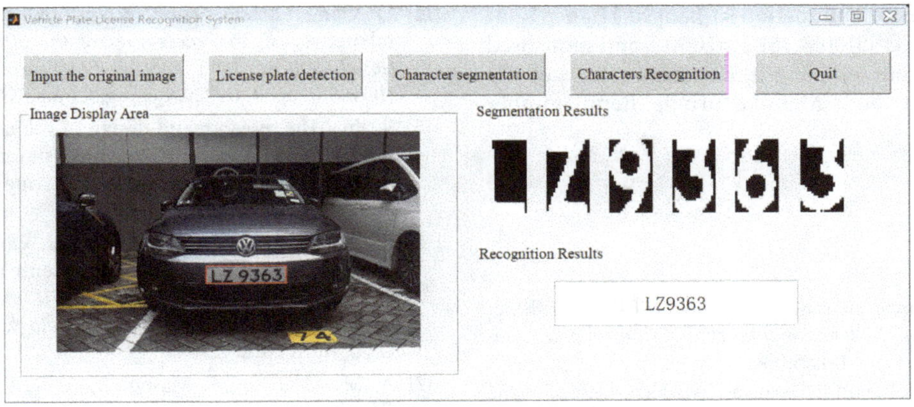

(a) A pale blue family car in Hong Kong

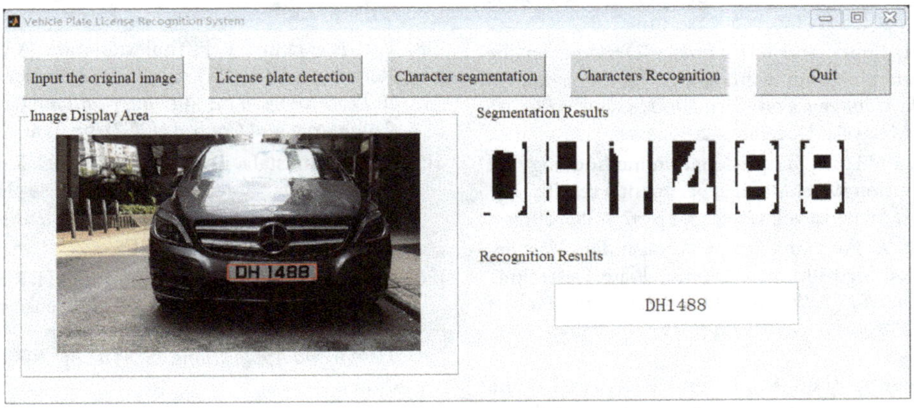

(b) A grey family car in Hong Kong

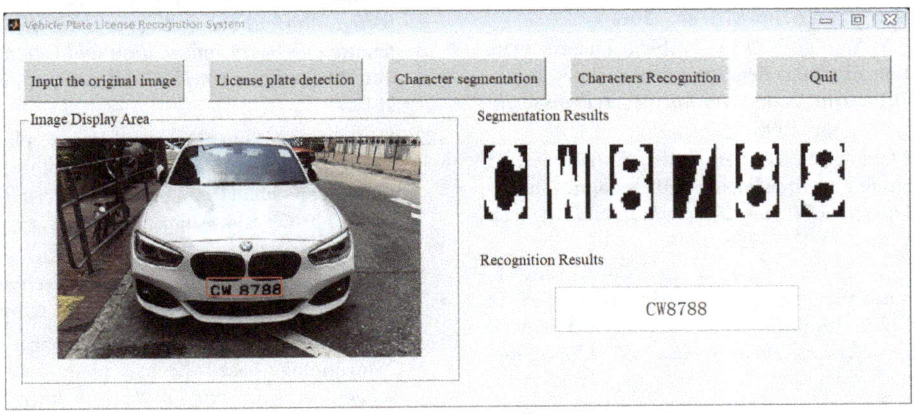

(c) A white family car in Hong Kong

FIGURE 18: Recognition results of several Hong Kong vehicle license plates.

the future work, extending the VPLR system would be further considered with RFID device and Bluetooth equipment combination to improve recognition accuracy and fit more complex requirements.

Conflicts of Interest

The authors declare that they have no conflicts of interest.

Acknowledgments

This work is partially supported by the National Natural Science Foundation of China (Grant No. 61601228, 61803208, 61403210), the Natural Science Foundation of Jiangsu (BK20161021, BK20180726), National Key Research and Development Program of China (Grant No. 2017YFB1103200), and the Natural Science Foundation of Jiangsu Higher Education Institution (17KJB510031, 18KJB120005) and sponsored by the Open Program of Jiangsu Key Laboratory of 3D Printing Equipment and Manufacturing, Item number 3DL201607.

References

[1] Y. Wiseman, "Vehicle identification by OCR, RFID and Bluetooth for toll roads," *International Journal of Control and Automation*, vol. 11, no. 9, pp. 1–12, 2018.

[2] A. Saini, S. Chandok, and P. Deshwal, "Advancement of traffic management system using RFID," in *Proceedings of the 2017 International Conference on Intelligent Computing and Control Systems (ICICCS)*, pp. 1254–1260, Madurai, June 2017.

[3] A. Sagahyroon, M. Eqbal, and F. Khamisi, "Drawing on the benefits of RFID and bluetooth technologies," in *Proceedings of the 2010 Asia Pacific Conference on Circuit and System, APCCAS 2010*, pp. 180–183, Malaysia, December 2010.

[4] Z. Youting, Y. Zhi, and L. Xiying, "Evaluation methodology for license plate recognition systems and experimental results," *IET Intelligent Transport Systems*, vol. 12, no. 5, pp. 375–385, 2018.

[5] A. Shahbahrami, B. A. Foomani, and A. Akoushideh, "A style-free and high speed algorithm for License Plate Detection," in *Proceedings of the 2017 10th Iranian Conference on Machine Vision and Image Processing (MVIP)*, pp. 76–81, Isfahan, Iran, November 2017.

[6] Y. Wen, Y. Lu, J. Yan, Z. Zhou, K. M. von Deneen, and P. Shi, "An algorithm for license plate recognition applied to intelligent transportation system," *IEEE Transactions on Intelligent Transportation Systems*, vol. 12, no. 3, pp. 830–845, 2011.

[7] C. Gou, K. Wang, Y. Yao, and Z. Li, "Vehicle License Plate Recognition Based on Extremal Regions and Restricted Boltzmann Machines," *IEEE Transactions on Intelligent Transportation Systems*, vol. 17, no. 4, pp. 1096–1107, 2016.

[8] Y.-P. Huang, T.-W. Chang, Y.-R. Chen, and F. E. Sandnes, "A back propagation based real-time license plate recognition system," *International Journal of Pattern Recognition and Artificial Intelligence*, vol. 22, no. 2, pp. 233–251, 2008.

[9] M. H. T. Brugge, J. H. Stevens, J. A. G. Nijhuis, and L. Spaanenburg, "License plate recognition using DTCNNs," in *Proceedings of the 1998 5th IEEE International Workshop on Cellular Neural Networks and Their Applications, CNNA*, pp. 212–217, April 1998.

[10] Y. Yuan, W. Zou, Y. Zhao, X. Wang, X. Hu, and N. Komodakis, "A robust and efficient approach to license plate detection," *IEEE Transactions on Image Processing*, vol. 26, no. 3, pp. 1102–1114, 2017.

[11] X. Chi, J. Dong, A. Liu, and H. Zhou, "A simple method for Chinese license plate recognition based on support vector machine," in *Proceedings of the 2006 International Conference on Communications, Circuits and Systems, ICCCAS*, pp. 2141–2145, China, June 2006.

[12] M. H. ter Brugge, J. A. Nijhuis, and L. Spaanenburg, "Transformational DT-CNN design from morphological specifications," *IEEE Transactions on Circuits and Systems I: Fundamental Theory and Applications*, vol. 45, no. 9, pp. 879–888, 1998.

[13] M. H. TerBrugge, L. C. Jain, and B. Lazzerrini, *Knowledge-Based Intelligent Techniques in Character Recognition: License Plate Recognition*, CRC Press, 1999.

[14] Q. Wang, "License plate recognition via convolutional neural networks," in *Proceedings of the 2017 8th IEEE International Conference on Software Engineering and Service Science (ICSESS)*, pp. 926–929, Beijing, China, November 2017.

[15] R. Fu, "The research and design of vehicle license plate recognition system in traffic management system," *International Journal of Signal Processing, Image Processing and Pattern Recognition*, vol. 9, no. 3, pp. 445–456, 2016.

[16] K. Almustafa, R. N. Zantout, H. R. Obeid, and F. N. Sibai, "Recognizing characters in Saudi license plates using character boundaries," in *Proceedings of the 2011 International Conference on Innovations in Information Technology, IIT 2011*, pp. 415–420, UAE, April 2011.

[17] A. M. Al-Ghaili, S. Mashohor, A. R. Ramli, and A. Ismail, "Vertical-edge-based car-license-plate detection method," *IEEE Transactions on Vehicular Technology*, vol. 62, no. 1, pp. 26–38, 2013.

[18] P. P. Afeefa and P. P. Thulasidharan, "Automatic License Plate Recognition (ALPR) using HTM cortical learning algorithm," in *Proceedings of the 2017 International Conference on Intelligent Computing and Control (I2C2)*, pp. 1–4, Coimbatore, June 2017.

[19] K. M. Almustafa, R. N. Zantout, and H. R. Obeid, "Peak position recognizing characters in Saudi license plates," in *Proceedings of the 2011 IEEE GCC Conference and Exhibition, GCC 2011*, pp. 186–189, UAE, February 2011.

[20] K. M. Almustafa, R. N. Zantout, and H. R. Obeid, "Pixel density: Recognizing characters in Saudi license plates," in *Proceedings of the 2010 10th International Conference on Intelligent Systems Design and Applications, ISDA'10*, pp. 308–313, Egypt, December 2010.

[21] S. Al-Shami, A. El-Zaart, R. Zantout, A. Zekri, and K. Almustafa, "A new feature extraction method for license plate recognition," in *Proceedings of the 2015 5th International Conference on Digital Information and Communication Technology and Its Applications, DICTAP 2015*, pp. 64–69, Lebanon, May 2015.

[22] H. Bai and C. Liu, "A hybrid license plate extraction method based on edge statistics and morphology," in *Proceedings of the 17th International Conference on Pattern Recognition (ICPR '04)*, vol. 2, pp. 831–834, August 2004.

[23] C. L. Tan, W. Huang, Z. Yu, and Y. Xu, "Imaged document text retrieval without OCR," *IEEE Transactions on Pattern Analysis and Machine Intelligence*, vol. 24, no. 6, pp. 838–844, 2002.

[24] . Jun-Wei Hsieh, . Shih-Hao Yu, and . Yung-Sheng Chen, "Morphology-based license plate detection from complex scenes," in *Proceedings of the 16th International Conference on Pattern Recognition*, pp. 176–179, Quebec City, Que., Canada.

[25] H. Bai, J. Zhu, and C. Liu, "A fast license plate extraction method on complex background," in *Proceedings of the 2003 IEEE International Conference on Intelligent Transportation Systems, ITSC 2003*, pp. 985–987, China, October 2003.

[26] G.-S. Hsu, J.-C. Chen, and Y.-Z. Chung, "Application-oriented license plate recognition," *IEEE Transactions on Vehicular Technology*, vol. 62, no. 2, pp. 552–561, 2013.

[27] N. Otsu, "A threshold selection method from gray-level histograms," *IEEE Transactions on Systems, Man, and Cybernetics*, vol. 9, no. 1, pp. 62–66, 1979.

[28] M. Zhang, F. Xie, J. Zhao, R. Sun, L. Zhang, and Y. Zhang, "Chinese license plates recognition method based on a robust and efficient feature extraction and bpnn algorithm," *Journal of Physics: Conference Series*, vol. 1004, p. 012022, 2018.

[29] F. Li and S. Gao, "Character recognition system based on back-propagation neural network," in *Proceedings of the 2010 International Conference on Machine Vision and Human-Machine Interface (MVHI '10)*, pp. 393–396, IEEE, China, April 2010.

[30] J. Dong, M. Sun, G. Liang, and K. Jin, "The improved neural network algorithm of license plate recognition," *International Journal of Signal Processing, Image Processing and Pattern Recognition*, vol. 8, no. 5, pp. 49–54, 2015.

[31] Z. Qu, Q. Chang, C. Chen et al., "An improved character recognition algorithm for license plate based on BP neural network," *Open Electrical & Electronic Engineering Journal*, vol. 8, pp. 202–207, 2014.

17

Emergency Situation Prediction Mechanism: A Novel Approach for Intelligent Transportation System Using Vehicular Ad Hoc Networks

P. Ganeshkumar[1] and P. Gokulakrishnan[2]

[1] *Department of Information Technology, PSNA College of Engineering and Technology, Dindigul, Tamil Nadu 624622, India*
[2] *Department of Computer Science and Engineering, PSNA College of Engineering and Technology, Dindigul, Tamil Nadu 624622, India*

Correspondence should be addressed to P. Gokulakrishnan; gokulakrishnan_77@yahoo.com

Academic Editor: Baozhen Yao

In Indian four-lane express highway, millions of vehicles are travelling every day. Accidents are unfortunate and frequently occurring in these highways causing deaths, increase in death toll, and damage to infrastructure. A mechanism is required to avoid such road accidents at the maximum to reduce the death toll. An Emergency Situation Prediction Mechanism, a novel and proactive approach, is proposed in this paper for achieving the best of Intelligent Transportation System using Vehicular Ad Hoc Network. ESPM intends to predict the possibility of occurrence of an accident in an Indian four-lane express highway. In ESPM, the emergency situation prediction is done by the Road Side Unit based on (i) the Status Report sent by the vehicles in the range of RSU and (ii) the road traffic flow analysis done by the RSU. Once the emergency situation or accident is predicted in advance, an Emergency Warning Message is constructed and disseminated to all vehicles in the area of RSU to alert and prevent the vehicles from accidents. ESPM performs well in emergency situation prediction in advance to the occurrence of an accident. ESPM predicts the emergency situation within 0.20 seconds which is comparatively less than the statistical value. The prediction accuracy of ESPM against vehicle density is found better in different traffic scenarios.

1. Introduction

This work is motivated by the statistics of increased highway accidents and rise in death toll every day. In India, 1, 37,423 people were killed and 4, 69,900 people were injured in road traffic crashes in 2013 (NCRB 2013). Traffic fatalities increased by about 54.3 percent during the period 2003–2013 [1]. The World Health Organization (WHO) reports say that the road accident is the fifth cause of death in the world. Nearly 1.3 million people die in road accidents every year. The accident injury rate is 50 times greater than the death rate [2]. The total number of vehicles all over the world is estimated to be one billion in 2010 [3]. In 2013, the total number of vehicles is estimated to be 1.2 billion which leads to an increase of accidents and deaths [4].

Road traffic accidents have a high impact on the economy of the developing countries and affect the development of those countries [5]. Around 20% of Gross Domestic Product (GDP) of Indian economy suffers because of road accidents [6]. Statistics say that if the driver of the vehicle is provided emergency warning at least one-half second before an accident then 60% of highway accidents could be avoided [7].

The objective of this paper is to design a framework to implement Intelligent Transportation System (ITS) using Vehicular Ad Hoc Network (VANET) with an intention to reduce the number of road accidents and to decrease the death toll caused by road accidents. An Emergency Situation Prediction Mechanism (ESPM), a proactive framework for accident prediction and prevention, is proposed to meet the objective. ESPM is designed for a four-lane express highway of Indian traffic system. In the existing Indian four-lane express highway the vehicles move in two lanes in left-hand side in one direction against the two lanes of the other side as shown in Figure 1. Four-lane express highway in India has a physical median or divider and generally has a speed limit of 60 km/h to 90 km/h [8].

ESPM attempts to predict the possibility of occurrence of an accident in advance before it occurs. An accident is said as an emergency situation in this paper. The Road Side Unit (RSU), the vehicles in the range of the RSU, and the four-lane express highway are considered in ESPM for emergency situation prediction and prevention.

Once the vehicles enter into the range of RSU, periodically the vehicles send a Status Report (SR) to RSU simultaneously the RSU performs traffic flow (TF) analysis. A Four-Lane Sensor Grid (FLSG) is designed to support ESPM for traffic flow analysis. ESPM performs prediction based on both SR and TF values. As the prediction process involves both the SR and TF values, the prediction accuracy (PA) is promisingly increased.

The highway road accidents can be categorized as head-on accidents, lateral impact, and rear-end accidents [9]. Sudden acceleration or deceleration of the vehicle is a major reason for rear-end accidents (i.e., chain of accidents) and head-on accidents. Improper or abnormal lane change in highway might cause lateral impact accidents. ESPM performs prediction for rear-end accidents, head-on accidents, and lateral impact accidents as discussed in Sections 5.3.1 and 5.3.2.

The rest of the paper is organized as follows. In Section 2, background of ESPM is presented. Section 3 presents related systems. Section 4 gives an overview of ESPM. Section 5 presents detailed design of ESPM. Simulation setup and result evaluation are discussed in Sections 6 and 7, respectively. Finally, Section 8 concludes the ESPM framework.

2. Background

Intelligent Transportation System (ITS) is an emerging area of research over the past two decades. Transportation system becomes intelligent with emerging computing hardware, positioning system, sensor technologies, high speed telecommunication, and data processing.

ITS makes the travels of ease and comfort, it also provides reliable and improved travel safety. The environment gets benefit out of ITS because of low carbon emissions and air pollutions. Due to improvement in traffic efficiency and reduction of accident rate the economy of countries is improved [5, 10].

VANET is a technology that uses moving vehicles as nodes and forms a Mobile Network. Vehicles communicate with each other via Vehicle to Vehicle communication (V2V) and with roadside base stations via Vehicle to Infrastructure communication (V2I) [11]. Vehicle to Vehicle communication is performed using Dedicated Short Range Communication (DSRC) and Vehicle to Roadside communication is performed using IEEE 802.11p [12] as shown in Figure 2.

ITS in VANET leads to the concept of Automated Highway System (AHS). AHS provides Safer Highway Transportation by making a vehicle to predict the actions of neighboring vehicles [13]. Establishment and maintenance of

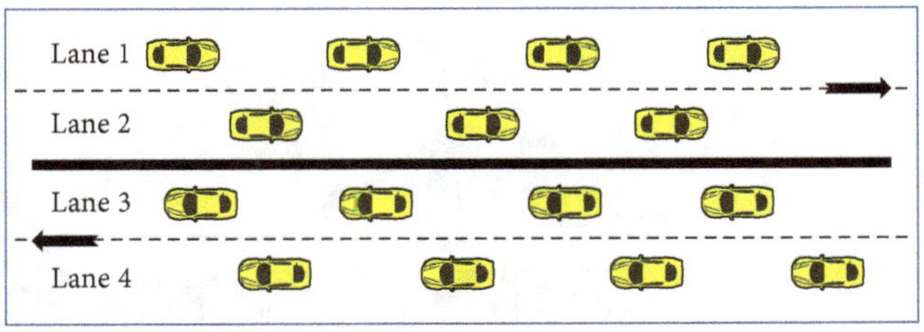

FIGURE 1: Indian four-lane express highway.

communication paths are more difficult in VANET due to the high mobility of the network. In order to prevent accidents, short reaction times from both system and driver are essential due to the high speed of the vehicles [14, 15]. Perception, modeling, decision making, and control are tough in real time safety analysis involving driver, vehicle, and the traffic environment itself [16].

3. Related Work

Caliendo et al. [17] considered vehicle density, traffic data, road conditions and weather for crash prediction in multi-lane roads. The authors have analyzed and classified severe accidents from total accidents. They have performed mathematical analyzes and provided a crash-prediction model for multilane roads. They have not considered important parameters such as vehicles and human (driver) behavior.

Hu et al. [18] have developed a traffic accident prediction scheme based on tracking of vehicles. Fuzzy self-organizing neural network method has been used for learning vehicle activity patterns. The authors have established a probability model for predicting traffic accidents. This model is suitable for urban vehicular scenario.

Sörstedt et al. [19] focus on the behavior or influence of the driver rather than vehicle and traffic conditions. A model with postulates is developed based on the behavior, actions, and influence of the driver on the vehicle. Other important parameters such as vehicle actions and road traffic conditions have not been considered.

Worrall et al. [20] propose a model named Intelligent Systems for Risk Assessment (ISRA). This model is developed with an aim to eliminate accidents by detecting risks and alerting other vehicles to take actions appropriately. Using artificial intelligence, this model attempts to assess the risk based on the dynamic behavior of the vehicle, environmental conditions, and ability of the human or driver. Risk assessment is made only based on less number of complex scenarios.

De Oña and Garach [21] used the Interactive Highway Safety Design Model (IHSDM) for analyzing the relationship between speed reduction and accidents in Spanish two-lane rural highways. IHSDM was actually developed for American two-lane rural highways. This mathematical model analyses the cause of accidents in tangent and curves of roads due to speed reduction. All scenarios of the rural highways were not considered.

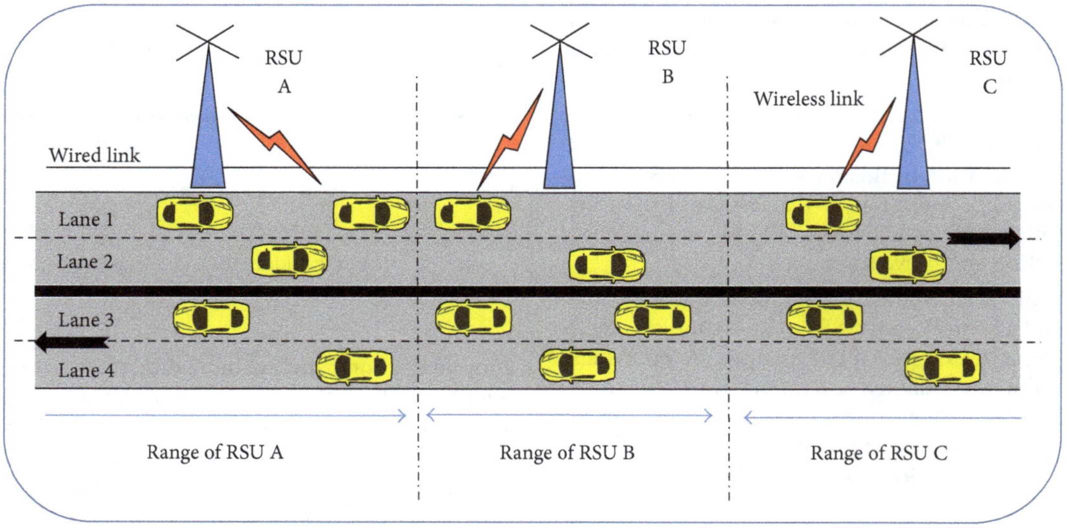

FIGURE 2: VANET Architecture.

Garcia-Lozano et al. [22] propose a warning service to the drivers in order to prevent from accidents due to dangerous road conditions and accidents caused by other vehicles. A part of the warning service, an alert message, is constructed based on the sensed environmental cum vehicle data and the message is distributed to the passing vehicles. The work needs to be extended for highly dense scenarios.

Rokade et al. [23] developed an accident prediction model using multiple linear regression analysis for Bhopal city in India. This model considers an urban multilane road conditions along with past traffic data for accident prediction. This model works on the statistical data and not on real time data.

Saravanan et al. [24] developed an accident risk prediction model using fuzzy approach. The road, driver, and pedestrians are considered as risk factors for accident. This model can be used to analyze or investigate the road accident causation. This is a reactive model used for after investigation of a road accident.

It is learned from the literature that most of the existing models for accident prediction is based on either the vehicle parameters or the traffic parameters and most models are reactive and based on statistical data rather than real time data. Sometime there is a possibility of failure in the process of prediction because of not considering either of the above parameters. Hence in ESPM, both the parameters of vehicle status and traffic conditions are considered for emergency situation prediction process to ensure the performance.

4. Overview of ESPM

The proposed Emergency Situation Prediction Mechanism (ESPM) is a framework for traffic accident prediction and prevention. Overall structure of ESPM is shown in Figure 3. In ESPM, the vehicles, a four-lane highway, and road side infrastructure are participating in the processes of reporting, monitoring, prediction, and prevention phases.

ESPM performs prediction of emergency situation in three phases and prevention in the fourth phase. First, the participating vehicles will send Status Report (SR) to the near-by Road Side Unit (RSU). Second, RSU monitors the traffic flow in its range. Third, the RSU performs emergency situation prediction based on the SR and TF values and in fourth phase the RSU constructs and disseminates Emergency Warning Message (EWM) to all vehicles in its range and to the near-by RSUs. As a result the vehicles have a chance to take preventive action against accidents either by reducing the speed of the vehicle or taking an alternate route [22].

In this paper much attention is given to the first three phases and the fourth phase will be a future work.

5. Detailed Design

The ESPM framework is broadly divided into four phases, namely, reporting, monitoring, prediction, and prevention. Each phase is assigned a distinct task. The pseudo code of ESPM is presented below followed by a detailed discussion.

Pseudocode of ESPM

Definitions

V_{ID}: vehicle ID;

Pos: position of the vehicle;

Sp: speed of the vehicle;

Yr: yaw rate (i.e., rotation angle);

SR: Status Report;

FLSG: Four-Lane Sensor Grid;

L_{MN}: Nth block in Mth lane of FLSG;

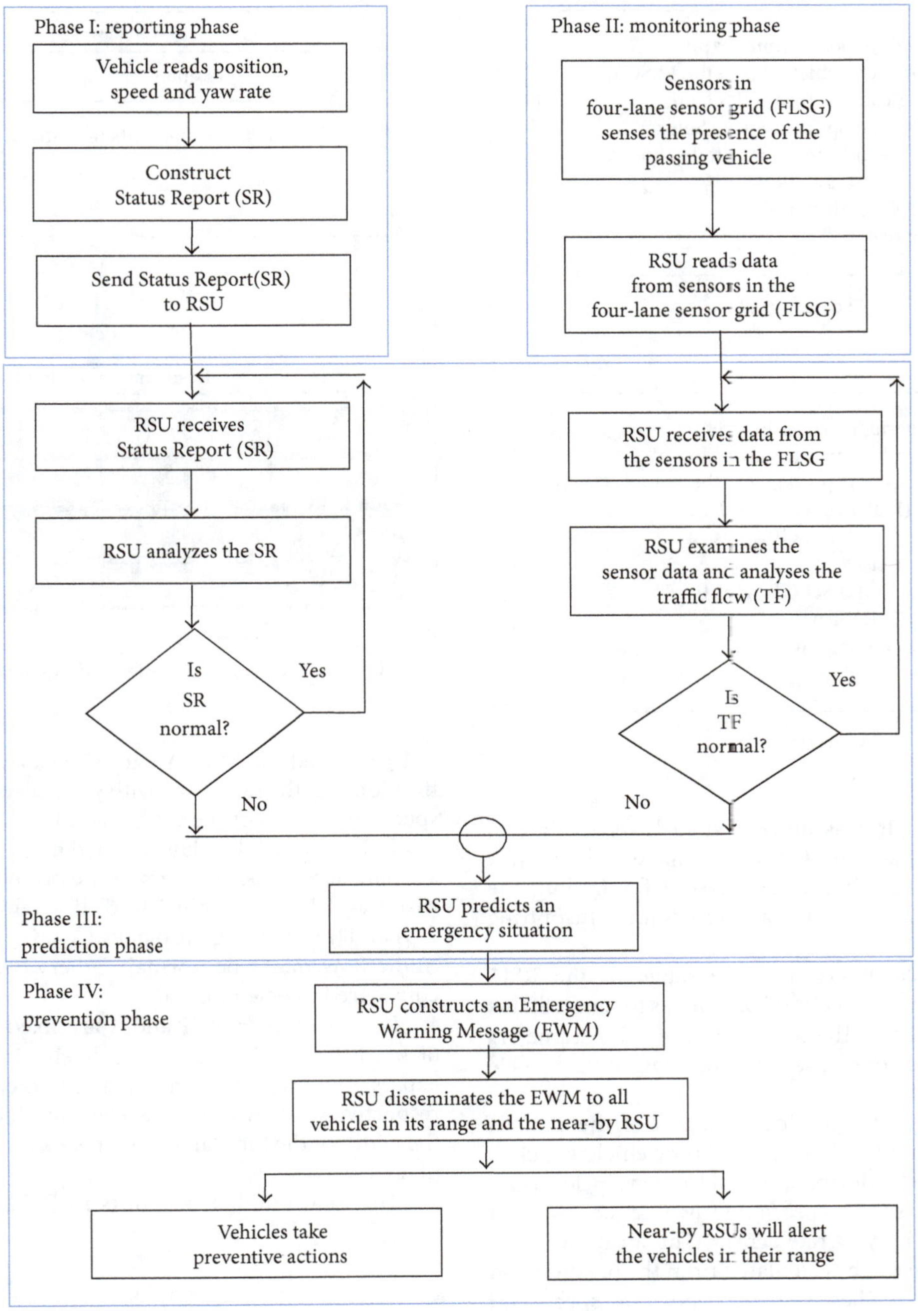

FIGURE 3: ESPM framework.

S_{MN}: road sensor in L_{MN};

Dm: decision matrix;

GP_{MN}: grid position in Dm (1 indicates presence of a vehicle);

RSU: Road Side Unit;

Periodic_Timer: set as 100 milliseconds;

Dp: decision parameter;

Td: traffic density.

Reporting Phase. See Algorithm 1.

Monitoring Phase. See Algorithm 2.

Prediction Phase

Module 1. See Algorithm 3.

Module 2. See Algorithm 4.

Module 3. See Algorithm 5.

(1) Sleep until *periodic _timer* expires
  (1) For each vehicle (V) in the FLSG do
    (1) Read Pos from GPS Unit.
    (2) Read Sp from Speed Sensor.
    (3) Read Yr from Yaw-Rate Sensor.
    (4) Construct SR with V_{ID}.
    (5) Send SR to RSU.
  (2) Return

ALGORITHM 1

(1) Sleep until *periodic _timer* expires
  (1) For each S_{MN} in L_{MN} do
    (1) Sense the presence of the Vehicle (V_{ID}).
      (1) If a vehicle is in L_{MN}
        (1) Set GP_{MN} as 1
      else
        (1) Set GP_{MN} as 0
      (2) Return
    (2) Send GP_{MN} to RSU.
  (2) Return

ALGORITHM 2

5.1. Reporting Phase. It is assumed that each vehicle (a car) is equipped with (i) sensors for measuring speed and yaw rate (ii) Global Positioning System (GPS) for finding the location of the vehicle and (iii) transceivers for transmitting and receiving data [25].

Once the vehicle enters into the range of the RSU, periodically (explained in Section 5.2.1), it has to send a Status Report (SR) to the RSU. The SR consists of the parameters such as *vehicle ID, position, speed*, and *yaw rate* of the vehicle as shown in Figure 4.

Vehicle			
ID	Position	Speed	Yaw rate

FIGURE 4: Components of Status Report (SR).

Where ID is used for the identification of the vehicle, position is the geographical position of the vehicle which is determined by the GPS device attached to the vehicle. Speed is determined by the speed sensor and yaw rate (i.e., rotation rate) is determined by yaw rate sensor. The location of the vehicle in a road lane can be calculated from the position and speed values of SR [20]. These vehicle ID, position, speed, and yaw rate are encapsulated in SR and reported to the RSU.

5.2. Monitoring Phase. In this phase, the RSU performs traffic flow monitoring based on the traffic flow data. The traffic flow data can be obtained with the help of sensors embedded in the four-lane express highway called Four-Lane Sensor Grid (FLSG).

5.2.1. Four-Lane Sensor Grid. A Four-Lane Sensor Grid (FLSG) is a setup, where each lane is divided into blocks [26] of length 6 meters as shown in Figure 5 and each block is fixed with sensors (like Sensys sensors [27]). These sensors are capable of detecting the presence and passing of vehicles in the lane. This sensor data can be passed to an Access Point (AP) attached the RSU. A pair of sensors can be installed in the blocks of the same lane with same distance to measure the speed and length of the vehicles [27].

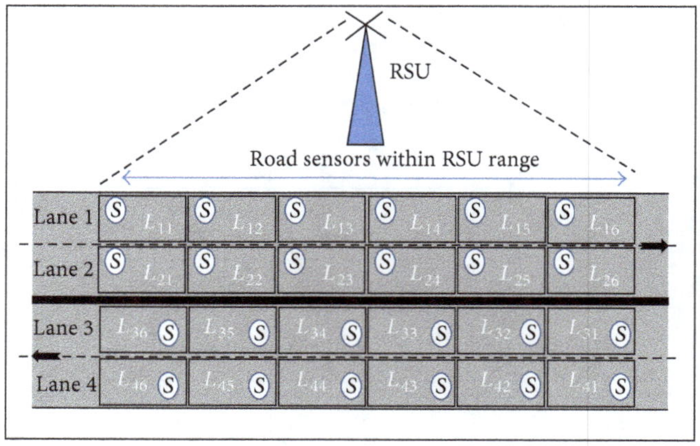

FIGURE 5: Organization of four-lane sensor grid (FLSG).

In FLSG, each block is denoted by L_{MN}, where M and N are lane and block numbers, respectively, and M lies within 1 to 4 and N lies within 1 to 6. If a vehicle enters in block L_{M1} and travels along the blocks $L_{M2}, L_{M3}, \ldots, L_{MN}$ then the traffic flow might be normal. Otherwise, the traffic flow is considered to be abnormal.

In real time, lanes 1 and 2 are fixed with travel speeds 60 km/h and 90 km/h, respectively, in Indian four-lane express highway. Lanes 1 and 2 are symmetric to lanes 4 and 3, respectively. The width of each lane in FLSG is 3.6 meters and the width of the four lanes is 17 meters including the highway divider.

The total number of blocks in FLSG can be computed using

$$\sum_{M=1}^{4} \sum_{N=1}^{6} L_{MN}. \tag{1}$$

The total number of Sensors in FLSG can be computed using

$$\sum_{M=1}^{4} \sum_{N=1}^{6} S_{MN}, \tag{2}$$

where M and N are lane and block number, respectively.

The length of blocks in each lane is assumed to be 6 meters and the minimum and maximum speed limits of the lanes are fixed to be 60 km/h and 90 km/h, respectively, in simulation. Therefore the period for sending the SR is fixed as 0.24 seconds that is the minimum period computed for the above speed limits. (The maximum period is 0.36 seconds).

(1) Sleep until *periodic _timer* expires
(2) For each vehicle (V_{ID}) in the FLSG do
    (1) Analyze the Pos, Sp and Yr.
    (2) If any of the following case arises
        (1) Case 1: pres_val (Pos) = prev_val (Pos)
        (2) Case 2: pres_val (Sp) *differs with* prev_val (Sp) by 30 Km/h
        (3) Case 3: pres_val (Yr) *differs with* prev_val (Yr) by 30 degrees.
    then
        (1) Set Dp as Abnormal.
    else
        (1) Set Dp as Normal.
(3) Return

ALGORITHM 3

(1) Sleep until *periodic _timer* expires
(2) For each S_{MN} in L_{MN} do
    (1) Read GP_{MN} value.
    (2) Construct Decision Matrix (Dm) with position values using GP_{MN}.
    (3) Analyze the Decision Matrix with Traffic Density (Td).
    (4) If V (GP_{MN}) wants to perform Lane Change
        (1) Case: Inner to Outer:
            (1) If $GP_{(M-1,N-1)}$, $GP_{(M-1,N)}$ and $GP_{(M-1,N+1)}$ values are 0.
                (1) Lane change is Normal.
                (2) Set Dp as Normal.
            (2) else
                (1) Lane change is Abnormal.
                (2) Set Dp as Abnormal
        (2) Case: Outer to Inner
            (1) If $GP_{(M+1,N-1)}$, $GP_{(M+1,N)}$ and $GP_{(M+1,N+1)}$ values are 0.
                (1) Lane change is Normal.
                (2) Set Dp as Normal.
            (2) else
                (1) Lane change is Abnormal.
                (2) Set Dp as Abnormal
(3) Return

ALGORITHM 4

In this paper, the size of FLSG is considered as 24 blocks (i.e., $M \times N$). The number of lanes can be increased dynamically up to 8 and the number of blocks can also be increased accordingly.

5.2.2. Working of Monitoring Phase. The RSU periodically monitors the traffic flow in its range. In VANET, traffic flow can be predicted using past and present traffic flow data (i.e., the location dynamics of the vehicles). Traffic flow prediction is not possible only by considering past traffic data due to on road traffic accidents, off road events, and unavailability of traffic data in all links of a traffic network because most roads/links are not equipped with traffic sensors [28]. Due to this reason, in ESPM the four-lane highway is equipped with traffic sensors [26].

RSU is programmed in such a way that it receives data from Sensys sensors [27] placed in the blocks of each lane of the four-lane highway. With these sensor data, it would be possible for the RSU to identify the presence of a vehicle in a particular block of the FLSG. Similarly all the sensors will send the sensed traffic data to the RSU. That is, if the sensor senses the presence of a vehicle then it sets the grid position (GP) value as 1. Otherwise, the sensor sets the GP value as 0.

5.3. Prediction Phase. The prediction phase is instrumental for emergency prediction based on the SR and TF data received from reporting and monitoring phases, respectively. If either SR or TF values are abnormal then RSU predicts an emergency situation (i.e., the possibility of occurrence of an accident).

5.3.1. Abnormality Based on the Status Report (SR). Once the RSU receives the SR, it checks whether speed, position, and yaw rate are within the specified limits. If any of the following cases arise then RSU concludes that the SR is abnormal.

(1) Sleep until *periodic_timer* expires
(2) Read Dp value in Module 1 and Module 2
    (1) If Dp is Abnormal then an Emergency Situation is predicted and report for prevention.
(3) Return

Algorithm 5

Case 1. If the position of the vehicle is the same in two continuous time periods then the vehicle is in halt state as represented by the following equation:

$$\text{Pos}(t_i) - \text{Pos}(t_{i-1}) = 0. \quad (3)$$

Case 2. If the difference in speed of the vehicle in two continuous time periods is greater than 30 km/h then the vehicle is with abnormal speed represented as

$$\begin{gathered} \text{Sp}(t_i) - \text{Sp}(t_{i-1}) > 30\,\text{km/h}, \\ \text{Sp}(t_{i-1}) - \text{Sp}(t_i) > 30\,\text{km/h}. \end{gathered} \quad (4)$$

Case 3. If the vehicle's change in yaw rate exceeds 30 degrees then the vehicle has performed an abnormal lane change represented as

$$\begin{gathered} \text{Yr}(t_i) - \text{Yr}(t_{i-1}) > 30\text{ degrees}, \\ \text{Yr}(t_{i-1}) - \text{Yr}(t_i) > 30\text{ degrees}. \end{gathered} \quad (5)$$

In (3), (4), and (5), Pos denote the position of the vehicle, Sp denotes the speed of the vehicle, and Yr denotes yaw rate of the vehicle and t_i is time.

5.3.2. Abnormality Based on the Traffic Flow or Sensor Data (TF). The RSU periodically receives the traffic data (i.e., GP value) through each sensor in FLSG; it transforms the traffic data into a decision matrix (DM) for traffic flow analysis as shown in Figure 6. The decision matrix is playing an important role in the prediction process.

If all the positions of DM are filled with 1s, then it means that the highway is filled with vehicles and the Traffic Density (TD) is high. If all the positions of DM are filled with 0s then the highway is free and TD is low. Three scenarios of abnormality are identified and explained below.

In FLSG, the sensors are fixed at the entry points of all the blocks. If a vehicle enters the block, immediately the sensor senses and reports to the RSU. The RSU records this by placing a "1" in the respective position of the decision matrix. The block size of FLSG is designed in such a way that, once a vehicle enters into a block means it automatically leaves the previous block (i.e., the sensor area in the previous block). Hence, if a vehicle is partially in two blocks the preceding sensor alone reports to the RSU about the presence of the vehicle.

Scenario 1 (abnormal deceleration of the vehicle). Once the speed of the moving vehicle decreases to 0 Km/h (i.e., deceleration to 0 Km/h) (or) the position of the moving vehicle remains the same for two continuous time periods then it is concluded that the vehicles resides in the same position (or) stops travelling.

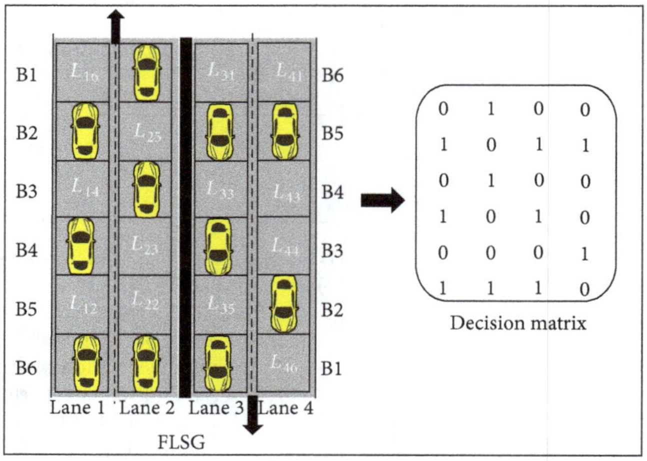

Figure 6: Transformation of four-lane sensor grid into decision matrix.

If a set of vehicles travelling in a lane one after the other, sudden stoppage of a vehicle might cause rear end collision [9, 29] as shown in Figure 7. Major mechanical problems, driver illness, and improper driving might cause the vehicle to suddenly stop [24].

A vehicle in L_{15} of DM has abnormally decelerated; as a cause, vehicles in L_{14}, L_{13}, and L_{24} have collided (i.e., rear end collision) as shown in Figure 7. The computation of ESPM is based on the values of the decision matrix. With respect to DM, the vehicle representing the value of GP_{15} halts; hence the vehicles representing the values of GP_{13}, GP_{14}, and GP_{24} collides.

In DM, if a particular column is filled with continuous 1s then the probability of an emergency situation is much higher and the change in GP values of DM need to be closely monitored against abnormality.

Scenario 2 (abnormal acceleration of the vehicle). If the moving vehicle suddenly accelerates its speed (or) the speed of the vehicle crosses the maximum lane speed limit then the vehicle might collide with the one before it (i.e., front end collision) [9]. This scenario might sometimes cause a series of front end collisions.

In case vehicles are moving closely one after the other, if a following vehicle suddenly accelerates its speed abnormally then the following vehicle collides with the one in front (front

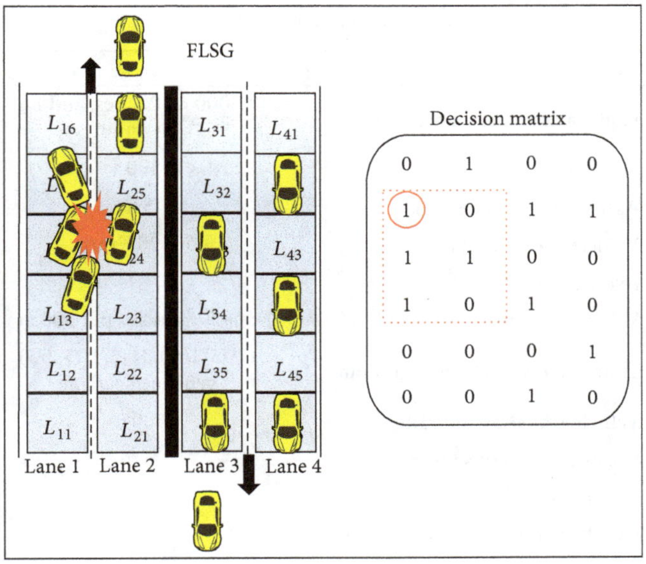

FIGURE 7: Scenario of abnormal deceleration with respective decision matrix.

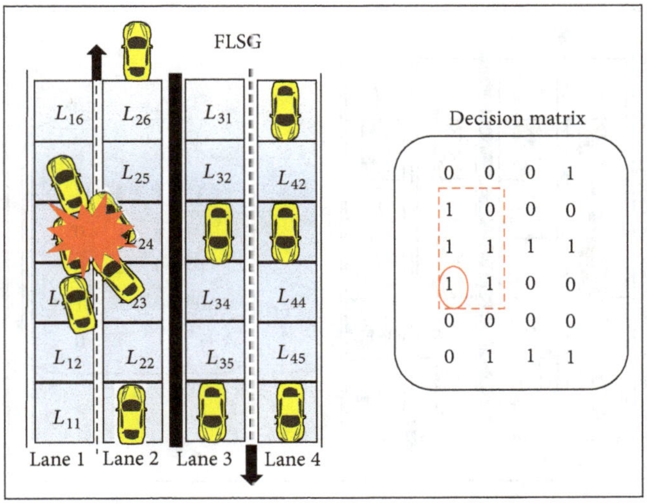

FIGURE 8: Scenario of abnormal acceleration with respective decision matrix.

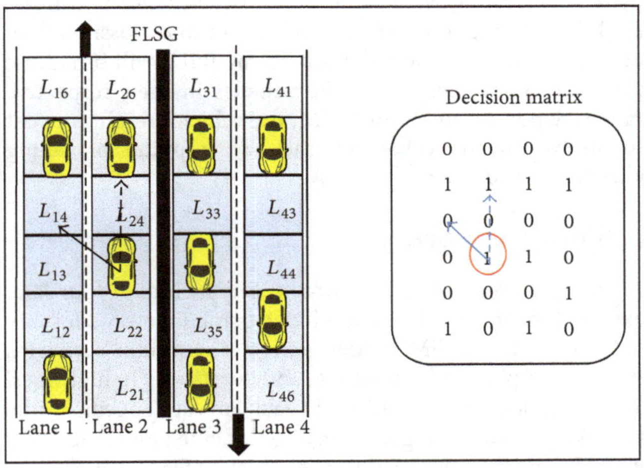

FIGURE 9: Abnormal inner to outer lane change with respective decision matrix.

end collision) and this might result in a chain of collisions as shown in Figure 8. The vehicle in L_{13} has accelerated abnormally which results in collision (front end collision) with the vehicles in L_{14}, L_{15}, L_{23}, and L_{24}. With respect to DM, the vehicle representing the value of GP_{13} has increased the speed abnormally. Hence the vehicles representing the values of GP_{14}, GP_{15}, GP_{23}, and GP_{24} collide.

Scenario 3 (abnormal lane change). The lane change the vehicle is normal only if the following condition is satisfied. Otherwise, the lane change is abnormal:

$$1 \geq V(M) \leq \frac{M_{\text{Max}}}{2}, \tag{6}$$

where M is the lane number, V is the vehicle, $M_{\max}$ is the maximum lane number in the highway, and $V(M)$ is vehicle's present lane.

The ESPM is designed for a four-lane express highway; the lane change is possible only between lane 1 and lane 2 in one direction and between lane 3 and lane 4 in the other direction (symmetric to the other direction). At par with expression (6), the vehicle can change its lane from lane 1 to lane 2 and vice versa. But a vehicle in lane 2 cannot perform lane change to either lane 3 or lane 4 because it causes head on accidents. Similarly a vehicle in lane 3 cannot perform lane change to either lane 1 or lane 2.

Two types of lane changes are possible such as inner to outer (ITO) and outer to inner (OTI). Consider the present position of the vehicle in DM as GP_{MN}; a smooth or normal ITO lane change is possible only if the position values $(M - 1, N - 1)$, $(M - 1, N)$, and $(M - 1, N + 1)$ of DM are zero. Figure 9 shows that a vehicle in L_{23} wants to change from lane 2 to lane 1. This lane change is normal only if GP_{12}, GP_{13}, and GP_{14} values of DM are zero.

Similarly, a smooth or normal OTI lane change is possible only if the position values $(M + 1, N - 1)$, $(M + 1, N)$, and $(M + 1, N + 1)$ of DM are zero. Figure 10 shows that a vehicle in L_{13} wants to change lane from lane 1 to lane 2. This lane change is normal only if GP_{22}, GP_{23}, and GP_{24} values of DM are zero.

Otherwise, both the lane changes would be abnormal and this leads to an emergency situation.

If any of the above three scenarios arise then it could be concluded that there is a possibility of occurrence of an accident and the prediction is successful (i.e., an emergency situation is predicted).

5.4. Prevention Phase. This prevention phase is purely dependent on the prediction phase. Once the emergency situation is predicted by the prediction phase, then the RSU constructs an Emergency Warning Message (EWM). This EWM is

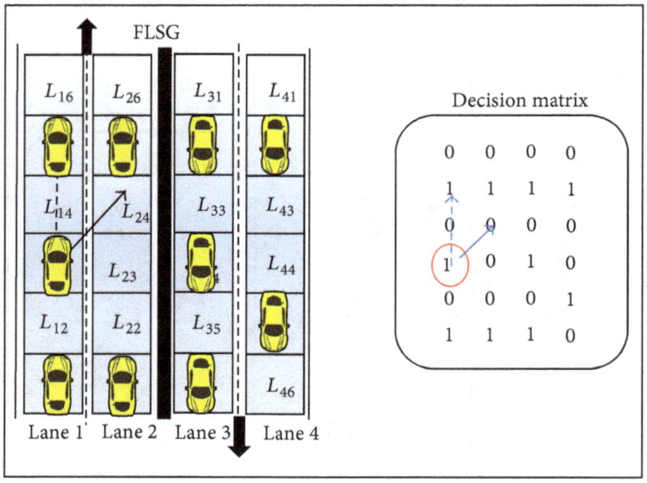

FIGURE 10: Abnormal outer to inner lane change with respective decision matrix.

then broadcasted by the RSU to all the vehicles and near-by RSUs in order to alert them about the possibility of occurrence of an accident. The near-by RSUs will broadcast this emergency message to the vehicles in their respective ranges. Upon reception of EWM, the vehicles might prevent themselves from accidents either by stopping or taking alternate lanes/routes.

6. Simulation Setup

A mobility scenario for highway traffic in Indian four-lane express highway has been developed using Freeway Mobility (FM) model. The FM model generates common scenarios such as stopping, lane changing, and overtaking in highways. The FM model injects vehicles in each lane at a specific traffic rate. As specified in Figure 2, vehicles of lane 1 and lane 2 will move in one direction against the vehicles of lane 3 and lane 4 in the opposite direction. During simulation the vehicles are placed randomly in four lanes. The speed of each vehicle in FM model [30] is based on the following:

$$\mathrm{Sp}\left(t_{i+1}\right) = \mathrm{Sp}\left(t_i\right) + \mathrm{random}\left(\right) \times A\left(t_i\right). \tag{7}$$

The FM model does not allows lane change; hence (7) has been changed with yaw rate and modified as

$$\mathrm{Sp}\left(t_{i+1}\right) = \left[\mathrm{Sp}\left(t_i\right) + \mathrm{random}\left(\right) \times A\left(t_i\right)\right] \times \mathrm{ROT}, \tag{8}$$

where $\mathrm{Sp}(t_i)$ is the speed of the vehicle at time "t," random() provides a random value in between $[-1, 1]$ used for both acceleration and deceleration of vehicles, and $A(t_i)$ is the vehicle's acceleration at time "t" and ROT is Radius Of Turn [31] and can be expressed as

$$\mathrm{ROT} = \frac{\mathrm{Sp}\left(t_i\right)}{\mathrm{Yr}}, \tag{9}$$

where Yr is yaw rate at maximum and expressed as

$$\mathrm{Yr} = \frac{\left(2\mathrm{Sp}\left(t_i\right)\right) \times \sin\theta}{d}, \tag{10}$$

where d is the distance of travel [32].

TABLE 1: Simulation parameters.

Parameter	Value
Highway scenario dimension	1000 m, 2 lanes, and 1 direction
Simulation time	300 s
Assumed vehicle length	5 m
Vehicle speed limits	60–90 Km/h
Average speed	70 Km/h
Vehicle density	50–70 vehicles per 500 m
Traffic rate/vehicle injection rate	(1/75, 1/60, 1/45, 1/15, 1/5) vehicles/sec/lane
Vehicle transmission range	500 m
Sensing radius of sensors	6 m
Number of vehicles	150
Number of road sensors	200
Number of lanes	2 lanes
Distance between two sensors in a lane	6 m
Distance between two sensors in two neighboring lanes	3.6 m
Period of message exchange	0.10 second (i.e., 100 milliseconds)

Network Simulator (ns 2.34) is used for simulation. The mobility traces of the scenarios (i) sudden stoppage (deceleration), (ii) sudden increase in speed (acceleration), and (iii) abnormal lane change of vehicles are developed using the FM model.

The simulation is performed on two layers of nodes. To simulate the fixed RSU and FLSG sensors nodes one (fixed) layer of nodes are used. The second layer of nodes is the moving vehicles whose mobility traces are opted from the FM model. The main simulation parameters are listed in Table 1. As two lanes in four-lane highway are symmetric to the other two lanes in the opposite direction, only two lanes are considered for simulation.

The moving vehicles will send SR to the fixed RSU nodes once in every 100 milliseconds (lesser that the required 240 milliseconds as explained in Section 5.2.1). Similarly the fixed sensor nodes in FLSG will send the data about presence of the vehicle in its coverage.

Based on both the SR and sensor data, the RSU node will analyze the abnormal behavior of vehicles and arrive at a conclusion about the possibility of occurrence of an accident (i.e., prediction). The vehicle might behave abnormally in any of the scenarios as mentioned in the previous Section 5.

7. Results

The results demonstrate the efficiency of the ESPM framework for emergency situation prediction. Prediction probability and vehicle density are selected as performance metrics for evaluating the performance of ESPM.

In order to demonstrate the accuracy of the ESPM approach the following three scenarios were simulated.

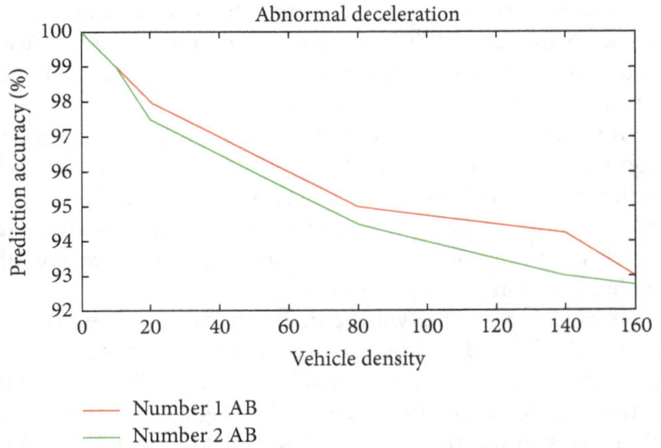

FIGURE 11: Prediction accuracy versus vehicle density for Scenario 1.

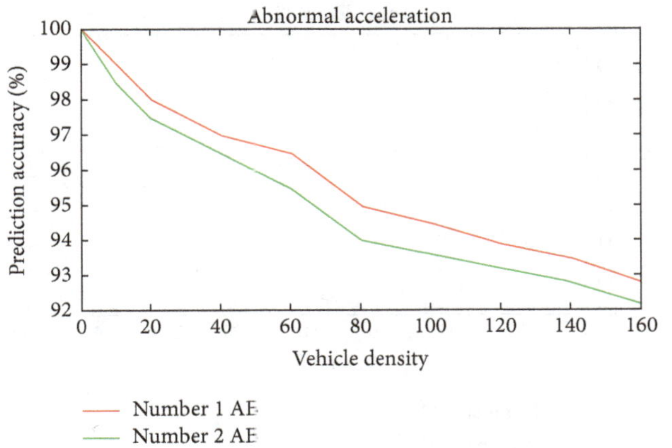

FIGURE 12: Prediction accuracy versus vehicle density for Scenario 2.

Scenario 1 (abnormal deceleration of the vehicle (sudden stoppage of a moving vehicle)). In this scenario, initially a moving vehicle (mobile node) in a lane has been forced to stop by making its speed reach 0 km/h instantly and the reaction in the system has been monitored. After receiving the SR of the respective vehicle, the RSU node has reported an abnormality within the specified time period and a successful prediction has been achieved.

The simulation was repeated for 10 runs by gradually increasing the vehicle density and randomly vehicles in the two lanes were made to stop suddenly. The reaction of both the vehicles and RSU has been monitored for accuracy and performance of ESPM. In Figure 11, the simulation results, prediction accuracy (PA) against vehicle density (VD) for one and two abnormal behaviors (AB) are above 92 percent.

Scenario 2 (abnormal acceleration of the vehicle (sudden increase in speed)). In this scenario, among the moving vehicles (mobile nodes) in two lanes a vehicle was forced to increase its speed to cross the maximum limit of 90 km/h and the systems' reaction has been monitored. The RSU node has reported an abnormality within the specified time period after reception of SR from the respective vehicle. Similarly the speed of the vehicles in two lanes was increased and the reaction of the RSU node has been monitored.

This simulation was repeated for 10 runs by gradually increasing the vehicle density (i.e., number of vehicles). The vehicle behavior and reaction of RSU was closely monitored for assessment of performance and the same is shown in Figure 12.

The simulation result shows that the prediction accuracy is close and above 92 percent both abnormal behaviors (AB). Comparing both Figures 11 and 12, prediction accuracy in Scenario 2 for two abnormal behaviors is slightly decreasing when comparing to Scenario 1.

Scenario 3 (abnormal lane change (inner to outer and outer to inner)). During simulation in this scenario, vehicles were made to change their lanes from inner to outer and outer to inner in different runs.

(a) A vehicle was allowed to perform a normal inner to outer (ITO) lane change and forced to perform an abnormal inner to outer lane change as explained in Section 5.

(b) Similar to the previous case a vehicle was allowed to perform a normal outer to inner (OTI) lane change and forced to perform an abnormal outer to inner lane change as explained in Section 5. Then the reaction of both the sensor nodes and RSU has been monitored. After receiving data from the sensor nodes, the RSU computed normal and abnormal lane changes within the specified time period. In the case of abnormality, RSU records the abnormality to report successful prediction of an emergency situation.

The prediction performance of ESPM in this scenario is promising and this is shown in Figures 13 and 14 for inner to outer (ITO) and outer to inner (OTI) lane changes, respectively. During simulation one and two abnormal lane changes (ABLC) were induced to measure the accuracy of prediction.

Figure 13 shows that the PA in Scenario 3(a) (ITO) is almost close to 91 percent for both one and two abnormal behaviors, respectively, whereas the prediction accuracy of Scenario 3(b) (OTI) is decreasing when comparing the previous one as in Figure 14.

The speed limit of inner lane is greater than the speed limit of outer lane. This leads to the increase of complexity in OTI lane change when comparing to ITO lane change. Due to this reason the prediction accuracy of OTI lane change might decrease.

A vehicle attempting for OTI lane change must increase its speed from 60 km/h range to 90 km/h range in order to meet the speed of vehicles in the inner lane. At par with (8) the speed of the vehicle performing lane change should cope up with the yaw rate and lane speed limit.

In all the three scenarios, the simulation results show that the successful prediction achieved within 0.2 seconds. This is much better than the statistical requirements of 0.5 seconds [7]. It is further noticed that if the vehicle density

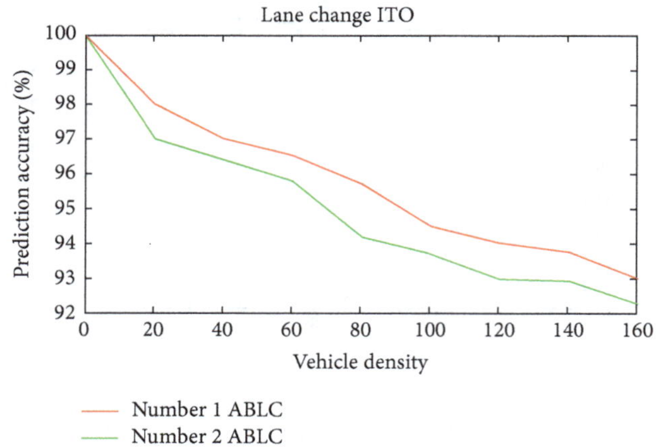

FIGURE 13: Prediction accuracy versus vehicle density for Scenario 3(a).

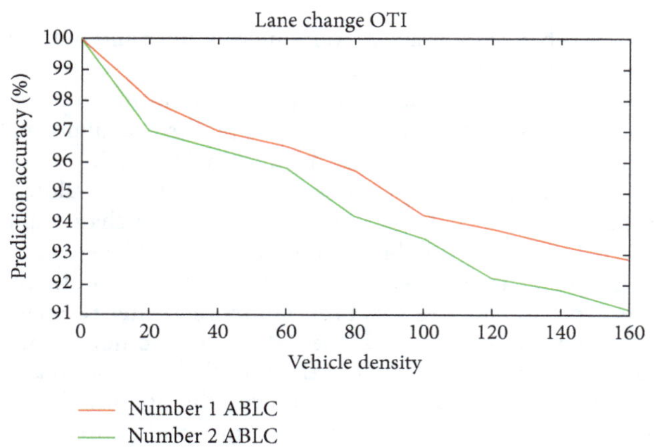

FIGURE 14: Prediction accuracy versus vehicle density for Scenario 3(b).

is less than the prediction accuracy is high but when the vehicle density increases the prediction accuracy decreases to close and above 92 percent. The average vehicle density is around 60 vehicles/second/lane in Indian four-lane express highways; hence this level of accuracy is acceptable to these road conditions.

8. Conclusion and Future Work

In this paper, an Emergency Situation Prediction Mechanism (ESPM) is proposed to predict the possibility of occurrence of an accident in Indian four-lane express highway. The primary objective of ESPM is to predict an emergency situation in advance, thereby preventing accidents and reducing the death toll. ESPM is used to perform prediction of an emergency situation in four phases. The first three phases (reporting, monitoring and prediction phases) are used for prediction and the fourth phase is used for prevention. In ESPM, the prediction accuracy is computed against vehicle density for three different scenarios. In all the three scenarios, it is noticed that there is a small gap between the analytical evaluation and simulation results. The results show that the performance of ESPM is promisingly increased towards prediction. Prediction accuracy of ESPM against vehicle density is almost above 92 percent.

Implementation of ESPM for Indian four-lane express highways is expensive because of the development and deployment of infrastructures such as FLSG and RSU even though most of the present day vehicles are automated vehicles. But in near future ESPM has a high scope of implementation for improving ITS.

In future, attempts will be made to improve the performance of ESPM. Based on prediction of emergency situation the prevention of accidents would be done by disseminating distress beacons (i.e., Emergency Warning Message) to alert all vehicles in the range of the RSU and the vehicles in the range of the near-by RSUs to prevent from accidents.

Conflict of Interests

The authors declare that there is no conflict of interests regarding the publication of this paper.

References

[1] NCRB, 2013, http://ncrb.gov.in/adsi2013/accidental%20deaths%202013.pdf.

[2] W. H. Who, World Health Organization, 2015, http://www.who.int/roadsafety/decade_of_action/plan/en/index.html.

[3] 2015, http://wardsauto.com/ar/world_vehicle_population_110815.

[4] Worldometers, 2013, http://www.worldometers.info/.

[5] F. N. Ogwueleka, S. Misra, T. C. Ogwueleka, and L. Fernandez-Sanz, "An artificial neural network model for road accident prediction: a case study of a developing country," *Acta Polytechnica Hungarica*, vol. 11, no. 5, pp. 177–197, 2014.

[6] 2015, http://nidm.gov.in/idmc2/PDF/Presentations/road_accidents/Pres4.pdf.

[7] C. D. Wang and J. P. Thompson, "Apparatus and method for motion detection and tracking of objects in a region for collision avoidance utilizing a real-time adaptive probabilistic neural network," US Patent no. 5,613,039, 1997.

[8] 2015, http://www.civil.iitb.ac.in/tvm/1111_nptel/553_MlHway/plain/plain.html.

[9] V. Milanés, J. Pérez, J. Godoy, and E. Onieva, "A fuzzy aid rear-end collision warning/avoidance system," *Expert Systems with Applications*, vol. 39, no. 10, pp. 9097–9107, 2012.

[10] K. N. Qureshi and A. H. Abdullah, "A survey on intelligent transportation systems," *Middle East Journal of Scientific Research*, vol. 15, no. 5, pp. 629–642, 2013.

[11] S. Panichpapiboon and W. Pattara-Atikom, "A review of information dissemination protocols for vehicular ad hoc networks," *IEEE Communications Surveys and Tutorials*, vol. 14, no. 3, pp. 784–798, 2012.

[12] K. A. Hafeez, L. Zhao, B. Ma, and J. W. Mark, "Performance analysis and enhancement of the DSRC for VANET's safety applications," *IEEE Transactions on Vehicular Technology*, vol. 62, no. 7, pp. 3069–3083, 2013.

[13] C. Ünsal, *Intelligent navigation of autonomous vehicles in an automated highway system: learning methods and interacting vehicles approach [Ph.D. thesis]*, Virginia Polytechnic Institute and State University, Blacksburg, Va, USA, 1998.

[14] G. Samara, T. Alhmiedat, and A. O. Abu Salem, "Dynamic safety message power control in VANET using PSO," *World of Computer Science and Information Technology Journal*, vol. 3, no. 10, pp. 176–184, 2013.

[15] K. Bilstrup, A. Böhm, K. Lidström, M. Jonsson, and T. Larsson, "VEHICLE ALERT SYSTEM," http://www.diva-portal.org/smash/get/diva2:239391/FULLTEXT01.pdf.

[16] H. Cheng, N. Zheng, X. Zhang, J. Qin, and H. van de Wetering, "Interactive road situation analysis for driver assistance and safety warning systems: framework and algorithms," *IEEE Transactions on Intelligent Transportation Systems*, vol. 8, no. 1, pp. 157–166, 2007.

[17] C. Caliendo, M. Guida, and A. Parisi, "A crash-prediction model for multilane roads," *Accident Analysis and Prevention*, vol. 39, no. 4, pp. 657–670, 2007.

[18] W. Hu, X. Xiao, D. Xie, T. Tan, and S. Maybank, "Traffic accident prediction using 3-D model-based vehicle tracking," *IEEE Transactions on Vehicular Technology*, vol. 53, no. 3, pp. 677–694, 2004.

[19] J. Sörstedt, L. Svensson, F. Sandblom, and L. Hammarstrand, "A new vehicle motion model for improved predictions and situation assessment," *IEEE Transactions on Intelligent Transportation Systems*, vol. 12, no. 4, pp. 1209–1219, 2011.

[20] S. Worrall, G. Agamennoni, J. Nieto, and E. Nebot, "A context-based approach to vehicle behavior prediction," *IEEE Intelligent Transportation Systems Magazine*, vol. 4, no. 3, pp. 32–44, 2012.

[21] J. De Oña and L. Garach, "Accidents prediction model based on speed reduction on Spanish two-lane rural highways," *Procedia—Social and Behavioral Sciences*, vol. 53, pp. 1010–1018, 2012.

[22] E. Garcia-Lozano, C. T. Barba, M. A. Igartua, and C. Campo, "A distributed bandwidth-efficient accident prevention system for interurban VANETs," in *Proceedings of the International Conference on Smart Communications in Network Technologies (SaCoNeT '13)*, pp. 1–5, Paris, France, June 2013.

[23] S. Rokade, K. Singh, S. K. Katiyar, and S. Gupta, "Development of accident prediction model," *International Journal of Advanced Engineering Technology*, vol. 1, no. 3, pp. 25–40, 2010.

[24] S. Saravanan, A. Sabari, and M. Geetha, "Fuzzy-based approach to predict accident risk on road network," *International Journal of Emerging Technology and Advanced Engineering*, vol. 4, no. 5, pp. 536–540, 2014.

[25] X. Yang, J. Liu, F. Zhao, and N. H. Vaidya, "A vehicle-to-vehicle communication protocol for cooperative collision warning," in *Proceedings of the 1st Annual International Conference on Mobile and Ubiquitous Systems: Networking and Services (MOBIQUITOUS '04)*, pp. 114–123, August 2004.

[26] A. Ahmed, D. Watling, and D. Ngoduy, "Significance of sensor location in real-time traffic state estimation," *Procedia Engineering*, vol. 77, pp. 114–122, 2014, Proceedings of the 4th International Symposium on Infrastructure Engineering in Developing Countries (IEDC '13).

[27] A. Haoui, R. Kavaler, and P. Varaiya, "Wireless magnetic sensors for traffic surveillance," *Transportation Research Part C: Emerging Technologies*, vol. 16, no. 3, pp. 294–306, 2008.

[28] A. Abadi, T. Rajabioun, and P. A. Ioannou, "Traffic flow prediction for road transportation networks with limited traffic data," *IEEE Transactions on Intelligent Transportation Systems*, vol. 16, no. 2, pp. 653–662, 2014.

[29] K. G. Manjunath and N. Jaisankar, "A survey on rear end collision avoidance system for automobiles," *International Journal of Engineering and Technology*, vol. 5, no. 2, pp. 1368–1372, 2013.

[30] G. Grilli, *Data dissemination in vehicular networks [Ph.D. dissertation]*, University of Rome "Tor Vergata", Rome, Italy, 2010.

[31] https://www.vboxautomotive.co.uk/downloads/Calculating%20Radius%20of%20Turn%20from%20Yaw.pdf.

[32] T. Gordon, A. Blankes Poor, M. Barnes, D. Blower, P. Green, and L. Kostyniuk, "YAW RATE ERROR—a dynamic measure of lane keeping control performance for the retrospective analysis of naturalistic driving data," 2015, http://www-nrd.nhtsa.dot.gov/pdf/esv/esv21/09-0326.pdf.

18

Experimental Evaluation of Cooperative Relative Positioning for Intelligent Transportation System

Suhua Tang,[1,2] Nao Kawanishi,[2] Rei Furukawa,[2,3] and Nobuaki Kubo[4]

[1] *Graduate School of Informatics and Engineering, The University of Electro-Communications, 1-5-1 Chofugaoka, Chofu, Tokyo 182-8585, Japan*
[2] *Advanced Telecommunications Research Institute International, 2-2-2 Hikaridai, Seika-cho, Soraku-gun, Kyoto 619-0288, Japan*
[3] *Kozo Keikaku Engineering Inc., 4-5-3 Chuo, Nakano-ku, Tokyo 164-0011, Japan*
[4] *Tokyo University of Marine Science and Technology, 2-1-6 Etchujima, Koto-ku, Tokyo 135-8533, Japan*

Correspondence should be addressed to Suhua Tang; shtang@uec.ac.jp

Academic Editor: Yuei-An Liou

Support system for safe driving heavily depends on global navigation satellite system. Pseudoranges between satellites and vehicles are measured to compute vehicles' positions and their relative positions. In urban areas, however, multipath errors (MPEs) in pseudoranges, caused by obstruction and reflection of roadside buildings, greatly degrade the precision of relative positions. On the other hand, simply removing all reflected signals might lead to a shortage of satellites in fixing positions. In our previous work, we suggested solving this dilemma by cooperative relative positioning (CoRelPos) which exploits spatial correlation of MPEs. In this paper, we collected the trace data of pseudoranges by driving cars in urban areas, analyzed the properties of MPEs (specifically, their dependency on signal strength, elevation angles of satellites, and receivers' speeds), and highlighted their spatial correlation. On this basis, the CoRelPos scheme is refined by considering the dynamics of MPEs. Evaluation results under practical vehicular scenarios confirm that properties of MPEs can be exploited to improve the precision of relative positions.

1. Introduction

The rapid and wide spread of motor vehicles after World War II has greatly changed human society. On one hand, it frees common people from the limitations of their geography and facilitates remote travel. On the other hand, it also brings some undesirable consequences such as vehicle accidents, which often lead to severe body injuries or even the loss of precious lives. To reduce vehicle accidents, many efforts have been devoted to support systems for safe driving, from two aspects. (i) Each vehicle independently detects the presence of nearby vehicles by using cameras or radars. But this fails to work when the line of sight (LOS) path between vehicles is obstructed. (ii) Vehicles cooperate to measure their distances, which relies on global navigation satellite system (GNSS) [1] and intervehicle communications (IVC) [2]. In this system, each vehicle receives positioning signals from satellites, measures the pseudoranges between itself and satellites, and on this basis computes its own position. Each vehicle further exchanges its own position and moving speed with nearby vehicles via IVC. Based on current positions and moving speeds of two vehicles, their distance (*relative position*) several seconds later is estimated, and the driver is warned in case a collision is predicted to happen. Such applications have been extensively evaluated in the ASV4 (advanced safety vehicle) experiments in Japan [3]. In our previous work, we found that although the real-time delivery of position information can be realized [4], the precision of position remains a big problem, especially in urban areas. How to improve the precision of relative position is the focus of this paper.

Positioning precision of GNSS might be degraded by several factors: clock and orbital errors of GNSS satellites, propagation delay caused by the atmosphere (ionosphere, troposphere), obstruction and reflection of roadside buildings, and clock error and thermal noise at a GNSS receiver.

Among these factors, propagation delay due to atmosphere is known to have spatial correlation and can be mitigated by ground based augmentation systems such as DGPS (differential global positioning system) [1] or space-based augmentation systems (SBAS).

Accurately measuring pseudoranges between vehicles and satellites usually requires a LOS path. In urban areas, however, the LOS path might be obstructed by roadside buildings, and then a GNSS receiver receives a reflected signal instead. The extended propagation length of a positioning signal due to reflections, compared with the LOS path, is called a *multipath error* (MPE). MPE, which only occurs in vehicles but not in reference stations installed in an open sky environment, can hardly be mitigated by DGPS and remains the largest error source in urban areas. Therefore, many efforts have been devoted to remove reflected signals before fixing absolute positions, including antenna design, correlator refinement (narrow correlator [5], early-late slope correlator [6], and strobe correlator [7]), modulation design (binary offset carrier in modern GPS [8]), carrier smoothing (Hatch filter [9]), signal separation (multipath estimating delay lock loop [10], spatial sampling via antenna array [11]), and detection of LOS path (using a 3D GIS database [12] or an omnidirectional infrared camera [13]). Removing all reflected signals, however, might lead to a shortage of satellites in fixing positions because few satellites are directly visible to vehicles in urban canyons.

Relative position between two vehicles plays a more important role than their *absolute* positions in avoiding collision accidents. In addition, computing relative positions needs the cooperation between vehicles. Vehicles, forming a network, can cooperate to improve positioning precision [14], using both intra- and intervehicle measurements. The latter requires a ranging capability in IVCs, which is absent in vehicular communication protocols such as 802.11p [15]. Instead of directly exchanging position information, it is suggested in [16] that each vehicle exchange its measured pseudoranges with nearby vehicles so that relative position is computed by using positioning signals from common satellites. But how to handle MPEs and how to deal with GNSS outage are left untouched. A GNSS outage occurs when the number of satellites is insufficient for fixing positions. This problem is usually solved by using an inertial navigation system (INS) with a gyroscope and accelerometers [17]. It is also possible to realize a tight integration of GNSS and INS by a single Kalman filter [18, 19]. As for vehicular networks, a car speed can be estimated from speed pulses instead of using accelerometers.

In our previous work [20], we assumed local *spatial correlation* of MPEs by the following reasoning: usually higher precision of relative positions is required when intervehicle distances get shorter; on the other hand, nearby receivers have similar spatial propagation environments and their MPEs are correlated; for example, MPEs in the pseudoranges at car A and car B in Figure 1 are almost the same. On this basis, we extended the DGPS concept and suggested a cooperative relative positioning (CoRelPos) scheme, using correlated information from common satellites, including reflected signals, to accurately compute relative positions. Initial results,

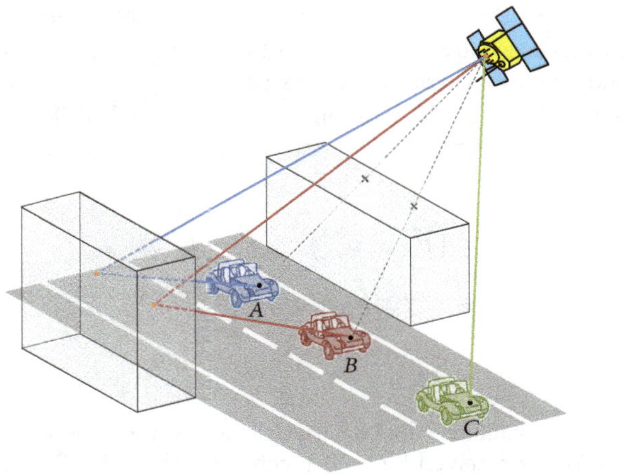

FIGURE 1: Correlation in MPEs of pseudoranges. MPEs at A and B are almost the same and are regarded as being spatially correlated.

where two receivers were installed on stationary cars or on one moving car, were reported.

In this work, we installed GNSS receivers on different cars and collected the trace data of pseudoranges by driving cars in urban areas. Based on these experimental data, we analyzed the properties of MPEs (specifically, their dependency on signal strength, elevation angles of satellites, and receivers' moving speed) and, most importantly, highlighted their *spatial correlation*. On this basis, the CoRelPos scheme is refined by considering the dynamics of MPEs. Evaluation results under practical vehicular scenarios confirm that properties of MPEs can be exploited to improve the precision of relative positions.

The rest of this paper is organized as follows: Section 2 briefly introduces the CoRelPos scheme [20]. Section 3 describes the experiment setup. Then, experimental results of MPEs are analyzed in Section 4. On this basis, the CoRelPos scheme is refined in Section 5, and positioning results are presented in Section 6. Finally, we conclude the paper with Section 7.

2. Brief Introduction of the CoRelPos Scheme

The basic idea of CoRelPos [20] is to use all correlated signals at two vehicles, including reflected ones, in fixing their relative position. To this end, each vehicle measures pseudoranges and exchanges with nearby vehicles via IVCs. Then, each vehicle selects signals with high correlation and uses them to compute relative positions.

A pseudorange $p_n^{(s)}$ between satellite $s = k, l$ and receiver $n = a, b$ is expressed as [1]

$$p_n^{(s)} = \rho_n^{(s)} + c \cdot \left(\Delta t_n - \Delta T^{(s)}\right) + d_{\text{ion},n}^{(s)} + d_{\text{trop},n}^{(s)} + m_n^{(s)} + \varepsilon_n^{(s)}, \quad (1)$$

where $\rho_n^{(s)}$ is the real distance between s and n, c is light speed, Δt_n and $\Delta T^{(s)}$ are clock errors of n and s, $d_{\text{ion},n}^{(s)}$ and $d_{\text{trop},n}^{(s)}$ are the delay effects of ionosphere and troposphere,

$m_n^{(s)}$ is a MPE, and $\varepsilon_n^{(s)}$ is the effect of thermal noise. When two receivers are close to each other, common errors ($\Delta T^{(s)}$) and spatially correlated errors ($d_{\text{ion},n}^{(s)}$, $d_{\text{trop},n}^{(s)}$) are cancelled out by differential operations. Then, the double difference of pseudoranges between two satellites k, l and two receivers a, b can be approximated as

$$\begin{aligned} p_{ab}^{(kl)} &= \left(p_a^{(k)} - p_b^{(k)}\right) - \left(p_a^{(l)} - p_b^{(l)}\right) \\ &\approx \rho_{ab}^{(kl)} + \left(m_a^{(k)} - m_b^{(k)}\right) - \left(m_a^{(l)} - m_b^{(l)}\right) \\ &\quad + \left(\varepsilon_a^{(k)} - \varepsilon_b^{(k)}\right) - \left(\varepsilon_a^{(l)} - \varepsilon_b^{(l)}\right). \end{aligned} \tag{2}$$

Assume satellite l has a large elevation angle so its signal is directly received by both receivers without MPEs; that is, $m_a^{(l)} = m_b^{(l)} = 0$. Then, the MPE difference with respect to satellite k can be estimated by

$$m_a^{(k)} - m_b^{(k)} \approx p_{ab}^{(kl)} - \rho_{ab}^{(kl)}. \tag{3}$$

$m_a^{(k)}$ and $m_b^{(k)}$ are regarded as being *spatially correlated* if $|p_{ab}^{(kl)} - \rho_{ab}^{(kl)}|$ is less than a threshold. When computing $\rho_{ab}^{(kl)}$, real positions of two receivers are required to find the true distances $\rho_n^{(s)}$. In the actual protocol, a Kalman filter [21] is exploited to predict these positions.

The state of a Kalman filter, $\mathbf{X}_{n,t} = [\mathbf{r}_{n,t}; \mathbf{v}_{n,t}; \dot{\mathbf{v}}_{n,t}; \delta_{n,t}; \Delta_{n,t}]$ (column vectors are stacked by the operator ";"), consists of receiver position $\mathbf{r}_{n,t}$, speed $\mathbf{v}_{n,t}$, acceleration $\dot{\mathbf{v}}_{n,t}$, and two clock errors ($\delta_{n,t}$ for GPS and $\Delta_{n,t}$ for GLONASS) at time t and evolves according to a matrix $\mathbf{\Phi}$, with a random variation $\mathbf{w}_{n,t-1}$, as follows:

$$\mathbf{X}_{n,t} = \mathbf{\Phi}\mathbf{X}_{n,t-1} + \mathbf{w}_{n,t-1}. \tag{4}$$

The measurement data $\mathbf{Y}_{n,t} = [\mathbf{p}_{n,t}; \mathbf{v}_{n,t}]$ contains both pseudoranges $\mathbf{p}_{n,t} = [p_{n,t}^{(1)}; p_{n,t}^{(2)}; \ldots]$ and receiver speed $\mathbf{v}_{n,t}$ and is indirectly associated with the state $\mathbf{X}_{n,t}$ by the following equation ($\boldsymbol{\rho}_{n,t} = [\rho_{n,t}^{(1)}; \rho_{n,t}^{(2)}; \ldots]$ is the true distance, $\mathbf{g}_{n,t}$ and $\mathbf{G}_{n,t}$ are flags indicating whether a satellite is GPS or GLONASS, and $\boldsymbol{\xi}_{n,t}$ is a measurement error):

$$\mathbf{Y}_{n,t} = \left[\boldsymbol{\rho}_{n,t} + \mathbf{g}_{n,t} \cdot \delta_{n,t} + \mathbf{G}_{n,t} \cdot \Delta_{n,t}; \mathbf{v}_{n,t}\right] + \boldsymbol{\xi}_{n,t}. \tag{5}$$

$\mathbf{Y}_{n,t}$ can be linearized near a state $\mathbf{X}_{n,t}$, where a small change $\partial\mathbf{X}_{n,t}$ leads to a change $\mathbf{H}_{n,t} \cdot \partial\mathbf{X}_{n,t}$ in $\mathbf{Y}_{n,t}$.

With the variance matrix of $\mathbf{w}_{n,t-1}$ and $\boldsymbol{\xi}_{n,t}$ being $\mathbf{Q}$ and $\mathbf{R}$, respectively, the process of updating state $\mathbf{X}_{n,t}$ and its variance $\mathbf{P}_{n,t}$ for a receiver $n = a, b$ using pseudoranges of satellites in $\mathbf{S}$ is briefly described below as three main steps [17].

(1) Receiver state prediction (the predicted position of receiver n, $\mathbf{r}_{n,t}^-$, is obtained from $\mathbf{X}_{n,t}^-$):

$$\mathbf{X}_{n,t}^- = \mathbf{\Phi}\mathbf{X}_{n,t-1}^+, \mathbf{P}_{n,t}^- = \mathbf{\Phi}\mathbf{P}_{n,t-1}^+\mathbf{\Phi}^T + \mathbf{Q}. \tag{6}$$

(2) Pseudorange estimation: $p_{n,t}^{(s)-}$ is estimated from $\rho_{n,t}^{(s)} = |\mathbf{r}_{n,t}^- - \mathbf{r}_t^{(s)}|$ by adding clock errors.

(3) Receiver state update: a Kalman gain $\mathbf{K}_{n,t}$ is computed based on the variance ($\mathbf{P}_{n,t}^-$) of predicted information and the variance $\mathbf{R}$ of measured information. Then, $\mathbf{X}_{n,t}^-$ is updated to $\mathbf{X}_{n,t}^+$ by adding the prediction error $\mathbf{Y}_{n,t} - (\mathbf{p}_{n,t}^-; \mathbf{v}_{n,t}^-)$ weighted by the Kalman gain using all satellites in $\mathbf{S}$, and $\mathbf{r}_{n,t}^+$ is obtained from $\mathbf{X}_{n,t}^+$

$$\begin{aligned} \mathbf{K}_{n,t} &= \mathbf{P}_{n,t}^- \cdot \mathbf{H}_{n,t}^T \cdot \left[\mathbf{H}_{n,t} \cdot \mathbf{P}_{n,t}^- \cdot \mathbf{H}_{n,t}^T + \mathbf{R}\right], \\ \mathbf{X}_{n,t}^+ &= \mathbf{X}_{n,t}^- + \mathbf{K}_{n,t} \cdot \left[\mathbf{Y}_{n,t} - \left(\mathbf{p}_{n,t}^-; \mathbf{v}_{n,t}^-\right)\right], \\ \mathbf{P}_{n,t}^+ &= \left(\mathbf{I} - \mathbf{K}_{n,t}\mathbf{H}_{\text{n},t}\right)\mathbf{P}_{n,t}^-. \end{aligned} \tag{7}$$

The above Kalman filtering process is divided into two parts to accomplish both correlation detection of MPEs and position computation.

Correlation detection of MPEs involves three steps.

(i) For each common satellite in $\mathbf{S}$, if its elevation angle is above a threshold (e.g., 50°) or the SNR at both receivers is above a threshold (e.g., 44 dB), it is added to a set $\mathbf{S}_R$; otherwise, it is added to a set $\mathbf{S}_C$. The thresholds are selected based on some initial analysis of the experiment data. Satellites in $\mathbf{S}_R$ are regarded as being reliable without MPEs and used in estimating current receiver positions, which are further used to detect correlation of MPEs of satellites in $\mathbf{S}_C$.

(ii) Step 1–3 of Kalman filtering are performed for each receiver: the predicted position is $\mathbf{r}_{n,t}^-$ in Step 1 and the updated position is $\mathbf{r}_{n,t}^+$ using $\mathbf{S}_R$ instead of $\mathbf{S}$ in Step 3.

(iii) For each satellite $k \in \mathbf{S}_C$, the metric in (3) is computed, using $\mathbf{r}_{n,t}^+$ as the estimated receiver position. Satellites whose metrics are beyond a threshold are removed from $\mathbf{S}_C$. After the correlation detection, $\mathbf{S}_C$ becomes $\mathbf{S}_C'$. MPEs of satellites in $\mathbf{S}_C'$ are regarded as being spatially correlated.

In position computation, Step 3 of Kalman filtering is performed again, but using $\mathbf{S}_R \cup \mathbf{S}_C'$ instead of $\mathbf{S}$. With the updated positions $\mathbf{r}_{a,t}^+$ and $\mathbf{r}_{b,t}^+$, their difference is regarded as the relative position.

3. Experiment Setup

In the experiments, we used two types of receivers, N-receiver (NovAtel OEM6) and U-receiver (u-blox EVK-6T), and two cars. As shown in Figure 2, two N-receivers (N_1, N_3) and two U-receivers (U_1, U_3) are installed on car 1; one N-receiver (N_2) and one U-receiver (U_2) are installed on car 2. Most of the time, two cars are driven along the same lane in a platoon, although they might temporarily change lanes, wait for signals at intersections, or be separated by other vehicles. In this way, we investigate the spatial correlation of MPEs at different interreceiver distances under practical vehicular scenarios.

In the experiment, car speed is obtained from the vehicle engine via the CAN-BUS (http://en.wikipedia.org/wiki/CAN_bus), and a fibre optic gyroscope is used to estimate

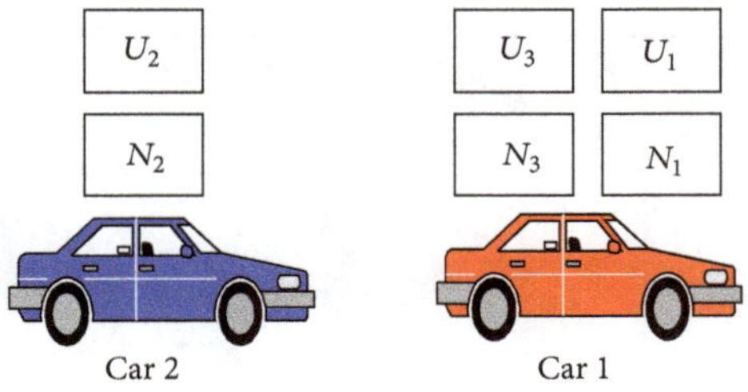

FIGURE 2: Experiment setup: two N-receivers and two U-receivers on car 1, one N-receiver and one U-receiver on car 2. Two cars are driven in a platoon in urban areas with buildings on roadsides.

car's moving direction. From GNSS receivers, both pseudoranges and carrier phases are obtained and recorded. Receivers' positions, fixed by offline RTK-GPS processing with carrier phases [22], are used as ground truth. In the evaluation, pseudorange data of L1-code is used to compute relative positions between receivers.

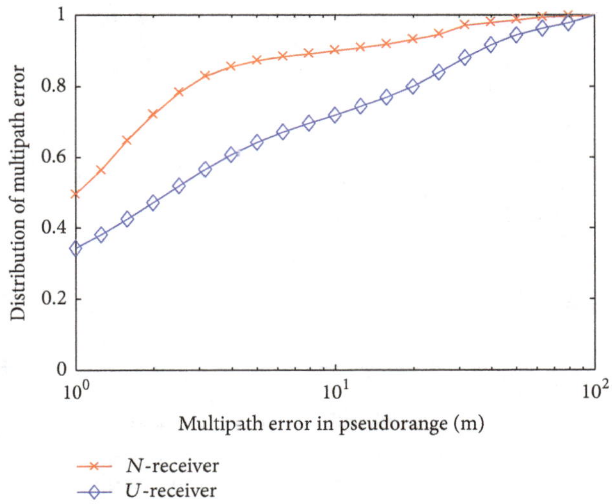

FIGURE 3: CDFs of MPEs in pseudoranges at two types of receivers.

4. Statistics of Multipath Error

In the analysis of MPE $m_a^{(k)}$ (in the pseudorange between receiver a and satellite k), a nearby reference station is used as receiver b. As the reference station is set up in an open sky environment, its MPE of satellite k, $m_b^{(k)}$, is zero. Then, (3) is simplified as

$$m_a^{(k)} \approx p_{ab}^{(kl)} - \rho_{ab}^{(kl)}, \tag{8}$$

by which $m_a^{(k)}$ can be estimated.

In the following, basic properties of MPE are first presented, and then, spatial correlation between MPEs at two receivers is analyzed.

4.1. Basic Properties of MPE. CDFs (cumulative distribution function [23]) of MPEs of two types of receivers are shown in Figure 3. For either receiver, MPEs in pseudoranges can be as large as 100 m. The two types of receivers have different properties: N-receivers try to improve the precision of pseudoranges while U-receivers aim to improve the sensitivity. Accordingly, their MPEs also have different properties. Compared with N-receivers, U-receivers generally have larger MPEs. With a threshold 2.5 m, 78% MPEs of N receivers are less than this threshold, while this probability decreases to 0.52 for U-receivers.

MPEs in pseudoranges are mainly caused by the reflection of positioning signals. In addition, a reflected signal, attenuated by each reflection, tends to have lower SNR (signal-to-noise ratio). Therefore, there is some correlation between MPEs and SNR. CDFs of MPEs, under different SNRs, are shown in Figure 4. MPEs at high SNR (SNR $\geq$ 45) are very small. In comparison, MPEs at low SNR (SNR $\leq$ 40) may have large values.

The elevation angles of satellites affect the HDOP (horizontal dilution of precision), which is further associated with the positioning precision. In addition, the probability that a positioning signal is obstructed and reflected increases when the elevation angle decreases. Therefore, MPEs are correlated with elevation angles of satellites. CDFs of MPEs, under different elevation angles, are shown in Figure 5. Generally, when elevation angle of a satellite is high (elevation angle $\geq$ 60), its MPE is very small. Hence, a satellite with a high elevation angle can be used as a reference satellite in estimating MPEs of other satellites. MPEs get larger as elevation angles decrease. At low elevation angles, U-receivers tend to have larger MPEs than N-receivers. This is because U-receivers have higher sensitivity. Then, positioning signals with multiple reflections can be received as well, which contributes to large MPEs.

There is also some correlation between MPEs and moving speeds. Figure 6 shows CDFs of MPEs under different speeds. MPEs tend to have large values at low speeds. This can be explained as follows: GNSS receivers use tracking loop to track positioning signals; at low speeds, even a reflected positioning signal is relatively stable for some period, and it can enter the tracking loop of a receiver; at higher speeds, however, this becomes less possible.

4.2. Correlation between MPEs. Next, we investigate the correlation between MPEs at different interreceiver distances. Here, $N_1 - N_3$, $U_1 - U_3$ are installed on the same car and regarded as pairs of receivers with a short distance (about 1.5 m). $N_1 - N_2$, $U_1 - U_2$ are installed on different cars and regarded as pairs of receivers with a long distance, whose CDF is shown in Figure 7. The median distance is aound 10 m and the average distance is 19.4 m.

In the following analysis, we focus on the satellites where *either or both receivers have nonnegligible MPEs* (MPE $\geq$ 5 m). In other words, satellites that are directly visible to both receivers (MPE $\approx$ 0) are not involved in the analysis.

As for MPEs of noncommon satellites, common satellites, and MPE difference of common satellites, their CDFs are

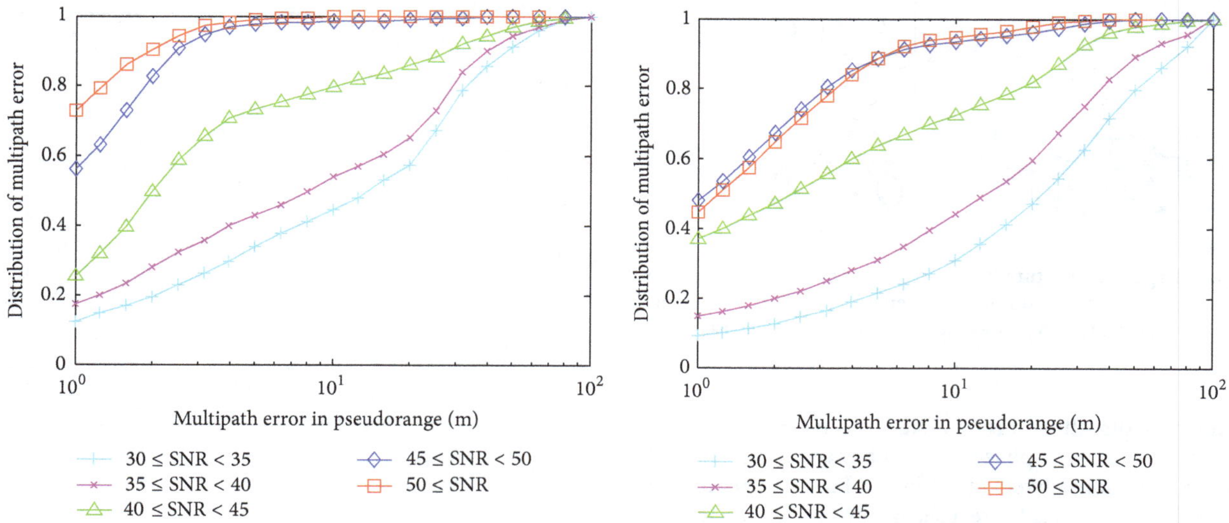

(a) N-receiver (b) U-receiver

FIGURE 4: CDFs of MPEs in pseudoranges under different SNR of received signals.

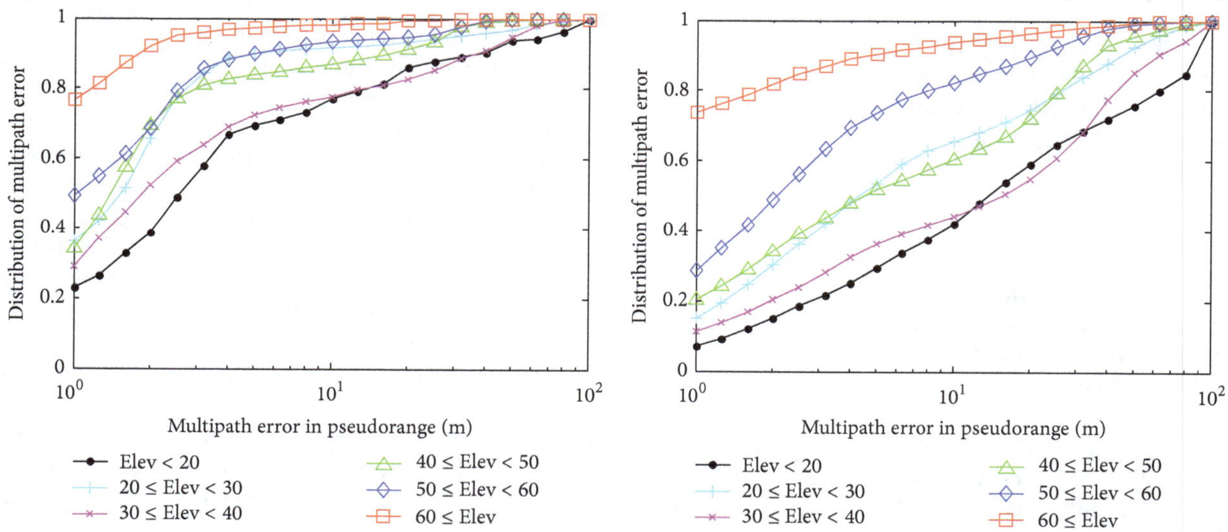

(a) N-receiver (b) U-receiver

FIGURE 5: CDFs of MPEs in pseudoranges under different elevation angles of satellites.

computed and shown in Figure 8. This figure reveals the following points.

(i) Under all cases, a MPE difference is much less than that of original MPEs. In other words, the concept of DGPS applies to two adjacent receivers since their MPEs can be cancelled by the differential operation.

(ii) The probability that a MPE difference is below a threshold (e.g., 10 m in Figure 8) is less than 1. Therefore, noncorrelated MPEs do exist, and some mechanisms are necessary to mitigate them.

(iii) As the interreceiver distance increases (e.g., (a) to (b) for N-receiver, (c) to (d) for U-receiver), the probability that a MPE difference is below a threshold (e.g., 10 m) decreases. In other words, correlation between MPEs is inversely related with the inter-receiver distance.

(iv) Compared with N-receivers, U-receivers tend to have lower correlation in MPEs (e.g., (a) to (c) where receivers are on the same car, (b) to (d) where receivers are on different cars).

(v) MPEs of common satellites are less than that of noncommon satellites. This is because a positioning signal is more likely a reflected one when it is only received by one receiver.

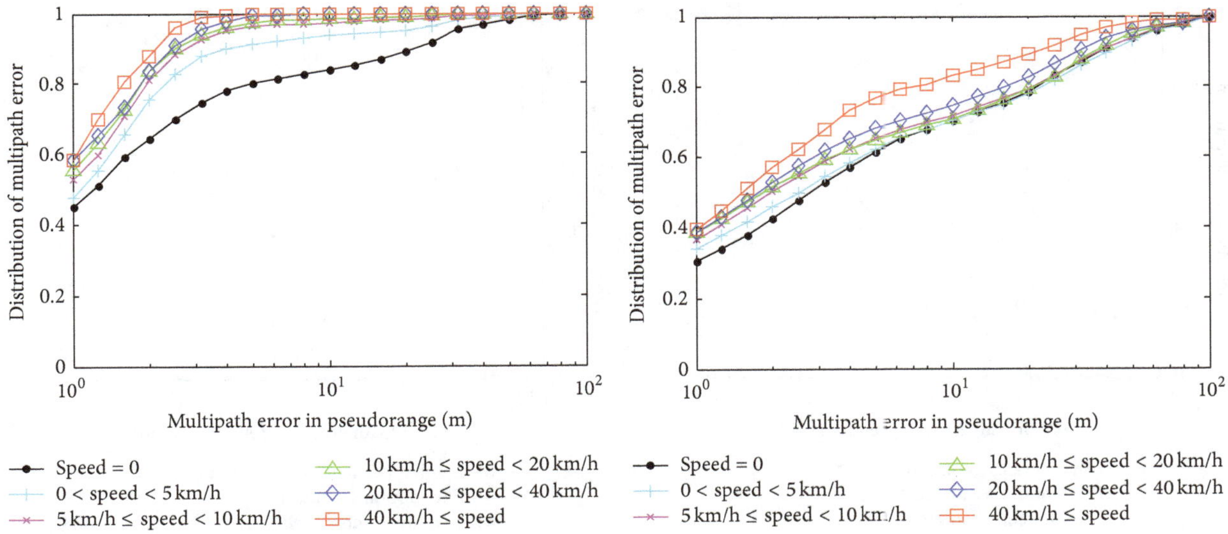

(a) N-receiver

(b) U-receiver

FIGURE 6: CDFs of MPEs in pseudoranges under different moving speeds of receivers.

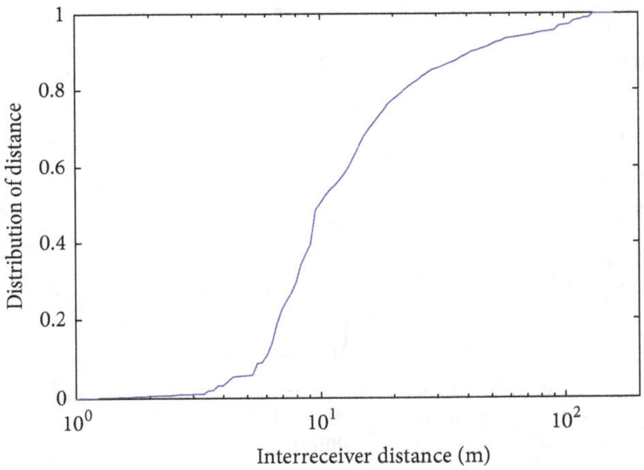

FIGURE 7: CDF of interreceiver distance when receivers are installed on top of separate cars.

TABLE 1: Probability of MPEs under different scenarios.

	N-receiver	U-receiver
Prob(MPE diff <5), short distance	0.537	0.329
Prob(MPE diff <5), long distance	0.364	0.203
Prob(MPE diff <10), short distance	0.743	0.607
Prob(MPE diff <10), long distance	0.579	0.440
Prob(single MPE), short distance	0.093	0.126
Prob(single MPE), long distance	0.184	0.234

The result in Figure 8 is summarized in Table 1. The probabilities that a MPE difference is below a certain threshold (5 m or 10 m) are computed for different interreceiver distances (short distance, long distance) and different types of receivers (N-receiver, U-receiver). With a relatively high probability, two receivers (especially N-receivers) have a correlation in their MPEs, although this correlation decreases at longer distances.

We further study the correlation between MPEs by investigating the distribution of MPE pairs (two MPEs at two receivers). Their normalized histograms are shown in Figure 9. In Figures 9(a) and 9(c), two receivers are installed in the same car with a short distance, and most MPE pairs lie along the diagonal of the horizontal axis. As the interreceiver distance increases (from (a) to (b) or from (c) to (d)), the peaks of histogram decrease and points are spread in a wider area. Even in this case, there is still a high probability that MPE pairs are near the diagonal, which confirm the correlation between MPEs, though the probability is less compared with that at a shorter distance. Again, MPEs of N receivers tend to have higher correlation than those of U receivers.

There are some points on either axis, which indicates that MPE is only found at one receiver, but with a negligible value at the other receiver. This corresponds to the case of car B and car C in Figure 1. In this case, the MPEs have low correlation, and the associated common satellite should be removed before fixing relative positions. The probabilities associated with this case are shown in the last two rows of Table 1.

As discussed before, the distribution of MPEs depends on moving speeds of receivers. A natural question is whether the correlation in MPEs is affected by speeds as well. We computed MPE difference of two receivers, when both speeds are less than 0.1 km/h or both are greater than 10 km/h (we did so because two receivers have different speeds in the long distance case). The CDFs of MPE of N receivers are shown in Figure 10, which shows a clear distinction between the MPE distributions under low and high speeds.

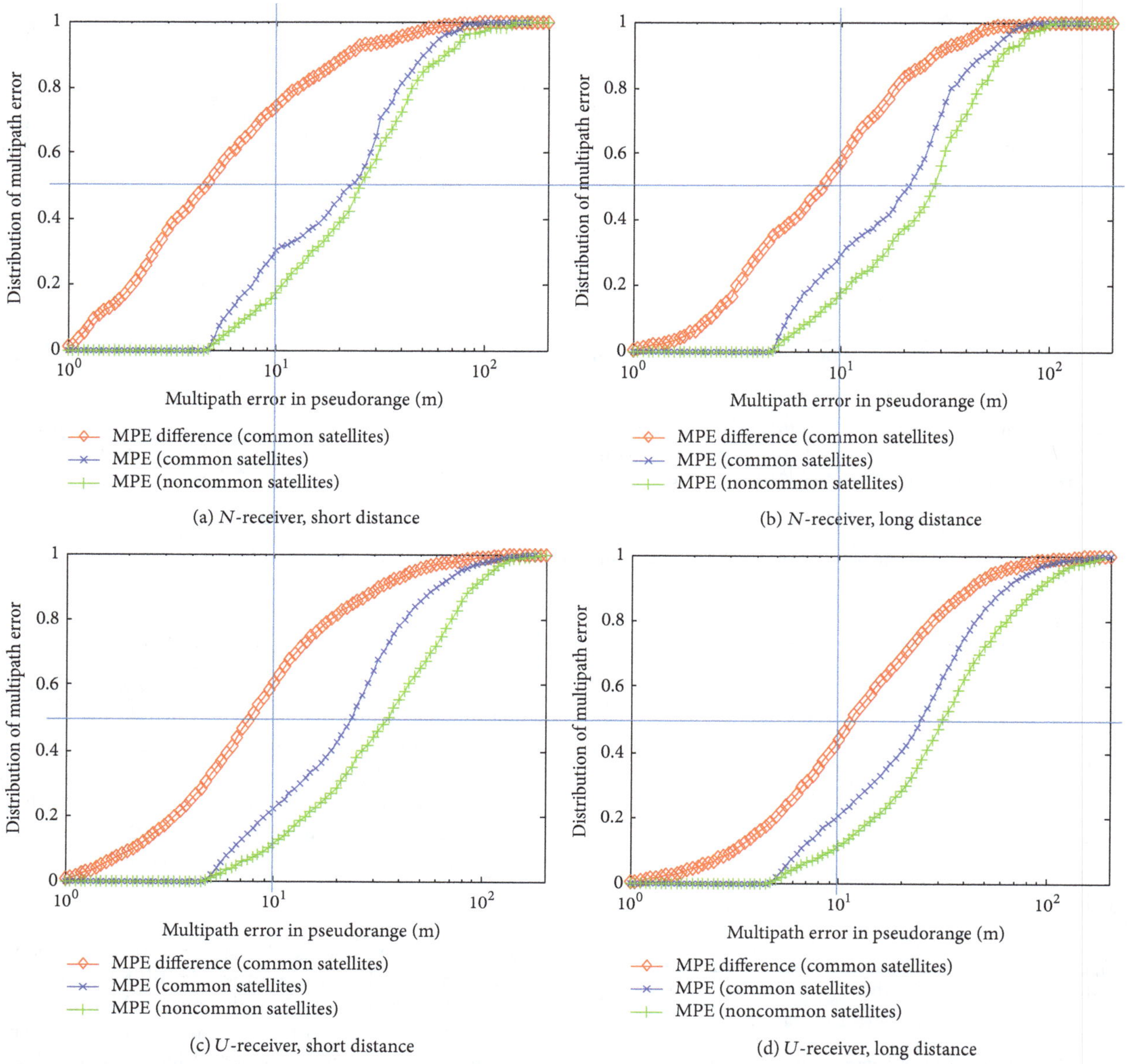

FIGURE 8: CDFs of MPEs of noncommon satellites, common satellites, and MPE difference of common satellites. (a) Two N-receivers with a short distance, (b) two N-receivers with a long distance, (c) two U-receivers with a short distance, and (d) two U-receivers with a long distance.

5. Refining the CoRelPos Scheme

Here, we further refine the CoRelPos scheme by considering the dynamics of MPEs.

A Kalman filter is used to combine the predicted position (based on previous position and speed) and measured pseudoranges to estimate the receiver position. Both predicted position and pseudoranges have a variance indicating their potential errors. From these variances, a Kalman gain is computed, and the new position is computed as the weighted sum of predicted position and the prediction error. The precision of predicted position depends on the accuracy of moving speed. It will have a larger weight when the prediction variance (indirectly, the speed variance) is small; on the other hand, the prediction error (pseudoranges) should have a large weight when the pseudoranges have small variances.

Typically, fixed variances are assigned for pseudoranges and moving speeds. But in the above analysis, we find that MPEs are related with speeds, besides elevation angles of satellites. Therefore, the variances of pseudoranges and speeds are adjusted as follows.

(i) A small variance for pseudoranges at a high speed and a large variance at a low speed: in other words, reflected positioning signals will be more leveraged at higher speeds. This is based on the relationship between MPE difference and speeds shown in Figure 10.

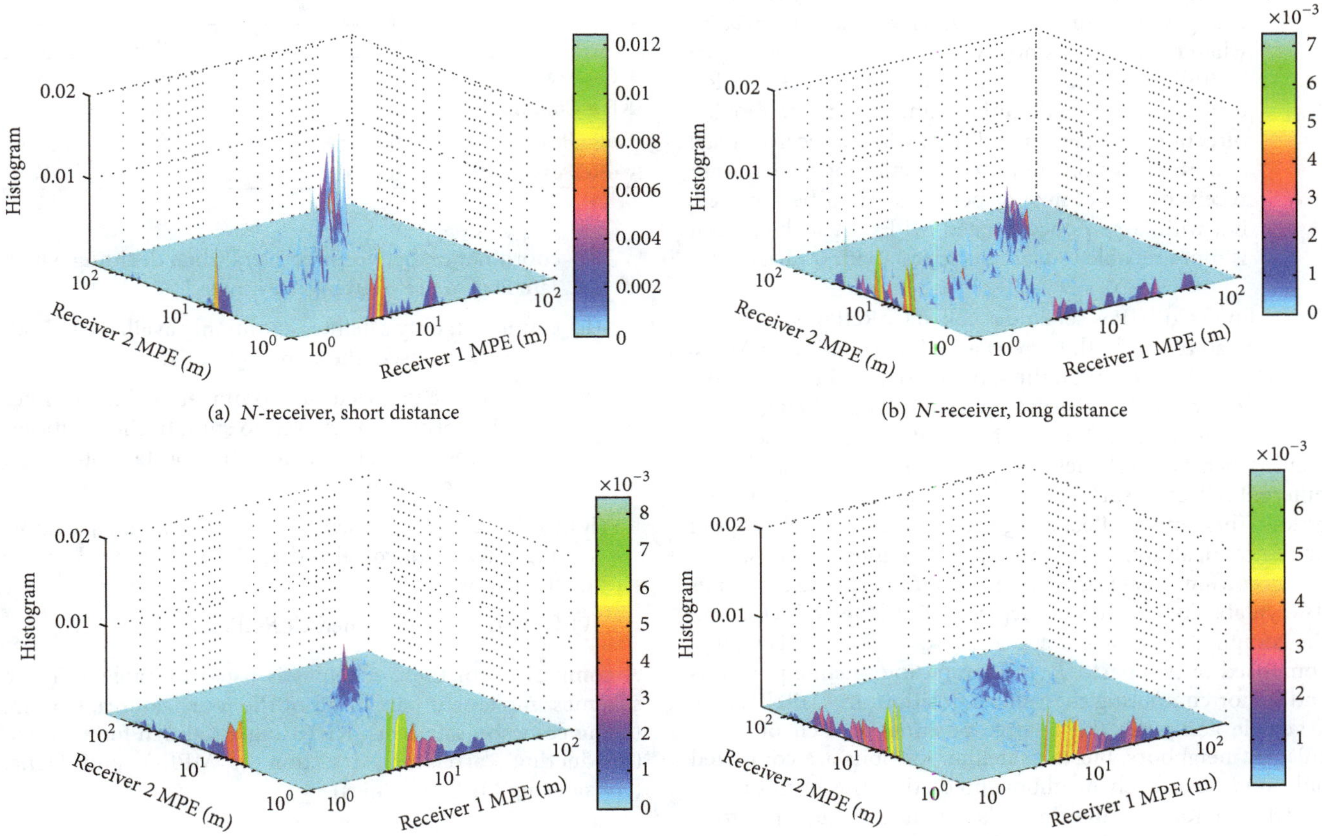

(a) N-receiver, short distance

(b) N-receiver, long distance

(c) U-receiver, short distance

(d) U-receiver, long distance

FIGURE 9: Histograms of MPE pairs. (a) Two N-receivers with a short distance, (b) two N-receivers with a long distance, (c) two U-receivers with a short distance, and (d) two U-receivers with a long distance.

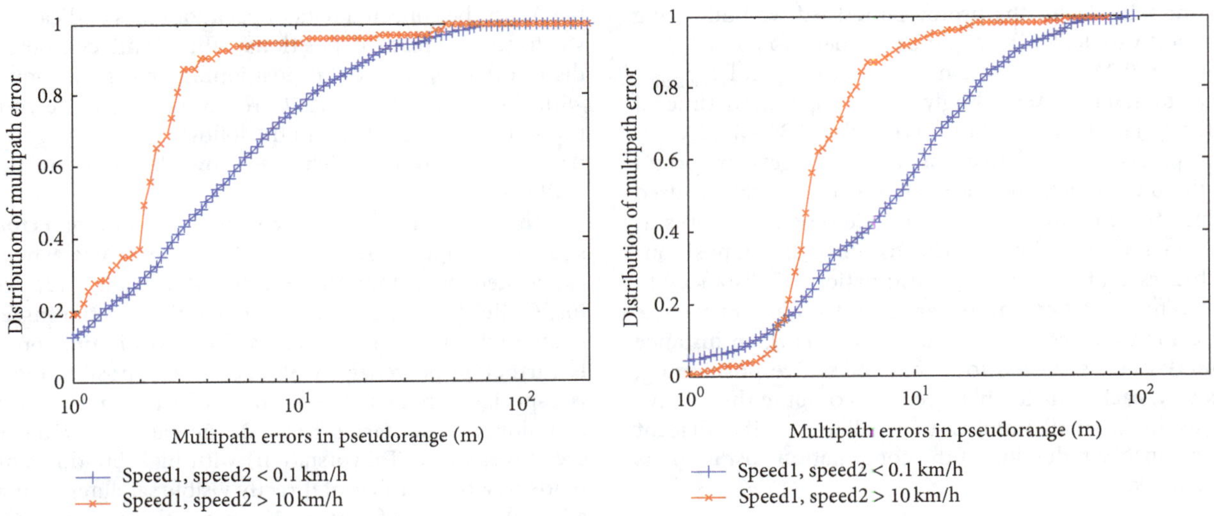

(a) N-receiver, short distance

(b) N-receiver, long distance

FIGURE 10: CDFs of MPE difference under different moving speeds. (a) Two N-receivers with a short distance and (b) two N-receivers with a long distance.

(ii) A large variance for speed when the moving direction changes, a small variance otherwise, and 0 variance when the receiver stops (when speed pulse gets equal to 0): in our analysis, we found that the estimation error of speed tends to be higher when the moving direction changes. In addition, when a receiver (car) comes to a stop, reflected positioning signals not received in the moving period start to be received, but with large noncorrelated MPEs. In such cases, it is more reliable to trust the speed (with 0 variance).

This refined CoRelPos scheme is called CoRelPos+.

As analyzed in [20], there is a high correlation in MPEs when two vehicles run on the same road and share the same reflective surface. In this case, reflected signals, being correlated, can be leveraged in relative positioning. On the other hand, when two vehicles run on different roads and have different reflective surfaces, the correlation in their MPEs will be low. Then, most reflected signals, being noncorrelated, will be removed, although the shortage of satellites remains.

In conventional positioning schemes, a vehicle computes its own absolute position independent of other vehicles, and the computation time per epoch, $T_{\mathrm{CONV}} = T_1 + T_2$, is mainly composed of two parts, T_1 for computing satellite positions and T_2 for computing the vehicle position. In CoRelPos(+), a vehicle computes the relative position of each of its N adjacent neighbors, but the satellite positions are computed only once. For each neighbor, predicting the position per vehicle via Kalman filtering takes a time T_3 and the total time for a pair of vehicles is $2T_3$; the correlation detection in MPEs takes a time T_4, and finally computing the pair of positions again using correlated signals roughly takes a time $2T_3$. Altogether, the computation cost per pair of vehicles is $T_{\mathrm{CoRelPos}} = T_1 + 4T_3 + T_4$, where T_4 is relatively small compared with T_3 and can be neglected. Then, the total computation cost per vehicle per epoch will be $T_1 + N \cdot 4T_3$, linearly increasing with N. In the prototype testbed we built using the free software RTKLIB [22], the parameters T_1 and T_{CONV} are equal to 0.34 ms and 0.73 ms, respectively, and T_{CoRelPos} is equal to 1.51 ms. Accordingly, the computation time of CoRelPos(+) under N neighbors is equal to $0.39 + 1.12N$ ms.

The precision of relative positions only gets important when the intervehicle distance gets short. This can be used to reduce the computation cost and CoRelPos(+) works as follows: each vehicle always performs conventional positioning schemes and gets a rough estimation of distances to nearby vehicles; when the distance to an adjacent vehicle gets less than a threshold (e.g., less than 50 m), an instance of CoRelPos(+) is started for accurate relative positioning; in this way, each vehicle only needs to compute the relative positions for a small number (a small N) of its adjacent vehicles. Further reduction of the computation cost is left as a future work.

5.1. A Short Comparison. In the experiment, we will evaluate five schemes, as follows.

(i) N-receiver: the relative position is directly computed from the positions output by N-receivers. There is no coordination between receivers when deciding which satellites to use in fixing positions.

(ii) CommSat: only satellites commonly available to both receivers are used, the same as in DGPS.

(iii) KF+CommSat: based on CommSat, a Kalman filter (speed information) is used to smooth the positions. This represents an implementation of state-of-the-art positioning methods.

(iv) CoRelPos [20]: based on KF+CommSat, pseudoranges with noncorrelated MPEs are removed before fixing positions.

(v) CoRelPos+: the refined CoRelPos scheme.

TABLE 2: Comparison of different schemes.

	Comm.	KF	Corr.	Dyna.
CommSat	Yes	No	No	No
KF + CommSat	Yes	Yes	No	No
CoRelPos	Yes	Yes	Yes	No
CoRelPos+	Yes	Yes	Yes	Yes

A comparison of these schemes is listed in Table 2. These schemes implement part or all of the four features, Comm. (using common satellites), KF (using a Kalman filter), Corr. (conducting correlation detection of MPEs), and Dyna. (considering MPE dynamics).

6. Results of Positioning

The evaluation of CoRelPos was conducted in [20] by setting two receivers on stationary cars or on one car with a short interreceiver distance. In this work, we installed two receivers on two separate cars with a longer interreceiver distance, aiming to make the experiment scenario as practical as possible. With the collected data of pseudoranges and car speeds, we fixed the relative positions offline and computed the distribution of horizontal positioning errors by comparing with the ground truth. As U-receivers have quite complex properties in their MPEs, in the following, we mainly use the data of N-receivers, which have shown higher correlation in MPEs.

Three scenarios and their relative positioning errors are shown in Figures 11, 12, and 13, respectively, where the experiment was done in January 2014. In all three scenarios, CoRelPos can effectively reduce the positioning errors compared with other schemes. The positioning precision is further improved by CoRelPos+. The positioning error is especially small in scenario 3, which can be explained as follows: in scenarios 1 and 2, the cars are driven in an area (Yaesu near Tokyo station) with high buildings on the roadsides; the number of directly visible satellites is small; in addition, there are frequent changes in the moving direction of receivers; in scenario 3 cars are driven from Yaesu to the Etchujima Island, and there are fewer tall buildings on roadsides and fewer changes in moving directions.

Positioning precision depends on the constellation of satellites, which is time-variant. Figure 14 shows the CDFs of

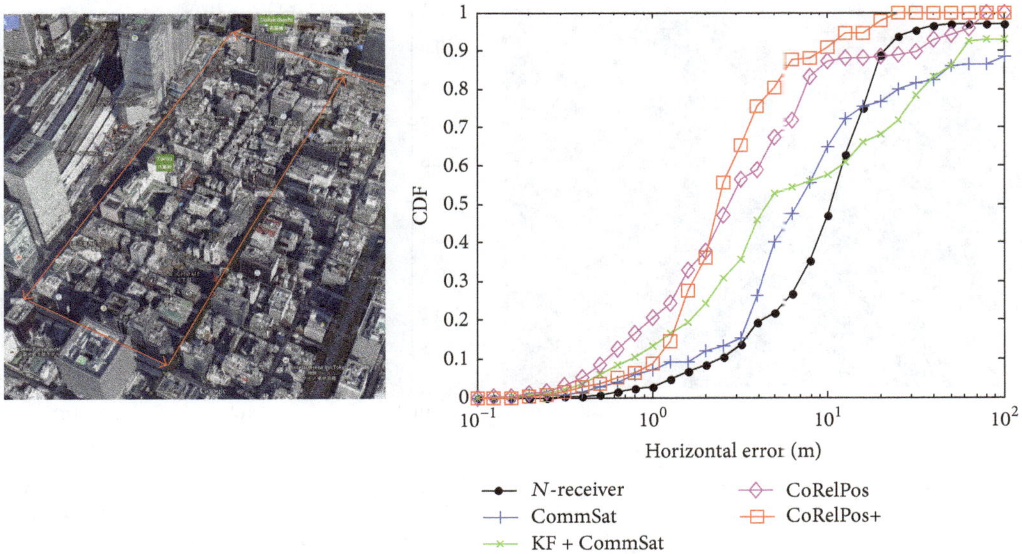

FIGURE 11: Scenario 1. CDFs of horizontal, relative positioning errors by different schemes.

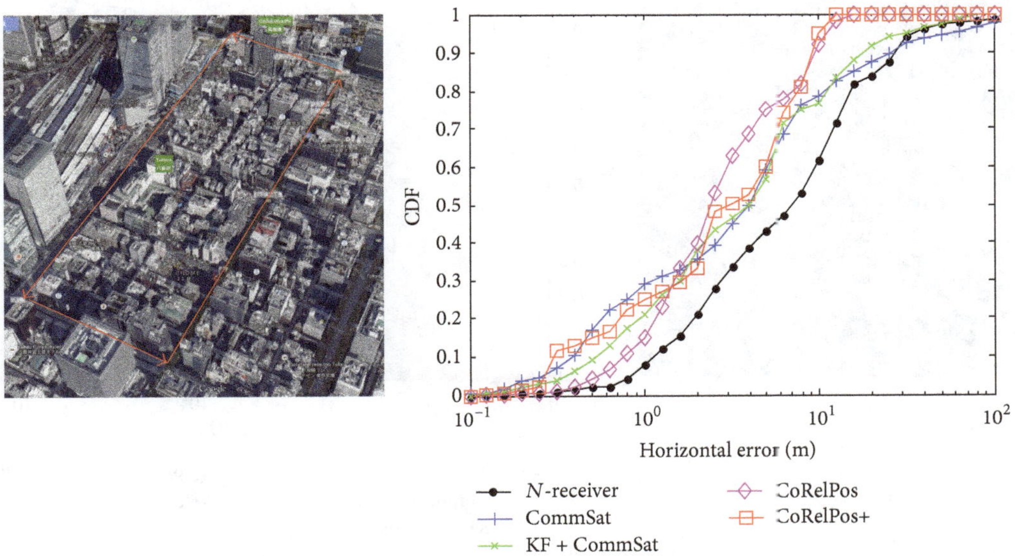

FIGURE 12: Scenario 2. CDFs of horizontal, relative positioning errors by different schemes.

horizontal, relative positioning errors around Yaesu, but on a different day in October 2013. This reflects a similar trend as in Figures 11, 12, and 13.

The RMS (root mean square) errors of five schemes, under four scenarios, are further summarized in Table 3. Compared with directly using N-receivers, coordinating receivers to use common satellites usually can reduce the errors. Exceptions do exist (e.g., scenario 1) because fewer satellites are available in the latter case. This is also confirmed by Table 4, which indicates that CommSat usually has a low rate of fixing positions. Combining position prediction with satellite positioning via KF+CommSat can reduce the outage and further reduce the errors, but its effect is limited by the MPEs. Removing noncorrelated MPEs, CoRelPos can greatly reduce the errors. These errors can be further reduced by considering the dynamics of MPE correlations in CoRelPos+.

TABLE 3: RMS errors of horizontal, relative positioning errors by different schemes (unit: m).

	Scen. 1	Scen. 2	Scen. 3	Scen. 4
N-receiver	53.03	32.96	35.59	29.33
CommSat	157.63	29.34	18.71	14.81
KF + CommSat	49.03	17.61	12.19	9.57
CoRelPos	22.76	6.44	5.44	5.42
CoRelPos+	7.63	6.40	1.54	3.48

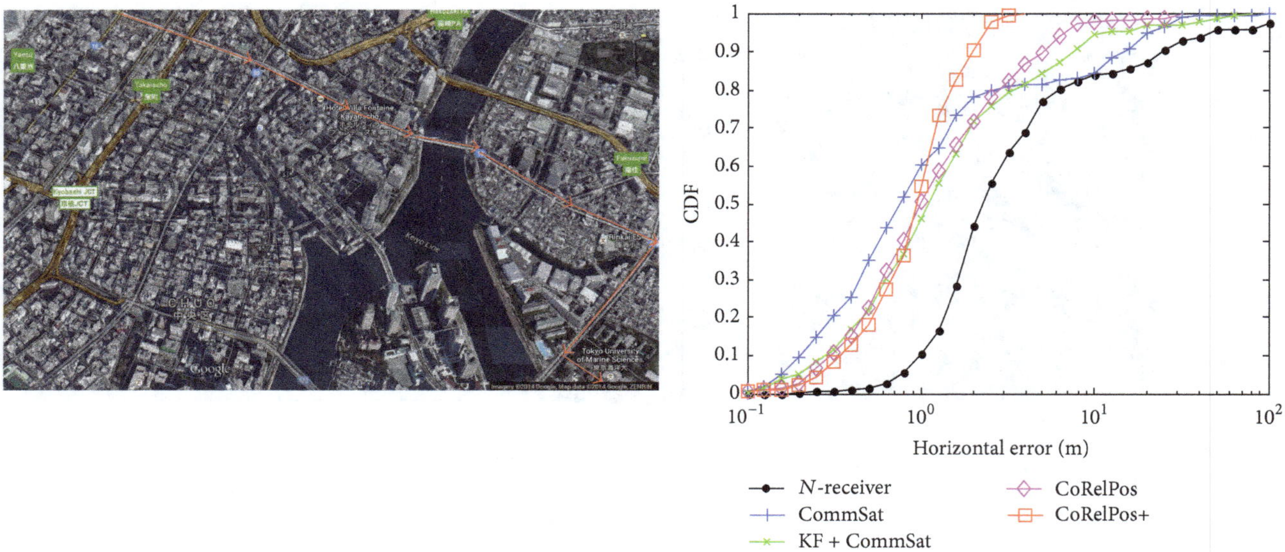

FIGURE 13: Scenario 3. CDFs of horizontal, relative positioning errors by different schemes.

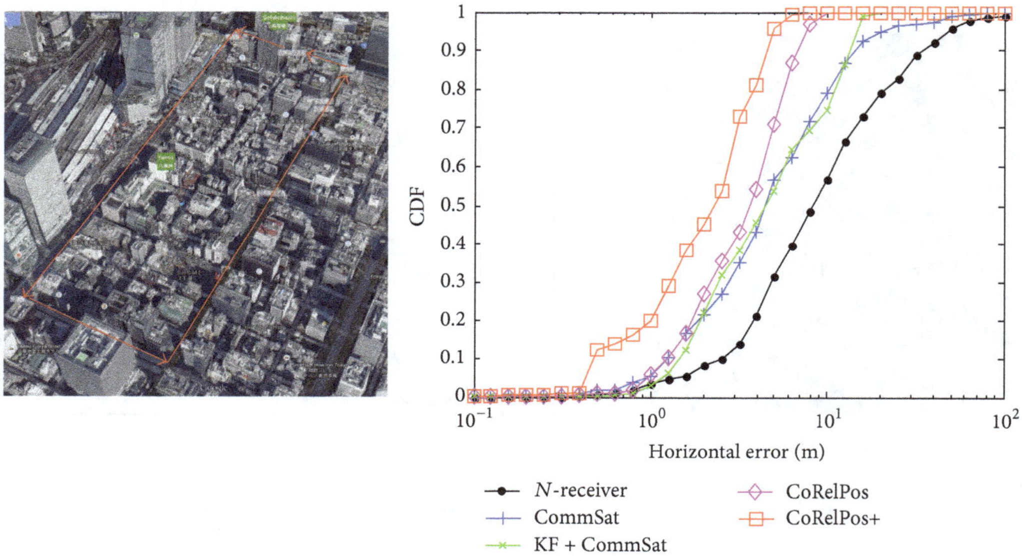

FIGURE 14: Scenario 4. CDFs of horizontal, relative positioning errors by different schemes.

TABLE 4: Percentage of epochs that relative positions can be fixed.

	Scen. 1	Scen. 2	Scen. 3	Scen. 4
N-receiver	85.7%	68.5%	63.4%	75.6%
CommSat	82.9%	42.6%	24.8%	63.3%
KF + CommSat	100%	99.1%	100%	95.3%
CoRelPos	100%	99.1%	100%	95.3%
CoRelPos+	100%	99.1%	100%	95.3%

7. Conclusion

Precision positioning in urban canyons remains a challenge due to the obstructions and reflections of roadside buildings. In this paper, analysis of experiment data reveals different properties of MPEs, especially their spatial correlation in local areas and the dependency of MPE correlations on moving speeds. Considering the dynamics of MPE helps further improve positioning precision. The evaluation of relative positioning by using two receivers on two cars makes the proposed scheme one step towards practical applications. In the future, we will continue the studies of the U-receivers which have more complex properties of MPEs.

Conflict of Interests

The authors declare that there is no conflict of interests regarding the publication of this paper.

Acknowledgment

This work is supported by the Strategic Information and Communications R&D Promotion Programme (SCOPE) funded by Ministry of Internal Affairs and Communications, Japan.

References

[1] P. Misra and P. Enge, *Global Positioning System: Signals, Measurements, and Performance*, Ganga-Jamuna, 2nd edition, 2006.

[2] M. L. Sichitiu and M. Kihl, "Inter-vehicle communication systems: a survey," *IEEE Communications Surveys and Tutorials*, vol. 10, no. 2, pp. 88–105, 2008.

[3] ASV4, http://www.soumu.go.jp/main_sosiki/joho_tsusin/policyreports/chousa/its/pdf/ 081219_2_si2-4.pdf.

[4] O. Shagdar, S. Tang, M. N. Shirazi, R. Suzuki, and S. Obana, "Reliable cut-through forwarding in CDMA inter-vehicle networks," in *Proceedings of the 2nd International Conference on Ubiquitous Information Management and Communication (ICUIMC '08)*, pp. 491–495, February 2008.

[5] A. J. Van Dierendonck, P. Fenton, and T. Ford, "Theory and performance of narrow correlator spacing in a GPS receiver," *Navigation, Journal of the Institute of Navigation*, vol. 39, no. 3, pp. 265–283, 1992.

[6] B. Townsend and P. Fenton, "A practical approach to the reduction of pseudorange multipath errors in a L1 GPS receiver," in *Proceedings of the 7th International Technical Meeting of the Satellite Division of the Institute of Navigation*, pp. 143–148, September 1994.

[7] V. A. Veitsel, A. V. Zhdanov, and M. I. Zhodzishsky, "The mitigation of multipath errors by strobe correlators in GPS/GLONASS receivers," *GPS Solutions*, vol. 2, no. 2, pp. 38–45, 1998.

[8] J. W. Betz, "Binary offset carrier modulations for radionavigation," *Navigation, Journal of the Institute of Navigation*, vol. 48, no. 4, pp. 227–246, 2001.

[9] R. Hatch, "The synergism of GPS code and carrier measurements," in *Proceedings of the 3rd International Geodetic Symposium on Satellite Doppler Positioning*, pp. 1213–1231, 1983.

[10] R. D. J. van Nee, J. Siereveld, P. C. Fenton, and B. R. Townsend, "Multipath estimating delay lock loop: approaching theoretical accuracy limits," in *Proceedings of the IEEE Position Location and Navigation Symposium*, pp. 246–251, Las Vegas, Nev, USA, April 1994.

[11] S. Rougerie, G. Carrié, F. Vincent, L. Ries, and M. Monnerat, "A new multipath mitigation method for GNSS receivers based on an antenna array," *International Journal of Navigation and Observation*, vol. 2012, Article ID 804732, 11 pages, 2012.

[12] Y. Suh and R. Shibasaki, "Evaluation of satellite-based navigation services in complex urban environments using a three-dimensional GIS," *IEICE Transactions on Communications*, vol. 90, no. 7, pp. 1816–1825, 2007.

[13] J.-I. Meguro, T. Murata, J.-I. Takiguchi, Y. Amano, and T. Hashizume, "GPS multipath mitigation for urban area using omnidirectional infrared camera," *IEEE Transactions on Intelligent Transportation Systems*, vol. 10, no. 1, pp. 22–30, 2009.

[14] M. Z. Win, A. Conti, S. Mazuelas et al., "Network localization and navigation via cooperation," *IEEE Communications Magazine*, vol. 49, no. 5, pp. 56–62, 2011.

[15] IEEE, "Wireless LAN medium access protocol (MAC) and physical layer (PHY) specification. Ammendment 6: wireless access in vehicular environment," IEEE Std 802.11p, 2010.

[16] N. Alam, A. T. Balaei, and A. G. Dempster, "Positioning enhancement with double differencing and DSRC," in *Proceedings of the 23rd International Technical Meeting of the Satellite Division of the Institute of Navigation (ION GNSS '10)*, pp. 1210–1218, September 2010.

[17] P. D. Groves, *Principles of GNSS, Inertial, and Multi Sensor Integrated Navigation Systems*, Artech House, 2007.

[18] A. Taha, C. Hancock, G. Roberts, and X. Meng, "The use of GPS and INS for centimeter precision during large GPS outages," in *Proceedings of the International Symposium on GPS/GNSS*, 2008.

[19] E. D. Martí, D. Martín, J. García, A. de la Escalera, J. M. Molina, and J. M. Armingol, "Context-aided sensor fusion for enhanced urban navigation," *Sensors*, vol. 12, no. 12, pp. 16802–16837, 2012.

[20] S. Tang, N. Kubo, N. Kawanishi, R. Furukawa, A. Hasegawa, and Y. Takeuchi, "Cooperative relative positioning for intelligent transportation system," *International Journal of Intelligent Transportation Systems Research*, 2014.

[21] H. Musoff and P. Zarchan, *Fundamentals of Kalman Filtering: A Practical Approach* American Institute of Aeronautics and Astronautics, 2nd edition, 2005.

[22] T. Takasu, N. Kubo, and A. Yasuda, "Development, evaluation and application of RTKLIB: a program library for RTK-GPS," in *Proceedings of the International Symposium on GPS/GNSS*, November 2007.

[23] D. P. Bertsekas and J. N. Tsitsiklis, *Introduction to Probability*, Athena Scientific, Belmont, Mass, USA, 2002.

19

Calculating the Contribution Rate of Intelligent Transportation System in Improving Urban Traffic Smooth Based on Advanced DID Model

Ming-wei Li,[1] Jun Yun,[1] and Na Liu[2]

[1]*Management School, Wuhan University of Technology, Wuhan 430070, China*
[2]*The Ministry of Transport Research Institute, Beijing 100029, China*

Correspondence should be addressed to Ming-wei Li; 467827811@qq.com

Academic Editor: Julien Bruchon

Recent years have witnessed the rapid development of intelligent transportation system around the world, which helps to relieve urban traffic congestion problems. For instance, many mega-cities in China have devoted a large amount of money and resources to the development of intelligent transportation system. This poses an intriguing and important issue: how to measure and quantify the contribution of intelligent transportation system to the urban city, which is still a puzzle. This paper proposes a matching difference-in-difference model to calculate the contribution rate of intelligent transportation system on traffic smoothness. Within the model, the main effect indicators of traffic smoothness are first identified, and then the evaluation index system is built, and finally the ideas of the matching pool are introduced. The proposed model is illustrated in Guangzhou, China (capital city of Guangdong province). The results show that introduction of ITS contributes 9.25% to the improvement of traffic smooth in Guangzhou. Also, the research explains the working mechanism of how ITS improves urban traffic smooth. Eventually, some strategy recommendations are put forward to improve urban traffic smooth.

1. Introduction

In August 2009, Premier Wen Jiabao clearly indicated the importance of establishing the "Experience China" center and seizing the initiative of the new round global network technology development. Since then, a new wave of Internet of Things (IOT) researching is rising in China. Nowadays, China is on the rapid development and construction of the Internet of Things, which has brought unprecedented opportunities for the management and service of the transportation industry. In 2011, the Ministry of Transport focused on promoting the development of transportation through IOT on the basis of the present development industry situation. During the research, aimed at guaranteeing the better development of transport industry, Guangzhou was chosen as the demonstration area and a "city intelligent transportation application demonstration project based on IOT" was organized. Moreover, as a new IOT product technology, Intelligence Transportation System (ITS) was applied in the traffic industry. Compared with the traditional transportation projects, the system mentioned above shows dynamic, novel, complex, extensive characteristics and so forth. In recent years, ITS has been constructed and applied in China while at the same time it has made certain achievements on improving the urban traffic operation and service so as to make the traffic smooth, safe, green, and convenient.

However, urban intelligent traffic is a complex integrated system, whose development and construction are affected by many factors, such as the urban traffic environment, population structure, geographical location, and economic development level. With the rapid development of urban intelligent transportation, the transportation situation has gained certain improvement. However, considerable attention has been currently paid to the evaluation index system and methods regarding the application effects of ITS. Besides, there are relatively rare researches on the application effects in the micro level of ITS. For example, researches about the contribution of ITS in proving urban traffic environment and travel quality, traffic smoothness, and security are relatively blank. Moreover, European Union issued the Guidebook for

Assessment of Transport Telemetric Applications: Updated Version in September 1998, which systematically expounded the evaluation thought of ITS project in EU. Besides, it had been used to evaluate the ITS project in London, Paris, and other cities. However, the evaluation is just related to the development and construction effects rather than the involvement with the application effects. In 2000, the United States issued "National ITS Architecture" (version 3.0) and comprehensively expounded the evaluation contents and methods of ITS. The evaluation content stipulated includes evaluation design, performance, cost and profitability analysis, risk analysis, and evaluation results. However, the micro-level research of ITS acting on urban traffic is not involved.

Although in the later research Underwood and Gehring [1] evaluated the ITS application effects from social, individual, and enterprise benefits, the research content and result are still in the macro level. Furthermore, Novak and McDnald [2] studied the potential macroeconomic effects of ITS in the United States, in which direct employment, economic multiplier, productivity, technology diffusion, and competitiveness are included. Moreover, Lee [3, 4], Zavergiu [5], Brand [6], and Beimborn et al. [7] established user efficiency analysis models, cost benefit analysis related models, and some other relevant models, so as to research the economic value of ITS projects. However, these researches are more from the perspective of ITS economic value instead of the improvement of traffic environment; besides, their evaluation indicators are different. Then, Xiao [8] who used the analytic hierarchy process and fuzzy mathematics theory to evaluate the effects of ITS in improving urban comprehensive transportation system once argued that the effects of ITS in improving traffic environment cannot be ignored. In addition, Zhao et al. [9] researched from public transport facilities, service, benefit, and social responsibility four aspects in total so as to establish the comprehensive evaluation index system of urban public traffic system. Additionally, Chen and Liao [10] evaluated ITS application effects from six aspects, namely, economic effect, social effect, resource consumption, environmental impact, and development ability. In all these researches, ITS application effects have been analyzed to a certain extent. However, more researches are still focusing on the economic and overall social effects. Besides, there has not yet been in-depth analysis on the specific effects of ITS to urban traffic. Until 2011, Cao et al. [11] defined and quantified the evaluation index system of ITS application effects from five aspects: comfort, safety, convenience, timeliness, and coordination, which has achieved a certain innovation and laid the foundation for the further study. However, it appeared that there have been some deficiencies with the evaluation method. The limitations of the methods led to the complexity of the evaluation process and affected the authenticity of the results. Therefore, in order to ensure the authenticity of the evaluation results, more suitable innovation methods should be developed and employed.

However during this period certain results in terms of the researches of evaluation methods have been obtained. For instance, Leviäkangas and Lähesmaa [12] discussed the application of AHP to analyze the cost benefits of ITS. It is too subjective to determine the weights of the efficiency indicators and this might widen the gap between evaluation results and actual situation. Moreover, Mattingly [13] constructed a model to show the value of ITS project in the monetary form, which was integrated through the analytic hierarchy process and multiple attribute value function method. After that, the model was used to evaluate ITS effects of Anaheim in California, thus confirming the feasibility of the model. However, the model cannot evaluate the application effects of the micro-level object effectively. Wang and Walton [14] discussed the grey correlation analysis (GRA) in evaluating ITS project. Zhang et al. [15] combined analytic hierarchy process with grey correlation method to evaluate the effects of the subsystems of ITS. Nevertheless, all these researches still cannot avoid the influence by the model characteristics; namely, the subjective evaluation results are stronger, which can seriously affect the veracity of the practical evaluation results.

In order to get a better evaluation model, some scholars proposed DID model due to the mature theoretical basis and practical verification of the model. Moreover, DID is more suitable for the evaluation of the benefits of new policy and the results are also more robust. For example, Wooldridge [16] established corresponding DID models according to different data types, while Ai and Norton [17] and Puhani [18] studied the parameter estimation computing method of generalized linear equation in DID model. Moreover, Drykker [19] used Wooldridge's test to verify the correlation between random errors and then corrected the model by generalized least squares method. By studying the minimum wage system in Britain from 1991 to 2000, Galiani et al. [20] evaluated the effects of minimum wage system on employment. Besides, Conti [21] calculated the contribution rate of euro in improving per capita GDP and labor productivity level by DID model. Lewis and Reiley [22] researched the contribution of offline sales products advertising for profit change. In addition, Liang and Zhou [23], Wu [24], Wang [25], and Deng and Wang [26] innovatively selected DID model to analyze the contribution of soil and water conservation engineering, new agriculture insurance policy and foreign strategic investment, and so forth. However, most researches focus on the fields of medical, equity, and economic reform, which could be inflicted with great short-time changes and easiness to control group. However, there are rare researches in the transportation field, while at the same time the applications are lacking innovation. Furthermore, researches on effects of ITS in improving urban traffic congestion are relatively blank. With more difficulties in choosing the control group, there would be more limitation in terms of the application of DID. Besides, the complexity of ITS also determines the difficulties in selecting control group. Thus, it is necessary to improve the DID model, and the innovative introduction of MDID mode is very important.

Despite the achievements acquired by the researches in the traffic smooth field [27–31], more researches are focusing on the cause and influence factors of traffic congestion or other fields. Besides, there are relatively scarce quantitative researches between transportation operation service and ITS. Although some scholars have studied the relationship between traffic smooth and ITS, there are still limited qualitative researches. All of these may lead to the fact that specific functions of ITS in improving traffic smooth cannot be expressed clearly. Therefore, strengthening quantitative research of ITS in improving urban traffic congestion is particularly important and necessary [32].

In order to acquire more scientific ITS investment and technology-integrated application with higher efficiency, this paper takes Guangzhou (capital city of Guangdong province, China) as an example (Guangzhou city is the ITS demonstration city in China) based on comprehensive analysis. After that, urban traffic smooth is chosen as the research object. Based on the literature analysis from domestic and foreign research results, characteristics of ITS and its influence factors have been refined. Then, relevant analysis has been conducted on the function of ITS to urban traffic smooth. In addition, a set of evaluation index system has been built so as to comprehensively, objectively, and accurately reflect the benefits of ITS in improving urban traffic smooth. Additionally, the indexes related to urban traffic smooth are extracted firstly, then the indexes which are greatly influenced by ITS are stripped out, and finally the influencing factors of these indexes are figured out. Finally, with advanced DID model (MDID, matching difference-in-difference) and other related econometric models, the contribution rate is calculated. Through the research, the contribution rate of ITS in improving Guangzhou's traffic smooth has been acquired, and the mechanism of ITS in perfecting urban traffic smooth has been proposed.

2. Model Formulation

2.1. Introduction to DID Model

2.1.1. DID Model. General DID estimation model is

$$y_i = \beta + \alpha_i d_i + u_i. \tag{1}$$

In the formula, $d_i = 1$ means experimental object, and $d_i = 0$ means control object. Average treatment effect on the treated ATT measures the average return of experimental group and control group:

$$\alpha^{\mathrm{ATT}} = E(\alpha_i \mid d_i = 1). \tag{2}$$

Investigating dynamic change and introducing time variable to the estimation model:

$$y_{it} = \beta + \alpha_i d_{it} + u_{it}. \tag{3}$$

The virtual variable $t = t_0$ is time section before policy implemented, and $t = t_1$ is the time section after policy implemented. Moreover, policy implementation effect is only acting on the experimental group. Observation time meets $t_0 < k < t_1$, and the formula is acquired as follows:

$$E[y_{it} \mid d_i, t] = \begin{cases} \beta + E[\alpha \mid d_i = 1] + \varepsilon_t & \text{if } d_i = 1,\ t = t_1 \\ \beta + \varepsilon_t & \text{others.} \end{cases} \tag{4}$$

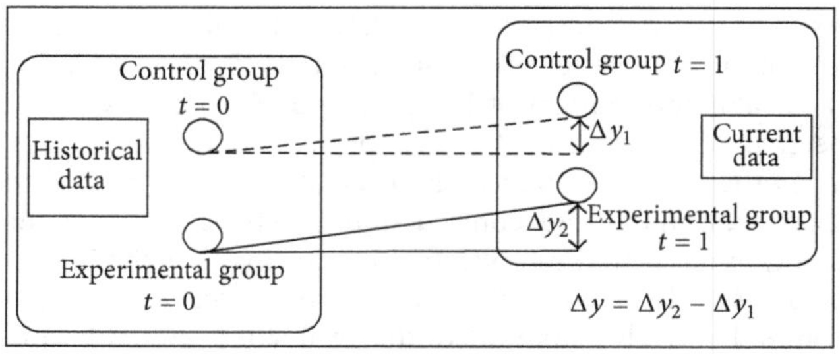

FIGURE 1: Application ideas of DID model.

Here the value of ATT is shown:

$$\begin{aligned} \alpha^{\mathrm{ATT}} &= E[\alpha \mid d_i = 1] \\ &= \{E[y_{it} \mid d_i = 1, t = t_1] - E[y_{it} \mid d_i = 1, t = t_0]\} \\ &\quad - \{E[y_{it} \mid d_i = 0, t = t_1] - E[y_{it} \mid d_i = 0, t = t_0]\}. \end{aligned} \tag{5}$$

Because $E[\widehat{\mathrm{ATT}}_{\mathrm{DID}}] = \alpha^{\mathrm{ATT}}$, the estimator of DID is

$$\widehat{\mathrm{ATT}}_{\mathrm{DID}} = [\overline{y}^1_{t_1} - \overline{y}^1_{t_0}] - [\overline{y}^0_{t_1} - \overline{y}^0_{t_0}]. \tag{6}$$

DID is the unbiased estimators of ATT. The estimator can not only measure the effects before and after the policy implemented but also measure the policy differences between experimental group and control group. After that, the policy implementation effects are obtained.

Here, setting the area that has built the ITS as "experimental group," the area without ITS is developed as "control group." DID model can effectively control the effects of other possible factors. Besides, it can not only distinguish the individual difference in experimental group after using ITS but also separate time differences of different areas. Then it comes to the identification of the application effect of ITS (Figure 1).

From the perspective of the DID model research and application, the selection of control group mainly relies on subjective judgment and lack of objective and quantitative evaluation standard. When control group is a single individual, the results will depend on the individual data to a great extent and result in poor robustness model. Therefore, it is necessary to improve the DID model.

2.2. Introduction to Matching Algorithm. Due to the strong subjectivity of DID model in selecting control group, the authenticity of evaluation results will be affected. Thus, matching algorithm is introduced in the paper to solve the problem. In terms of matching algorithm, adopting an approximation experimental algorithm is implemented to distinguish experimental group and control group effectively. Matching algorithm believes that individuals share the same characteristics, due to which they also have the same response to the same policy variables. That is to say, the decision-making behaviors of experimental group and control group are not affected by unobservable variables. Moreover, matching algorithm can enlarge the select range of control group, which can avoid the subjective randomness in selecting control group and eliminate the estimation deviation caused by nonmutual support domain and nonidentical distribution [33].

Matching model assumes that the selection biases are entirely determined by observable variables and it also rules out the selection biases caused by unobservable factors. It means that when the control variables are determined, whether affected by policy or not, they will have nothing to do with the value of index Y. Call it conditional independence assumption (CIA), and mathematical formula is

$$E(Y \mid X, d = 1) = E(Y \mid X, d = 0), \quad Y \perp d \mid X. \quad (7)$$

The CIA assumption in matching model would limit some variables selected into control group, but the restriction does not cause interference to the precision of results. It is for the reason that DID model allows the existence of observation factors.

Here, setting α_0^{ATT} is the ATT of control variable, and α_1^{ATT} is the ATT of experimental group. When CIA assumption is satisfied, formulas are acquired as

$$\begin{aligned}\alpha_0^{\mathrm{ATT}} &= E\left[y_{it} \mid d_i = 0, t = t_1\right] - E\left[y_{it} \mid d_i = 0, t = t_0\right] \\ &= 0.\end{aligned} \quad (8)$$

That is to say, the change of control group is 0 before and after the policy implemented. Therefore, α^{ATT} is

$$\begin{aligned}\alpha^{\mathrm{ATT}} &= \alpha_1^{\mathrm{ATT}} - \alpha_0^{\mathrm{ATT}} = \alpha_1^{\mathrm{ATT}} - 0 \\ &= E\left[y_{it} \mid d_i = 1, t = t_1\right] - E\left[y_{it} \mid d_i = 1, t = t_0\right].\end{aligned} \quad (9)$$

However, there will be multidimensional problems due to the increasing matching strategy with the rise of dimension X. In the early method, covariate (CV) is used for the calculation, which would result in "dimension disaster (when the dimension increased, the volume of space will increase quickly, which lead to the fact that the available data will become very sparse)" and the calculation amount is too large. In order to solve the multidimensional problem, Rosenbaum and Rubin [33] put forward the propensity score (PS) as a covariate function. If CV meets CIA hypothesis, then PS could also meet the requirement. Besides, the dimension reduction can be completed successfully by transferring the information contained in CV to PS. Therefore, PS(X) is selected to replace the CV's X, and formula $Y \perp d \mid X \Rightarrow Y \perp d \mid ps(X)$ is employed to achieve the dimension reduction:

$$\begin{aligned}\Pr(D = 1 \mid Y(0), Y(1), \mathrm{PS}(X)) &= E(D = 1 \mid Y(0), \\ &\quad Y(1), \mathrm{PS}(X)) \\ &= E[E(D = 1 \mid Y(0), Y(1), \mathrm{PS}(X), X) \mid Y(0), \\ &\quad Y(1), \mathrm{PS}(X)] = E[E(D = 1 \mid Y(0), Y(1), X) \mid \\ &\quad Y(0), Y(1), \mathrm{PS}(X)] = E[E(D = 1 \mid X) \mid Y(0), \\ &\quad Y(1), \mathrm{PS}(X)] = E[\mathrm{PS}(X) \mid Y(0), Y(1), \mathrm{PS}(X)] \\ &= \mathrm{PS}(X), \\ \Pr(D = 1 \mid Y(0), Y(1), \mathrm{PS}(X)) \\ &= \Pr(D = 1 \mid \mathrm{PS}(X)) = \mathrm{PS}(X).\end{aligned} \quad (10)$$

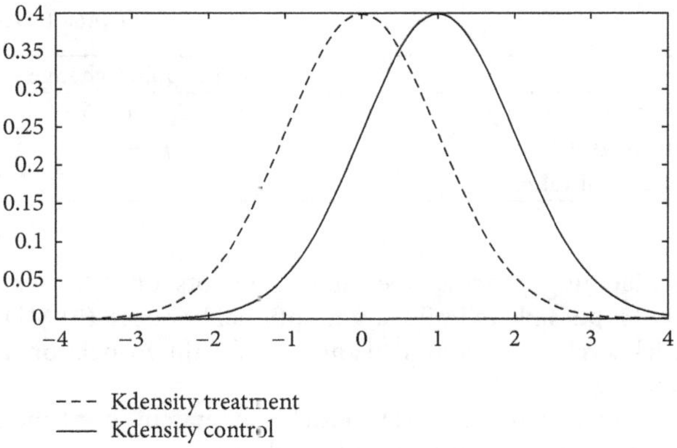

FIGURE 2: Common support domains.

When dealing with PS, Heckman believed that Logist (Logist model uses cumulative Logist distribution function to describe the relationship between selection probability and explanatory variables and limited the value of explained variables ranging from 0 to 1: $p_i = E(Y = 1 \mid X_i) = e^{\beta_0+\beta_1 X_i}/(1+e^{\beta_0+\beta_1 X_i})$; the odds ratio selection or not selection is $p_i/(1-p_i)$, taking logarithm of the odds ratio to obtain the measurement model: $L_i = \ln(p_i/(1-p_i)) = \beta_0+\beta_1 X_i+\varepsilon_i$) has more advantages than Probit (Probit model uses the standard normal distribution function to describe the relationship between the selection probabilities and the explanatory variables and limited the value of explained variables ranging from 0 to 1: $p(Y_i = 1) = \Phi(z_i) = \int_{-\infty}^{z_i}(1/\sqrt{2\pi})\exp(-s_i^2/2)ds$; then it implemented simple inverse function transformation and got the following: $z_i = \Phi^{-1}(p_i) = \beta_0 + \beta_1 X_i + \varepsilon_i$). Therefore, the estimation values of matching can be obtained with the following formula:

$$\widehat{\mathrm{ATT}}_{\mathrm{Matching}} = \frac{1}{N_1}\sum_{i=1}^{N_1}\left(y_{1i} - \sum_{j=1}^{N_0} w_{ij} y_{0j}\right). \quad (11)$$

N_0 and N_1 refer to the total number of individuals. w_{ij} means the differences between experimental group and control group, and it meets $w_{ij} \in [0, 1]$, $\sum_{j=1}^{N_0} w_{ij} = 1$.

Meanwhile, in order to select optimal control group, common support hypothesis is introduced here as $0 < P(d = 1 \mid X) < 1$. The cross region between experimental group and control group is shown in Figure 2.

2.3. Induction to MDID Model. MDID model aims to find optimal control group on the basis of DID model through bringing in matching module and matching algorithm. Then, virtual processing individuals have been developed before the intervention implemented. Last, the virtual processing individual comes to the implementation of the double difference based on the matching selection of control group.

MDID model has the following advantages when it comes to calculating the contribution rate: (1) MDID could solve the human factors generated by DID model. (2) Matching algorithm could not solve time fixed effects, but DID could. (3) MDID allows the existence of invisible factors.

TABLE 1: Basic principle of MDID.

	Before the policy change	After the policy change	Differential value
Experimental group	$\beta_0 + \beta_2 + \alpha + \delta$	$\beta_0 + \beta_1 + \beta_2 + \gamma + \alpha + \delta$	$\Delta y_t = \beta_1 + \gamma$
Control group	$\beta_0 + \beta_2 + \delta$	$\beta_0 + \beta_1 + \beta_2 + \delta$	$\Delta y_c = \beta_1$
Differential value	—	—	$\Delta\Delta y = \gamma$

(4) Matching process does not need lots of data, which makes it possible to build larger optional libraries. (5) MDID model is relatively complete and easy for the inspection and inference.

According to the DID model and matching estimator, MDID estimator is established as follows:

$$\widehat{\text{ATT}}_{\text{MDID}} = \frac{1}{N_1}\sum_{i=1}^{N_1}\left\{\left(y_{1it_1} - y_{1it_0}\right) - \sum_{j=1}^{N_0} w_{ij}\left(y_{0jt_1} - y_{0jt_0}\right)\right\}. \quad (12)$$

In view of the above estimators, considering the other influencing factors in control variables, the following panel model is constructed here to estimate ATT:

$$w_{ij}y_{it} = \beta_0 + \beta_1 t_{it} + \beta_2 x_{it} + \alpha_i d_{it} + \gamma t_{it} d_{it} + \delta_i + \mu_{it}. \quad (13)$$

Among them, y_{it} is random variable, w_{ij} is weight, and x_{it} is control variable. β_0, β_1, and β_2 are coefficient vectors, and β_1 means how the economic behaviors of the experimental group and control group change with time under the policy intervention. δ_i and u_{it} are random disturbance items, which are not changing with time. α_i is the difference between control group and experiment group and it does not change with time. $t_{it}d_{it}$ is the cross term of virtual variable; γ is the target variable and it is the change degree of the experiment group after accepting the policy.

The original hypothesis of DID is $H_0 : \gamma = 0$; when there is no policy intervention, $\gamma = 0$, and DID effects of experimental group and control group are 0.

Table 1 summarizes the basic principle of MDID method.

3. Numerical Study

3.1. Analysis of Smooth Indicators

3.1.1. Qualitative Analysis. Traffic smooth can be explained from two aspects: smooth state and smooth efficiency. Smooth state can be understood as road unimpeded, which could be represented by urban average traffic running speed. Urban traffic smooth depends on traffic conditions of each section or road intersection in the whole traffic network. Generally speaking, lots of congestion reasons resulted in poor traffic smooth, with the main reasons being the mismatching and incompatibility of traffic demands and supplies.

Traffic demands are mainly influenced by urban economy, traffic composition, motorized travelling, and other factors. Among them, urbanization process affects travel demands, for the reason that more urban population and advanced economy mean greater travel demands. Besides, traffic forms could directly determine the motorized travel demands and the smooth state of urban traffic. Basic operation directly determines urban traffic intensity, such as travel consumption and distance.

TABLE 2: Main impact indicators of traffic smooth.

Category	Indicator contents	Relevance to the smooth
The social economic	Population GDP	Negative correlation
Traffic demand	Motor vehicle ownership Civilian car ownership	Negative correlation
Traffic supply	Road mileage, road area	Positive correlation
	Public vehicle number	Positive correlation
	Public transportation passenger	Positive correlation

Traffic supply mainly contains urban traffic management and planning, road infrastructure, related transportation service facilities, and so forth. Urban traffic management and planning determine the spatial layout of urban transportation, such as road network structure and travel direction. Moreover, road infrastructure includes road mileage, road area, road network structure, public transportation, rail transportation, and other facilities, while the transportation service facilities mainly include vehicle parking and vehicle transfer.

Based on the above analysis and modeling necessity, factors whose data are quantifiable and accessible are selected to form the indicators, as shown in Table 2.

(1) Population and GDP. Economic development will bring substantial increase in urban population and will also lead to a great travel demand. Moreover, economic development could lead to the production of more purchasing power of motor vehicles. Therefore, the more development the urban economy has gained, the greater the traffic travel demand will be, which also means higher requirement of transportation supply and carrying capacity.

(2) Number of Motor Vehicles. As the visual expression of road traffic and parking demands, vehicle amount largely determines the requirements of motor vehicle travel. In terms of the slowly growing urban roads and tight urban land, the rapid growth of motor vehicles will provide the road with more carrying capacity, which would quickly reach saturation. Besides, the insufficient parking supply capacity will directly result in severe urban traffic congestion.

(3) Road Mileage and Road Area. Urban road is the carrier of urban traffic, and road mileage as well as road areas can directly determine the carrying capacity of the road network.

TABLE 3: Related data of traffic smooth in Beijing.

Year	ASPD	GDP	POP	CVECH	RODM	RODA	PVECH	PPER
2000	18.9	3161	1357	104.1	2471	3502	10353	34.8
2001	19.0	3711	1383	114.5	2493	3701	12945	39.2
2002	19.8	4330	1423	133.9	2504	3857	15046	43.5
2003	20.1	5024	1456	163.1	3347	4315	16753	37.1
2004	20.0	6033	1493	182.4	4064	7287	18451	43.9
2005	23.4	6970	1538	209.7	4073	7437	18503	45.0
2006	22.5	8118	1581	239.1	4419	7286	19522	39.8
2007	20.5	9847	1633	273.4	4460	7632	19395	42.3
2008	23.1	11115	1695	313.7	6186	8940	21507	47.1
2009	23.1	12153	1755	368.1	6247	9179	21716	51.7
2010	22.2	14114	1961	449.7	6355	9395	21548	50.5

Data sources: Beijing Transportation Development Annual Report.

Longer road and larger areas mean more driving smooth of urban traffic. However, due to the restriction of urban planning and land, construction funds, and many other factors, the growth rate of urban roads is lower than that of motor vehicle. Taking Beijing as an example, the average annual growth of road mileage is 10.12% from 2005 to 2010. Excluding the large-scale construction during the 2008 Olympic Games, the average annual growth rate is only 3%, which is less than 13% of the average annual growth of motor vehicle.

(4) Number of Public Vehicles. After years of exploration, optimizing the traffic travel structure has become more and more important. Besides, improving the carrying capacity and service level of public transport also becomes an important way to solve the urban traffic congestion problem. The number of public vehicles, which plays a positive role in easing urban traffic congestion, could reflect the supply capacity of urban public transport.

(5) Number of Public Transport Passengers. Public transport passenger can describe supply capacity of public transport from the perspective of transportation production. Generally, people believed that, with greater amount of public transport passenger, urban transport operation would turn better. At present, the main public transport travel modes are public electric vehicle (regular bus), mass transit, and taxi.

3.1.2. Quantitative Analysis. The purpose of the quantitative analysis is to clarify the correlation among indicators and the relationship between indicators and traffic smooth. As showed in foregoing analysis, the paper defines average traffic speed of road network as the characterizing indicator of urban traffic smooth. Considering the regularity between indicators and data availability, the paper collects the data of Beijing from 2000 to 2010 for further analysis, as shown in Table 3.

Here, the following symbols are used to represent the name of each indicator:

ASPD: average traffic speed of road network (km/h).

GDP: GDP (hundred million Yuan).

POP: resident population (ten thousand people).

CVECH: number of civilian vehicles (ten thousand vehicles).

RODM: road mileage (km).

RODA: road area (ten thousand m^2).

PVECH: number of public vehicles.

PPER: number of public transport passengers.

Utilizing EViews 6.0 to implement correlation test, the specific results are shown in Table 4.

The results showed that the correlation coefficient of GDP, POP, other factors, and ASPD is all above 0.7. With high correlation, these factors are main factors affecting average traffic speed of road network.

Indicators are screened by referring to the existing evaluation achievements about traffic smooth both at home and abroad and following the metering modeling principles of "from general to simple." In terms of screening the independent variables, the CIA assumption in MDID model could provide some supporting functions: independent variables are influenced by "intervention," but it cannot influence the "intervention." According to MDID assumption, variables which cannot meet CIA assumptions will be eliminated. Finally, Granger test is used as the specific implementation method.

Step 1. Invoke the analysis results of speed's factors and establish timing database.

Step 2. Conduct Granger test between influencing factors and ASPD one by one.

Step 3. Strip out the factors of ASPD Granger reason.

The results show that Granger test result of traffic speed is significant (P value < 0.05). Hence, the number of public transport passengers' indicator is not suitable for the independent variable of speed.

3.2. Building the Matching Pool. DID model needs a comparison city, while at the same time Guangzhou city is the only ITS demonstration project city in China. Therefore, the control group cannot be the ITS demonstration project city. However, there are two major drawbacks for the manual selection of the control group cities. First, how much difference the control group cities can tolerate? It may be 10% or 30%. There is no objective standard. Second, in case of one single selected city, the results will heavily depend on the data of the chosen city. Under the circumstances, the introduction of matching process can solve these two issues effectively and provide theoretical support for selecting optimal control group. Before the execution of the matching algorithms, the first thing is to select the cities that can be allowed to enter into the matching pool

According to the model theory, it should be assumed that policy is exogenous in terms of selecting control group. That is to say, ITS only affects Guangzhou and has no effect on control group cities; or the influences can be ignored.

Table 4: Correlation between indicators.

Probability	Correlation							
	ASPD	GDP	POP	CVECH	RODM	RODA	PVECH	PPER
ASPD	1.000	0.748	0.703	0.727	0.805	0.830	0.826	0.704
	—	0.008	0.016	0.011	0.003	0.002	0.002	0.016
GDP	0.748	1.000	0.982	0.992	0.967	0.929	0.893	0.828
	0.008	—	0.000	0.000	0.000	0.000	0.000	0.002
POP	0.703	0.982	1.000	0.995	0.934	0.888	0.851	0.822
	0.016	0.000	—	0.000	0.000	0.000	0.001	0.002
CVECH	0.727	0.992	0.995	1.000	0.955	0.905	0.867	0.832
	0.011	0.000	0.000	—	0.000	0.000	0.001	0.002
RODM	0.805	0.967	0.934	0.955	1.000	0.956	0.916	0.827
	0.003	0.000	0.000	0.000	—	0.000	0.000	0.002
RODA	0.830	0.929	0.888	0.905	0.956	1.000	0.937	0.824
	0.002	0.000	0.000	0.000	0.000	—	0.000	0.002
PVECH	0.826	0.893	0.851	0.867	0.916	0.937	1.000	0.800
	0.002	0.000	0.001	0.001	0.000	0.000	—	0.003
PPER	0.704	0.828	0.822	0.832	0.827	0.824	0.800	1.000
	0.016	0.002	0.002	0.002	0.002	0.002	0.003	—

This hypothesis implied a few requirements: (1) the model allows observation factors; (2) in addition to the policy, the influences of remaining factors are the same; (3) the characteristics of control and experimental group are steady.

According to the above requirements and combined with the situation of China's administrative division, 36 center cities are selected to make up the matching pool. However, considering the outlier features of Tibet's data, the investigation only uses Suzhou as the replacement of Lhasa.

MDID model requires that the cross-section data must be implemented before the policy. Therefore, we can choose the data of the 36 cities in 2010 as alternative library cross-section data, as shown in Table 5.

3.3. Analysis of Matching Process. Based on the above data, Probit model is employed to do the robust regression. The results are shown in Table 6.

The results show that in matching Probit regression process the robust test of PVECH's coefficient is nonsignificant, and there is higher colinearity between PVECH and CVEH. After the removal of PVECH, the values of R^2 and AIC/SIC are experiencing obvious change.

The colinearity problems of RODM and RODA cannot be ignored, and it is unsuitable for them to be put into the same model. Thus, only RODM is retained. Therefore, the statistical magnitude of the 4 variances' Wald χ^2 is 19.89, the value of P is 0.0005, the likelihood of log pseudo is −2.2760393, and pseudo R^2 is 0.5019, with the fitting degree being acceptable.

Then, the matching process comes to the implementation of Probit regression and calculating the value of PS. Under common support domain, radius matching algorithm method (radius value is 0.1%) is used to select control group cities and calculate the weights of each city. The significant control group cities are Shenzhen, Shanghai, and Suzhou. Then, the Gaussian kernel density function $K(\|p_i - p_j\|_{j\in\{d=0\}}) = e^{-(p_i-p_j)^2/2\sigma^2}$ is employed (setting Gaussian width parameter as 0.1). Last, the weight of control group cities is calculated according to the following formula:

$$\omega_i = \frac{K_i\left(\left\|p_i - p_j\right\|_{j\in\{d=0\}}\right) = e^{-(p_i-p_j)^2/2\sigma^2}}{\sum K_i\left(\left\|p_i - p_j\right\|_{j\in\{d=0\}}\right) = e^{-(p_i-p_j)^2/2\sigma^2}}. \tag{14}$$

The results are shown in Table 7.

PS value of the remaining 32 cities is less than 0.001, so it is not listed.

For guaranteeing the robustness of model, doing logarithm process to data, and recalculating PS value, 0.2265 (Guangzhou), 0.4052 (Shenzhen), 0.1697 (Shanghai), and 0.1371 (Suzhou), the results show that the model is provided with good stability.

Therefore, a virtual city composed of 0.403 Shenzhen, 0.139 Shanghai, and 0.458 Suzhou has been set: city A = 0.403 ∗ Shenzhen + 0.139 ∗ Shanghai + 0.458 ∗ Suzhou. It is supposed that each indicator of city A is the same to Guangzhou. Then, city A is set as the control group to calculate the contribution rate.

After that, contribution rate is calculated by collecting relevant data of the experimental group (Guangzhou) and control group (Shenzhen, Shanghai, and Suzhou) from 2008 to 2012, as shown in Table 8.

Contribution rate can be calculated by the changes of y stemming from the changes of x divided by the total changes of y. According to Solow Growth model and making a difference of logarithm, the coefficient of virtual variable is the contribution rate. Meanings of variables and indicators are shown in Table 9. Therefore, GRP is set as the group virtual variable. If GRP = 0, it means that the variable should be in control group. And if GRP = 1, it means that the variable should be in the experimental group. Set ITS as policy virtual

TABLE 5: Cities and corresponding data in matching pool.

City	GDP	CVECH	POP	RODM	RODA	PVECH
Shijiazhuang	34010186.00	901653.00	1002.81	1475	4147	90657.00
Taiyuan	17780539.00	605048.00	354.02	1780	2432	39353.00
Hohhot	18657116.00	321551.00	278.76	720	1609	10189.00
Shenyang	50175400.00	984312.00	804.65	2895	5706	10275.00
Dalian	51581600.00	944885.00	663.82	2899	4135	7145.00
Changchun	33290329.00	664845.00	757.70	3659	6260	23935.00
Harbin	36648538.00	652435.00	991.81	1427	3296	21618.00
Nanjing	51306491.82	830524.00	786.03	5599	9576	19539.00
Hangzhou	59491687.00	1248056.00	851.97	2194	4754	70304.00
Ningbo	51630017.00	877434.00	744.29	1439	2379	16689.00
Hefei	29616700.00	386060.00	540.49	2013	4323	16707.00
Fuzhou	31234092.00	433734.00	699.27	1101	2327	11108.00
Xiamen	20600737.00	385391.00	353.13	1213	2994	7987.00
Nanchang	22001059.00	362429.00	502.66	965	1806	6303.00
Jinan	39105271.00	797359.00	674.62	4498	5907	12297.00
Qingdao	56661900.00	975571.00	860.77	3409	5893	15544.00
Zhengzhou	40408926.00	963010.00	809.04	1338	3158	21572.00
Wuhan	55659300.00	1046500.00	944.00	2682	7273	9838.00
Changsha	45470573.00	1008677.00	684.35	1781	3618	28502.00
Guangzhou	107482827.92	1598934.00	1228.96	6986	9731	7396.00
Shenzhen	95815101.00	1669674.00	1016.10	12613	8941	3099.00
Nanning	18002613.00	432199.00	702.62	1307	3205	21681.00
Haikou	6171948.00	238038.00	159.35	111	316	18808.00
Chengdu	55513336.00	2599300.00	1144.37	2610	6460	5819.00
Guiyang	11218174.00	604425.00	430.52	872	1348	11416.00
Kunming	21203700.00	1324215.00	632.00	1420	2815	2067.00
Xi'an	32416900.00	957162.00	845.45	2428	5342	18986.00
Lanzhou	11003898.00	246100.00	323.56	906	2162	9146.00
Xining	6282800.00	309900.00	221.01	433	737	7448.00
Yinchuan	7694227.00	225597.00	185.31	506	1652	6202.00
Beijing	141135800.00	4497200.00	1910.96	6355	9395	11584.00
Urumqi	13385172.00	1271400.00	311.00	1632	2005	15085.00
Chongqing	79255800.00	1143000.00	2872.00	5130	9931	14303.00
Tianjin	92244600.00	1582400.00	1263.73	5439	9159	7297.00
Shanghai	171659800.00	1755100.00	2256.48	4713	9299	27114.00
Suzhou	92289100.00	1261001.00	991.90	2904	5928	7937.00

Data source: urban passenger statistical reports of the Ministry of Transport.

variables and GRP ∗ ITS as cross terms, and the final model is shown as

$$w * \text{ASPD} = \beta_0 + \beta_1 * \text{RODM} + \beta_2 * \text{CVECH} + \beta_3 * \text{GDP} + \alpha_1 * \text{GRP} + \alpha_2 * \text{ITS} + \gamma * \text{GRP} * \text{ITS} + \mu. \tag{15}$$

3.4. Analyzing the Calculated Results. The data above are not in the symmetrical distribution, so it is difficult for the conditional expectation of least squares model to accurately reflect the real situation. Therefore, the paper establishes the following fixed effects estimation model and used quantile regression to achieve fitting.

The estimation and test results of coefficient and variance are shown in Table 10.

The paper can obtain the following conclusions based on the estimation results.

(1) The coefficient of GRP ∗ ITS could reflect the net contribution rate of ITS project to Guangzhou's ASPD, and the value is about 0.0925. Hence, the net contribution rate of ITS project to the ASPD increasing of Guangzhou is 9.25%. Moreover, 9.25% of the effects are contributed by ITS project in terms of improving the traffic smooth.

(2) Variance $R^2 = 0.7041$, which means that 70.41% of the total variation can be explained by the model

Table 6: Regression results.

Variable	Coef.	Robust std. err.	z	$P > z$	[95% conf. interval]	
GDP	$1.37E-07$	$5.07E-08$	2.7	0.007	$3.74E-08$	$2.36E-07$
CVECH	$7.01E-08$	$2.97E-07$	0.24	0.814	$-5.13E-07$	$6.53E-07$
RODM	−0.0000411	0.0001591	−0.26	0.796	−0.0003529	0.0002707
POP	−0.0091995	0.0039181	−2.35	0.019	−0.0168789	−0.0015201
_cons	−3.895322	0.6351935	−6.13	0	−5.140279	−2.650366

Table 7: PS values and the weights.

Significant city	PS	Weight
Guangzhou	0.2465	1
Shenzhen	0.2917	40.28%
Shanghai	0.1034	13.89%
Suzhou	0.3310	45.83%

with four variables and two free dummy variables, while the other 30% is caused by factors that are not included in the model.

(3) Table 10 shows that the intercept coefficient _cons = 0.054003. Due to the fact that the control group cannot completely lose connection with ITS project, the value indicating the effect proportion of ITS project to control group is 5.4%, and this value can be ignored. Therefore, _cons = 0.054003 also fully illustrates the scientificity in selecting control group and supporting the authenticity of the contribution results.

(4) The coefficient of policy virtual variables is −0.031374, which means that traffic smooth of control group is also affected by time factor and other related factors. Besides, traffic smooth level decreased by 3.14% from the year 2008 to 2012, which shows that traffic smooth of control group is relatively stable during the period, and it is beneficial to use MDID contribution assessment model.

(5) The regression coefficient of group virtual variable is −0.035766. It means that before the implementation of ITS project, the difference of traffic smooth between control group and experimental group is only 3.58%. It also shows that the selection of control group and experimental group in this paper is quite reasonable.

(6) The coefficients of RODM, CVECH, and GDP are 1.085374, −1.867530, and 1.865249, respectively, proving that the RODM and GDP have a positive correlation with traffic smooth, and the correlation value of GDP is higher than RODM. In contrast, CVECH has a negative correlation with traffic smooth.

The applications of ITS will inevitably exert a positive influence on the urban transportation development. However, with the rapid development of the city and coupled with the effects of many other factors on ITS, the urban traffic smooth has not been significantly improved. Therefore, based on these conclusions, the paper puts forward policy recommendations to improve urban traffic smooth from several aspects.

(1) The paper enhances the management level of urban traffic and improve the existing road network structure. City should figure out the requirements of its market economy, take into consideration the current situation of its own traffic, and promote reform of its management system. For example, the city could actively promote the separation of enterprise and government, so as to preferentially develop public transport, strengthen the using of information technology, and develop multicore urban spatial structure. Moreover, in terms of doing road network planning by the government, they should predict the traffic environment in a long period in the future. The situation where the occurrence of the later construction would deny the preoutcomes will be avoided. Last, perfect the static traffic facilities, such as establishing three-dimensional garage.

(2) It improves traffic smooth under the condition of ensuring effective urban development process. The development of economy and increasing population is bound to increase the number of motor vehicles. Thus, the traffic smooth state will be affected. Then, traffic demand will be guided reasonably and the traffic congestion will be eased by adopting the price mechanism, such as improving vehicle taxes and parking fees, charging road congestion fees, and implementing public transport travel concessions.

(3) It optimizes urban transportation infrastructure operational state. At present, China's motor vehicle ownership has increased by 15%, while the urban road has just grown by 3%. This ratio is extremely unreasonable. Due to the limitation of urban land resources, it is necessary to maximize the limited land resources and road utilization rate. For example, common cross overpass will be established and intelligent traffic lights will be used to control the intersection. Firstly, set intersection spacing of different road reasonable levels and then strengthen interconnection rates between roads. After that, shorten the travel distance; furthermore, install intelligent traffic display and allocate the traffic flow to be more reasonable. Finally, it comes to reducing travel time.

(4) It reasonably configures traffic composition mode. Firstly, improve facilities and services of public

TABLE 8: Data of experimental group and control group.

City	Year	ASPD	RODM	POP	CVECH	GDP
Shenzhen	2008	29.90	5849	933.3250	1252747	77,867,920.00
	2009	27.40	6035	974.6450	1419005	82,013,176.00
	2010	26.40	6184	1016.1050	1669674	95,815,101.00
	2011	28.00	5977	1041.9718	1939653	115,055,298.00
	2012	26.50	6015	1050.7432	2210821	129,501,000.00
Suzhou	2008	31.30	5013	897.3807	840431	70,780,900.00
	2009	25.70	5178	924.7993	1027058	77,402,000.00
	2010	25.80	5348	991.8973	1261001	92,289,100.00
	2011	25.90	5524	1049.3582	1518866	107,169,900.00
	2012	26.00	5672	1053.3855	1778977	120,116,500.00
Shanghai	2008	16.98	15844	2102.1141	1321229	140,698,700.00
	2009	17.70	16071	2175.4742	1471052	150,464,500.00
	2010	16.64	16687	2256.4845	1755100	171,659,800.00
	2011	15.64	16792	2325.0594	1947515	191,956,900.00
	2012	14.70	17316	2363.9464	2126637	201,817,200.00
Guangzhou	2008	21.00	5434	1084.1750	1171731	82,873,816.00
	2009	18.70	5519	1151.1550	1340540	91,382,135.00
	2010	22.70	6986	1228.9649	1598934	107,482,827.92
	2011	22.90	7072	1273.0500	1857740	124,234,390.00
	2012	25.30	7100	1279.5143	2041592	135,512,000.00

Data source: statistical yearbook of the cities and traffic statistics yearbook.

TABLE 9: Meanings of variables and indicators.

Main variable	Variable declaration
Decision variable	ASPD
Explanatory variable	CVECH
Control variable	RODM, GDP
Policy variable	$t_{it} = 1$ means after the implementation of ITS project (2011-2012) $t_{it} = 0$ means before the implementation of ITS project (2008~2010)
Category variable	$d_{it} = 1$ means the objects of experimental group $d_{it} = 0$ means the objects of control group

transport, which then can attract more people to take public transportation. For example, the layout of public transport, developing various forms of rapid transit buses, implementing different discount bus cards at peak period, and improving public transport share ratio may be optimizing.

4. Conclusions and Further Study

This paper reveals the influence factors and evaluation indexes of ITS on urban traffic smooth, based on the analysis of ITS in terms of improvement of urban traffic smooth. Combining the improved DID model, we calculate the contribution rate of ITS in improving urban traffic smooth. Through innovative study of the DID model, the research overcomes the application limitations of the DID model, which can serve as a useful tool for policy making of transportation department and evaluation of demonstration project. Also, the study provides new insights for quantitative research of urban transportation industry in terms of investment and decision. The result reveals that ITS works to improve urban traffic smooth to some extent. Specifically, ITS helps to relieve the traffic congestion through taking full advantage of the urban road area, reduce vehicle travel time and distance, and improve the urban traffic smooth. In a nutshell, the proposed quantitative analysis methodology in this paper can provide some new insights for ITS decision makers for strategic planning purposes. Moreover, this study can provide theoretical reference for the performance analysis of city's large demonstration projects and innovative ideas for the technical processing of DID model.

Although ITS has improved the traffic smooth to some extent, the improvement of average travel speed is still not obvious as a result of the increase of motor vehicles. In view of the above, the contributions of ITS cannot be ignored in terms of improving the average travel speed. Accordingly, it is necessary to accelerate the construction of ITS and make a practical action in improving urban traffic smooth. Although the modeling methodology proposed in this study provides a new avenue for evaluating the effects of ITS on urban traffic smooth, further extensions are necessary.

(1) The paper identifies the main indicators that affect traffic smooth. However, it is difficult to quantify the related degree between the indicators and ITS, since there are plenty of factors that influence traffic smooth. As such, some high correlation indicators

Table 10: Estimation results of the model.

Variable	Coef.	Bootstrap std. err.	t	$P > \|t\|$	[95% conf. interval]	
RODM	1.085374	0.201156	5.400000**	0.000	0.6303271	1.54042
CVECH	−1.867530	0.890793	−2.100000*	0.065	−3.882644	0.1475842
GDP	1.865249	0.568868	3.280000*	0.010	0.5783794	3.152118
GRP	−0.035766	0.050119	−0.710000	0.494	−0.1491443	0.0776114
ITS	−0.031374	0.049013	−0.640000	0.538	−0.1422493	0.0795012
GRP ∗ ITS	0.092490	0.063868	1.450000	0.182	−0.0519902	0.2369696
_cons	0.054003	0.094830	0.570000	0.583	−0.1605178	0.2685238

t statistics in parentheses: $^{*}P < 0.05$ and $^{**}P < 0.01$.

may be ignored, and this may further cause a significant bias in the evaluation capability of the model. Accordingly, it is worthy of more attention in a further study.

(2) Because of the special characteristics of ITS such as complexity, dynamism, and the hysteresis of application effects, one key problem is to determine the starting time-point of evaluation firstly. An appropriate starting time-point can guarantee that the evaluation results reflect the application benefits of ITS accurately. Therefore, in the future research, it is necessary to study the evaluation time-point of ITS application effects. Also, it is necessary to choose the point data to evaluate the contribution rate and to strengthen the scientific rationality of the results.

(3) Although the MDID model works well in the selection of control group, it has high requirements of the individuals that enter the matching pool. Therefore, it needs to strengthen the discussion about the selection of control group in further research. As such, DID model can be preferably used in traffic fields.

Conflict of Interests

All the authors have declared that they have no conflict of interests to this work.

References

[1] S. E. Underwood and S. G. Gehring, "Framework for evaluating intelligent vehicle-highway systems," *Transportation Research Record*, vol. 1453, pp. 16–22, 1994.

[2] D. C. Novak and M. McDnald, "A general overview of the potential macroeconomic impacts of ITS investment in the United States," in *Proceedings of the Annual Meeting of the Transportation Research Board*, Washington, DC, USA, January 1998.

[3] D. B. Lee, *Benefit-Cost Evaluation of ITS Projects: Seattle WSDOT Web Site (SE-26)*, FHWA/FTA Joint Program Office, USDOT, 1999.

[4] D. B. Lee and L. R. Klein, *Benefit-Cost Evaluation of ITS Traveler Information: Kiosks*, Volpe Center, USDOT, 1997.

[5] R. Zavergiu, "Intelligent transportation systems—an approach to benefit studies," Tech. Rep. TP 12695E, Transportation Development Centre, Montréal, Canada, 1996.

[6] D. Brand, "Applying benefit/cost analysis to identify and measure the benefits of intelligent transportation systems," *Transportation Research Record*, no. 1651, pp. 23–29, 1998.

[7] E. A. Beimborn, Z. Ren-peng, and M. Neluheni, "A framework and methods for evaluation of benefits of intelligent transportation systems," Transportation Research Board Record 1777, 2001.

[8] H.-B. Xiao, "Evaluation index system and evaluation methods on city integrated transport system," *Transportation Science & Technology*, vol. 234, no. 3, pp. 87–90, 2009.

[9] Q. Zhao, S.-M. Huang, S.-X. Di, and G.-S. Lu, "Comprehensive evaluation index system of urban public traffic system," *Transport Standardization*, vol. 208, no. 11, pp. 83–86, 2009.

[10] B.-W. Chen and X.-F. Liao, "Research on evaluation index system of highway transportation sustainable development," *China Journal of Highway and Transport*, vol. 22, no. 5, pp. 112–116, 2009.

[11] L. Cao, W.-X. Zha, W.-T. Zhao, and S.-B. Wang, "Establishing transfer evaluation index system of BRT," *Transport Engineering and Safety*, vol. 246, no. 11, pp. 126–128, 2011.

[12] P. Leviäkangas and J. Lähesmaa, "Profitability evaluation of intelligent transport system investments," *Journal of Transportation Engineering*, vol. 128, no. 3, pp. 276–286, 2002.

[13] S. P. Mattingly, *Decision Theory for Performance Evaluation of New Technologies Incorporating Institutional Issues: APP Lineation to Traffic Control Implementation*, University of California, Berkeley, Calif, USA, 2000.

[14] Z. Wang and C. M. Walton, "A multicriteria approach for evaluation and alternative selection of ITS deployments," in *Proceedings of the 11th Annual World Congress on Intelligent Transportation Systems (ITS '04)*, Nagoya, Japan, 2004.

[15] Q. Zhang, J.-L. Shen, and C. Li, "Bidding appraisal mechanism and synthetic appraisal method in ITS projects," *Journal of Transportation Systems Engineering and Information Technology*, vol. 5, no. 4, pp. 18–29, 2005.

[16] J. M. Wooldridge, *Introductory Econometrics: A Modern Approach*, South-Western College Publishing, 2nd edition, 2002.

[17] C. Ai and E. C. Norton, "Interaction terms in logit and probit models," *Economics Letters*, vol. 80, no. 1, pp. 123–129, 2003.

[18] P. A. Puhani, "The treatment effect, cross difference and interaction term in nonlinear 'difference-in-differences' models," IZA Discussion Paper 2, 2008.

[19] D. M. Drykker, "Testing for serial correlation in linear panel-data models," *The Stata Journal*, vol. 3, no. 2, pp. 168–177, 2003.

[20] S. Galiani, P. Gertler, and E. Schargrodsky, "Water for life: the impact of the privatization of water services on child mortality," *Journal of Political Economy*, vol. 113, no. 1, pp. 83–120, 2005.

[21] M. Conti, "The introduction of the euro and economic growth: some panel data evidence," *Journal of Applied Economics*, vol. 17, no. 2, pp. 199–211, 2014.

[22] R. A. Lewis and D. H. Reiley, "Online ads and offline sales: measuring the effect of retail advertising via a controlled experiment on Yahoo!," *Quantitative Marketing and Economics*, vol. 12, no. 3, pp. 235–266, 2014.

[23] J. Liang and Y. Zhou, "An empirical study on GEM and enterprises innovation," *Science Research Management*, vol. 34, no. 2, pp. 89–92, 2013.

[24] Y.-L. Wu, "Soil and water conservation projects and national food safety—evidence from the panel data of northeast black soil areas," *China Population, Resources and Environment*, vol. 24, no. 5, pp. 304–308, 2014.

[25] H. Wang, "New farming maintain to evaluate the contribution of the rural residents' consumption growth in China—based on DID model," *Modern Business*, no. 2, 2014.

[26] M. Deng and Q.-H. Wang, "Relationship between China's strategic introduction of overseas investment and X-efficiency of Chinese-funded banks," *International Business*, no. 2, pp. 21–30, 2014.

[27] A. Tzedakis, "Different vehicle speeds and congestion costs," *Journal of Transport Economics and Policy*, vol. 14, no. 1, pp. 81–103, 2005.

[28] C. Wright and P. Roberg, "The conceptual structure of traffic jams," *Transport Policy*, vol. 5, no. 1, pp. 23–35, 1998.

[29] H. Rehborn and M. Koller, "A study of the influence of severe environmental conditions on common traffic congestion features," *Journal of Advanced Transportation*, vol. 48, no. 8, pp. 1107–1120, 2014.

[30] A. Waadt, "Traffic congestion estimation service mobile assisted positioning schemes in GSM networks," *Procedia Earthand Plaetary Science*, vol. 1, no. 1, pp. 1385–1392, 2009.

[31] Y.-L. Li and C. Wang, "Urban planning and population density: factors affecting urban traffic development research," *Xinjiang Social Science*, no. 3, pp. 42–46, 2013 (Chinese).

[32] M. Anisimova, "Darwin and Fisher meet at biotech: on the potential of computational molecular evolution in industry," *BMC Evolutionary Biology*, vol. 15, article 76, 2015.

[33] P. R. Rosenbaum and D. B. Rubin, "The central role of the propensity score in observational studies for causal effects," *Biometrika*, vol. 70, no. 1, pp. 41–55, 1983.

Permissions

All chapters in this book were first published in JAT, by Hindawi Publishing Corporation; hereby published with permission under the Creative Commons Attribution License or equivalent. Every chapter published in this book has been scrutinized by our experts. Their significance has been extensively debated. The topics covered herein carry significant findings which will fuel the growth of the discipline. They may even be implemented as practical applications or may be referred to as a beginning point for another development.

The contributors of this book come from diverse backgrounds, making this book a truly international effort. This book will bring forth new frontiers with its revolutionizing research information and detailed analysis of the nascent developments around the world.

We would like to thank all the contributing authors for lending their expertise to make the book truly unique. They have played a crucial role in the development of this book. Without their invaluable contributions this book wouldn't have been possible. They have made vital efforts to compile up to date information on the varied aspects of this subject to make this book a valuable addition to the collection of many professionals and students.

This book was conceptualized with the vision of imparting up-to-date information and advanced data in this field. To ensure the same, a matchless editorial board was set up. Every individual on the board went through rigorous rounds of assessment to prove their worth. After which they invested a large part of their time researching and compiling the most relevant data for our readers.

The editorial board has been involved in producing this book since its inception. They have spent rigorous hours researching and exploring the diverse topics which have resulted in the successful publishing of this book. They have passed on their knowledge of decades through this book. To expedite this challenging task, the publisher supported the team at every step. A small team of assistant editors was also appointed to further simplify the editing procedure and attain best results for the readers.

Apart from the editorial board, the designing team has also invested a significant amount of their time in understanding the subject and creating the most relevant covers. They scrutinized every image to scout for the most suitable representation of the subject and create an appropriate cover for the book.

The publishing team has been an ardent support to the editorial, designing and production team. Their endless efforts to recruit the best for this project, has resulted in the accomplishment of this book. They are a veteran in the field of academics and their pool of knowledge is as vast as their experience in printing. Their expertise and guidance has proved useful at every step. Their uncompromising quality standards have made this book an exceptional effort. Their encouragement from time to time has been an inspiration for everyone.

The publisher and the editorial board hope that this book will prove to be a valuable piece of knowledge for researchers, students, practitioners and scholars across the globe.

List of Contributors

Muhammad Alam, José Fonseca and Joaquim Ferreira
Instituto de Telecomunicações, Universidade de Aveiro, Portugal

Davide Moroni, Gabriele Pieri, Marco Tampucci and Giuseppe Riccardo Leone
Institute of Information Science and Technologies, National Research Council of Italy, Pisa, Italy

Miguel Gomes
Micro I/O, Sistemas Electrónicos, Lda, Aveiro, Portugal

Jiandong Zhao, Jinjin Tang and Lingxi Zhu
School of Traffic and Transportation, Beijing Jiaotong University, Beijing, China

Jiandong Zhao
MOE Key Laboratory for Urban Transportation Complex Systems Theory and Technology, Beijing Jiaotong University, Beijing, China

Yuan Gao
School of Mechanical and Electronic Control Engineering, Beijing Jiaotong University, Beijing, China

Jiaqi Ma
Department of Civil and Architectural Engineering and Construction Management, University of Cincinnati, Cincinnati, OH, USA

Chao Qian, Meng Yang, Peiqi Li and Shuguang Li
School of Electronic and Control Engineering, Chang'an University, Middle-Section of Nan'er Huan Road, Xi'an, Shaanxi 710064, China

Jiawei Gui and Qunqi Wu
School of Economics and Management, Chang'an University, Xi'an, Shanxi 710064, China
Center of Comprehensive Transportation Economic Management, Chang'an University, Xi'an, Shanxi 710064, China

Krishna Raghuwaiya, Bibhya Sharma and Jito Vanualailai
School of Computing, Information & Mathematical Sciences, The University of the South Pacific, Suva, Fiji

Carlos Fernández
Karlsruhe Institute of Technology, 76131 Karlsruhe, Germany

Jesús Muñoz-Bulnes
ALTRAN Innovation, Madrid, Spain

David Fernández-Llorca, Ignacio Parra, Iván García-Daza, Rubén Izquierdo and Miguel Á. Sotelo
Computer Engineering Department, University of Alcalá, Madrid, Spain

Yusor Rafid Bahar Al-Mayouf, Omar Adil Mahdi and Namar A. Taha
Department of Computer Science, College of Education for Pure Science (Ibn Al-Haitham), University of Baghdad, Iraq

Nor Fadzilah Abdullah
Centre of Advanced Electronic and Communication Engineering, Faculty of Engineering and Built Environment, Universiti Kebangsaan Malaysia, Malaysia

Suleman Khan
School of Information Technology, Monash University, Bandar Sunway 47500, Malaysia

Muhammad Alam
Department of Computer Science and Software Engineering, Xi'an Jiaotong-Liverpool University, Suzhou, China

Guang Chen, Canbo Ye, Kai Chen and Xingbo Liu
College of Automotive Engineering, Tongji University, China

Guang Chen, Muhammad Aafaque, Florian Röhrbein, Zhenshan Bing, Gereon Hinz and Alois Knoll
Robotics and Embedded Systems, Technische Universität München, Germany

Hu Cao
State Key Laboratory of Advanced Design and Manufacturing for Vehicle Body, Hunan University, China

Jieneng Chen
College of Electronics and Information Engineering, Tongji University, China

Jörg Conradt
Department of Computational Science and Technology, KTH Royal Institute of Technology, Sweden

Walter Stechele
Integrated Systems, Technische Universität München, Germany

Mushtaq Ahmad and Zahid Khan
Southwest Jiaotong University, Chengdu, Sichuan 611756, China

Qingchun Chen
Guangzhou University, Guangzhou, Guangdong 510006, China

Yu Yao
School of Information Engineering, East China Jiaotong University, Nanchang 330031, China

Lenan Wu
School of Information Science and Engineering, Southeast University, Nanjing 210096, China

Wei Liu, Longsheng Wei and Yongbo Li
School of Automation, China University of Geosciences, Wuhan, Hubei 430074, China

Muhammed Tahsin Rahman and Aboelmagd Noureldin
Edward S. Rogers Sr. Department of Electrical and Computer Engineering, University of Toronto, Toronto, ON, Canada

Tashfeen Karamat
Department of Electrical and Computer Engineering, Queen's University, Kingston, ON, Canada

Sidney Givigi and Aboelmagd Noureldin
Department of Electrical and Computer Engineering, RMCC, Kingston, ON, Canada

Hyun-ho Chang
Graduate School of Environmental Studies, Seoul National University, Seoul, Republic of Korea

Byoung-jo Yoon
Department of Urban Engineering, Incheon National University, Incheon, Republic of Korea

Balázs Németh and Péter Gáspár
Systems and Control Laboratory, Institute for Computer Science and Control, Hungarian Academy of Sciences, Kende Utca 13-17, H-1111 Budapest, Hungary

Tamás Hegedűs
Department of Control for Transportation and Vehicle Systems, Budapest University of Technology and Economics, Budapest, Hungary

Lin-heng Li, Jing Gan and Wen-quan Li
School of Transportation, Southeast University, China

Ahmed Hussein, Pablo Marín-Plaza, Fernando García and José María Armingol
Intelligent Systems Lab (LSI) Research Group, Universidad Carlos III de Madrid (UC3M), Calle Butarque 15, Leganés, 28911 Madrid, Spain

Ferran Mochol Belenguer
Traffic Control Systems Group, ITACA Institute, Universitat Politècnica de València, Valencia 46022, Spain

Antonio Mochol Salcedo
Department of Electronic Engineering, ITACA Institute, Universitat Politècnica de València, Valencia 46022, Spain

Victor Milián Sánchez
Chemical and Nuclear Engineering Department, Institute of Industrial, Radiological and Environmental Safety, Universitat Politècnica de València, Valencia 46022, Spain

José Humberto Arroyo Núñez
Electronic and Telecommunications Engineering Department, Universidad Politécnica de Tulancingo, Tulancingo 43629, Mexico

Sanchai Jaktheerangkoon, Kulit Na Nakorn and Kultida Rojviboonchai
Chulalongkorn University Big Data Analytics and IoT Center (CUBIC), Department of Computer Engineering, Faculty of Engineering, Chulalongkorn University, Bangkok, Thailand

Fei Xie, Jing Zhao, Jiquan Yang, Yijian Liu and Xinyue Yuan
School of Electrical and Automation Engineering, Nanjing Normal University, Nanjing 210042, China

Fei Xie, Jing Zhao, Jiquan Yang, Yijian Liu and Xinyue Yuan
Jiangsu Key Laboratory of 3D Printing Equipment and Manufacturing, Nanjing 210042, China

Ming Zhang
Department of Electronic Engineering, City University of Hong Kong, Hong Kong

Jing Zhao
School of Automation, Nanjing University of Posts and Telecommunications, Nanjing 210023, China

Index